Network Analysis and Synthesis

Second Edition

Wiley International Edition

Network Analysis and Synthesis

Second Edition

by Franklin F. Kuo

Bell Telephone Laboratories, Inc.

John Wiley & Sons, Inc., New York | London | Sydney

Toppan Company. Ltd., Tokyo, Japan

Wiley International Edition

Library of Congress Catalog Card Number : 66-16127
Printed in Singapore by Toppan Printing Co. (S) Pte. Ltd.

To My Father and Mother

Preface

In the second edition, I have tried to keep the organization of the first edition. Most of the new material are additions aimed at strengthening the weaknesses of the original edition. Some specific changes deserve mention. The most important of these is a new chapter on computer applications (Chapter 15). In the past five years, digital computers have brought about many significant changes in the content of engineering subject matter concerned with both analysis and design. In analysis, computation has become an important adjunct to theory. Theory establishes the foundation of the subject matter; computation provides clarity, depth, and insight. In design, the computer has not only contributed precision and speed to existing procedures but has made practicable design methods that employ iteration and simulation. The importance of computer-aided design cannot be overemphasized. In Chapter 15 I have attempted to survey some digital computer applications in the areas of network analysis and design. I strongly encourage all students to read this chapter for cultural interest, if not for survival.

Another new section contains a rigorous treatment of the unit impulse. It was difficult to decide whether to incorporate this material in Chapter 2 in the discussion of signals or in a separate appendix. By putting generalized functions in an appendix, I have left the decision of whether to teach the rigorous treatment up to the individual instructor.

Other changes worth mentioning are: (1) two new sections on the Fourier integral in Chapter 3; (2) a section on initial and final conditions in Chapter 5; (3) a section on Bode plots in Chapter 8; (4) revised material on two-port parameters in Chapter 9; and (5) new sections on frequency and transient responses of filters in Chapter 13. Major or minor changes may be found in every chapter, with the exception of Chapters 11 and 12. In addition, many new problems are included at the end of each chapter.

A brief description of the subject matter follows. Chapters 1 and 2 deal with signal representation and certain general characteristics of linear systems. Chapter 3 deals with Fourier analysis, and includes the impulse method for evaluating Fourier coefficients. Chapters 4 and 5 discuss solutions of network differential equations in the time domain. In Chapters 6 and 7, the goals are the same as those of the two preceding chapters, except that the viewpoint here is that of the frequency domain. Chapter 8 deals with the amplitude, phase, and delay of a system function.

The final seven chapters are concerned with network synthesis. Chapter 9 deals with two-ports. In Chapter 10, the elements of realizability theory are presented. Chapters 11 and 12 are concerned with elementary driving-point and transfer function synthesis procedures. In Chapter 13, some fundamental concepts in modern filter design are introduced. Chapter 14 deals with the use of scattering matrices in network analysis and design. And, as mentioned earlier, Chapter 15 contains a brief survey of digital computer techniques in system analysis. In addition, there are five appendices covering the rudiments of matrix algebra, generalized functions, complex variables, proofs of Brune's theorems, and a visual aid to filter approximation.

The book is intended for a two-semester course in network theory. Chapters 1 through 8 may be used in a one-semester undergraduate or beginning graduate course in transient analysis or linear system analysis. Chapters 9 and 15 are to be used in a subsequent course on network synthesis.

The second edition is largely a result of the feedback from the professors who have used this book and from their students, who have discovered errors and weaknesses in the original edition. I wish to express my sincere appreciation to those who provided this feedback.

Special thanks are due to the following people: Robert Barnard of Wayne State, Charles Belove and Peter Dorato of the Polytechnic Institute of Brooklyn, James Kaiser and Philip Sherman of the Bell Telephone Laboratories, Evan Moustakas of San Jose State College, A. J. Welch of the University of Texas, and David Landgrebe of Purdue. I am particularly indebted to Mac Van Valkenburg of Princeton University and Robert Tracey of Illinois for editorial advice and to Donald Ford of Wiley for help and encouragement.

In addition, I wish to thank Elizabeth Jenkins, Lynn Zicchino, and Joanne Mangione of the Bell Telephone Laboratories for their efficient and careful typing of the manuscript.

Berkeley Heights,
New Jersey,
December, 1965

F. F. Kuo

Preface to the First Edition

This book is an introduction to the study of electric networks based upon a *system theoretic* approach. In contrast to many present textbooks, the emphasis is not on the form and structure of a network but rather on its *excitation-response* properties. In other words, the major theme is concerned with how a linear network behaves as a signal processor. Special emphasis is given to the descriptions of a linear network by its system function in the frequency domain and its impulse response in the time domain. With the use of the system function as a unifying link, the transition from network analysis to synthesis can be accomplished with relative ease.

The book was originally conceived as a set of notes for a second course in network analysis at the Polytechnic Institute of Brooklyn. It assumes that the student has already had a course in steady-state circuit analysis. He should be familiar with Kirchhoff's laws, mesh and node equations, standard network theorems, and, preferably, he should have an elementary understanding of network topology.

A brief description of the subject matter follows. Chapters 1 and 2 deal with signal representation and certain characteristics of linear networks. Chapters 3, 4, 5, and 6 discuss transient analysis from both a time domain viewpoint, i.e., in terms of differential equations and the impulse response, and a frequency domain viewpoint using Fourier and Laplace transforms. Chapter 7 is concerned with the use of poles and zeros in both transient and steady-state analysis. Chapter 8 contains a classical treatment of network functions.

The final five chapters deal with network synthesis. In Chapter 9, the elements of realizability theory are presented. Chapters 10 and 11 are concerned with elementary driving-point and transfer function synthesis procedures. In Chapter 12, some fundamental concepts in modern filter design are introduced. Chapter 13 deals with the use of scattering matrices in network analysis and synthesis. In addition, there are three appendices covering the rudiments of matrix algebra, complex variables, and proofs of Brune's realizability theorems.

The book is intended for a two-semester course in network theory. Chapters 1 through 7 can be used in a one-semester undergraduate or beginning graduate course in transient analysis or linear system analysis. Chapters 8 through 13 are to be used in a subsequent course on network synthesis.

It was my very good fortune to have studied under Professor M. E. Van Valkenburg at the University of Illinois. I have been profoundly influenced by his philosophy of teaching and writing, which places strong emphasis upon clarity of exposition. In keeping with this philosophy, I have tried to present complicated material from a simple viewpoint, and I have included a large number of illustrative examples and exercises. In addition, I have tried to take a middle ground between mathematical rigor and intuitive understanding. Unless a proof contributes materially to the understanding of a theorem, it is omitted in favor of an intuitive argument. For example, in the treatment of unit impulses, a development in terms of a *generalized function* is first introduced. It is stressed that the unit impulse is not really a function but actually a sequence of functions whose limit point is undefined. Then, the less rigorous, intuitive notion of an impulse "function" is presented. The treatment then proceeds along the nonrigorous path.

There are a number of topics which have been omitted. One of these is network topology, which seems to be in vogue at present. I have purposely omitted topology because it seems out of place in a book that de-emphasizes the form and structure approach to network analysis.

In an expository book of this nature, it is almost impossible to reference adequately all the original contributors in the vast and fertile field of network theory. I apologize to those whose names were omitted either through oversight or ignorance. At the end of the book, some supplementary textbooks are listed for the student who either wishes to fill in some gaps in his training or wants to obtain a different point of view.

I acknowledge with gratitude the help and advice given to me by my colleagues at the Bell Telephone Laboratories and by my former colleagues at the Polytechnic Institute of Brooklyn. I wish to express my sincere appreciation to the many reviewers whose advice and criticism were invaluable in revising preliminary drafts of the manuscript. Professors R. D. Barnard of Wayne State University and R. W. Newcomb of Stanford University deserve specific thanks for their critical reading of the entire manuscript and numerous helpful suggestions and comments.

In addition, I wish to thank Mrs. Elizabeth Jenkins and Miss Elizabeth

La Jeunesse of the Bell Telephone Laboratories for their efficient and careful typing of the manuscript.

Finally, to my wife Dora, I owe a special debt of gratitude. Her encouragement and cooperation made the writing of this book an enjoyable undertaking.

F. F. KUO

Murray Hill, New Jersey,
January, 1962

Contents

chapter 1

Signals and systems

This book is an introduction to electric network theory. The first half of the book is devoted to network analysis and the remainder to network synthesis and design. What *are* network analysis and synthesis? In a generally accepted definition of network analysis and synthesis, there are three key words: the *excitation*, the *network*, and the *response* as depicted in Fig. 1.1. Network analysis is concerned with determining the response, given the excitation and the network. In network synthesis, the problem is to design the network given the excitation and the desired response. In this chapter we will outline some of the problems to be encountered in this book without going into the actual details of the problems. We will also discuss some basic definitions.

1.1 SIGNAL ANALYSIS

For electric networks, the excitation and response are given in terms of voltages and currents which are functions of time, t. In general, these functions of time are called *signals*. In describing signals, we use the two universal languages of electrical engineering—*time* and *frequency*. Strictly speaking, a signal is a function of time. However, the signal can be described equally well in terms of *spectral* or *frequency* information. As between any two languages, such as French and German, translation is needed to render information given in one language comprehensible in the

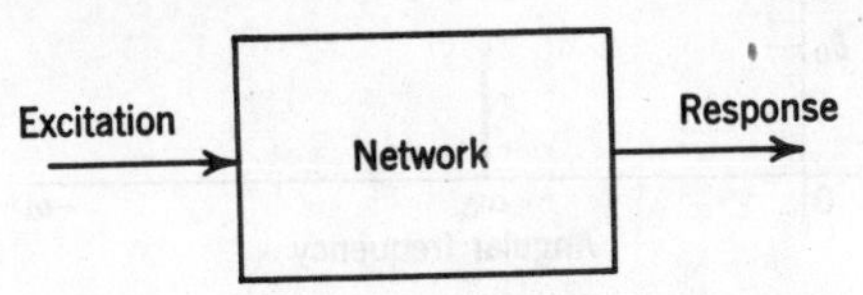

FIG. 1.1. The objects of our concern.

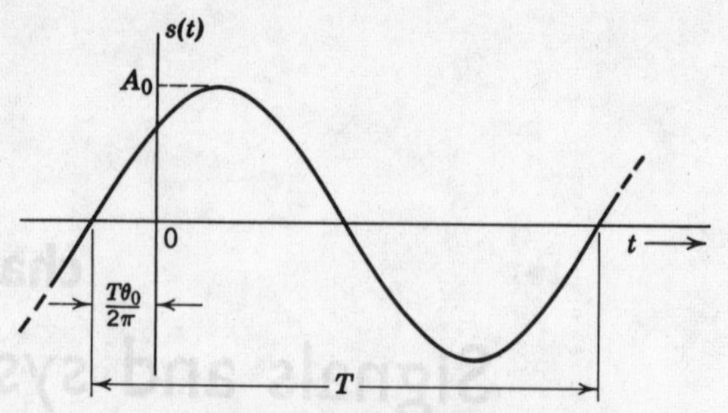

FIG. 1.2. Sinusoidal signal.

other. Between time and frequency, the translation is effected by the *Fourier series*, the *Fourier integral*, and the *Laplace transform*. We shall have ample opportunity to define and study these terms later in the book. At the moment, let us examine how a signal can be described in terms of both frequency and time. Consider the sinusoidal signal

$$s(t) = A_0 \sin(\omega_0 t + \theta_0) \tag{1.1}$$

where A_0 is the *amplitude*, θ_0 is the *phase shift*, and ω_0 is the *angular frequency* as given by the equation

$$\omega_0 = \frac{2\pi}{T} \tag{1.2}$$

where T is the period of the sinusoid. The signal is plotted ag..inst time in Fig. 1.2. An equally complete description of the signal is obtained if we

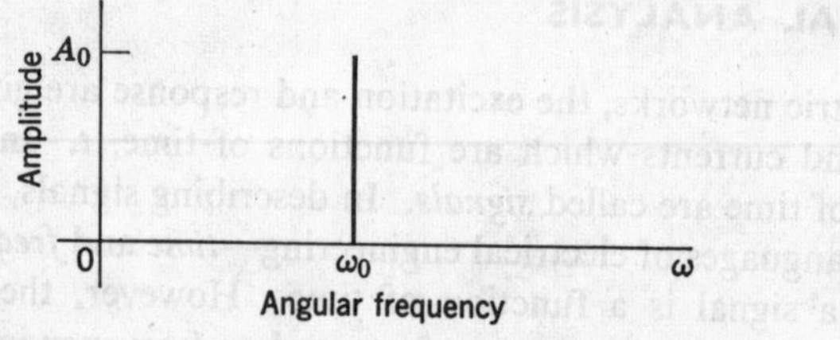

FIG. 1.3a. Plot of amplitude A versus angular frequency ω.

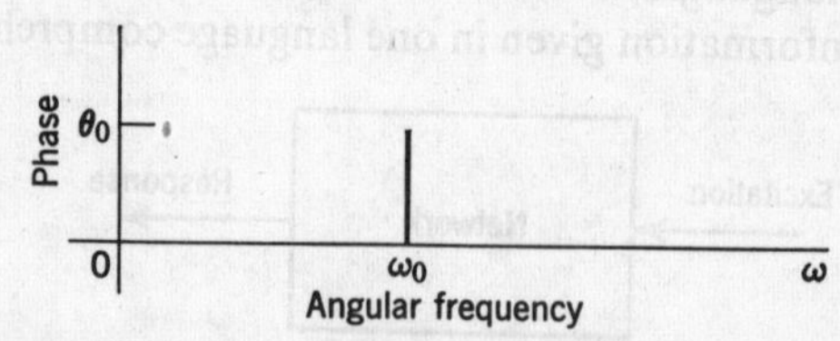

FIG. 1.3b. Plot of phase θ versus angular frequency ω.

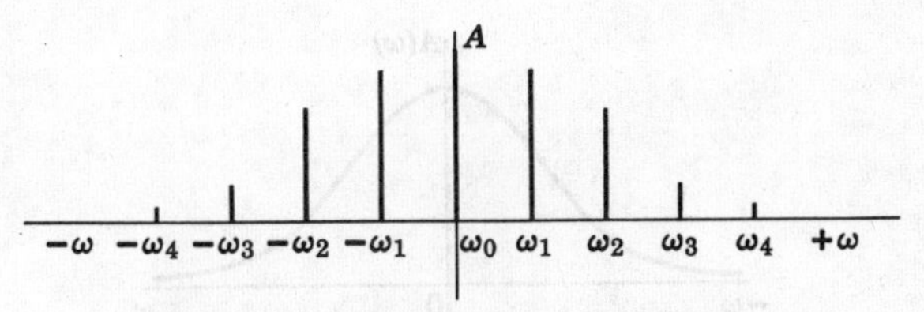

FIG. 1.4a. Discrete amplitude spectrum.

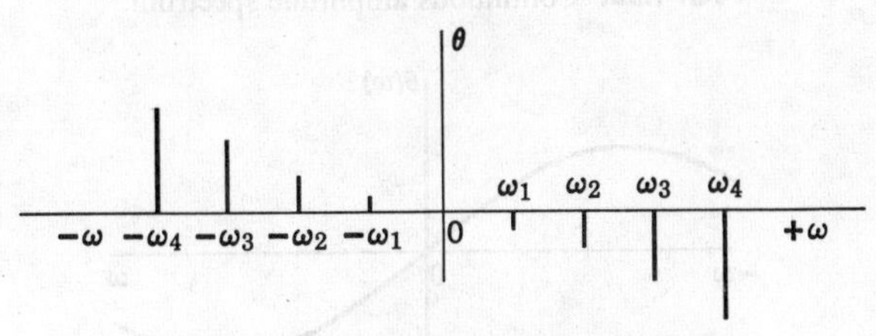

FIG. 1.4b. Discrete phase spectrum.

let the angular frequency ω be the independent variable. In this case, the signal is described in terms of A_0, ω_0, and θ_0, as shown in Fig. 1.3*a*, where amplitude is plotted against frequency, and in Fig. 1.3*b*, where phase shift is plotted.

Now suppose that the signal is made up of $2n + 1$ sinusoidal components

$$s(t) = \sum_{i=-n}^{n} A_i \sin (\omega_i t + \theta_i) \tag{1.3}$$

The spectral description of the signal would then contain $2n + 1$ lines at $\pm\omega_1, \pm\omega_2, \ldots, \pm\omega_n$, as given in Figs. 1.4*a* and *b*. These discrete spectra of amplitude A versus ω and phase shift θ versus ω are sometimes called *line spectra*. Consider the case when the number of these spectral lines become infinite and the intervals $\omega_{i+1} - \omega_i$ between the lines approach zero. Then there is no longer any discrimination between one frequency and another, so that the discrete line spectra fuse into a *continuous* spectra, as shown by the example in Figs. 1.5*a* and *b*. In the continuous case, the sum in Eq. 1.3 becomes an integral

$$s(t) = \int_{-\infty}^{\infty} A(\omega) \sin [\omega t + \theta(\omega)]\, d\omega \tag{1.4}$$

where $A(\omega)$ is known as the *amplitude spectrum* and $\theta(\omega)$ as the *phase spectrum.*

As we shall see later, periodic signals such as the sine wave in Fig. 1.2 can be described in terms of discrete spectra through the use of Fourier series. On the other hand, a nonperiodic signal such as the triangular

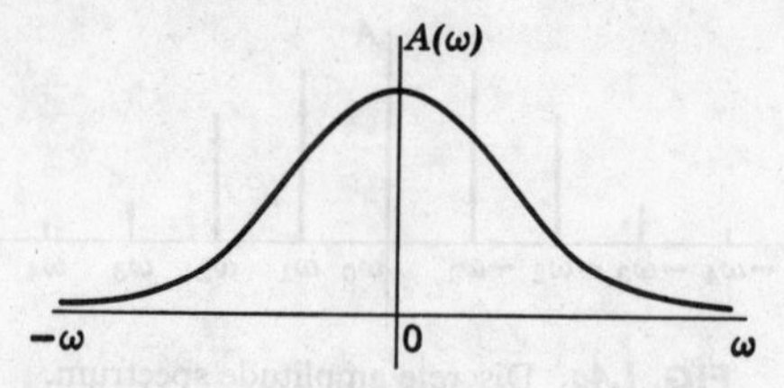

FIG. 1.5*a*. Continuous amplitude spectrum.

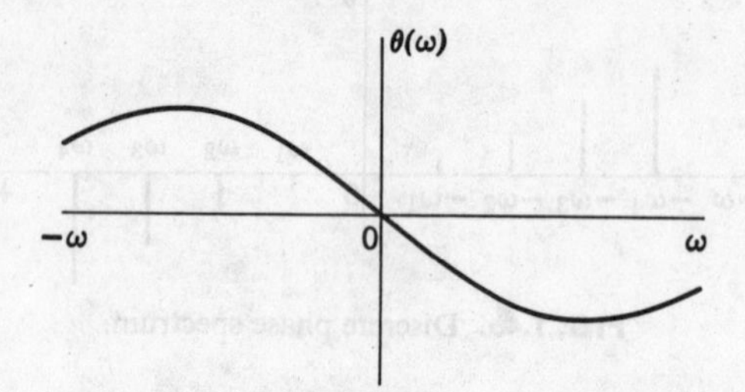

FIG. 1.5*b*. Continuous phase spectrum.

pulse in Fig. 1.6 can only be described in terms of continuous spectra through the Fourier integral transform.

1.2 COMPLEX FREQUENCY

In this section, we will consider the concept of *complex frequency*. As we shall see, the complex frequency variable

$$s = \sigma + j\omega \tag{1.5}$$

is a generalized frequency variable whose real part σ describes growth and decay of the amplitudes of signals, and whose imaginary part $j\omega$ is angular frequency in the usual sense. The idea of complex frequency is developed by examining the cisoidal signal

$$\mathrm{S}(t) = Ae^{j\omega t} \tag{1.6}$$

FIG. 1.6. Triangular signal.

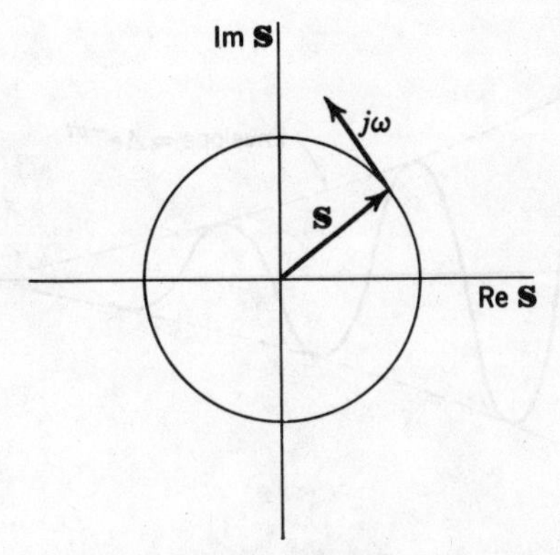

FIG. 1.7. Rotating phasor.

when $\mathbf{S}(t)$ is represented as a rotating phasor,[1] as shown in Fig. 1.7. The angular frequency ω of the phasor can then be thought of as a *velocity* at the end of the phasor. In particular the velocity ω is always at right angles to the phasor, as shown in Fig. 1.7. However, consider the general case when the velocity is inclined at any arbitrary angle ψ as given in Figs. 1.8*a* and 1.8*b*. In this case, if the velocity is given by the symbol s, we see that s is composed of a component ω at right angle to the phasor $\mathbf{S}$ as well as a component σ, which is parallel to $\mathbf{S}$. In Fig. 1.8*a*, s has a component $-\sigma$ toward the origin. As the phasor $\mathbf{S}$ spins in a counterclockwise fashion, the phasor decreases in amplitude. The resulting wave for the real and imaginary parts of $\mathbf{S}(t)$ are *damped sinusoids* as given by

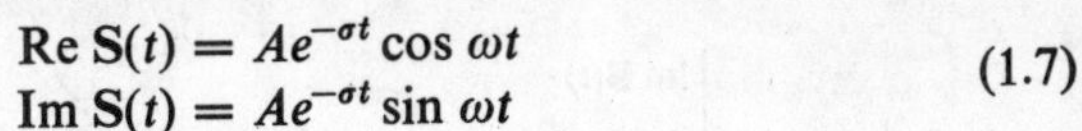

$$\begin{aligned} \text{Re}\,\mathbf{S}(t) &= Ae^{-\sigma t}\cos\omega t \\ \text{Im}\,\mathbf{S}(t) &= Ae^{-\sigma t}\sin\omega t \end{aligned} \qquad (1.7)$$

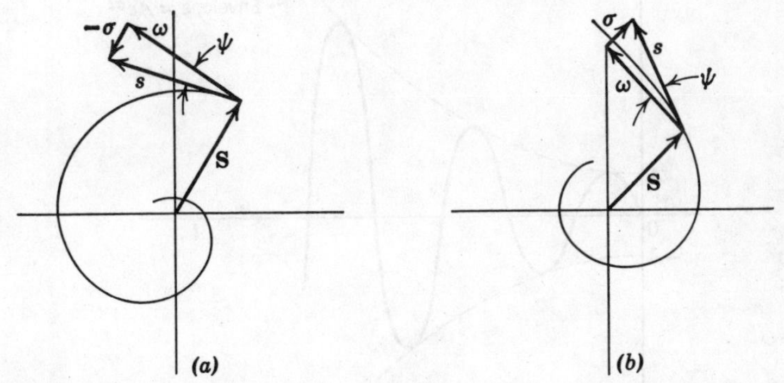

FIG. 1.8. (*a*) Rotating phasor with exponentially decreasing amplitude. (*b*) Rotating phasor with exponentially increasing amplitude.

[1] A phasor $\mathbf{S}$ is a complex number characterized by a magnitude and a phase angle (see Appendix C).

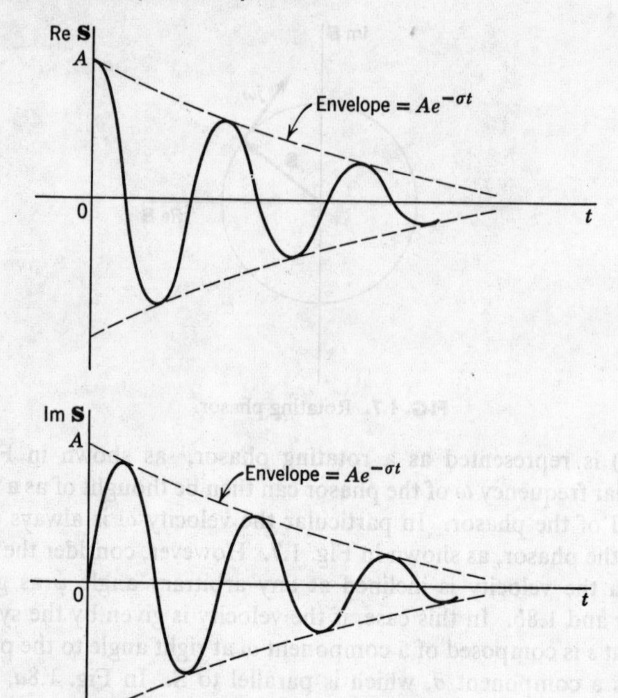

FIG. 1.9. Damped sinusoids.

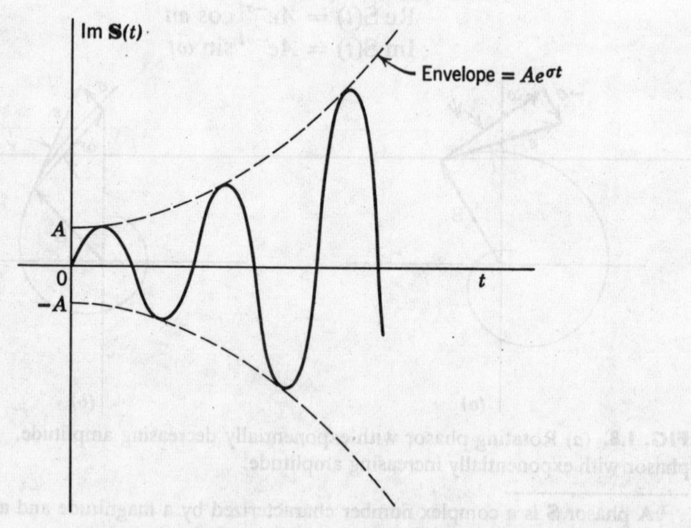

FIG. 1.10. Exponentially increasing sinusoid.

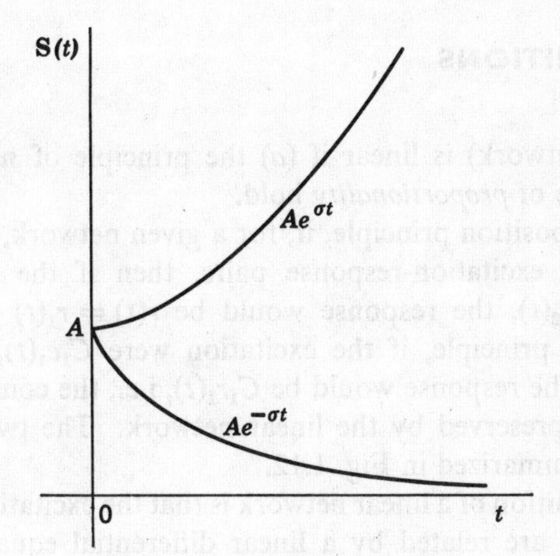

FIG. 1.11. Exponential signals.

which are shown in Fig. 1.9. Note that the damped sinusoid has an exponential envelope decay, $Ae^{-\sigma t}$. In Fig. 1.8*b*, the phasor is shown with a positive real component of velocity $+\sigma$. Therefore, as the phasor spins, the amplitudes of the real and imaginary parts increase exponentially with an envelope $Ae^{+\sigma t}$, as shown by Im S(t) in Fig. 1.10.

From this discussion, it is apparent that the generalized cisoidal signal

$$S(t) = Ae^{st} = Ae^{(\sigma + j\omega)t} \tag{1.8}$$

describes the growth and decay of the amplitudes in addition to angular frequency in the usual sense. When $\sigma = 0$, the sinusoid is undamped, and when $j\omega = 0$, the signal is an exponential signal

$$S(t) = Ae^{\pm\sigma t} \tag{1.9}$$

as shown in Fig. 1.11. Finally, if $\sigma = j\omega = 0$, then the signal is a constant A. Thus we see the versatility of a complex frequency description.

1.3 NETWORK ANALYSIS

As mentioned before, the characterization of the excitation and response signals in time and frequency makes up only part of the analysis problem. The other part consists of characterizing the network itself in terms of time and frequency, and determining how the network behaves as a signal processer. Let us turn our attention now to a brief study of the properties of linear networks and the general characteristics of signal processing by a linear system.

BASIC DEFINITIONS

Linear

A system (network) is linear if (*a*) the principle of *superposition* and (*b*) the principle of *proportionality* hold.

By the superposition principle, if, for a given network, $[e_1(t), r_1(t)]$ and $[e_2(t), r_2(t)]$ are excitation-response pairs, then if the excitation were $e(t) = e_1(t) + e_2(t)$, the response would be $r(t) = r_1(t) + r_2(t)$. By the proportionality principle, if the excitation were $C_1e_1(t)$, where C_1 is a constant, then the response would be $C_1r_1(t)$, i.e., the constant of proportionality C_1 is preserved by the linear network. The two conditions of linearity are summarized in Fig. 1.12.

Another definition of a linear network is that the excitation and response of the network are related by a linear differential equation. We shall discuss this definition in Chapter 4 on differential equations.

Passive

A linear network is *passive*[2] if (*a*) the energy delivered to the network is nonnegative for any arbitrary excitation, and (*b*) if no voltages or currents appear between any two terminals before an excitation is applied.

Reciprocal

A network is said to be *reciprocal* if when the points of excitation and measurement of response are interchanged, the relationship between

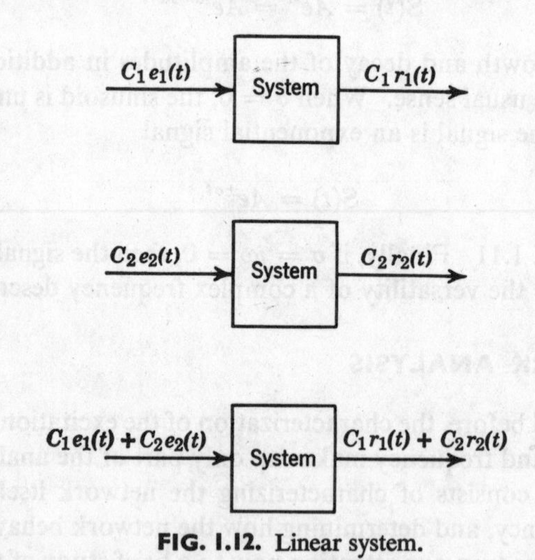

FIG. 1.12. Linear system.

[2] G. Raisbeck, "A Definition of Passive Linear Networks in Terms of Time and Energy," *J. Appl. Phys.*, 25 (Dec. 1954), 1510–1514.

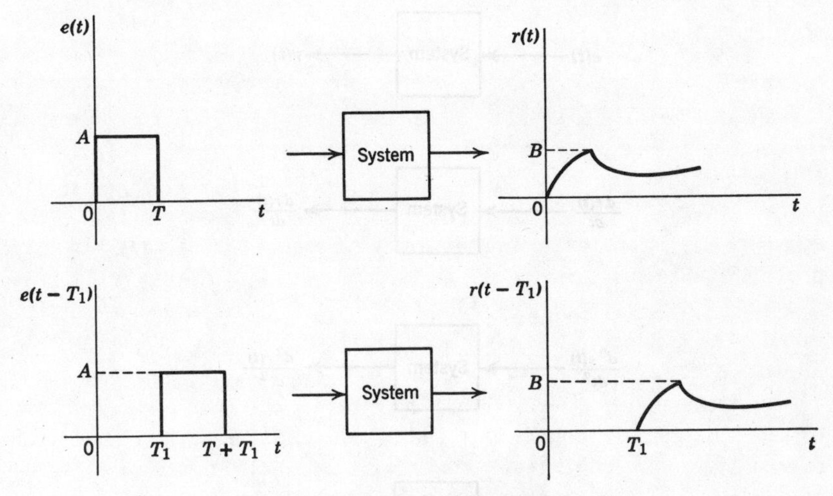

FIG. 1.13. Time-invariant system.

excitation and response remains the same. Thus must be true for any choice of points of excitation and response.

Causal

We say a system is *causal* if its response is nonanticipatory, i.e., if

$$e(t) = 0 \qquad t < T$$

then

$$r(t) = 0 \qquad t < T \tag{1.10}$$

In other words, a system is causal if before an excitation is applied at $t = T$, the response is zero for $-\infty < t < T$.

Time invariant

A system is *time invariant* if $e(t) \rightarrow r(t)$ implies that $e(t \pm T) \rightarrow r(t \pm T)$, where the symbol $\rightarrow$ means "gives rise to." To understand the concept of time invariance in a linear system, let us suppose that initially the excitation is introduced at $t = 0$, which gives rise to a response $r(t)$. If the excitation were introduced at $t = T$, and if the shape of the response waveform were the same as in the first case, but delayed by a time T (Fig. 1.13), then we could say the system is time invariant. Another way of looking at this concept is through the fact that time-invariant systems contain only elements that do not vary with time. It should be mentioned here that linear systems need not be time invariant.

Derivative property

From the time-invariant property we can show that, if $e(t)$ at the input gives rise to $r(t)$ at the output (Fig. 1.14), then, if the input were $e'(t)$,

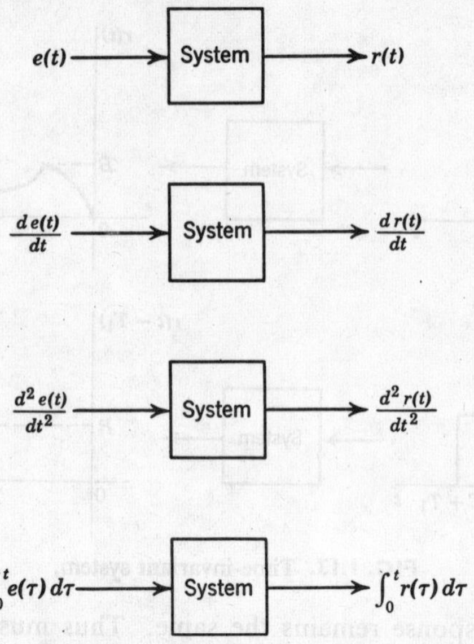

FIG. 1.14. Some implications of linear time-invariant systems.

i.e., the derivative of $e(t)$, the response would be $r'(t)$. The proof is quite simple. Consider an excitation $e(t + \epsilon)$ where ϵ is a real quantity. By the time-invariant property, the response would be $r(t + \epsilon)$. Now suppose the excitation were

$$e_1(t) = \frac{1}{\epsilon}[e(t + \epsilon) - e(t)] \tag{1.11}$$

then according to the linearity and time-invariant properties, the response would be

$$r_1(t) = \frac{1}{\epsilon}[r(t + \epsilon) - r(t)] \tag{1.12}$$

Taking the limit as $\epsilon \to 0$, we see that

$$\begin{aligned} \lim_{\epsilon \to 0} e_1(t) &= \frac{d}{dt} e(t) \\ \lim_{\epsilon \to 0} r_1(t) &= \frac{d}{dt} r(t) \end{aligned} \tag{1.13}$$

We can extend this idea to higher derivatives as well as for the integrals of $e(t)$ and $r(t)$, as shown in Fig. 1.14.

Ideal models

Let us now examine some idealized models of linear systems. The systems given in the following all have properties which make them very useful in signal processing.

FIG. 1.15. Amplifier.

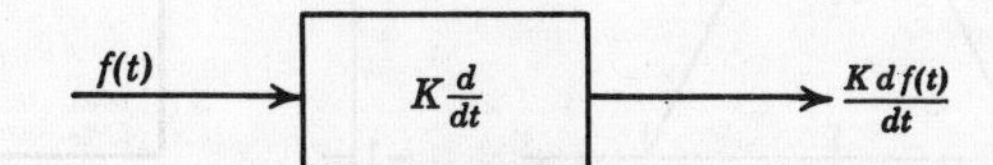

FIG. 1.16. Differentiator.

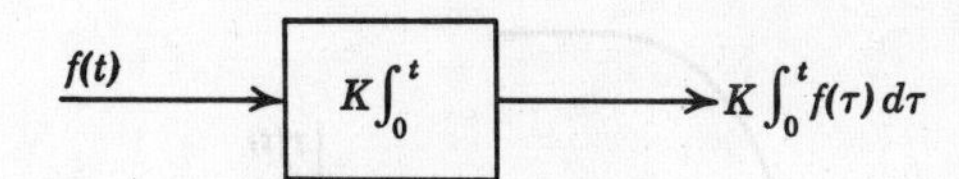

FIG. 1.17. Integrator.

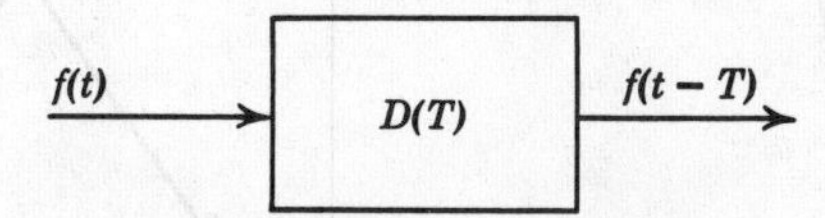

FIG. 1.18. Time-delay network.

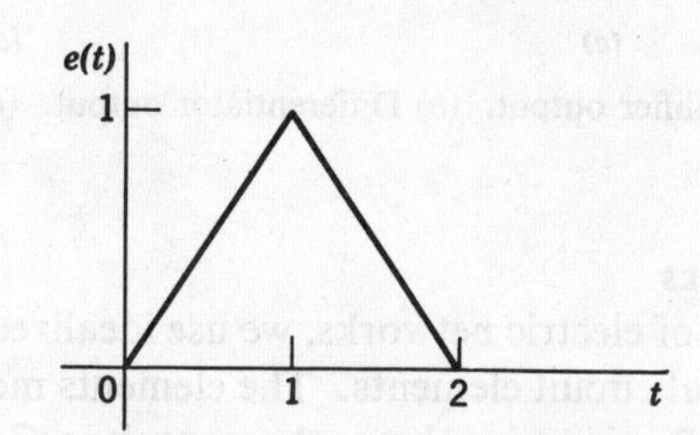

FIG. 1.19. Excitation function.

1. *Amplifier:* An amplifier scales up the magnitude of the input, i.e., $r(t) = Ke(t)$, where K is a constant (Fig. 1.15).
2. *Differentiator:* The input signal is differentiated and possibly scaled up or down (Fig. 1.16).
3. *Integrator:* The output is the integral of the input, as shown in Fig. 1.17.
4. *Time delayer:* The output is delayed by an amount T, but retains the same wave shape as the input (Fig. 1.18).

Suppose we take the triangular pulse in Fig. 1.19 as the input signal. Then the outputs for each of the four systems just described are shown in Figs. 1.20*a*–1.20*d*.

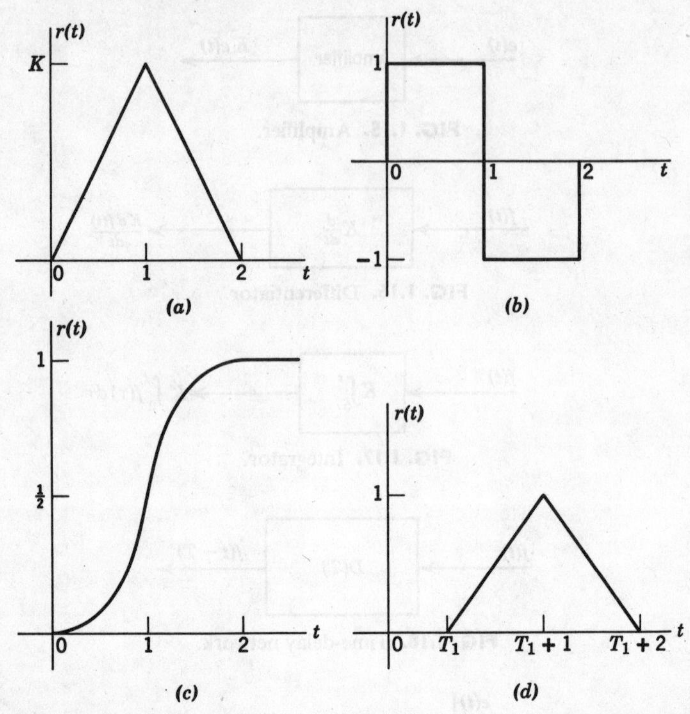

FIG. 1.20. (*a*) Amplifier output. (*b*) Differentiator output. (*c*) Integrator output. (*d*) Delayed output.

Ideal elements

In the analysis of electric networks, we use idealized linear mathematical models of physical circuit elements. The elements most often encountered are the resistor R, given in ohms, the capacitor C, given in farads, and the inductor L, expressed in henrys. The endpoints of the elements are called *terminals*. A *port* is defined as any pair of two terminals into which energy is supplied or withdrawn or where network variables may be measured or observed. In Fig. 1.21 we have an example of a two-port network.

The energy sources that make up the excitation functions are ideal *current* or *voltage sources*, as shown in Figs. 1.22*a* and *b*. The polarities

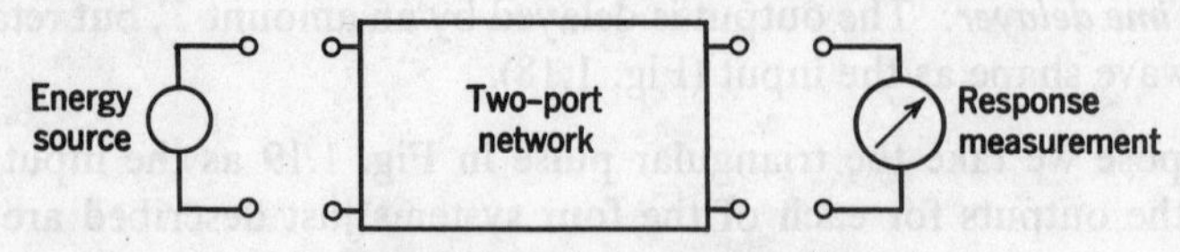

FIG. 1.21. Two-port network.

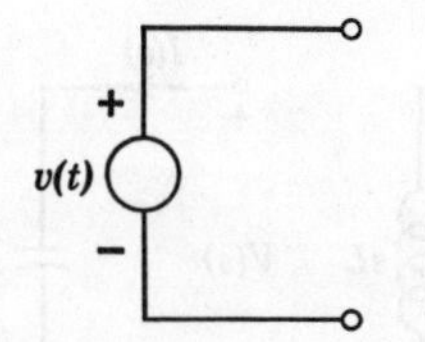

FIG. 1.22a. Voltage source.

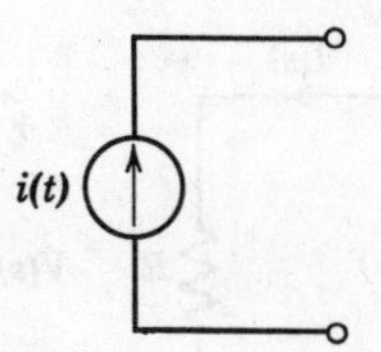

FIG. 1.22b. Current source.

indicated for the voltage source and the direction of flow for the current source are arbitrarily assumed for reference purposes only. An ideal voltage source is an energy source that provides, at a given port, a voltage signal that is independent of the current at that port. If we interchange the words "current" and "voltage" in the last definition, we then define an ideal current source.

In network analysis, the principal problem is to find the relationships that exist between the currents and voltages at the ports of the network. Certain simple voltage-current relationships for the network elements also serve as defining equations for the elements themselves. For example, when the currents and voltages are expressed as functions of time, then the R, L, and C elements, shown in Fig. 1.23, are defined by the equations

$$v(t) = Ri(t) \qquad \text{or} \qquad i(t) = \frac{1}{R}\,v(t)$$

$$v(t) = L\,\frac{di(t)}{dt} \qquad \text{or} \qquad i(t) = \frac{1}{L}\int_0^t v(x)\,dx + i(0) \qquad (1.14)$$

$$v(t) = \frac{1}{C}\int_0^t i(x)\,dx + v(0) \qquad \text{or} \qquad i(t) = C\,\frac{dv(t)}{dt}$$

where the constants of integration $i(0)$ and $v(0)$ are *initial conditions* to be discussed in detail later.

Expressed as a function of the complex frequency variable s, the equations

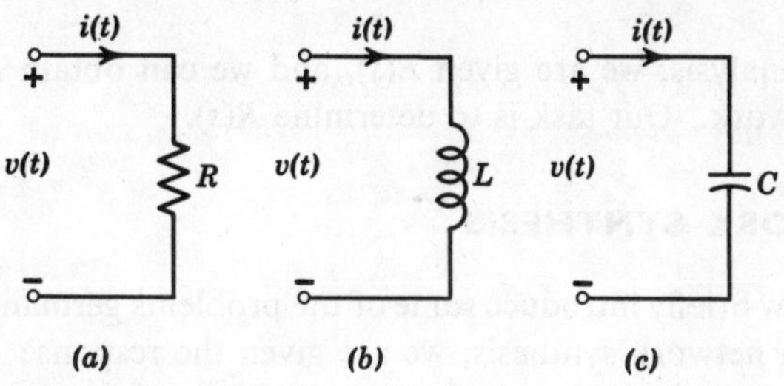

FIG. 1.23. (a) Resistor. (b) Inductor. (c) Capacitor.

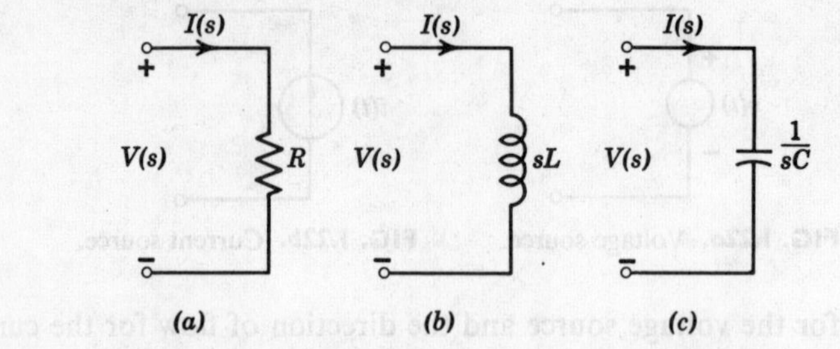

FIG. 1.24. (*a*) Resistor. (*b*) Inductor. (*c*) Capacitor.

defining the R, L, and C elements, shown in Fig. 1.24, are (ignoring initial conditions for the moment)

$$V(s) = RI(s) \quad \text{or} \quad I(s) = \frac{1}{R}V(s)$$

$$V(s) = sLI(s) \quad \text{or} \quad I(s) = \frac{1}{sL}V(s) \tag{1.15}$$

$$V(s) = \frac{1}{sC}I(s) \quad \text{or} \quad I(s) = sCV(s)$$

We see that in the *time domain*, i.e., where the independent variable is t, the voltage-current relationships are given in terms of differential equations. On the other hand, in the *complex-frequency domain*, the voltage-current relationships for the elements are expressed in *algebraic* equations. Algebraic equations are, in most cases, more easily solved than differential equations. Herein lies the *raison d'être* for describing signals and networks in the frequency domain as well as in the time domain.

When a network is made up of an interconnection of linear circuit elements, the network is described by its *system* or *transfer function* $H(s)$. The response $R(s)$ and the excitation $E(s)$ are related by the equation

$$R(s) = H(s)\,E(s). \tag{1.16}$$

In network analysis, we are given $E(s)$, and we can obtain $H(s)$ directly from the network. Our task is to determine $R(s)$.

1.4 NETWORK SYNTHESIS

We will now briefly introduce some of the problems germane to network synthesis. In network synthesis, we are given the response $R(s)$ and the excitation $E(s)$, and we are required to synthesize the network from the

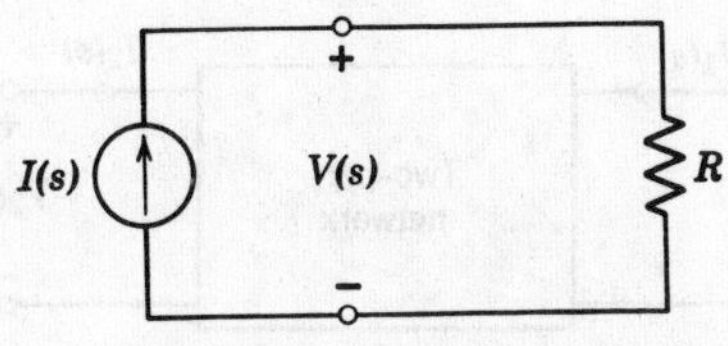

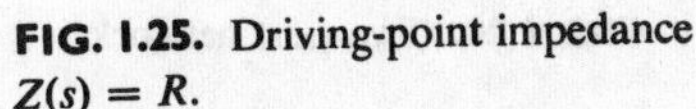

FIG. 1.25. Driving-point impedance $Z(s) = R$.

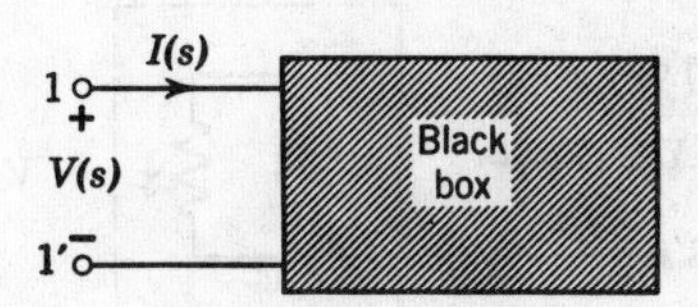

FIG. 1.26. Black box.

system function

$$H(s) = \frac{R(s)}{E(s)} \tag{1.17}$$

Since $R(s)$ and $E(s)$ are voltages or currents, then $H(s)$ is denoted generally as an *immittance* if $R(s)$ is a voltage and $E(s)$ is a current, or vice versa. A *driving-point immittance*[3] is defined to be a function for which the variables are measured at the same port. Thus a *driving-point impedance* $Z(s)$ at a given port is the function

$$Z(s) = \frac{V(s)}{I(s)} \tag{1.18}$$

where the excitation is a current $I(s)$ and the response is a voltage $V(s)$, as shown in Fig. 1.25. When we interchange the words "current" and "voltage" in the last definition, we then have a *driving-point admittance*. An example of a driving-point impedance is the network in Fig. 1.25, where

$$Z(s) = \frac{V(s)}{I(s)} = R \tag{1.19}$$

Now suppose the resistor in Fig. 1.25 were enclosed in a "black box." We have no access to this black box, except at the terminals 1-1′ in Fig. 1.26. Our task is to determine the network in the black box. Suppose we are given the information that, for a given excitation $I(s)$, the voltage response $V(s)$ is proportional to $I(s)$ by the equation

$$V(s) = K\,I(s) \tag{1.20}$$

An obvious solution, though not unique, is that the network consists of a resistor of value $R = K\,\Omega$. Suppose next that the excitation is a voltage $V(s)$, the response is a current $I(s)$, and that

$$Y(s) = \frac{I(s)}{V(s)} = 3 + 4s \tag{1.21}$$

[3] IRE Standards on Circuits "Linear Passive Networks," *Proc. IRE*, **48**, No. 9 (Sept. 1960), 1608–1610.

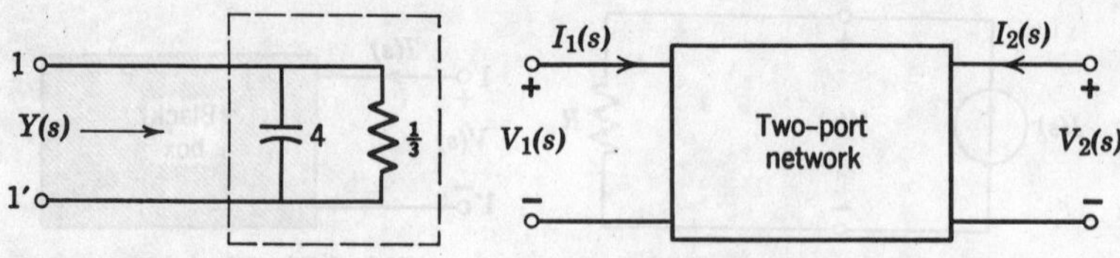

FIG. 1.27. Network realization for $Y(s)$.

FIG. 1.28. Two-port network.

Our task is to synthesize a network equivalent to the network in the black box. From a close scrutiny of the driving-point admittance $Y(s)$, we see that a possible solution might consist of a resistor of $\frac{1}{3}\ \Omega$ in parallel with a capacitor of 4 farads, as seen in Fig. 1.27.

The problem of driving-point synthesis, as shown from the examples just given, consists of decomposing a given immittance function into basic recognizable parts (such as $3 + 4s$). Before we proceed with the mechanics of decomposition, we must first determine whether the function is *realizable*, i.e., can it be synthesized in terms of positive resistances, inductances, and capacitances? It will be shown that realizable driving-point immittances belong to a class of functions known as *positive real* or, simply, p.r. functions. From the properties of p.r. functions, we can test a given driving-point function for realizability. (The Appendices present a short introduction to complex variables as well as the proofs of some theorems on positive real functions.) With a knowledge of p.r. functions, we then go on to examine special driving-point functions. These include functions which can be realized with two kinds of elements only—the *L-C*, *R-C*, and *R-L* immittances.

Next we proceed to the synthesis of transfer functions. According to the IRE Standards on passive linear networks,[4] a *transfer function* or *transmittance* is a system function for which the variables are measured at different ports. There are many different forms which a transfer function might take. For example, consider the two-port network in Fig. 1.28. If the excitation is $I_1(s)$ and the response $V_2(s)$, the transfer function is a *transfer impedance*

$$Z_{21}(s) = \frac{V_2(s)}{I_1(s)} \tag{1.22}$$

On the other hand, if $V_1(s)$ were the excitation and $V_2(s)$ the response, then we would have a *voltage-ratio* transfer function

$$H(s) = \frac{V_2(s)}{V_1(s)} \tag{1.23}$$

[4] *Loc. cit.*

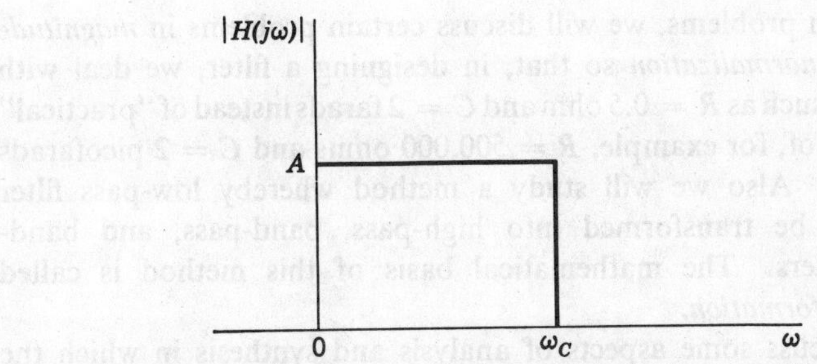

FIG. 1.29. Ideal amplitude spectrum for low-pass filter.

As for driving-point functions, there are certain properties which a transfer function must satisfy in order to be realizable. We shall study these realizability conditions and then proceed to the synthesis of some simple transfer functions.

The most important aspect of transfer function synthesis is *filter design*. A filter is defined as a network which passes a certain portion of a frequency spectrum and blocks the remainder of the spectrum. By the term "blocking," we imply that the magnitude response $|H(j\omega)|$ of the filter is approximately zero for that frequency range. Thus, an ideal *low-pass* filter is a network which passes all frequencies up to a *cutoff* frequency ω_C, and blocks all frequencies above ω_C, as shown in Fig. 1.29.

One aspect of filter design is to synthesize the network from the transfer function $H(s)$. The other aspect deals with the problem of obtaining a realizable transmittance $H(s)$ given the specification of, for example, the magnitude characteristic in Fig. 1.29. This part of the synthesis is generally referred to as the *approximation* problem. Why the word "approximation?" Because frequency response characteristics of the R, L, and C elements are continuous (with the exception of isolated points called *resonance* points), a network containing these elements cannot be made to cut off abruptly at ω_C in Fig. 1.29. Instead, we can realize low-pass filters which have the magnitude characteristics of Fig. 1.30. In connection with

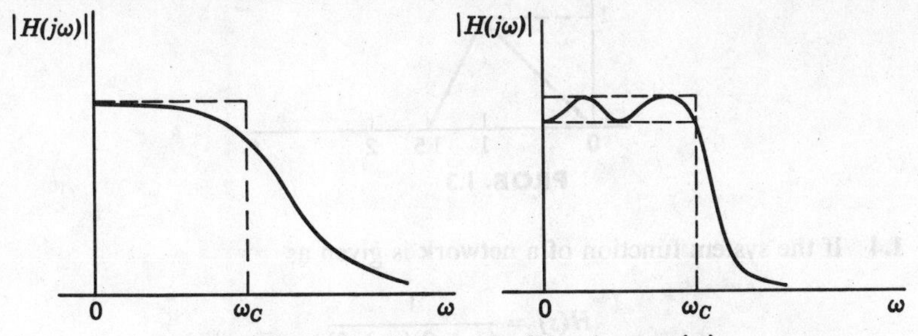

FIG. 1.30. Realizable low-pass filter characteristics.

the filter design problems, we will discuss certain problems in *magnitude* and *frequency normalization* so that, in designing a filter, we deal with element values such as $R = 0.5$ ohm and $C = 2$ farads instead of "practical" element values of, for example, $R = 500{,}000$ ohms and $C = 2$ picofarads (pico $= 10^{-12}$). Also we will study a method whereby low-pass filter designs might be transformed into high-pass, band-pass, and band-elimination filters. The mathematical basis of this method is called *frequency transformation.*

We next discuss some aspects of analysis and synthesis in which the excitation and response functions are given in terms of *power* rather than of voltage and current. We will examine the power-transfer properties of linear networks, using *scattering parameters*, which describe the incident and reflected power of the network at its ports.

Finally, in Chapter 15, we will examine some of the many uses of high-speed digital computers in circuit analysis and design. In addition to a general survey of the field, we will also study some specific computer programs in circuit analysis.

Problems

1.1 Draw the line spectra for the signal

$$s(t) = 3 \sin\left(t + \frac{\pi}{4}\right) + 4 \sin\left(2t - \frac{\pi}{8}\right) + 6 \sin 3t$$

1.2 Find the response to the excitation $\sin t$ into a *sampler* that closes every $K\pi/4$ seconds where $K = 0, 1, 2, \ldots$. Draw the response for $0 \leq t \leq 2\pi$.

1.3 Find the response to the excitation shown in the figure when the network is (*a*) an ideal differentiator; (*b*) an ideal integrator.

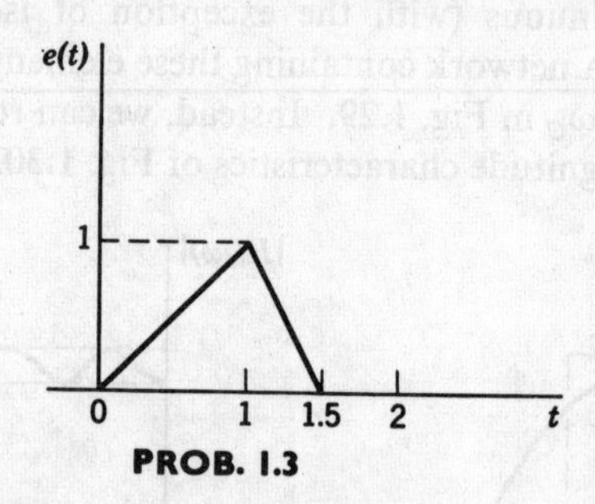

PROB. 1.3

1.4 If the system function of a network is given as

$$H(s) = \frac{1}{(s+2)(s+3)}$$

find the response $R(s)$ if the excitation is

$$E(s) = \frac{3}{s}$$

1.5 Given the driving-point functions find their simplest network realizations.

(*a*) $$Z(s) = 3 + 2s + \frac{1}{3s}$$

(*b*) $$Y(s) = 2s + \frac{3s}{s + 2}$$

(*c*) $$Z(s) = 3 + \frac{s}{s^2 + 2}$$

(*d*) $$Y(s) = \frac{1}{3s + 2} + \frac{2s}{s^2 + 4}$$

1.6 For the network shown, write the mesh equation in terms of (*a*) differential equations and (*b*) the complex-frequency variable s.

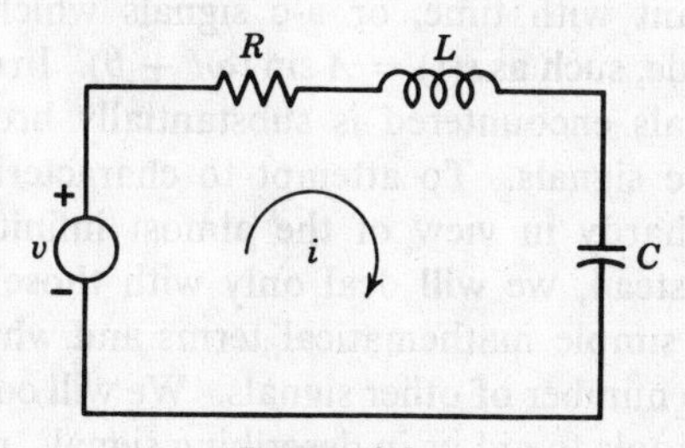

PROB. 1.6

1.7 For the network shown, write the node equation in terms of (*a*) differential equations and (*b*) complex-frequency form.

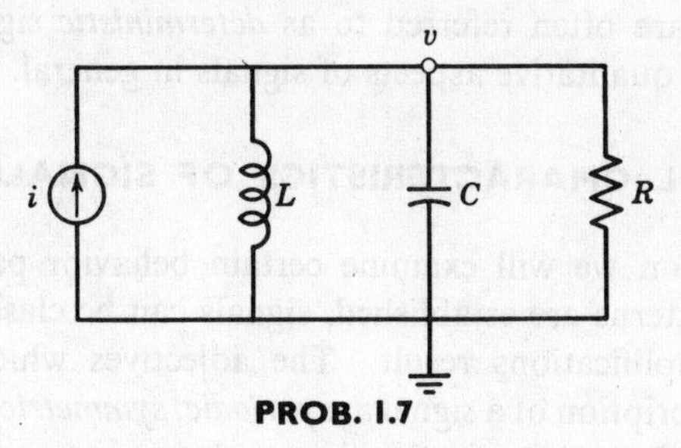

PROB. 1.7

1.8 Suppose the response of a linear system to an excitation $e(t)$ were $r(t) = 3e^{-4t}$. What would the response be to an excitation of $e(t - 2)$?

chapter 2
Signals and waveforms

Our main concern in this chapter is the characterization of signals as functions of time. In previous studies we have dealt with d-c signals that were constant with time, or a-c signals which were sinusoids of constant amplitude, such as $s(t) = A \sin (\omega t + \theta)$. In engineering practice, the class of signals encountered is substantially broader in scope than simple a-c or d-c signals. To attempt to characterize each member of the class is foolhardy in view of the almost infinite variety of signals encountered. Instead, we will deal only with those signals that can be characterized in simple mathematical terms and which serve as *building blocks* for a large number of other signals. We will concentrate on formulating analytical tools to aid us in describing signals, rather than deal with the representation of specific signals. Because of time and space limitations, we will cover only signals which do not exhibit random behavior, i.e., signals which can be explicitly characterized as functions of time. These signals are often referred to as *deterministic* signals. Let us first discuss certain qualitative aspects of signals in general.

2.1 GENERAL CHARACTERISTICS OF SIGNALS

In this section we will examine certain behavior patterns of signals. Once these patterns are established, signals can be classified accordingly, and some simplifications result. The adjectives which give a general qualitative description of a signal are *periodic*, *symmetrical*, and *continuous*. Let us discuss these terms in the given order.

First, signals are either *periodic* or *aperiodic*. If a signal is periodic, then it is described by the equation

$$s(t) = s(t \pm kT) \qquad k = 0, 1, 2, \ldots \tag{2.1}$$

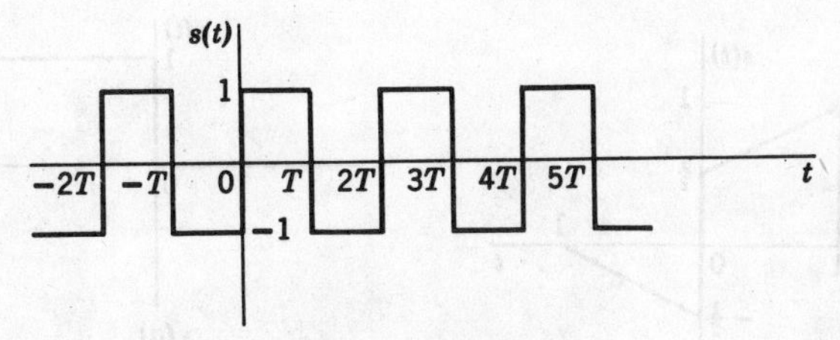

FIG. 2.1. Square wave.

where T is the period of the signal. The sine wave, $\sin t$, is periodic with period $T = 2\pi$. Another example of a periodic signal is the *square wave* given in Fig. 2.1. On the other hand, the signals given in Fig. 2.2 are aperiodic, because the pulse patterns do not repeat after a certain finite interval T. Alternatively, these signals may be considered "periodic" with an infinite period.

Next, consider the *symmetry* properties of a signal. The key adjectives here are *even* and *odd*. A signal function can be even or odd or neither. An even function obeys the relation

$$s(t) = s(-t) \tag{2.2}$$

For an odd function
$$s(t) = -s(-t) \tag{2.3}$$

For example, the function $\sin t$ is odd, whereas $\cos t$ is even. The square pulse in Fig. 2.2*a* is even, whereas the triangular pulse is odd (Fig. 2.2*b*).

Observe that a signal need not be even or odd. Two examples of signals of this type are shown in Figs. 2.3*a* and 2.4*a*. It is significant to note, however, that any signal $s(t)$ can be resolved into an even component $s_e(t)$ and an odd component $s_0(t)$ such that

$$s(t) = s_e(t) + s_0(t) \tag{2.4}$$

For example, the signals in Figs. 2.3*a* and 2.4*a* can be decomposed into odd and even components, as indicated in Figs. 2.3*b*, 2.3*c*, 2.4*b*, and 2.4*c*.

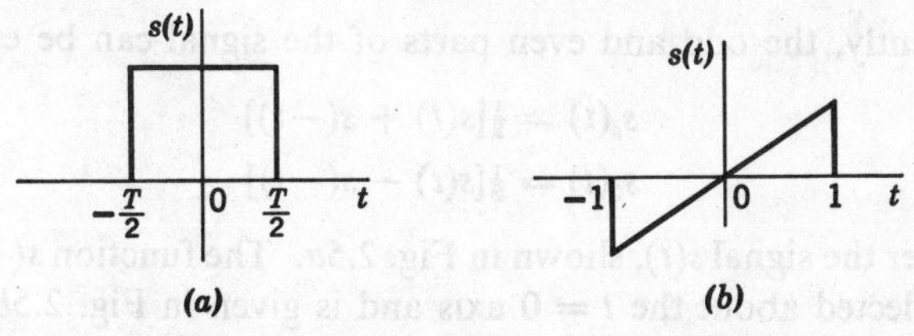

FIG. 2.2. (*a*) Even function. (*b*) Odd function.

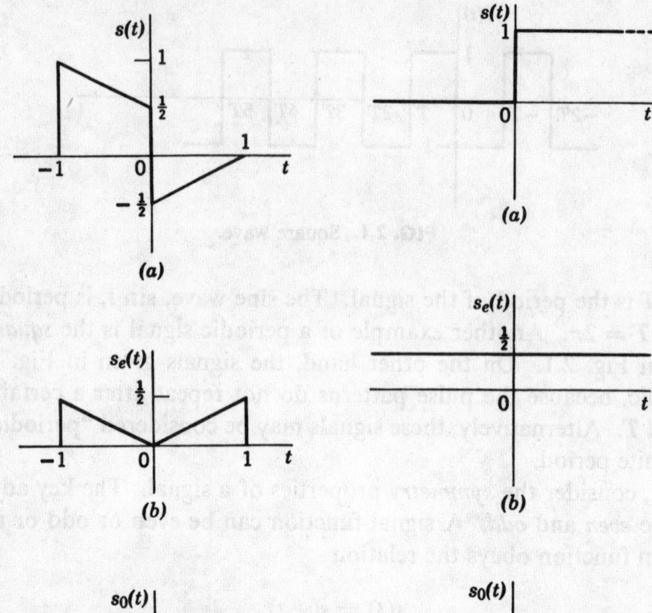

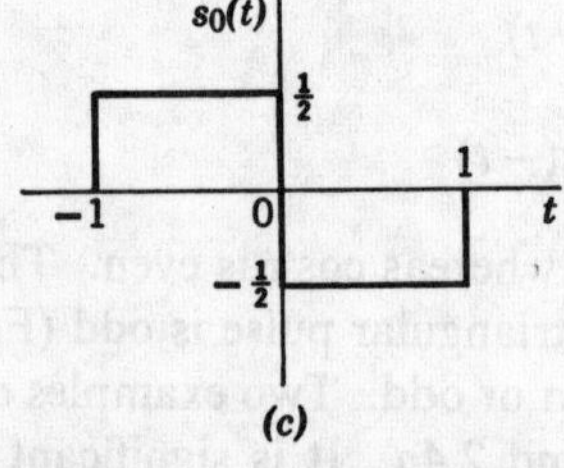

FIG. 2.3. Decomposition into odd and even components. (*a*) Original function. (*b*) Even part. (*c*) Odd part.

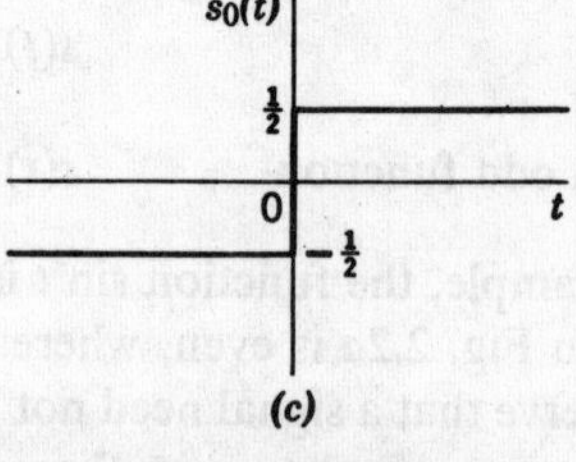

FIG. 2.4. Decomposition into even and odd components. (*a*) Unit step function. (*b*) Even part of unit step. (*c*) Odd part of unit step.

From Eq. 2.4 we observe that

$$\begin{aligned} s(-t) &= s_e(-t) + s_0(-t) \\ &= s_e(t) - s_0(t) \end{aligned} \tag{2.5}$$

Consequently, the odd and even parts of the signal can be expressed as

$$\begin{aligned} s_e(t) &= \tfrac{1}{2}[s(t) + s(-t)] \\ s_0(t) &= \tfrac{1}{2}[s(t) - s(-t)] \end{aligned} \tag{2.6}$$

Consider the signal $s(t)$, shown in Fig. 2.5*a*. The function $s(-t)$ is equal to $s(t)$ reflected about the $t = 0$ axis and is given in Fig. 2.5*b*. We then obtain $s_e(t)$ and $s_0(t)$ as shown in Figs. 2.5*c* and *d*, respectively.

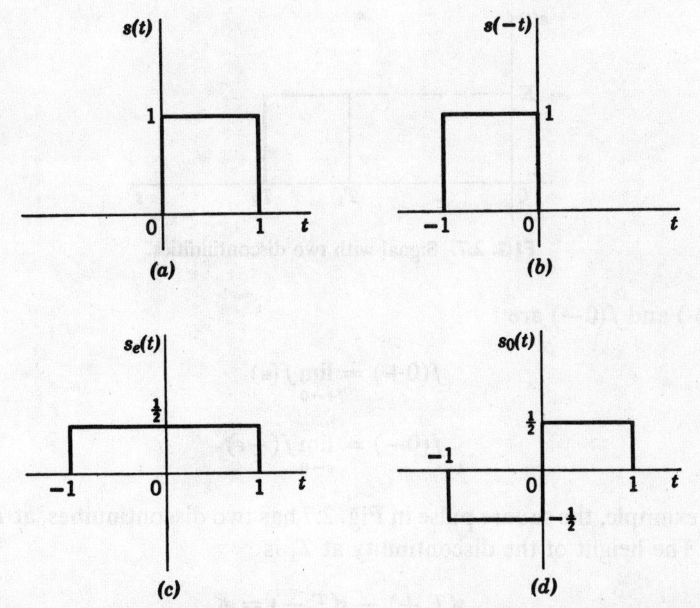

FIG. 2.5. Decomposition into odd and even components from $s(t)$ and $s(-t)$.

Now let us turn our attention to the *continuity* property of signals. Consider the signal shown in Fig. 2.6. At $t = T$, the signal is *discontinuous*. The height of the discontinuity is

$$f(T+) - f(T-) = A \tag{2.7}$$

where

$$f(T+) = \lim_{\epsilon \to 0} f(T + \epsilon)$$
$$f(T-) = \lim_{\epsilon \to 0} f(T - \epsilon) \tag{2.8}$$

and ϵ is a real positive quantity. In particular, we are concerned with discontinuities in the neighborhood of $t = 0$. From Eq. 2.8, the points

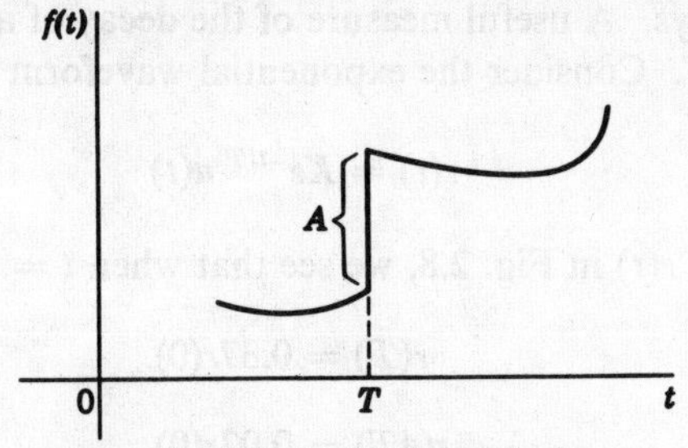

FIG. 2.6. Signal with discontinuity.

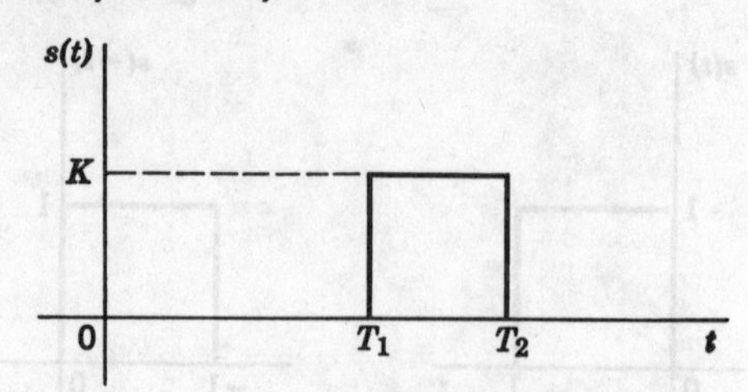

FIG. 2.7. Signal with two discontinuities.

$f(0+)$ and $f(0-)$ are

$$f(0+) = \lim_{\epsilon \to 0} f(\epsilon)$$
$$f(0-) = \lim_{\epsilon \to 0} f(-\epsilon) \tag{2.9}$$

For example, the square pulse in Fig. 2.7 has two discontinuities, at T_1 and T_2. The height of the discontinuity at T_1 is

$$s(T_1+) - s(T_1-) = K \tag{2.10}$$

Similarly, the height of the discontinuity at T_2 is $-K$.

2.2 GENERAL DESCRIPTIONS OF SIGNALS

In this section we consider various time domain descriptions of signals. In particular, we examine the meanings of the following terms: *time constant, rms value, d-c value, duty cycle,* and *crest factor*. The term, time constant, refers only to exponential waveforms; the remaining four terms describe only periodic waveforms.

Time constant

In many physical problems, it is important to know how quickly a waveform decays. A useful measure of the decay of an exponential is the *time constant T*. Consider the exponential waveform described by

$$r(t) = Ke^{-t/T}\, u(t) \tag{2.11}$$

From a plot of $r(t)$ in Fig. 2.8, we see that when $t = T$,

$$r(T) = 0.37r(0) \tag{2.12}$$

Also
$$r(4T) = 0.02r(0) \tag{2.13}$$

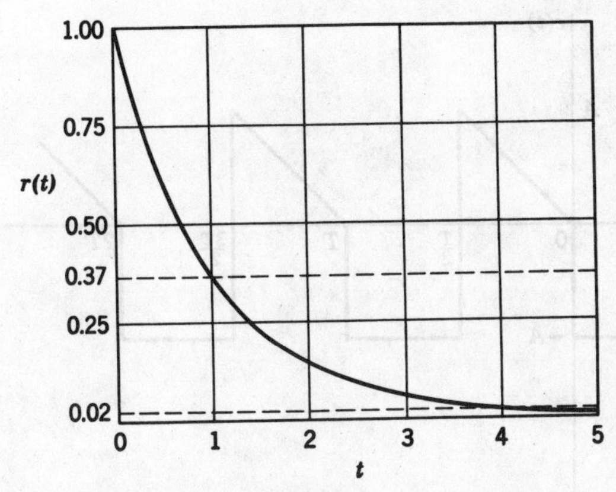

FIG. 2.8. Normalized curve for time constant $T = 1$.

Observe that the larger the time constant, the longer it requires for the waveform to reach 37% of its peak value. In circuit analysis, common time constants are the factors RC and R/L.

RMS Value

The *rms* or *root mean square* value of a periodic waveform $e(t)$ is defined as

$$e_{\text{rms}} = \left[\frac{1}{T}\int_0^T e^2(t)\,dt\right]^{1/2} \tag{2.14}$$

where T is the period. If the waveform is not periodic, the term rms does not apply. As an example, let us calculate the rms voltage for the periodic waveform in Fig. 2.9.

$$\begin{aligned} e_{\text{rms}} &= \left\{\frac{1}{T}\left[\int_0^{T/2}\left(\frac{2A}{T}t\right)^2 dt + \int_{T/2}^{T} A^2\,dt\right]\right\}^{1/2} \\ &= \left\{\frac{1}{T}\left[\frac{4A^2}{T^2}\frac{t^3}{3}\bigg|_0^{T/2} + A^2t\bigg|_{T/2}^{T}\right]\right\}^{1/2} \\ &= \left(\frac{A^2}{6} + \frac{A^2}{2}\right)^{1/2} \\ &= \sqrt{2/3}\,A \text{ v} \end{aligned} \tag{2.15}$$

D-C Value

The d-c value of a waveform has meaning only when the waveform is periodic. It is the average value of the waveform over one period

$$e_{\text{d-c}} = \frac{1}{T}\int_0^T e(t)\,dt \tag{2.16}$$

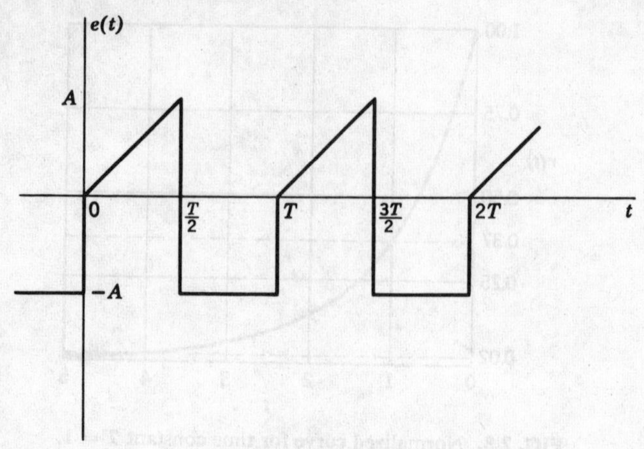

FIG. 2.9. Periodic waveform.

The square wave in Fig. 2.1 has zero d-c value, whereas the waveform in Fig. 2.9 has a d-c value of

$$e_{\text{d-c}} = \frac{1}{T}\left[\frac{AT}{4} - \frac{AT}{2}\right] = -\frac{A}{4}\ \text{v} \tag{2.17}$$

Duty cycle

The term *duty cycle, D,* is defined as the ratio of the time duration of the *positive* cycle t_0 of a periodic waveform to the period, T, that is,

$$D = \frac{t_0}{T} \tag{2.18}$$

The duty cycle of a pulse train becomes important in dealing with waveforms of the type shown in Fig. 2.10, where most of the energy is concentrated in a narrow pulse of width t_0. The rms voltage of the waveform in Fig. 2.10 is

$$\begin{aligned} e_{\text{rms}} &= \left(\frac{1}{T}\int_0^{t_0} A^2\, dt\right)^{1/2} \\ &= A\sqrt{t_0/T} \\ &= A\sqrt{D} \end{aligned} \tag{2.19}$$

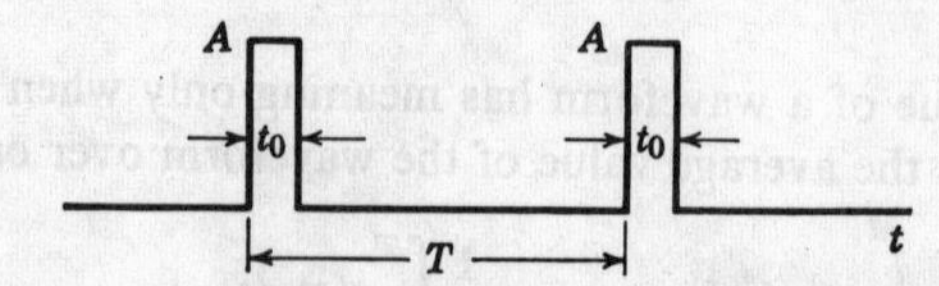

FIG. 2.10. Periodic waveform with small duty cycle.

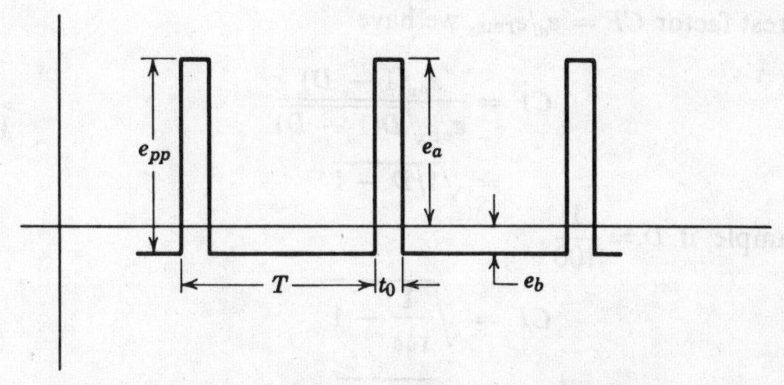

FIG. 2.11. Periodic waveform with zero d-c and small duty cycle.

We see that the smaller the duty cycle, the smaller the rms voltage. The square wave in Fig. 2.1 has a 50% duty cycle.

Crest factor

Crest factor[1] is defined as the ratio of the peak voltage of a periodic waveform to the rms value (with the d-c component removed). Explicitly, for any waveform with zero d-c such as the one shown in Fig. 2.11—crest factor, *CF*, is defined as

$$CF = \frac{e_a}{e_{\text{rms}}} \quad \text{or} \quad \frac{e_b}{e_{\text{rms}}} \tag{2.20}$$

whichever is greater. For the waveform in Fig. 2.11, the peak-to-peak voltage is defined as

$$e_{pp} = e_a + e_b \tag{2.21}$$

Since the waveform has zero d-c value

$$e_a t_0 = e_b(T - t_0) \tag{2.22}$$

Also,

$$e_b = e_{pp} D \tag{2.23}$$

and

$$e_a = e_{pp}(1 - D) \tag{2.24}$$

The rms value of the waveform is

$$\begin{aligned} e_{\text{rms}} &= \left(\frac{e_{pp}^{\,2}(1 - D)^2\, t_0 + e_{pp}^{\,2} D^2 (T - t_0)}{T} \right)^{1/2} \\ &= e_{pp}\sqrt{D(1 - D)} \end{aligned} \tag{2.25}$$

[1] G. Justice, "The Significance of Crest Factor," Hewlett-Packard Journal, **15**, No. 5 (Jan., 1964), 4–5.

Since crest factor $CF = e_a/e_{\mathrm{rms}}$, we have

$$CF = \frac{e_{pp}(1 - D)}{e_{pp}\sqrt{D(1 - D)}} = \sqrt{1/D - 1} \tag{2.26}$$

For example, if $D = \dfrac{1}{100}$,

$$CF = \sqrt{\frac{1}{\frac{1}{100}} - 1} = \sqrt{100 - 1} \simeq 10 \tag{2.27}$$

If $D = \dfrac{1}{10{,}000}$,

$$CF = \sqrt{10{,}000 - 1} \simeq 100 \tag{2.28}$$

A voltmeter with high crest factor is able to read accurately rms values of signals whose waveforms differ from sinusoids, in particular, signals with low duty factor. Note that the smallest value of crest factor occurs for the maximum value of D, that is, $D_{\max} = 0.5$,

$$CF_{\min} = \sqrt{1/D_{\max} - 1} = 1 \tag{2.29}$$

2.3 THE STEP FUNCTION AND ASSOCIATED WAVEFORMS

The unit step function $u(t)$ shown in Fig. 2.12 is defined as

$$u(t) = 0 \qquad t < 0$$
$$= 1 \qquad t \geq 0 \tag{2.30}$$

The physical analogy of a unit step excitation corresponds to a switch S, which closes at $t = 0$ and connects a d-c battery of 1 volt to a given circuit, as shown in Fig. 2.13. Note that the unit step is zero whenever the

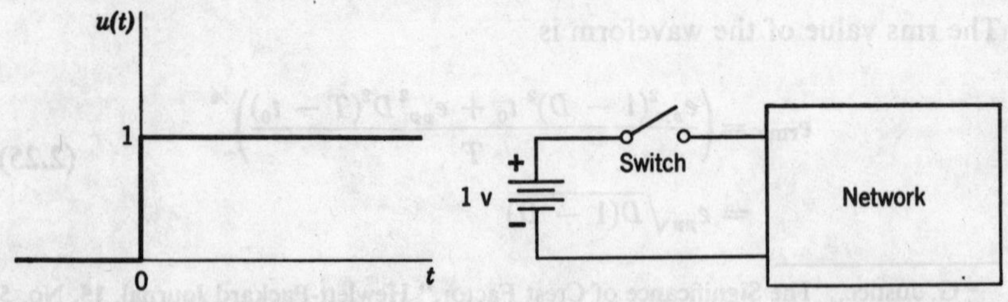

FIG. 2.12. Unit step function.

FIG. 2.13. Network analog of unit step.

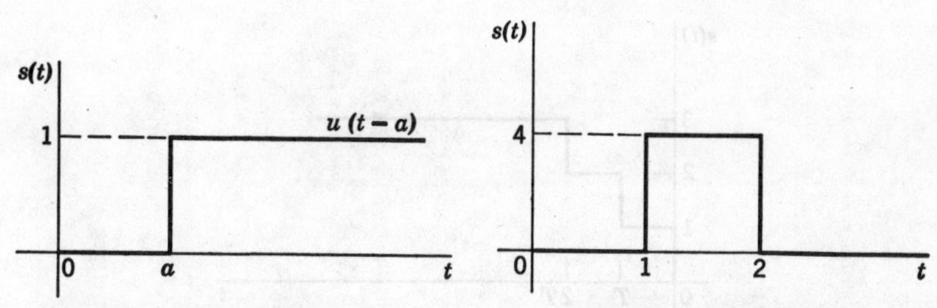

FIG. 2.14. Shifted step function. **FIG. 2.15.** Square pulse.

argument (t) within the parentheses is negative, and is unity when the argument (t) is greater than zero. Thus the function $u(t - a)$, where $a > 0$, is defined by

$$\begin{aligned} u(t - a) &= 0 \qquad t < a \\ &= 1 \qquad t \geq a \end{aligned} \tag{2.31}$$

and is shown in Fig. 2.14. Note that the jump discontinuity of the step occurs when the argument within the parentheses is zero. This forms the basis of the *shifting* property of the step function. Also, the height of the jump discontinuity of the step can be scaled up or down by the multiplication of a constant K.

With the use of the change of amplitude and the shifting properties of the step function, we can proceed to construct a family of pulse waveforms. For example, the square pulse in Fig. 2.15 can be constructed by the sum of two step functions

$$s(t) = 4u(t - 1) + (-4)\,u(t - 2) \tag{2.32}$$

as given in Fig. 2.16. The "staircase" function, shown in Fig. 2.17, is characterized by the equation

$$s(t) = \sum_{k=0}^{2} u(t - kT) \tag{2.33}$$

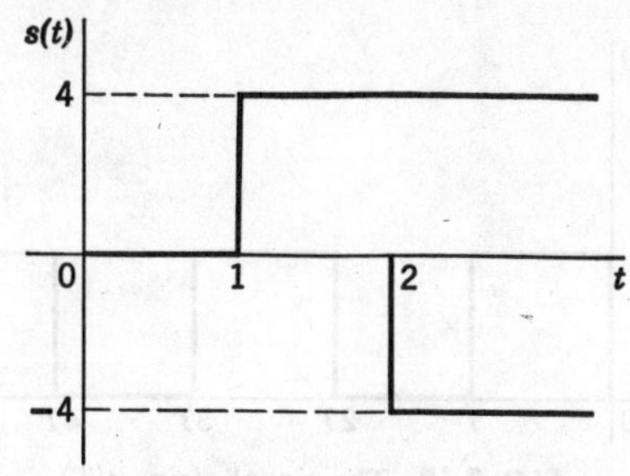

FIG. 2.16. Construction of square pulse by step function.

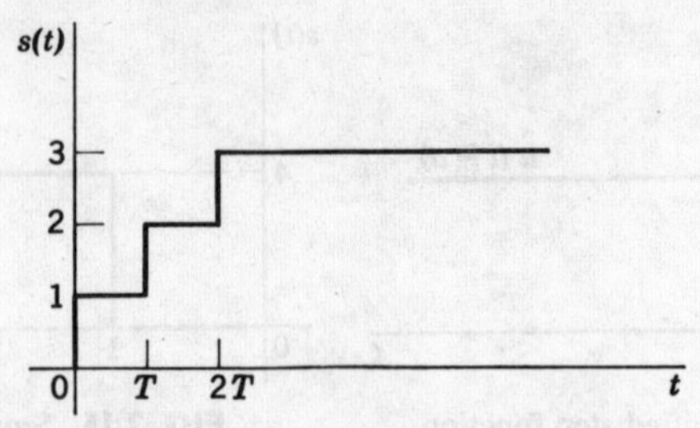

FIG. 2.17. Staircase function.

Finally, let us construct the square wave in Fig. 2.1. Using the shifting property, we see that the square wave is given by (for $t \geq 0$)

$$s(t) = u(t) - 2u(t - T) + 2u(t - 2T) - 2u(t - 3T) + \cdots \quad (2.34)$$

A simpler way to represent the square wave is by using the property that the step function is zero whenever its argument is negative. Restricting ourselves to the interval $t \geq 0$, the function

$$s(t) = u\left(\sin \frac{\pi t}{T}\right) \quad (2.35)$$

is zero whenever $\sin(\pi t/T)$ is negative, as seen by the waveform in Fig. 2.18. It is now apparent that the square wave in Fig. 2.1 can be represented as

$$s(t) = u\left(\sin \frac{\pi t}{T}\right) - u\left(-\sin \frac{\pi t}{T}\right) \quad (2.36)$$

Another method of describing the square wave is to consider a generalization of the step function known as the *sgn function* (pronounced signum). The sgn function is defined as

$$\begin{aligned} \operatorname{sgn}[f(t)] &= 1 \qquad f(t) > 0 \\ &= 0 \qquad f(t) = 0 \\ &= -1 \qquad f(t) < 0 \end{aligned} \quad (2.37)$$

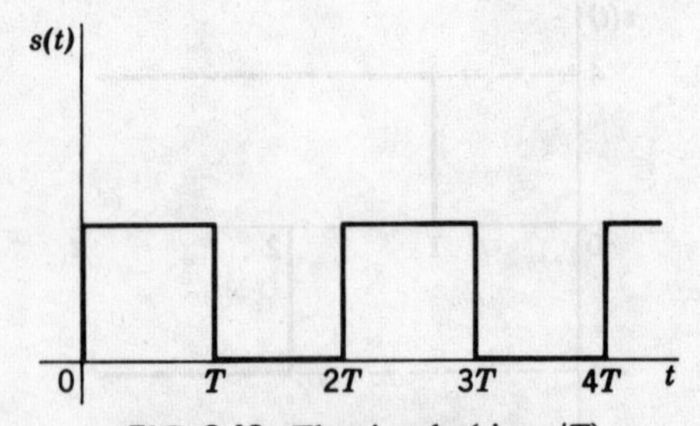

FIG. 2.18. The signal $u(\sin \pi t/T)$.

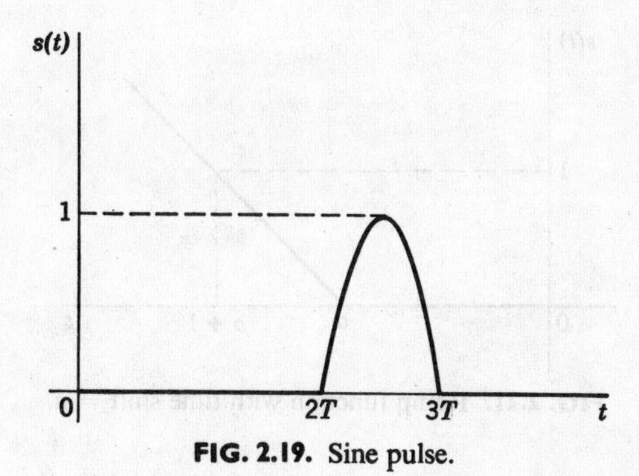

FIG. 2.19. Sine pulse.

Thus the square wave in Fig. 2.1 is simply expressed as

$$s(t) = \text{sgn}\left(\sin\frac{\pi t}{T}\right) \tag{2.38}$$

Returning to the shifting property of the step function, we see that the single sine pulse in Fig. 2.19 can be represented as

$$s(t) = \sin\frac{\pi t}{T}[u(t - 2T) - u(t - 3T)] \tag{2.39}$$

The step function is also extremely useful in representing the shifted or delayed version of any given signal. For example, consider the *unit ramp function*

$$\rho(t) = t\,u(t) \tag{2.40}$$

shown in Fig. 2.20. Suppose the ramp is delayed by an amount $t = a$, as shown in Fig. 2.21. How do we represent the delayed version of ramp?

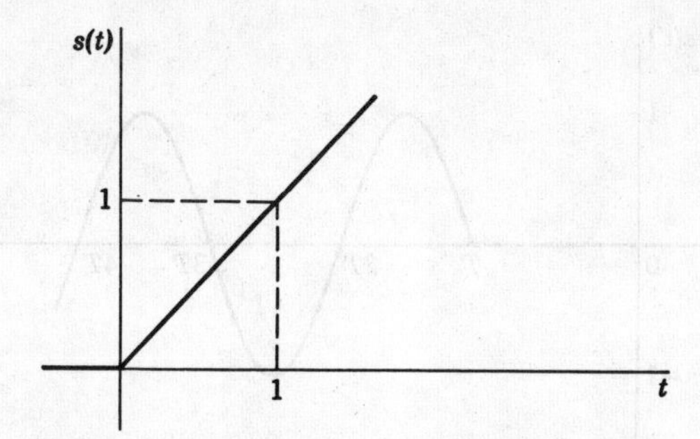

FIG. 2.20. Ramp function with zero time shift.

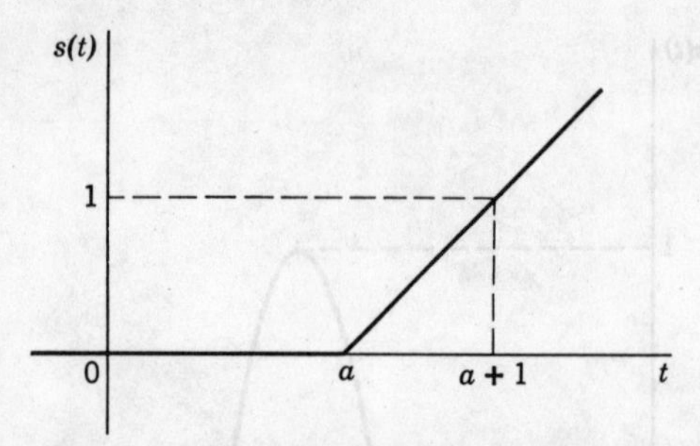

FIG. 2.21. Ramp function with time shift = a.

First, let us replace the variable t by a new variable $t' = t - a$. Then

$$\rho(t') = t'\, u(t') \tag{2.41}$$

When $\rho(t')$ is plotted against t', the resulting curve is identical to the plot of $s(t)$ versus t in Fig. 2.20. If, however, we substitute $t - a = t'$ in $\rho(t')$, we then have

$$\rho(t') = (t - a)\, u(t - a) \tag{2.42}$$

When we plot $\rho(t')$ against t, we have the delayed version of $\rho(t)$ shown in Fig. 2.21.

From the preceding discussion, it is clear that if any signal $f(t)\,u(t)$ is delayed by a time T, the delayed or shifted signal is given by

$$f(t') = f(t - T)\, u(t - T) \tag{2.43}$$

For example, let us delay the function $(\sin \pi t/T)\, u(t)$ by a period T. Then the delayed function $s(t')$, shown in Fig. 2.22, is

$$s(t') = \left[\sin \frac{\pi}{T}(t - T)\right] u(t - T) \tag{2.44}$$

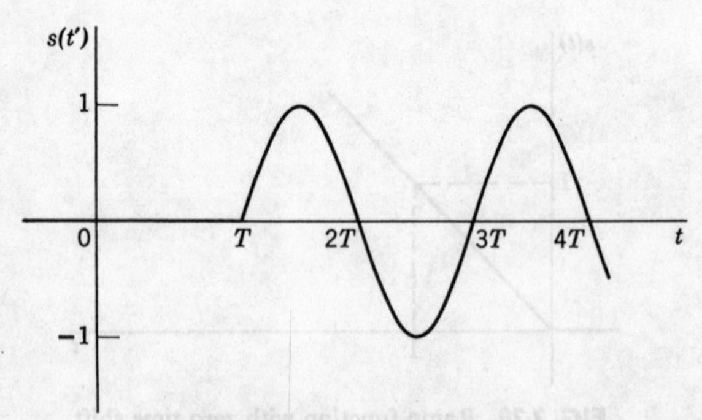

FIG. 2.22. Shifted sine wave.

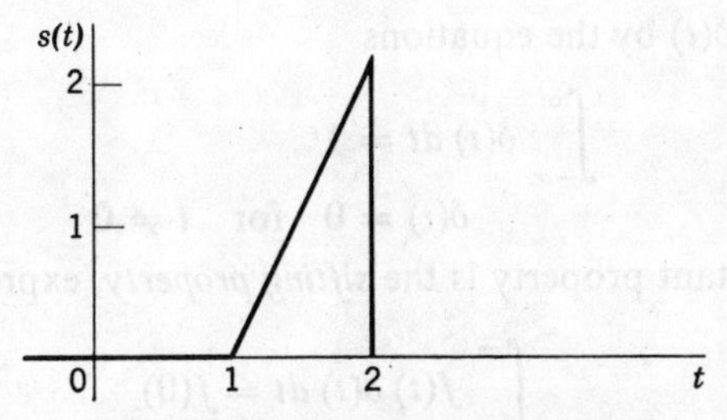

FIG. 2.23. Triangular pulse.

As a final example, consider the waveform in Fig. 2.23, whose component parts are given in Fig. 2.24. For increasing t, the first nonzero component is the function $2(t-1)\,u(t-1)$, which represents the straight line of slope 2 at $t = 1$. At $t = 2$, the rise of the straight line is to be arrested, so we add to the first component a term equal to $-2(t-2)\,u(t-2)$ with a slope of -2. The sum is then a constant equal to 2. We then add a term $-2u(t-2)$ to bring the level down to zero. Thus,

$$s(t) = 2(t-1)\,u(t-1) - 2(t-2)\,u(t-2) - 2u(t-2) \qquad (2.45)$$

2.4 THE UNIT IMPULSE

The *unit impulse*, or *delta function*, is a mathematical anomaly. P. A. M. Dirac first used it in his writings on quantum mechanics.[2] He defined the

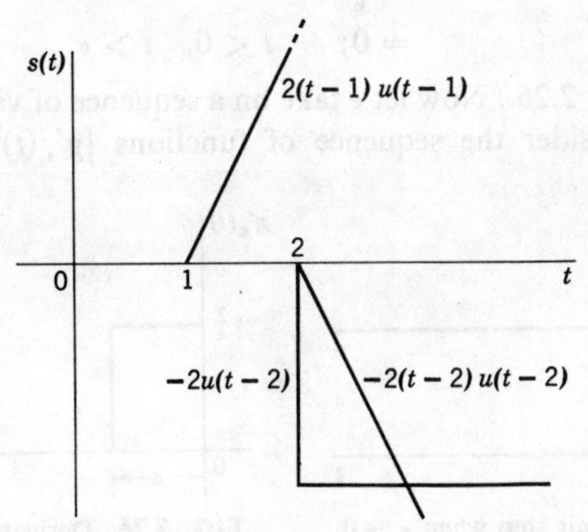

FIG. 2.24. Decomposition of the triangular pulse in Fig. 2.23.

[2] P. A. M. Dirac, *The Principles of Quantum Mechanics* Oxford University Press, 1930.

delta function $\delta(t)$ by the equations

$$\int_{-\infty}^{\infty} \delta(t)\, dt = 1 \tag{2.46}$$

$$\delta(t) = 0 \quad \text{for} \quad t \neq 0 \tag{2.47}$$

Its most important property is the *sifting property*, expressed by

$$\int_{-\infty}^{\infty} f(t)\, \delta(t)\, dt = f(0) \tag{2.48}$$

In this section we will examine the unit impulse from a nonrigorous approach. Those who prefer a rigorous treatment should refer to Appendix B for development of this discussion. The material in that appendix is based on the theory of generalized functions originated by G. Temple.[3] In Appendix B it is shown that the unit impulse is the derivative of the unit step

$$\delta(t) = u'(t) \tag{2.49}$$

At first glance this statement is doubtful. After all, the derivative of the unit step is zero everywhere except at the jump discontinuity, and it does not even exist at that point! However, consider the function $g_\epsilon(t)$ in Fig. 2.25. It is clear that as ϵ goes to zero, $g_\epsilon(t)$ approaches a unit step, that is,

$$\lim_{\epsilon \to 0} g_\epsilon(t) = u(t) \tag{2.50}$$

Taking the derivative of $g_\epsilon(t)$, we obtain $g'_\epsilon(t)$, which is defined by the equations

$$\begin{aligned} g'_\epsilon(t) &= \frac{1}{\epsilon}; \qquad 0 \leq t \leq \epsilon \\ &= 0; \qquad t < 0, \quad t > \epsilon \end{aligned} \tag{2.51}$$

as shown in Fig. 2.26. Now let ϵ take on a sequence of values ϵ_i such that $\epsilon_i > \epsilon_{i+1}$. Consider the sequence of functions $\{g'_{\epsilon_i}(t)\}$ for decreasing

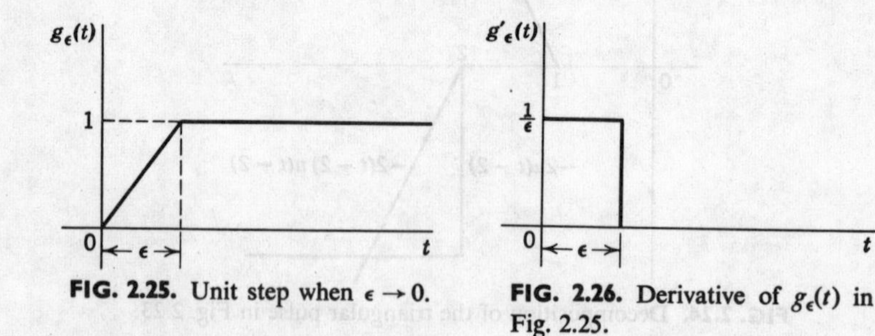

FIG. 2.25. Unit step when $\epsilon \to 0$.

FIG. 2.26. Derivative of $g_\epsilon(t)$ in Fig. 2.25.

[3] G. Temple, "The Theory of Generalized Functions," *Proc. Royal Society*, *A*, **228**, 1955, 175–190.

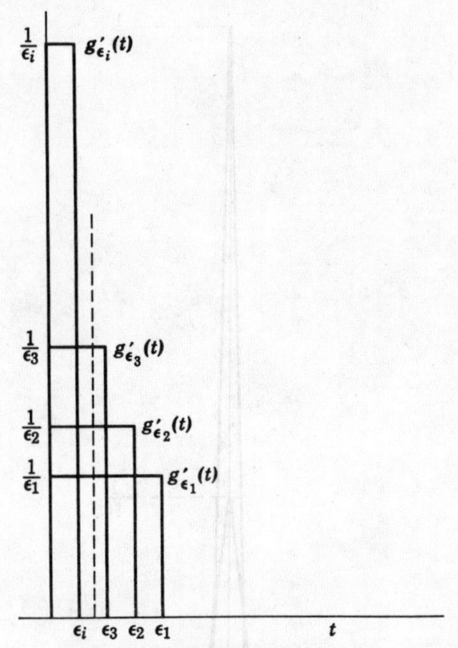

FIG. 2.27. The sequence $\{g_{\epsilon_i}(t)\}$.

values of ϵ_i, as shown in Fig. 2.27. The sequence has the following property:

$$\lim_{\epsilon_i \to 0} \int_{t_1<0}^{t_2>0} g'_{\epsilon_i}(t)\, dt = 1 \tag{2.52}$$

where t_1 and t_2 are arbitrary real numbers. For every nonzero value of ϵ, there corresponds a well-behaved function (i.e., it does not "blow up") $g'_{\epsilon_i}(t)$. As ϵ_i approaches zero,

$$g'_{\epsilon_i}(0+) \xrightarrow[\epsilon_i \to 0]{} \infty \tag{2.53}$$

so that the limit of the sequence is not defined in the classical sense. Another sequence of functions which obeys the property given in Eq. 2.52 is the sequence $\{f_{\epsilon_i}(t)\}$ in Fig. 2.28. We now define the *unit impulse* $\delta(t)$ as the class of *all* sequences of functions which obey Eq. 2.52. In particular, we define

$$\begin{aligned} \int_{t_1<0}^{t_2>0} \delta(t)\, dt &\triangleq \lim_{\epsilon_i \to 0} \int_{t_1<0}^{t_2>0} g'_{\epsilon_i}(t)\, dt \\ &\triangleq \lim_{\epsilon_i \to 0} \int_{t_1<0}^{t_2>0} f_{\epsilon_i}(t)\, dt \end{aligned} \tag{2.54}$$

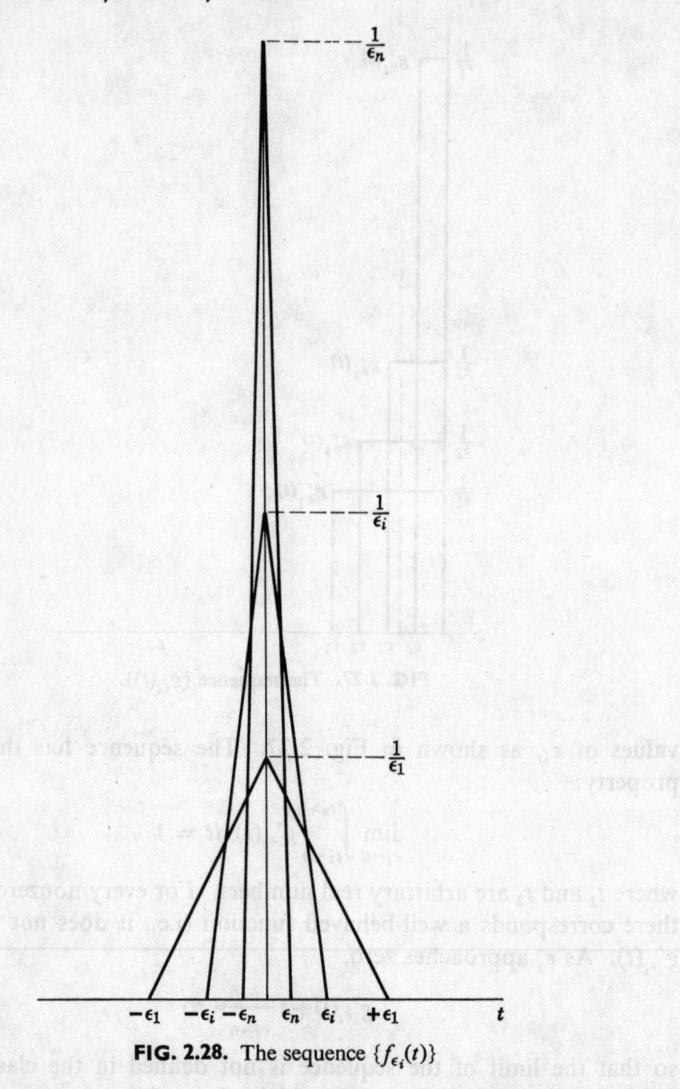

FIG. 2.28. The sequence $\{f_{\epsilon_i}(t)\}$

It should be stressed that this is not a rigorous definition (which, as stated previously, is found in Appendix B) but merely a heuristic one. From the previous definition we can think of the delta "function" as having the additional properties,

$$\begin{aligned} \delta(0) &= \infty \\ \delta(t) &= 0 \quad \text{for} \quad t \neq 0 \end{aligned} \tag{2.55}$$

Continuing with this heuristic treatment, we say that the area "under" the impulse is unity, and, since the impulse is zero for $t \neq 0$, we have

$$\int_{-\infty}^{\infty} \delta(t)\, dt = \int_{0-}^{0+} \delta(t)\, dt = 1 \tag{2.56}$$

Thus the entire area of the impulse is "concentrated" at $t = 0$. Consequently, any integral that does not integrate through $t = 0$ is zero, as seen by

$$\int_{-\infty}^{0-} \delta(t)\, dt = \int_{0+}^{+\infty} \delta(t)\, dt = 0 \tag{2.57}$$

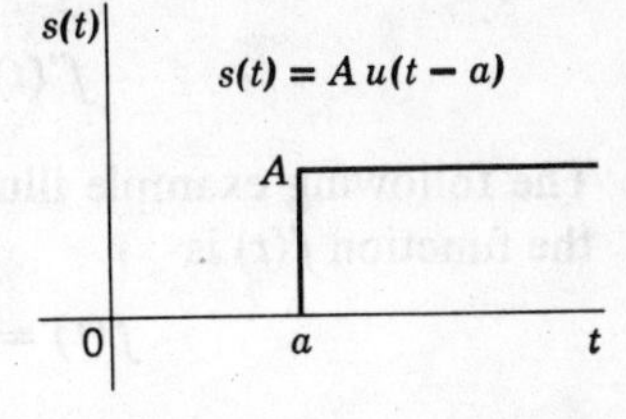

The change of scale and time shift properties discussed earlier also apply for the impulse function. The derivative of a step function

$$s(t) = A\, u(t - a) \tag{2.58}$$

yields an impulse function

$$s'(t) = A\, \delta(t - a) \tag{2.59}$$

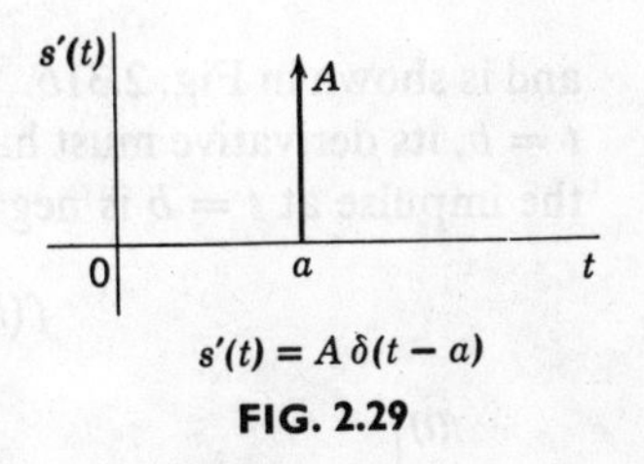

FIG. 2.29

which is shown in Fig. 2.29. Graphically, we represent an impulse function by an arrowhead pointing upward, with the constant multiplier A written next to the arrowhead. Note that A is the area under the impulse $A\, \delta(t - a)$.

Consider the implications of Eqs. 2.58 and 2.59. From these equations we see that the derivative of the step at the jump discontinuity of height A yields an impulse of area A at that same point $t = T$. Generalizing on this argument, consider any function $f(t)$ with a jump discontinuity at $t = T$. Then the derivative, $f'(t)$ must have an impulse at $t = T$. As an example, consider $f(t)$ in Fig. 2.30. At $t = T, f(t)$ has a discontinuity of height A,

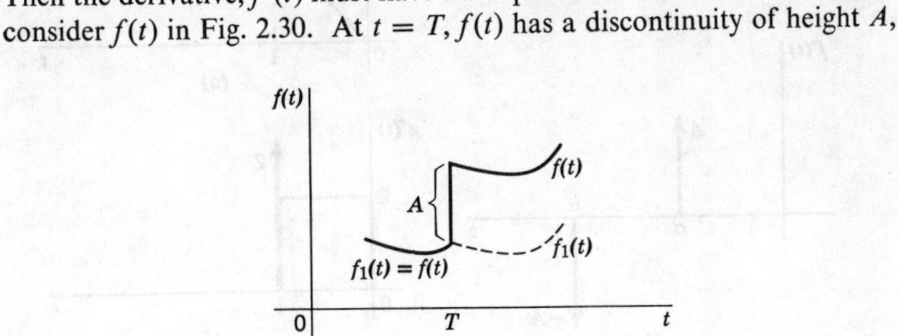

FIG. 2.30. Function with discontinuity at T.

which is given as

$$A = f(T+) - f(T-) \tag{2.60}$$

Let us define $f_1(t)$ as being equal to $f(t)$ for $t < T$, and having the *same shape* as $f(t)$, but without the discontinuity for $t > T$, that is,

$$f_1(t) = f(t) - A\,u(t - T) \tag{2.61}$$

The derivative $f'(t)$ is then

$$f'(t) = f'_1(t) + A\,\delta(t - T) \tag{2.62}$$

The following example illustrates this point more clearly. In Fig. 2.31*a*, the function $f(t)$ is

$$f(t) = A\,u(t - a) - A\,u(t - b) \tag{2.63}$$

Its derivative is $$f'(t) = A\,\delta(t - a) - A\,\delta(t - b) \tag{2.64}$$

and is shown in Fig. 2.31*b*. Since $f(t)$ has two discontinuities, at $t = a$ and $t = b$, its derivative must have impulses at those points. The coefficient of the impulse at $t = b$ is negative because

$$f(b+) - f(b-) = -A \tag{2.65}$$

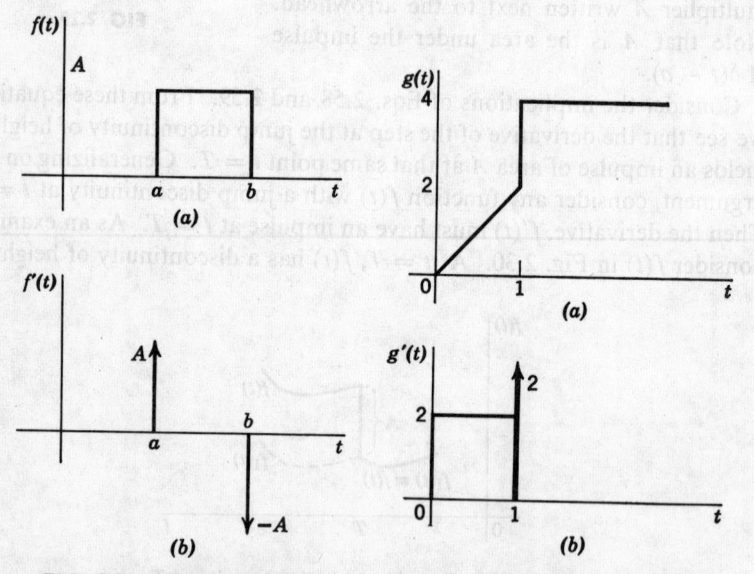

FIG. 2.31. (*a*) Square pulse. (*b*) Derivative of square pulse.

FIG. 2.32. (*a*) Signal. (*b*) Derivative.

As a second example, consider the function $g(t)$ shown in Fig. 2.32*a*. We obtain $g'(t)$ by inspection, and note that the discontinuity at $t = 1$ produces the impulse in $g'(t)$ of area

$$g(1+) - g(1-) = 2, \tag{2.66}$$

as given in Fig. 2.32*b*.

Another interesting property of the impulse function is expressed by the integral

$$\int_{-\infty}^{+\infty} f(t)\,\delta(t - T)\,dt = f(T) \tag{2.67}$$

This integral is easily evaluated if we consider that $\delta(t - T) = 0$ for all $t \neq T$. Therefore, the product

$$f(t)\,\delta(t - T) = 0 \qquad \text{all } t \neq T \tag{2.68}$$

If $f(t)$ is single-valued at $t = T$, $f(T)$ can be factored from the integral so that we obtain

$$f(T)\int_{-\infty}^{\infty} \delta(t - T)\,dt = f(T) \tag{2.69}$$

Figure 2.33 shows $f(t)$ and $\delta(t - T)$, where $f(t)$ is continuous at $t = T$. If $f(t)$ has a discontinuity at $t = T$, the integral

$$\int_{-\infty}^{+\infty} f(t)\,\delta(t - T)\,dt$$

is not defined because the value of $f(T)$ is not uniquely given. Consider the following examples.

Example 2.1

$$f(t) = e^{j\omega t}$$

$$\int_{-\infty}^{+\infty} e^{j\omega t}\delta(t - T)dt = e^{j\omega T} \tag{2.70}$$

Example 2.2

$$f(t) = \sin t$$

$$\int_{-\infty}^{+\infty} \sin t\,\delta\left(t - \frac{\pi}{4}\right)dt = \frac{1}{\sqrt{2}} \tag{2.71}$$

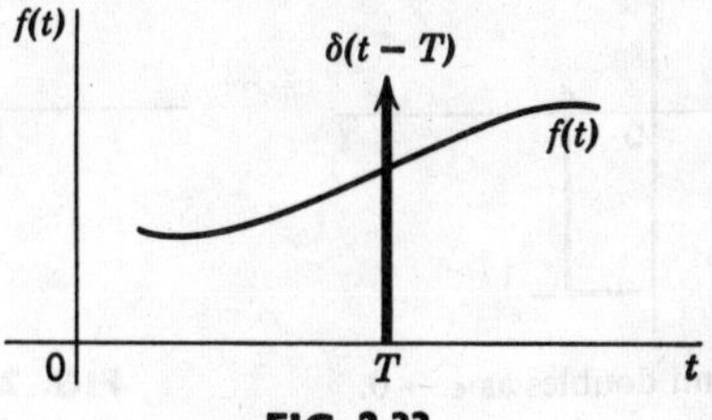

FIG. 2.33

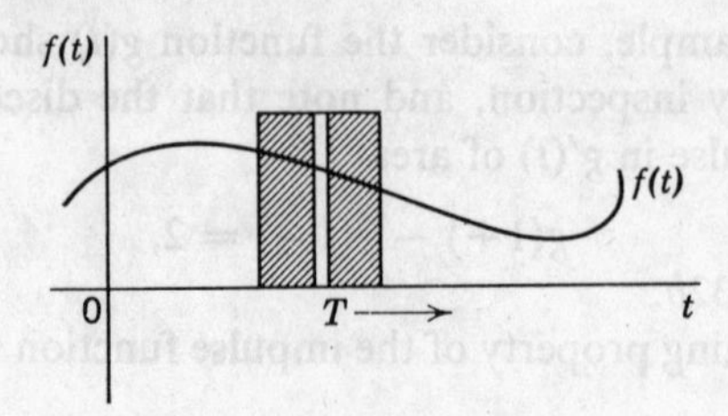

FIG. 2.34. Impulse scanning.

Consider next the case where $f(t)$ is continuous for $-\infty < t < \infty$. Let us direct our attention to the integral

$$\int_{-\infty}^{\infty} f(t)\,\delta(t-T)\,dt = f(T) \tag{2.72}$$

which holds for all t in this case. If T were varied from $-\infty$ to $+\infty$, then $f(t)$ would be reproduced in its entirety. An operation of this sort corresponds to *scanning* the function $f(t)$ by moving a sheet of paper with a thin slit across a plot of the function, as shown in Fig. 2.34.

Let us now examine higher order derivatives of the unit step function. Here we represent the unit impulse by the function $f_\epsilon(t)$ in Fig. 2.28, which, as $\epsilon \to 0$, becomes the unit impulse. The derivative of $f_\epsilon(t)$ is given in Fig. 2.35. As ϵ approaches zero, $f'_\epsilon(t)$ approaches the derivative of the unit impulse $\delta'(t)$, which consists of a pair of impulses as seen in Fig. 2.36. The area under $\delta'(t)$, which is sometimes called a *doublet*, is equal to zero. Thus,

$$\int_{-\infty}^{\infty} \delta'(t)\,dt = 0 \tag{2.73}$$

The other significant property of the doublet is

$$\int_{-\infty}^{\infty} f(t)\,\delta'(t-T)\,dt = -f'(T) \tag{2.74}$$

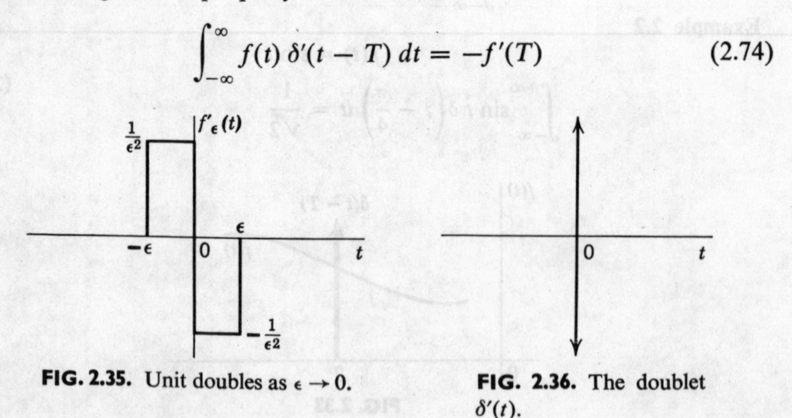

FIG. 2.35. Unit doubles as $\epsilon \to 0$.

FIG. 2.36. The doublet $\delta'(t)$.

where $f'(T)$ is the derivative of $f(t)$ evaluated at $t = T$ where, again, we assume that $f(t)$ is continuous. Equation 2.74 can be proved by integration by parts. Thus,

$$\int_{-\infty}^{\infty} f(t)\,\delta'(t - T)\,dt = f(t)\,\delta(t - T)\Big|_{-\infty}^{\infty} - \int_{-\infty}^{\infty} f'(t)\,\delta(t - T)\,dt$$
$$= -f'(T). \tag{2.75}$$

It can be shown in general that

$$\int_{-\infty}^{\infty} f(t)\,\delta^{(n)}(t - T)\,dt = (-1)^n f^{(n)}(T) \tag{2.76}$$

where $\delta^{(n)}$ and $f^{(n)}$ denote nth derivatives. The higher order derivatives of $\delta(t)$ can be evaluated in similar fashion.

Problems

2.1 Resolve the waveforms in the figure into odd and even components.

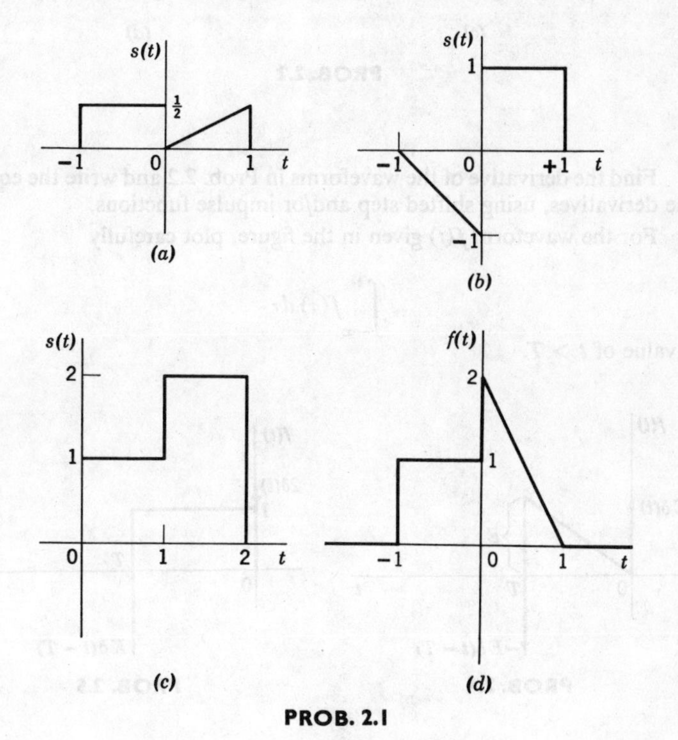

PROB. 2.1

2.2 Write the equation for the waveforms in the figure using shifted step functions.

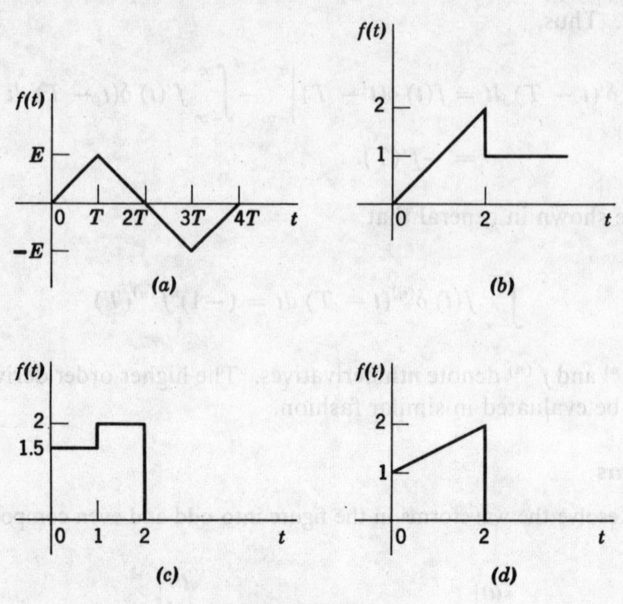

PROB. 2.2

2.3 Find the derivative of the waveforms in Prob. 2.2 and write the equations for the derivatives, using shifted step and/or impulse functions.

2.4 For the waveform $f(t)$ given in the figure, plot carefully

$$\int_{-\infty}^{t} f(\tau)\, d\tau$$

for a value of $t > T$.

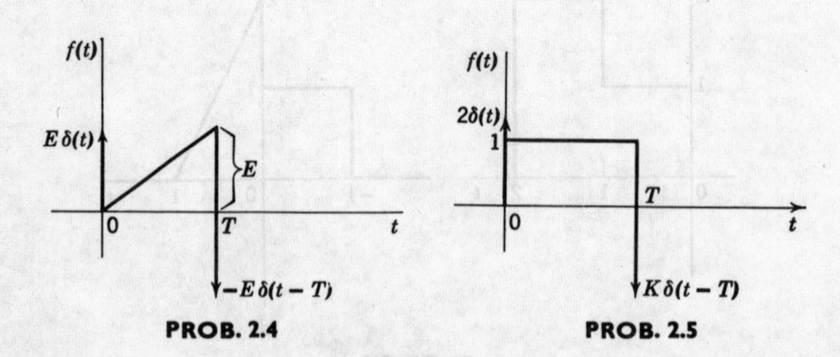

PROB. 2.4 PROB. 2.5

2.5 For the waveform $f(t)$, shown in the figure, determine what value K must be so that

(*a*)
$$\int_{-\infty}^{\infty} f(t)\, dt = 0$$

(*b*)
$$\int_{0+}^{\infty} f(t)\, dt = 0$$

2.6 For the waveforms shown in the figure, express in terms of elementary time functions, that is, t^n, $u(t - t_0)$, $\delta(t - t_0)$, (*a*) $f(t)$ (*b*) $f'(t)$ (*c*) $\int_{-\infty}^{t+} f(\tau)\, d\tau$

Sketch the waveforms for (*b*) and (*c*) neatly.

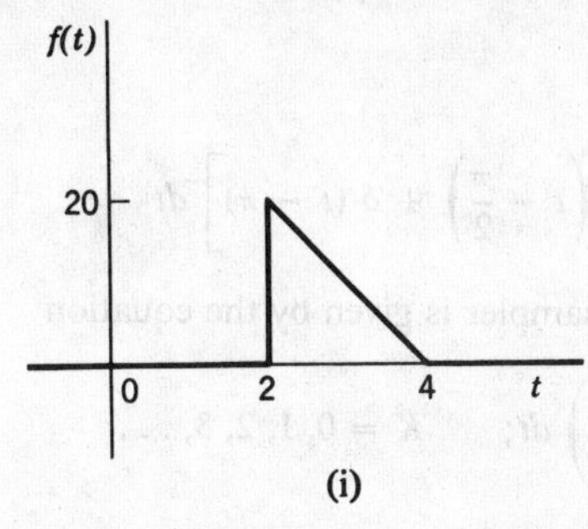

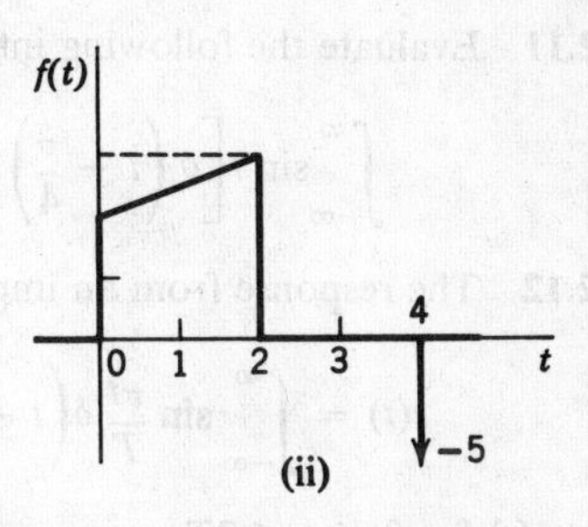

PROB. 2.6

2.7 Prove that

(*a*)
$$\delta'(x) = -\delta'(-x)$$

(*b*)
$$-\delta(x) = x\, \delta'(x)$$

(*c*)
$$\int_{-\infty}^{\infty} x\, \delta(x)\, dx = 0$$

2.8 The waveform $f(t)$ in the figure is defined as

$$f(t) = \frac{3}{\epsilon^3}(t - \epsilon)^2, \qquad 0 \le t \le \epsilon$$

$$= 0, \qquad \text{elsewhere}$$

Show that as $\epsilon \to 0$, $f(t)$ becomes a unit impulse.

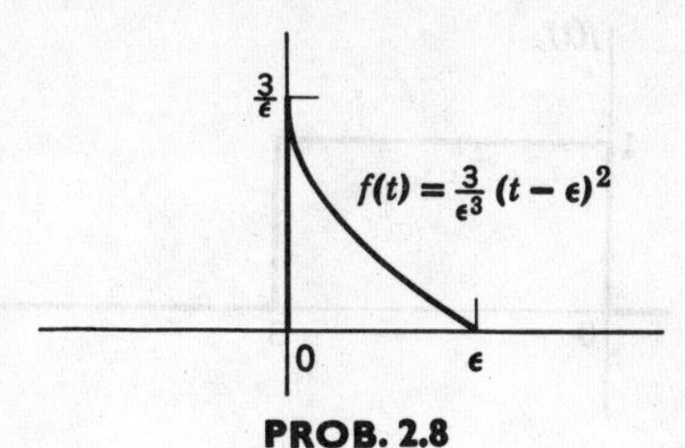

PROB. 2.8

2.9 Plot

(*a*) $$\delta(\cos t)$$

(*b*) $$t \operatorname{sgn}(\cos t); \qquad 0 \leq t \leq 2\pi$$

2.10 Evaluate the following integrals

(*a*) $$\int_{-\infty}^{\infty} \delta(t - T_1)\, u(t - T_2)\, dt; \qquad T_2 < T_1$$

(*b*) $$\int_{-\infty}^{\infty} \delta(\omega - \omega_0) \cos \omega t \, d\omega$$

(*c*) $$\int_{-\infty}^{\infty} [\delta(t) - A\,\delta(t - T_1) + 2A\,\delta(t - T_2)] e^{-jn\omega t}\, dt$$

2.11 Evaluate the following integral

$$\int_{-\infty}^{\infty} \sin t \left[\delta\left(t - \frac{\pi}{4}\right) + \delta'\left(t - \frac{\pi}{2}\right) + \delta''(t - \pi) \right] dt$$

2.12 The response from an impulse sampler is given by the equation

$$r(t) = \int_{-\infty}^{\infty} \sin \frac{\pi t}{T}\, \delta\left(t - K\frac{T}{4}\right) dt; \qquad K = 0, 1, 2, 3, \ldots$$

Plot $r(t)$ for $0 \leq t \leq 2T$.

2.13 If the step response of a linear, time-invariant system is $r_s(t) = 2e^{-t}\, u(t)$, determine the impulse response $h(t)$, and plot.

2.14 For the system in Prob. 2.13 determine the response due to a staircase excitation

$$s(t) = \sum_{k=0}^{3} u(t - kT)$$

Plot both excitation and response functions.

2.15 If the impulse response of a time-invariant system is $h(t) = e^{-t}\, u(t)$, determine the response due to an excitation

$$e(t) = 2\delta(t - 1) - 2\delta(t - 2)$$

Plot $e(t)$.

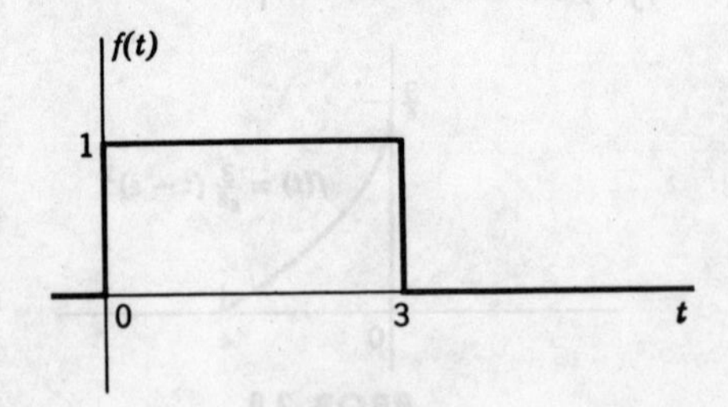

PROB. 2.16

2.16 The unit step response of a linear system is

$$\alpha(t) = (2e^{-2t} - 1)\, u(t)$$

(*a*) Find the response $r(t)$ to the input $f(t)$.
(*b*) Make a reasonably accurate sketch of the response. Show all pertinent dimensions.

chapter 3

The frequency domain: Fourier analysis

3.1 INTRODUCTION

One of the most common classes of signals encountered are periodic signals. If T is the period of the signal, then

$$s(t) = s(t \pm nT) \qquad n = 0, 1, 2, \ldots \tag{3.1}$$

In addition to being periodic, if $s(t)$ has only a finite number of discontinuities in any finite period and if the integral

$$\int_{\alpha}^{\alpha+T} |s(t)|\, dt$$

is finite (where α is an arbitrary real number), then $s(t)$ can be expanded into the infinite trigonometric series

$$\begin{aligned} s(t) = \frac{a_0}{2} &+ a_1 \cos \omega t + a_2 \cos 2\omega t + \cdots \\ &+ b_1 \sin \omega t + b_2 \sin 2\omega t + \cdots \end{aligned} \tag{3.2}$$

where $\omega = 2\pi/T$. This trigonometric series is generally referred to as the *Fourier series*. In compact form, the Fourier series is

$$s(t) = \frac{a_0}{2} + \sum_{n=1}^{\infty} (a_n \cos n\omega t + b_n \sin n\omega t) \tag{3.3}$$

It is apparent from Eqs. 3.2 and 3.3 that, when $s(t)$ is expanded in a Fourier series, we can describe $s(t)$ completely in terms of the coefficients of its harmonic terms, $a_0, a_1, a_2, \ldots, b_1, b_2, \ldots$. These coefficients constitute a *frequency domain* description of the signal. Our task now is to derive the equations for the coefficients a_i, b_i in terms of the given signal function $s(t)$. Let us first discuss the mathematical basis of Fourier series, the theory of *orthogonal sets.*

3.2 ORTHOGONAL FUNCTIONS

Consider any two functions $f_1(t)$ and $f_2(t)$ that are not identically zero. Then if

$$\int_{T_1}^{T_2} f_1(t) f_2(t)\, dt = 0 \tag{3.4}$$

we say that $f_1(t)$ and $f_2(t)$ are *orthogonal* over the interval $[T_1, T_2]$. For example, the functions *sin t* and *cos t* are orthogonal over the interval $n2\pi \leq t \leq (n+1)2\pi$. Consider next a set of real functions $\{\phi_1(t), \phi_2(t), \ldots, \phi_n(t)\}$. If the functions obey the condition

$$(\phi_i, \phi_j) \equiv \int_{T_1}^{T_2} \phi_i(t)\, \phi_j(t)\, dt = 0, \qquad i \neq j \tag{3.5}$$

then the set $\{\phi_i\}$ forms an *orthogonal set* over the interval $[T_1, T_2]$. In Eq. 3.5 the integral is denoted by the *inner product* (ϕ_i, ϕ_j). For convenience here, we use the inner product notation in our discussions.

The set $\{\phi_i\}$ is *orthonormal* over $[T_1, T_2]$ if

$$\begin{aligned} (\phi_i, \phi_j) &= 0 \qquad i \neq j \\ &= 1 \qquad i = j \end{aligned} \tag{3.6}$$

The *norm* of an element ϕ_k in the set $\{\phi_i\}$ is defined as

$$\|\phi_k\| = (\phi_k, \phi_k)^{1/2} = \left(\int_{T_1}^{T_2} \phi_k^{\,2}(t)\, dt\right)^{1/2} \tag{3.7}$$

We can normalize any orthogonal set $\{\phi_1, \phi_2, \ldots, \phi_n\}$ by dividing each term ϕ_k by its norm $\|\phi_k\|$.

Example 3.1. The Laguerre set,[1] which has been shown to be very useful in time domain approximation, is orthogonal over $[0, \infty]$. The first four terms

---

[1] W. H. Kautz, "Transient Synthesis in the Time Domain," *Trans. IRE on Circuit Theory*, CT-1, No. 3 (Sept. 1954), 29–39.

of the Laguerre set are

$$\phi_1(t) = e^{-at}$$

$$\phi_2(t) = e^{-at}[1 - 2(at)]$$

$$\phi_3(t) = e^{-at}[1 - 4(at) + 2(at)^2] \tag{3.8}$$

$$\phi_4(t) = e^{-at}[1 - 6at + 6(at)^2 - \tfrac{4}{3}(at)^3]$$

$$\cdots$$

To show that the set is orthogonal, let us consider the integral

$$\int_0^\infty \phi_1(t)\,\phi_3(t)\,dt = \int_0^\infty e^{-2at}[1 - 4(at) + 2(at)^2]\,dt \tag{3.9}$$

Letting $\tau = at$, we have

$$\begin{aligned}(\phi_1, \phi_3) &= \frac{1}{a}\int_0^\infty e^{-2\tau}(1 - 4\tau + 2\tau^2)\,d\tau \\ &= \frac{1}{a}\left[\frac{1}{2} - 4\left(\frac{1}{4}\right) + 2\left(\frac{2}{8}\right)\right] = 0\end{aligned} \tag{3.10}$$

The norms of $\phi_1(t)$ and $\phi_2(t)$ are

$$\|\phi_1\| = \left(\int_0^\infty e^{-2at}\,dt\right)^{1/2} = \frac{1}{\sqrt{2a}} \tag{3.11}$$

$$\begin{aligned}\|\phi_2\| &= \left\{\int_0^\infty e^{-2at}[1 - 4(at) + 4(at)^2]\,dt\right\}^{1/2} \\ &= \frac{1}{\sqrt{2a}}\end{aligned} \tag{3.12}$$

It is not difficult to verify that the norms of all the elements in the set are also equal to $1/\sqrt{2a}$. Therefore, to render the Laguerre set orthonormal, we divide each element ϕ_i by $1/\sqrt{2a}$.

3.3 APPROXIMATION USING ORTHOGONAL FUNCTIONS

In this section we explore some of the uses of orthogonal functions in the linear approximation of functions. The principal problem is that of approximating a function $f(t)$ by a sequence of functions $f_n(t)$ such that the *mean squared error*

$$\epsilon = \lim_{n\to\infty}\int_{T_1}^{T_2} [f(t) - f_n(t)]^2\,dt = 0 \tag{3.13}$$

When Eq. 3.13 is satisfied, we say that $\{f_n(t)\}$ *converges in the mean to* $f(t)$.

To examine the concept of convergence in the mean more closely, we must first consider the following definitions:

Definition 3.1 Given a function $f(t)$ and constant $p > 0$ for which

$$\int_{T_1}^{T_2} |f(t)|^p \, dt < \infty,$$

we say that $f(t)$ is *integrable* L^p *in* $[T_1, T_2]$, and we write $f(t) \in L^p$ in $[T_1, T_2]$.

Definition 3.2 If $f(t) \in L^p$ in $[T_1, T_2]$, and $\{f_n(t)\}$ is a sequence of functions integrable L^p in $[T_1, T_2]$, we say that if

$$\lim_{n\to\infty} \int_{T_1}^{T_2} |f(t) - f_n(t)|^p \, dt = 0$$

then $\{f_n(t)\}$ converges in the mean of order p to $f(t)$. Specifically, when $p = 2$ we say that $\{f_n(t)\}$ converges in the mean to $f(t)$.

The principle of least squares

Now let us consider the case when $f_n(t)$ consists of a linear combination of orthonormal functions $\phi_1, \phi_2, \ldots, \phi_n$.

$$f_n(t) = \sum_{i=1}^{n} a_i \, \phi_i(t) \tag{3.14}$$

Our problem is to determine the constants a_i such that the integral squared error

$$\|f - f_n\|^2 = \int_{T_1}^{T_2} [f(t) - f_n(t)]^2 \, dt \tag{3.15}$$

is a minimum. The *principle of least squares* states that in order to attain minimum squared error, the constants a_i must have the values

$$c_i = \int_{T_1}^{T_2} f(t) \, \phi_i(t) \, dt \tag{3.16}$$

Proof. We shall show that in order for $\|f - f_n\|^2$ to be minimum, we must set $a_i = c_i$ for every $i = 1, 2, \ldots, n$.

$$\begin{aligned} \|f - f_n\|^2 &= (f, f) - 2(f, f_n) + (f_n, f_n) \\ &= \|f\|^2 - 2\sum_{i=1}^{n} a_i (f, \phi_i) + \sum_{i=1}^{n} a_i^2 \|\phi_i\|^2 \end{aligned} \tag{3.17}$$

Since the set $\{\phi_i\}$ is orthonormal, $\|\phi_i\|^2 = 1$, and by definition, $c_i = (f, \phi_i)$. We thus have

$$\|f - f_n\|^2 = \|f\|^2 - 2\sum_{i=1}^{n} a_i c_i + \sum_{i=1}^{n} a_i^2 \tag{3.18}$$

Adding and subtracting $\sum_{i=1}^{n} c_i^2$ gives

$$\begin{aligned}\|f - f_n\|^2 &= \|f\|^2 - 2\sum_{i=1}^{n} a_i c_i + \sum_{i=1}^{n} c_i^2 + \sum_{i=1}^{n} a_i^2 - \sum_{i=1}^{n} c_i^2 \\ &= \|f\|^2 + \sum_{i=1}^{n} (c_i - a_i)^2 - \sum_{i=1}^{n} c_i^2\end{aligned} \tag{3.19}$$

We see that in order to attain minimum integral squared error, we must set $a_i = c_i$. The coefficients c_i, defined in Eq. 3.16, are called the *Fourier coefficients* of $f(t)$ with respect to the orthonormal set $\{\phi_i(t)\}$.

Parseval's equality

Consider $f_n(t)$ given in Eq. 3.14. We see that

$$\int_{T_1}^{T_2} [f_n(t)]^2\, dt = \sum_{i=1}^{n} c_i^2 \tag{3.20}$$

since ϕ_i are orthonormal functions. This result is known as *Parseval's equality*, and is important in determining the energy of a periodic signal.

3.4 FOURIER SERIES

Let us return to the Fourier series as defined earlier in this chapter,

$$s(t) = \frac{a_0}{2} + \sum_{n=1}^{\infty} (a_n \cos n\omega t + b_n \sin n\omega t) \tag{3.21}$$

From our discussion of approximation by orthonormal functions, we can see that the periodic function $s(t)$ with period T can be approximated by a Fourier series $s_n(t)$ such that $s_n(t)$ converges in the mean to $s(t)$, that is,

$$\lim_{n\to\infty} \int_{\alpha}^{\alpha+T} [s(t) - s_n(t)]^2\, dt \to 0 \tag{3.22}$$

where α is any real number. We know, moreover, that if n is finite, the mean squared error $\|s(t) - s_n(t)\|^2$ is minimized when the constants a_i, b_i are the Fourier coefficients of $s(t)$ with respect to the orthonormal set

$$\left\{\frac{\cos k\omega t}{(T/2)^{1/2}}, \frac{\sin k\omega t}{(T/2)^{1/2}}\right\}, \quad k = 0, 1, 2, \ldots$$

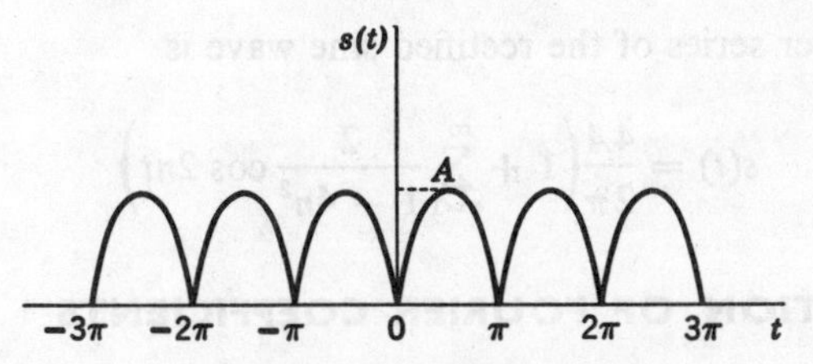

FIG. 3.1. Rectified sine wave.

In explicit form the Fourier coefficients, according to the definition given earlier, are obtained from the equations

$$a_0 = \frac{2}{T}\int_{\alpha}^{\alpha+T} s(t)\,dt \tag{3.23}$$

$$a_k = \frac{2}{T}\int_{\alpha}^{\alpha+T} s(t)\cos k\omega t\,dt \tag{3.24}$$

$$b_k = \frac{2}{T}\int_{\alpha}^{\alpha+T} s(t)\sin k\omega t\,dt \tag{3.25}$$

We should note that because the Fourier series $s_n(t)$ only converges to $s(t)$ *in the mean*, when $s(t)$ contains a jump discontinuity, for example, at t_0

$$s_n(t_0) = \frac{s(t_0+) + s(t_0-)}{2} \tag{3.26}$$

At any point t_1 that $s(t)$ is differentiable (thus naturally continuous) $s_n(t_1)$ converges to $s(t_1)$.[2]

As an example, let us determine the Fourier coefficients of the fully rectified sine wave in Fig. 3.1. As we observe, the period is $T = \pi$ so that the fundamental frequency is $\omega = 2$. The signal is given as

$$s(t) = A\,|\sin t| \tag{3.27}$$

Let us take $\alpha = 0$ and evaluate between 0 and π. Using the formula just derived, we have

$$b_n = \frac{2}{\pi}\int_0^{\pi} s(t)\sin 2nt\,dt = 0 \tag{3.28}$$

$$a_0 = \frac{2A}{\pi}\int_0^{\pi}\sin t\,dt = \frac{4A}{\pi} \tag{3.29}$$

$$a_n = \frac{2}{\pi}\int_0^{\pi} s(t)\cos 2nt\,dt$$

$$= \frac{1}{1-4n^2}\frac{4A}{\pi} \tag{3.30}$$

[2] For a proof see H. F. Davis *Fourier Series and Orthogonal Functions*, Allyn and Bacon, Boston, 1963, pp. 92–95.

Thus the Fourier series of the rectified sine wave is

$$s(t) = \frac{4A}{2\pi}\left(1 + \sum_{n=1}^{\infty} \frac{2}{1 - 4n^2} \cos 2nt\right) \tag{3.31}$$

3.5 EVALUATION OF FOURIER COEFFICIENTS

In this section we will consider two other useful forms of Fourier series. In addition, we will discuss a number of methods to simplify the evaluation of Fourier coefficients. First, let us examine how the evaluation of coefficients is simplified by symmetry considerations. From Eqs. 3.23–3.25 which give the general formulas for the Fourier coefficients, let us take $\alpha = -T/2$ and represent the integrals as the sum of two separate parts, that is,

$$\begin{aligned} a_n &= \frac{2}{T}\left[\int_0^{T/2} s(t) \cos n\omega t \, dt + \int_{-T/2}^{0} s(t) \cos n\omega t \, dt\right] \\ b_n &= \frac{2}{T}\left[\int_0^{T/2} s(t) \sin n\omega t \, dt + \int_{-T/2}^{0} s(t) \sin n\omega t \, dt\right] \end{aligned} \tag{3.32}$$

Since the variable (t) in the above integrals is a dummy variable, let us substitute $x = t$ in the integrals with limits $(0;\ T/2)$, and let $x = -t$ in the integrals with limits $(-T/2;\ 0)$. Then we have

$$\begin{aligned} a_n &= \frac{2}{T}\int_0^{T/2} [s(x) + s(-x)] \cos n\omega x \, dx \\ b_n &= \frac{2}{T}\int_0^{T/2} [s(x) - s(-x)] \sin n\omega x \, dx \end{aligned} \tag{3.33}$$

Suppose now the function is odd, that is, $s(x) = -s(-x)$, then we see that $a_n = 0$ for all n, and

$$b_n = \frac{4}{T}\int_0^{T/2} s(x) \sin n\omega x \, dx \tag{3.34}$$

This implies that, if a function is odd, its Fourier series will contain only sine terms. On the other hand, suppose the function is even, that is, $s(x) = s(-x)$, then $b_n = 0$ and

$$a_n = \frac{4}{T}\int_0^{T/2} s(x) \cos n\omega x \, dx \tag{3.35}$$

Consequently, the Fourier series of an even function will contain only cosine terms.

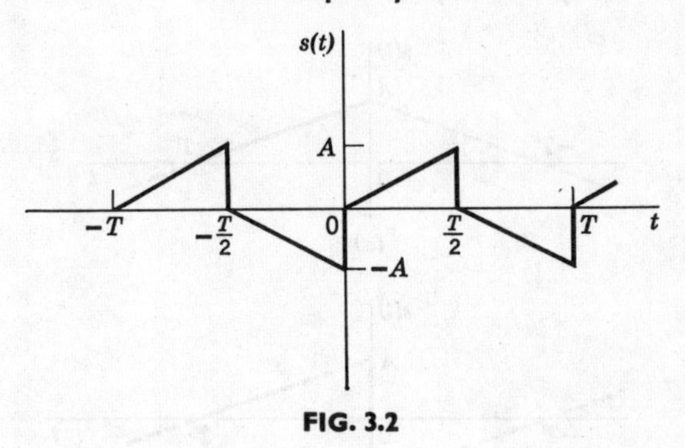

FIG. 3.2

Suppose next, the function $s(t)$ obeys the condition

$$s\left(t \pm \frac{T}{2}\right) = -s(t) \tag{3.36}$$

as given by the example in Fig. 3.2. Then we can show that $s(t)$ contains only odd harmonic terms, that is,

$$a_n = b_n = 0; \qquad n \text{ even}$$

and

$$\begin{aligned} a_n &= \frac{4}{T}\int_0^{T/2} s(t)\cos n\omega t\, dt \\ b_n &= \frac{4}{T}\int_0^{T/2} s(t)\sin n\omega t\, dt, \qquad n \text{ odd} \end{aligned} \tag{3.37}$$

With this knowledge of symmetry conditions, let us examine how we can approximate an arbitrary time function $s(t)$ by a Fourier series within an interval $[0, T]$. Outside this interval, the Fourier series $s_n(t)$ is not required to fit $s(t)$. Consider the signal $s(t)$ in Fig. 3.3. We can approximate $s(t)$ by any of the periodic functions shown in Fig. 3.4. Observe that each periodic waveform exhibits some sort of symmetry.

Now let us consider two other useful forms of Fourier series. The first is the *Fourier cosine series*, which is based upon the trigonometric identity,

$$C_n \cos(n\omega t + \theta_n) = C_n \cos n\omega t \cos \theta_n - C_n \sin n\omega t \sin \theta_n \tag{3.38}$$

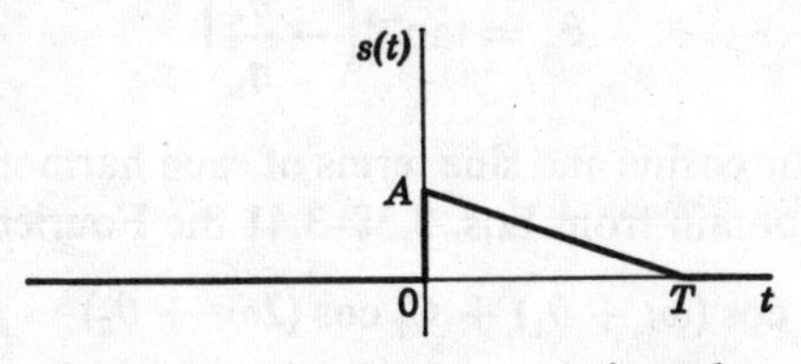

FIG. 3.3. Signal to be approximated.

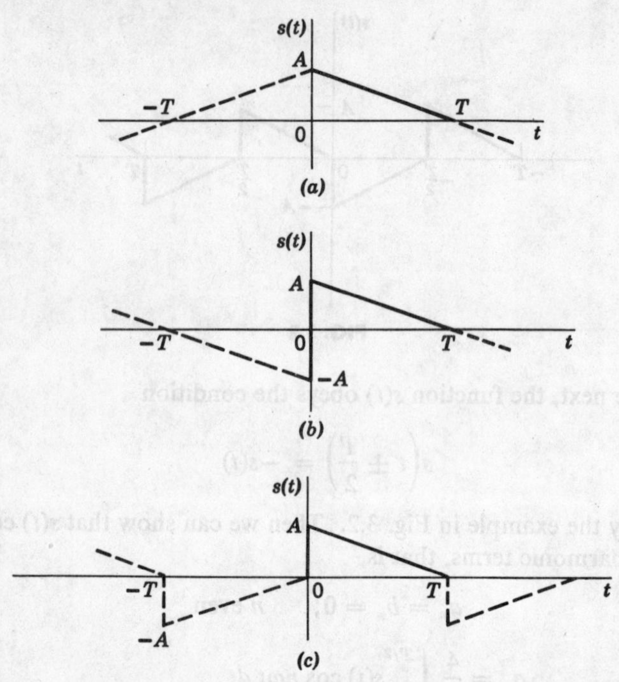

FIG. 3.4. (*a*) Even function cosine terms only. (*b*) Odd function sine terms only. (*c*) Odd harmonics only with both sine and cosine terms.

We can derive the form of the Fourier cosine series by setting

$$a_n = C_n \cos \theta_n \tag{3.39}$$

and

$$b_n = -C_n \sin \theta_n \tag{3.40}$$

We then obtain C_n and θ_n in terms of a_n and b_n, as

$$C_n = (a_n^2 + b_n^2)^{1/2}$$

$$C_0 = \frac{a_0}{2} \tag{3.41}$$

$$\theta_n = \tan^{-1}\left(-\frac{b_n}{a_n}\right)$$

If we combine the cosine and sine terms of each harmonic in the original series, we readily obtain from Eqs. 3.38–3.41 the Fourier cosine series

$$f(t) = C_0 + C_1 \cos(\omega t + \theta_1) + C_2 \cos(2\omega t + \theta_2)$$
$$+ C_3 \cos(3\omega t + \theta_3) + \cdots + C_n \cos(n\omega t + \theta_n) + \cdots \tag{3.42}$$

It should be noted that the coefficients C_n are usually taken to be positive. If however, a term such as $-3 \cos 2\omega t$ carries a negative sign, then we can use the equivalent form

$$-3 \cos 2\omega t = 3 \cos (2\omega t + \pi) \tag{3.43}$$

For example, the Fourier series of the fully rectified sine wave in Fig. 3.1 was shown to be

$$s(t) = \frac{4A}{2\pi}\left(1 + \sum_{n=1}^{\infty} \frac{2}{1 - 4n^2} \cos 2nt\right) \tag{3.44}$$

Expressed as a Fourier cosine series, $s(t)$ is

$$s(t) = \frac{4A}{2\pi}\left[1 + \sum_{n=1}^{\infty} \frac{2}{4n^2 - 1} \cos (2nt + \pi)\right] \tag{3.45}$$

Next we consider the *complex form* of a Fourier series. If we express $\cos n\omega t$ and $\sin n\omega t$ in terms of complex exponentials, then the Fourier series can be written as

$$\begin{aligned} s(t) &= \frac{a_0}{2} + \sum_{n=1}^{\infty}\left(a_n \frac{e^{jn\omega t} + e^{-jn\omega t}}{2} + b_n \frac{e^{jn\omega t} - e^{-jn\omega t}}{2j}\right) \\ &= \frac{a_0}{2} + \sum_{n=1}^{\infty}\left(\frac{a_n - jb_n}{2} e^{jn\omega t} + \frac{a_n + jb_n}{2} e^{-jn\omega t}\right) \end{aligned} \tag{3.46}$$

If we define

$$\beta_n = \frac{a_n - jb_n}{2}, \qquad \beta_{-n} = \frac{a_n + jb_n}{2}, \qquad \beta_0 = \frac{a_0}{2} \tag{3.47}$$

then the complex form of the Fourier series is

$$\begin{aligned} s(t) &= \beta_0 + \sum_{n=1}^{\infty}(\beta_n e^{jn\omega t} + \beta_{-n} e^{-jn\omega t}) \\ &= \sum_{n=-\infty}^{\infty} \beta_n e^{jn\omega t} \end{aligned} \tag{3.48}$$

We can readily express the coefficient β_n as a function of $s(t)$, since

$$\begin{aligned} \beta_n &= \frac{a_n - jb_n}{2} \\ &= \frac{1}{T}\int_0^T s(t)(\cos n\omega t - j \sin n\omega t)\, dt \\ &= \frac{1}{T}\int_0^T s(t) e^{-jn\omega t}\, dt \end{aligned} \tag{3.49}$$

Equation 3.49 is sometimes called the *discrete Fourier transform* of $s(t)$ and Eq. 3.48 is the *inverse transform* of $\beta_n(n\omega) = \beta_n$.

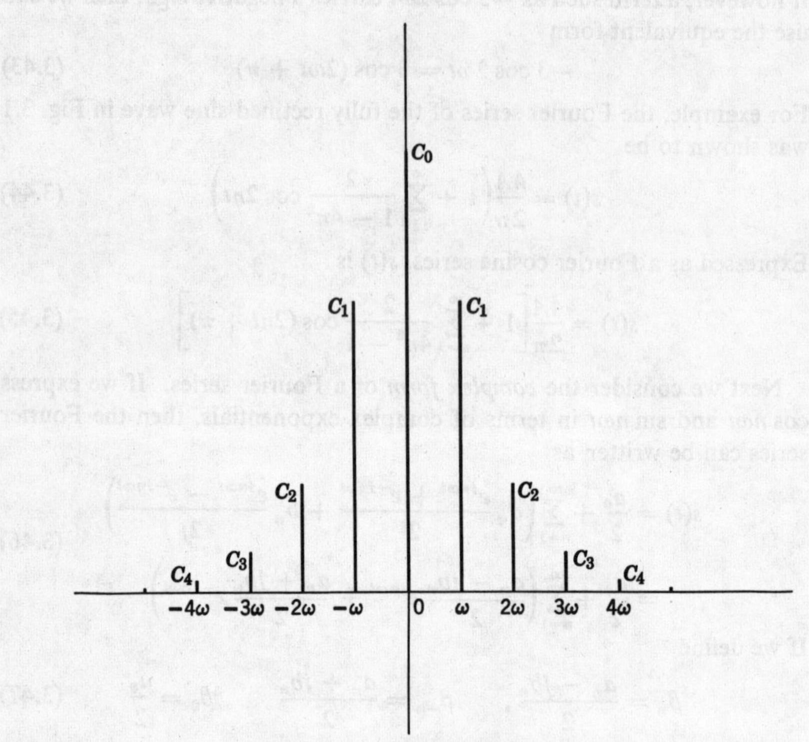

FIG. 3.5. Amplitude spectrum.

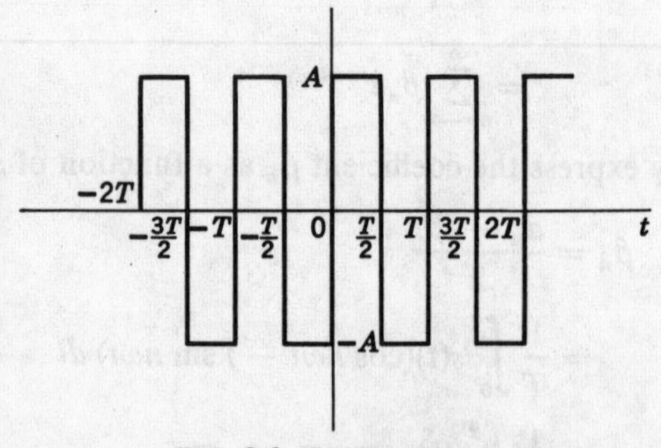

FIG. 3.6. Square wave.

Observe that β_n is usually complex and can be represented as

$$\beta_n = \text{Re}\,\beta_n + j\,\text{Im}\,\beta_n \tag{3.50}$$

The real part of β_n, Re β_n, is obtained from Eq. 3.49 as

$$\text{Re}\,\beta_n = \frac{1}{T}\int_0^T s(t)\cos n\omega t\,dt \tag{3.51}$$

and the imaginary part of β_n is

$$j\,\text{Im}\,\beta_n = \frac{-j}{T}\int_0^T s(t)\sin n\omega t\,dt \tag{3.52}$$

It is clear that Re β_n is an even function in n, whereas Im β_n is an odd function in n. The *amplitude spectrum* of the Fourier series is defined as

$$|\beta_n| = (\text{Re}^2\,\beta_n + \text{Im}^2\,\beta_n)^{1/2} \tag{3.53}$$

and the *phase spectrum* is defined as

$$\phi_n = \arctan\frac{\text{Im}\,\beta_n}{\text{Re}\,\beta_n} \tag{3.54}$$

It is easily seen that the amplitude spectrum is an even function and the phase spectrum is an odd function in n. The amplitude spectrum provides us with valuable insight as to where to *truncate* the infinite series and still maintain a good approximation to the original waveform. From a plot of the amplitude spectrum, we can almost pick out by inspection the nontrivial terms in the series. For the amplitude spectrum in Fig. 3.5, we see that a good approximation can be obtained if we disregard any harmonic above the third.

As an example, let us obtain the complex Fourier coefficients for the square wave in Fig. 3.6. Let us also find the amplitude and phase spectra of the square wave. From Fig. 3.6, we note that $s(t)$ is an odd function. Moreover, since $s(t - T/2) = -s(t)$, the series has only odd harmonics. From Eq. 3.49 we obtain the coefficients of the complex Fourier series as

$$\begin{aligned}\beta_n &= \frac{1}{T}\int_0^{T/2} Ae^{-jn\omega t}\,dt - \frac{1}{T}\int_{T/2}^{T} Ae^{-jn\omega t}\,dt \\ &= \frac{A}{jn\omega T}(1 - 2e^{-(jn\omega T/2)} + e^{-jn\omega T})\end{aligned} \tag{3.55}$$

Since $n\omega T = n2\pi$, β_n can be simplified to

$$\beta_n = \frac{A}{j2n\pi}(1 - 2e^{-jn\pi} + e^{-j2n\pi}) \tag{3.56}$$

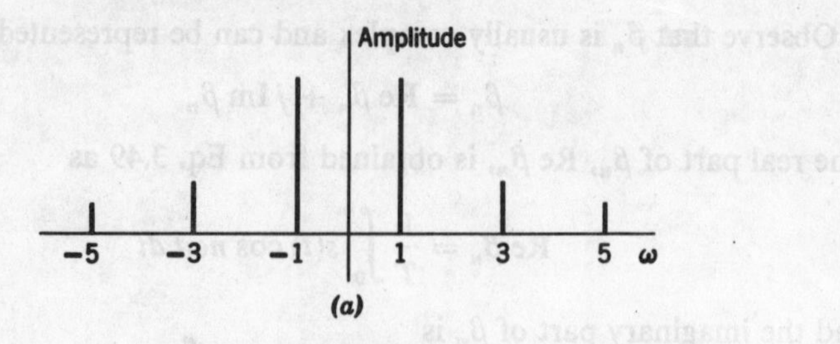

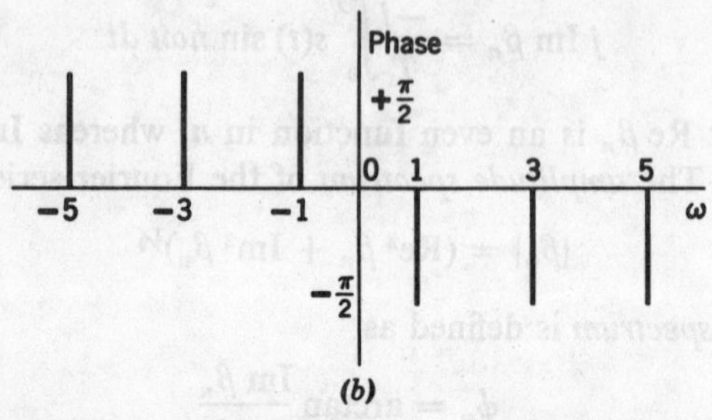

FIG. 3.7. Discrete spectra of square wave. (*a*) Amplitude. (*b*) Phase.

Simplifying β_n one step further, we obtain

$$\beta_n = \frac{2A}{jn\pi} \qquad n \text{ odd}$$
$$= 0 \qquad n \text{ even} \tag{3.57}$$

The amplitude and phase spectra of the square wave are given in Fig. 3.7.

3.6 EVALUATION OF FOURIER COEFFICIENTS USING UNIT IMPULSES

In this section we make use of a basic property of impulse functions to simplify the calculation of complex Fourier coefficients. This method is restricted to functions which are made up of *straight-line components* only. Thus the method applies for the square wave in Fig. 3.6. The method is based on the relation

$$\int_{-\infty}^{\infty} f(t)\,\delta(t - T_1)\,dt = f(T_1) \tag{3.58}$$

Let us use this equation to evaluate the complex Fourier coefficients for the impulse train in Fig. 3.8. Using Eq. 3.58 with $f(t) = e^{-jn\omega t}$, we have

$$\beta_n = \frac{A}{T}\int_0^T \delta\left(t - \frac{T}{2}\right)e^{-jn\omega t}\,dt = \frac{A}{T}\,e^{-(jn\omega T/2)} \tag{3.59}$$

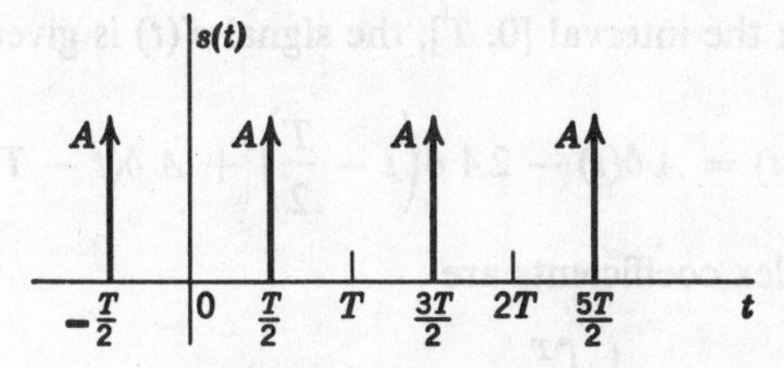

FIG. 3.8. Impulse train.

We see that the complex Fourier coefficients for impulse functions are obtained by simply substituting the time at which the impulses occur into the expression, $e^{-jn\omega t}$.

In the evaluation of Fourier coefficients, we must remember that the limits for the β_n integral are taken over *one period* only, i.e., we consider only a single period of the signal in the analysis. Consider, as an example, the square wave in Fig. 3.6. To evaluate β_n, we consider only a single period of the square wave, say, from $t = 0$ to $t = T$, as shown in Fig. 3.9*a*. Since the square wave is not made up of impulses, let us differentiate the single period of the square wave to give $s'(t)$, as shown in Fig. 3.9*b*. We can now evaluate the complex Fourier coefficients for the derivative $s'(t)$, which clearly is made up of impulses alone. Analytically, if $s(t)$ is given as

$$s(t) = \sum_{n=-\infty}^{\infty} \beta_n e^{jn\omega t} \tag{3.60}$$

then the derivative of $s(t)$ is

$$s'(t) = \sum_{n=-\infty}^{\infty} jn\omega\, \beta_n e^{jn\omega t} \tag{3.61}$$

Here, we define a new complex coefficient

$$\gamma_n = jn\omega\, \beta_n \tag{3.62}$$

or

$$\beta_n = \frac{\gamma_n}{jn\omega} \tag{3.63}$$

If the derivative $s'(t)$ is a function which consists of impulse components alone, then we simply evaluate γ_n first and then obtain β_n from Eq. 3.63. For example, the derivative of the square wave yields the impulse train

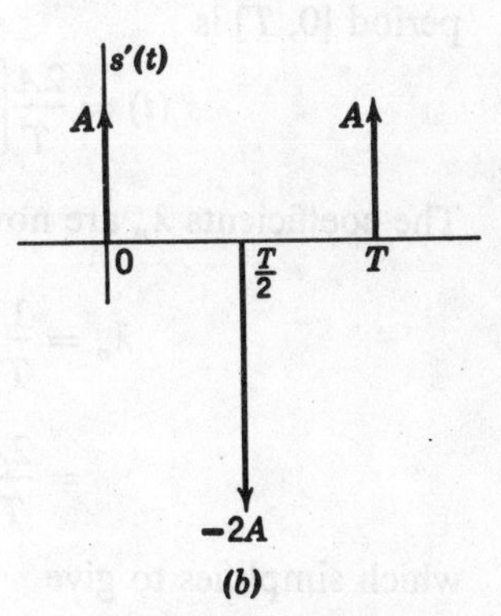

FIG. 3.9. (*a*) Square wave over period [0, *T*]. (*b*) Derivative of square wave over period [0, *T*].

in Fig. 3.9*b*. In the interval [0, *T*], the signal $s'(t)$ is given as

$$s'(t) = A\,\delta(t) - 2A\,\delta\left(t - \frac{T}{2}\right) + A\,\delta(t - T) \tag{3.64}$$

Then the complex coefficients are

$$\begin{aligned} \gamma_n &= \frac{1}{T}\int_0^T s'(t)e^{-jn\omega t}\,dt \\ &= \frac{A}{T}(1 - 2e^{-(jn\omega T/2)} + e^{-jn\omega T}) \end{aligned} \tag{3.65}$$

The Fourier coefficients of the square wave are

$$\begin{aligned} \beta_n &= \frac{\gamma_n}{jn\omega} \\ &= \frac{A}{jn\omega T}(1 - 2e^{-(jn\omega T/2)} + e^{-jn\omega T}) \end{aligned} \tag{3.66}$$

which checks with the solution obtained in the standard way in Eq. 3.55.

If the first derivative, $s'(t)$, does not contain impulses, then we must differentiate again to yield

$$s''(t) = \sum_{n=-\infty}^{\infty} \lambda_n e^{jn\omega t} \tag{3.67}$$

where

$$\lambda_n = jn\omega\gamma_n = (jn\omega)^2\,\beta_n \tag{3.68}$$

For the triangular pulse in Fig. 3.10, the second derivative over the period [0, *T*] is

$$s''(t) = \frac{2A}{T}\left[\delta(t) - 2\delta\left(t - \frac{T}{2}\right) + \delta(t - T)\right] \tag{3.69}$$

The coefficients λ_n are now obtained as

$$\begin{aligned} \lambda_n &= \frac{1}{T}\int_0^T s''(t)e^{-jn\omega t}\,dt \\ &= \frac{2A}{T^2}(1 - 2e^{-(jn\omega T/2)} + e^{-jn\omega T}) \end{aligned} \tag{3.70}$$

which simplifies to give

$$\begin{aligned} \lambda_n &= \frac{8A}{T^2} && n \text{ odd} \\ &= 0 && n \text{ even} \end{aligned} \tag{3.71}$$

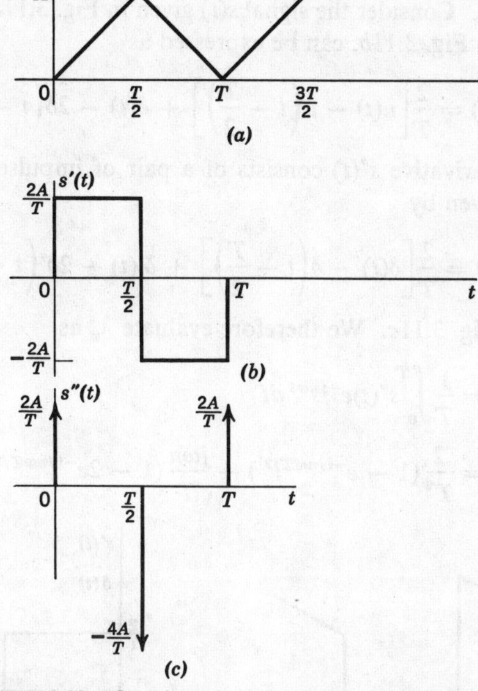

FIG. 3.10. The triangular wave and its derivatives.

From λ_n we obtain

$$\begin{aligned} \beta_n &= \frac{\lambda_n}{(j\omega n)^2} \\ &= -\frac{2A}{n^2\pi^2} \qquad n \text{ odd} \\ &= 0 \qquad\qquad n \text{ even} \end{aligned} \tag{3.72}$$

A slight difficulty arises if the expression for $s'(t)$ contains an impulse in addition to other straight-line terms. Because of these straight-line terms we must differentiate once more. However, from this additional differentiation, we obtain the derivative of the impulse as well. This presents no difficulty, however, because we know that

$$\int_{-\infty}^{\infty} s(t)\, \delta'(t - T) = -s'(T) \tag{3.73}$$

so that

$$\int_{-\infty}^{\infty} \delta'(t - T)e^{-jn\omega t}\, dt = jn\omega e^{-jn\omega T} \tag{3.74}$$

We can therefore tolerate doublets or even higher derivatives of impulses in the analysis. Consider the signal $s(t)$ given in Fig. 3.11a. Its derivative $s'(t)$, shown in Fig. 3.11b, can be expressed as

$$s'(t) = \frac{2}{T}\left[u(t) - u\left(t - \frac{T}{2}\right)\right] + \delta(t) - 2\delta\left(t - \frac{T}{2}\right) \tag{3.75}$$

The second derivative $s''(t)$ consists of a pair of impulses and a pair of doublets as given by

$$s''(t) = \frac{2}{T}\left[\delta(t) - \delta\left(t - \frac{T}{2}\right)\right] + \delta'(t) - 2\delta'\left(t - \frac{T}{2}\right) \tag{3.76}$$

as shown in Fig. 3.11c. We therefore evaluate λ_n as

$$\begin{aligned}\lambda_n &= \frac{1}{T}\int_0^T s''(t)e^{-jn\omega t}\,dt \\ &= \frac{2}{T^2}(1 - e^{-(jn\omega T/2)}) + \frac{j\omega n}{T}(1 - 2e^{-(jn\omega T/2)})\end{aligned} \tag{3.77}$$

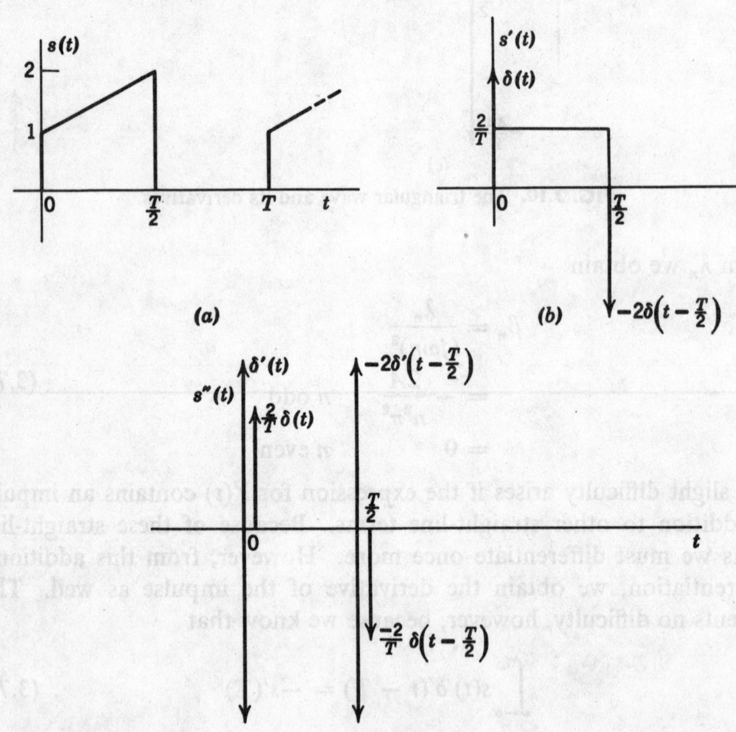

FIG. 3.11

The complex coefficients β_n are now obtained as

$$\begin{aligned}\beta_n &= \frac{\lambda_n}{(j\omega n)^2} \\ &= \frac{2}{(j\omega nT)^2}(1 - e^{-(jn\omega T/2)}) + \frac{1}{j\omega nT}(1 - 2e^{-(jn\omega T/2)})\end{aligned} \tag{3.78}$$

Simplifying, we have

$$\begin{aligned}\beta_n &= -\frac{1}{n^2\pi^2} + \frac{3}{j2\pi n} \qquad n \text{ odd} \\ &= -\frac{1}{j2\pi n} \qquad n \text{ even}\end{aligned} \tag{3.79}$$

In conclusion, it must be pointed out that the method of using impulses to evaluate Fourier coefficients does not give the d-c coefficient, $a_0/2$ or β_0. We obtain this coefficient through standard methods as given by Eq. 3.23.

3.7 THE FOURIER INTEGRAL

In this section we extend our analysis of signals to the aperiodic case. We show through a plausibility argument that generally, aperiodic signals have continuous amplitude and phase spectra. In our discussion of Fourier series, the complex coefficient β_n for periodic signals was also called the *discrete Fourier transform*

$$\beta(nf_0) = \frac{1}{T}\int_{-T/2}^{T/2} s(t)e^{-jn2\pi f_0 t}\,dt \tag{3.80}$$

and the inverse (discrete) transform was

$$s(t) = \sum_{n=-\infty}^{\infty} \beta(nf_0)e^{jn2\pi f_0 t} \tag{3.81}$$

From the discrete Fourier transform we obtain amplitude and phase spectra which consist of discrete lines. The spacing between adjacent lines in the spectrum is

$$\Delta f = (n + 1)f_0 - nf_0 = \frac{1}{T} \tag{3.82}$$

As the period T becomes larger, the spacing between the harmonic lines in the spectrum becomes smaller. For aperiodic signals, we let T approach infinity so that, in the limit, the discrete spectrum becomes *continuous*.

We now define the *Fourier integral* or *transform* as

$$S(f) = \lim_{\substack{T\to\infty \\ \Delta f\to 0}} \frac{\beta(nf_0)}{f_0} = \int_{-\infty}^{\infty} s(t)e^{-j2\pi f t}\, dt \tag{3.83}$$

The inverse transform is

$$s(t) = \int_{-\infty}^{\infty} S(f)e^{j2\pi ft}\, df \tag{3.84}$$

Equations 3.83 and 3.84 are sometimes called the *Fourier transform pair.* If we let $\mathcal{F}$ denote the operation of Fourier transformation and $\mathcal{F}^{-1}$ denote inverse transformation, then

$$\begin{aligned} S(f) &= \mathcal{F} \cdot s(t) \\ s(t) &= \mathcal{F}^{-1} \cdot S(f) \end{aligned} \tag{3.85}$$

In general, the Fourier transform $S(f)$ is complex and can be denoted as

$$S(f) = \text{Re}\, S(f) + j\,\text{Im}\, S(f) \tag{3.86}$$

The real part of $S(f)$ is obtained through the formula

$$\begin{aligned} \text{Re}\, S(f) &= \tfrac{1}{2}[S(f) + S(-f)] \\ &= \int_{-\infty}^{\infty} s(t) \cos 2\pi f t\, dt \end{aligned} \tag{3.87}$$

and the imaginary part through

$$\begin{aligned} \text{Im}\, S(f) &= \frac{1}{2j}[S(f) - S(-f)] \\ &= -\int_{-\infty}^{\infty} s(t) \sin 2\pi f t\, dt \end{aligned} \tag{3.88}$$

The amplitude spectrum of $S(f)$ is defined as

$$A(f) = [\text{Re}\, S(f)^2 + \text{Im}\, S(f)^2]^{1/2} \tag{3.89}$$

and the phase spectrum is

$$\phi(f) = \arctan \frac{\text{Im}\, S(f)}{\text{Re}\, S(f)} \tag{3.90}$$

Using the amplitude and phase definition of the Fourier transform, the inverse transform can be expressed as

$$s(t) = \int_{-\infty}^{\infty} A(f) \cos [2\pi ft - \phi(f)]\, df \tag{3.91}$$

Let us examine some examples.

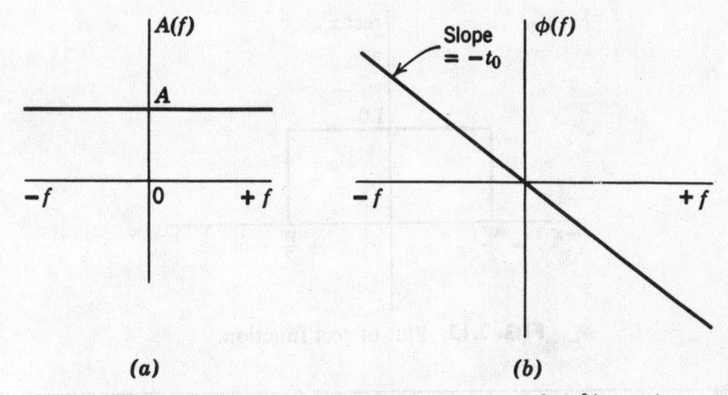

FIG. 3.12. Amplitude and phase spectrum of $A\ \delta(t - t_0)$.

Example 3.2.[3]

$$s(t) = A\ \delta(t - t_0)$$

$$S(f) = \int_{-\infty}^{\infty} A\ \delta(t - t_0)e^{-j2\pi ft}\ dt$$
$$= Ae^{-j2\pi ft_0} \tag{3.92}$$

Its amplitude spectrum is

$$A(f) = A \tag{3.93}$$

while its phase spectrum is

$$\phi(f) = -2\pi ft_0 \tag{3.94}$$

as shown in Fig. 3.12.

Example 3.3. Next consider the rectangular function plotted in Fig. 3.13. Formally, we define the function as the *rect* function.

$$\text{rect } x \begin{cases} = 1 & |x| \leq \dfrac{W}{2} \\ = 0 & |x| > \dfrac{W}{2} \end{cases} \tag{3.95}$$

The inverse transform of rect f is defined as sinc t (pronounced *sink*),

$$\mathcal{F}^{-1}[\text{rect } f] = \text{sinc } t$$
$$= \int_{-W/2}^{W/2} e^{j2\pi ft}\ df \tag{3.96}$$
$$= \frac{\sin \pi Wt}{\pi t}$$

[3] It should be noted here that the Fourier transform of a generalized function is also a generalized function. In other words, if $\phi \in C$, $(\mathcal{F} \cdot \phi) \in C$. For example, $\mathcal{F} \cdot \delta(t) = 1$, where 1 is described by a generalized function $1_n(f)$. We will not go into the formal details of Fourier transforms of generalized functions here. For an excellent treatment of the subject see M. J. Lighthill, *Fourier Analysis and Generalized Functions*, England, Cambridge University Press, 1955.

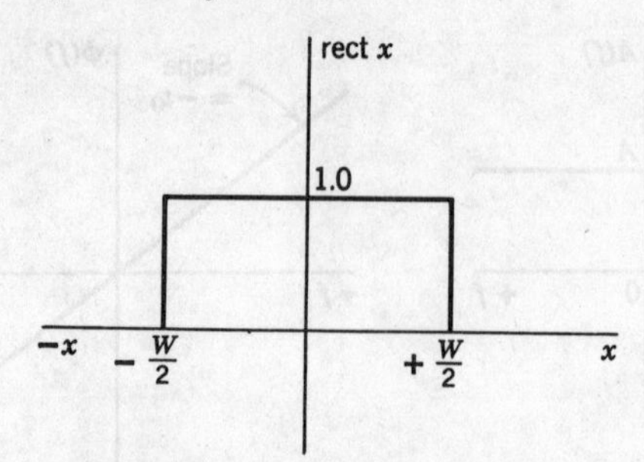

FIG. 3.13. Plot of rect function.

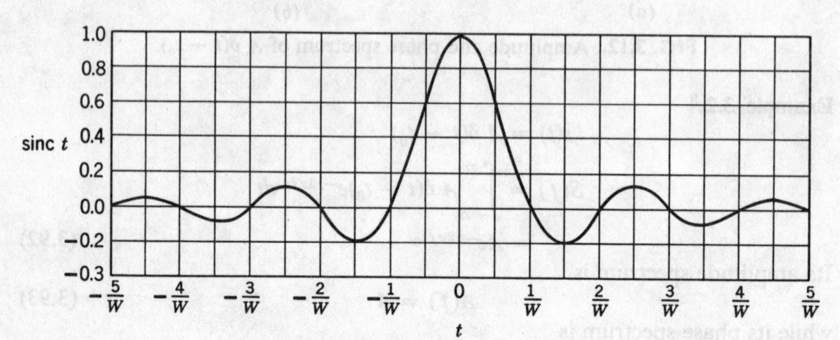

FIG. 3.14. The sinc *t* curve.

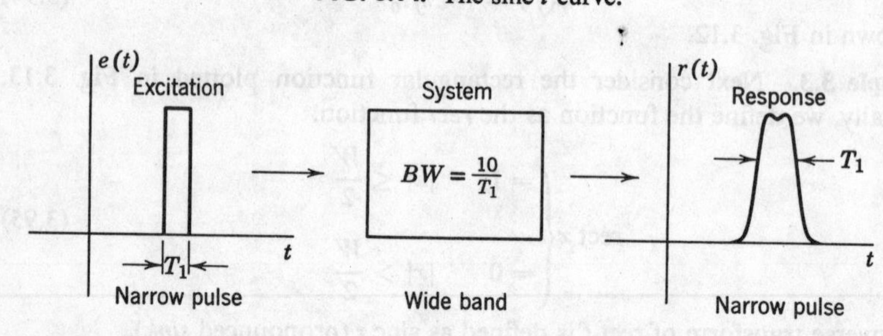

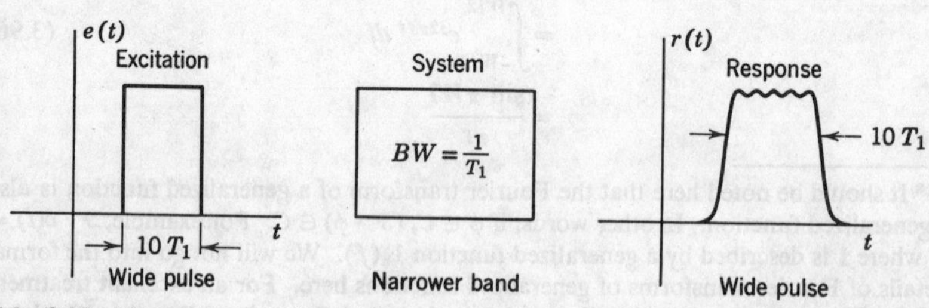

FIG. 3.15. Illustration of the reciprocity relationships between time duration and bandwidth.

From the plot of sinc t in Fig. 3.14 we see that sinc t falls as does $|t|^{-1}$, with zeros at $t = n/W$, $n = 1, 2, 3, \ldots$ We also note that most of the energy of the signal is concentrated between the points $-1/W < t < 1/W$. Let us define the *time duration* of a signal as that point, t_0, beyond which the amplitude is never greater than a specified value, for example, ϵ_0. We can effectively regard the time duration of the sinc function as $t_0 = \pm 1/W$. The value W, as we see from Fig. 3.13, is the spectral *bandwidth* of the rect function. We see that if W increases, t_0 decreases. The preceding example illustrates the reciprocal relationship between the time duration of a signal and the spectral bandwidth of its Fourier transform. This concept is quite fundamental. It illustrates why in pulse transmission, narrow pulses, i.e., those with small time durations, can only be transmitted through filters with large bandwidths; whereas pulses with longer time durations do not require such wide bandwidths, as illustrated in Fig. 3.15.

3.8 PROPERTIES OF FOURIER TRANSFORMS

In this section we consider some important properties of Fourier transforms.

Linearity

The linearity property of Fourier transforms states that the Fourier transform of a sum of two signals is the sum of their individual Fourier transforms, that is,

$$\mathcal{F}[c_1 s_1(t) + c_2 s_2(t)] = c_1 S_1(f) + c_2 S_2(f) \tag{3.97}$$

Differentiation

This property states that the Fourier transform of the derivative of a signal is $j2\pi f$ times the Fourier transform of the signal itself:

$$\mathcal{F} \cdot s'(t) = j2\pi f\, S(f) \tag{3.98}$$

or more generally,

$$\mathcal{F} \cdot s^{(n)}(t) = (j2\pi f)^n\, S(f) \tag{3.99}$$

The proof is obtained by taking the derivative of both sides of the inverse transform definition,

$$\begin{aligned} s'(t) &= \frac{d}{dt}\int_{-\infty}^{\infty} S(f)e^{j2\pi ft}\, df \\ &= \int_{-\infty}^{\infty} j2\pi f\, S(f)e^{j2\pi ft}\, df \end{aligned} \tag{3.100}$$

Similarly, it is easily shown that the transform of the integral of $s(t)$ is

$$\mathscr{F}\left[\int_{-\infty}^{t} s(\tau)\, d\tau\right] = \frac{1}{j2\pi f} S(f) \tag{3.101}$$

Consider the following example

$$s(t) = e^{-at}\, u(t) \tag{3.102}$$

Its Fourier transform is

$$\begin{aligned} S(f) &= \int_{-\infty}^{\infty} e^{-at}\, u(t) e^{-j2\pi f t}\, dt \\ &= \int_{0}^{\infty} e^{-at} e^{-j2\pi f t}\, dt = \frac{1}{a + j2\pi f} \end{aligned} \tag{3.103}$$

The derivative of $s(t)$ is

$$s'(t) = \delta(t) - ae^{-at}\, u(t) \tag{3.104}$$

Its Fourier transform is

$$\begin{aligned} \mathscr{F}[s'(t)] &= 1 - \frac{a}{a + j2\pi f} = \frac{j2\pi f}{a + j2\pi f} \\ &= j2\pi f\, S(f) \end{aligned} \tag{3.105}$$

Symmetry

The symmetry property of Fourier transforms states that if

$$\mathscr{F} \cdot x(t) = X(f) \tag{3.106}$$

then

$$\mathscr{F} \cdot X(t) = x(-f) \tag{3.107}$$

This property follows directly from the symmetrical nature of the Fourier transform pair in Eqs. 3.83 and 3.84.

Example 3.4. From the preceding section, we know that

$$\mathscr{F} \cdot \text{sinc}\, t = \text{rect}\, f \tag{3.108}$$

It is then simple to show that

$$\mathscr{F} \cdot \text{rect}\, t = \text{sinc}\,(-f) = \text{sinc}\, f \tag{3.109}$$

which conforms to the statement of the symmetry property. Consider next the Fourier transform of the unit impulse, $\mathscr{F} \cdot \delta(t) = 1$. From the symmetry property we can show that

$$\mathscr{F} \cdot 1 = \delta(f) \tag{3.110}$$

as shown in Fig. 3.16. The foregoing example is also an extreme illustration of the time-duration and bandwidth reciprocity relationship. It says that zero time duration, $\delta(t)$, gives rise to infinite bandwidth in the frequency domain; while zero bandwidth, $\delta(f)$ corresponds to infinite time duration.

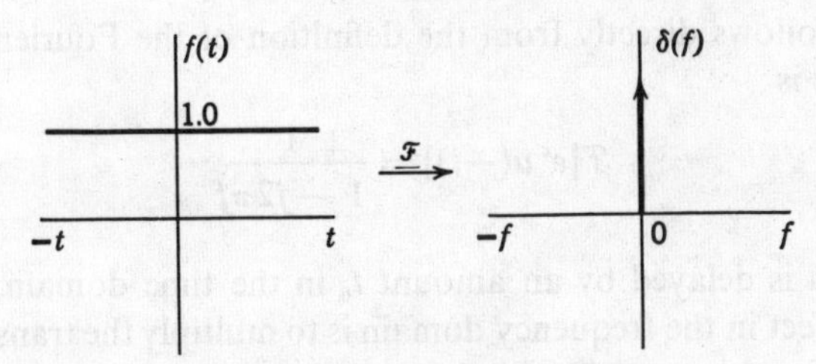

FIG. 3.16. Fourier transform of $f(t) = 1.0$.

Scale change

The scale-change property describes the time-duration and bandwidth reciprocity relationship. It states that

$$\mathcal{F}\left[s\left(\frac{t}{a}\right)\right] = |a|\, S(af) \tag{3.111}$$

Proof. We prove this property most easily through the inverse transform

$$\mathcal{F}^{-1}[|a|\, S(af)] = |a| \int_{-\infty}^{\infty} S(af) e^{j2\pi ft}\, df \tag{3.112}$$

Let $f' = af$; then

$$\begin{aligned}\mathcal{F}^{-1}[|a|\, S(f')] &= |a| \int_{-\infty}^{\infty} S(f') e^{j2\pi f'(t/a)} \frac{df'}{a} \\ &= s\left(\frac{t}{a}\right)\end{aligned} \tag{3.113}$$

As an example, consider

$$\mathcal{F}[e^{-at}\, u(t)] = \frac{1}{j2\pi f + a} \tag{3.114}$$

then

$$\begin{aligned}\mathcal{F}[e^{-t}\, u(t)] &= \frac{|a|}{j2\pi af + a} \\ &= \frac{1}{j2\pi f + 1}\end{aligned} \tag{3.115}$$

if $a > 0$.

Folding

The folding property states that

$$\mathcal{F}[s(-t)] = S(-f) \tag{3.116}$$

The proof follows directly from the definition of the Fourier transform. An example is

$$\mathcal{F}[e^t u(-t)] = \frac{1}{1 - j2\pi f} \tag{3.117}$$

Delay

If a signal is delayed by an amount t_0 in the time domain, the corresponding effect in the frequency domain is to multiply the transform of the undelayed signal by $e^{-j2\pi f t_0}$, that is,

$$\mathcal{F}[s(t - t_0)] = e^{-j2\pi f t_0}\, S(f) \tag{3.118}$$

For example,

$$\mathcal{F}[e^{-a(t-t_0)} u(t - t_0)] = \frac{e^{-j2\pi f t_0}}{a + j2\pi f} \tag{3.119}$$

Modulation

The *modulation* or *frequency shift* property of Fourier transforms states that if a Fourier transform is shifted in frequency by an amount f_0, the corresponding effect in time is described by multiplying the original signal by $e^{j2\pi f_0 t}$, that is,

$$\mathcal{F}^{-1}[S(f - f_0)] = e^{j2\pi f_0 t}\, s(t) \tag{3.120}$$

Example 3.5. Given $S(f)$ in Fig. 3.17*a*, let us find the inverse transform of $S_1(f)$ in Fig. 3.17*b* in terms of $s(t) = \mathcal{F}^{-1} S(f)$. We know that

$$S_1(f) = S(f - f_0) + S(f + f_0) \tag{3.121}$$

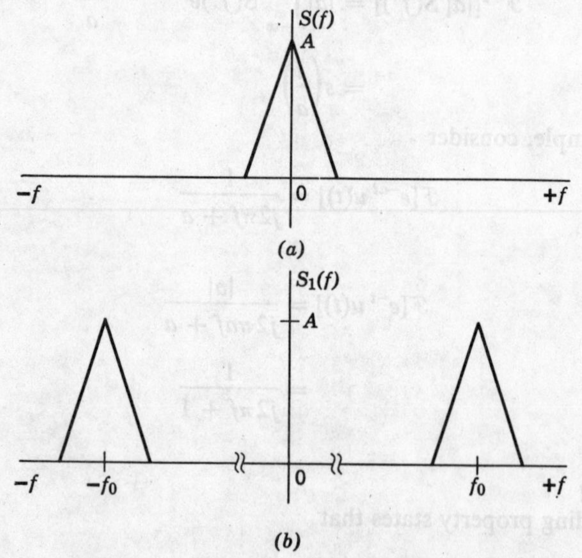

FIG. 3.17. Demonstration of amplitude modulation.

Then $$\mathcal{F}^{-1} S_1(f) = e^{j2\pi f_0 t}\, s(t) + e^{-j2\pi f_0 t}\, s(t) = 2s(t) \cos 2\pi f_0 t \tag{3.122}$$

Thus we see that multiplying a signal by a cosine or sine wave in the time domain corresponds to shifting its spectrum by an amount $\pm f_0$. In transmission terminology f_0 is the *carrier frequency*, and the process of multiplying $s(t)$ by $\cos 2\pi f_0 t$ is called *amplitude modulation.*

Parseval's theorem

An important theorem which relates energy in the time and frequency domains is *Parseval's theorem*, which states that

$$\int_{-\infty}^{\infty} s_1(t)\, s_2(t)\, dt = \int_{-\infty}^{\infty} S_1(f)\, S_2(-f)\, df \tag{3.123}$$

The proof is obtained very simply as follows:

$$\begin{aligned}\int_{-\infty}^{\infty} s_1(t)\, s_2(t)\, dt &= \int_{-\infty}^{\infty} s_2(t)\, dt \int_{-\infty}^{\infty} S_1(f) e^{j2\pi ft}\, df \\ &= \int_{-\infty}^{\infty} S_1(f)\, df \int_{-\infty}^{\infty} s_2(t) e^{j2\pi ft}\, dt \\ &= \int_{-\infty}^{\infty} S_1(f)\, S_2(-f)\, df\end{aligned} \tag{3.124}$$

In particular, when $s_1(t) = s_2(t)$, we have a corollary of Parseval's theorem known as *Plancheral's theorem.*

$$\int_{-\infty}^{\infty} s^2(t)\, dt = \int_{-\infty}^{\infty} |S(f)|^2\, df \tag{3.125}$$

If $s(t)$ is equal to the current through, or the voltage across a 1-ohm resistor, the total energy is

$$\int_{-\infty}^{\infty} s^2(t)\, dt$$

We see from Eq. 3.125 that the total energy is also equal to the area under the curve of $|S(f)|^2$. Thus $|S(f)|^2$ is sometimes called an *energy density* or *energy spectrum.*

Problems

3.1 Show that the set $\{1, \sin n\pi t/T, \cos n\pi t/T\}$, $n = 1, 2, 3, \ldots$, forms an orthogonal set over an interval $[\alpha, \alpha + 2T]$, where α is any real number. Find the norms for the members of the set and normalize the set.

3.2 Given the functions $f_1(t)$ and $f_2(t)$ expressed in terms of complex Fourier series

$$f_1(t) = \sum_{n=-\infty}^{\infty} \alpha_n e^{jn\omega t}$$

$$f_2(t) = \sum_{m=-\infty}^{\infty} \beta_m e^{jm\omega t}$$

where both $f_1(t)$ and $f_2(t)$ have the same period T, and

$$\alpha_n = |\alpha_n| e^{j\phi_n}, \qquad \beta_m = |\beta_m| e^{j\phi_m}$$

show that

$$P = \frac{1}{T}\int_0^T f_1(t)\bar{f}_2(t)\,dt$$

$$= \alpha_0\beta_0 + 2\sum_{n=1}^{\infty} |\alpha_n\beta_n| \cos(\theta_n - \phi_n)$$

Note that

$$\int_0^T e^{j(n+m)\omega t}\,dt = \begin{cases} T, & m = -n \\ 0, & m \neq -n \end{cases}$$

3.3 For the periodic signals in the figure, determine the Fourier coefficients a_n, b_n.

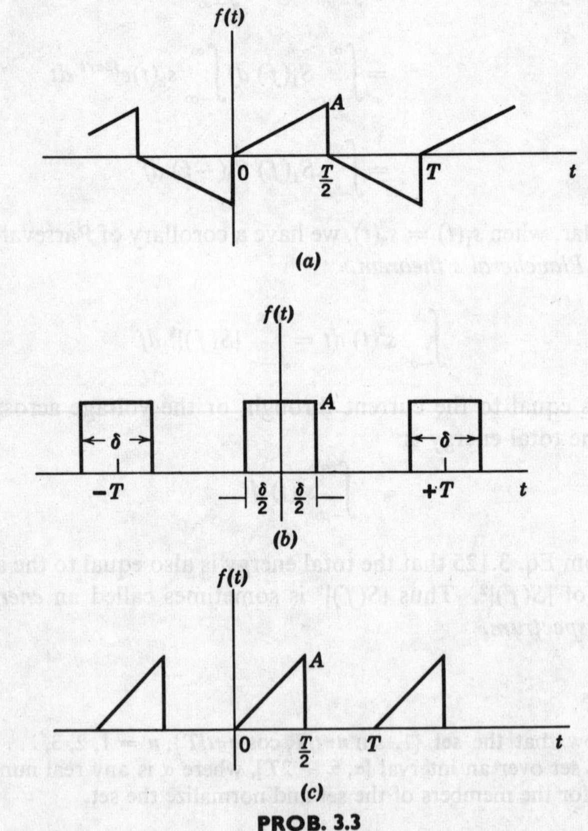

PROB. 3.3

3.4 For the waveforms in Prob. 3.3, find the discrete amplitude and phase spectra and plot.

3.5 For the waveforms in Prob. 3.3 determine the complex Fourier coefficients using the impulse function method.

3.6 Find the complex Fourier coefficients for the function shown in the figure.

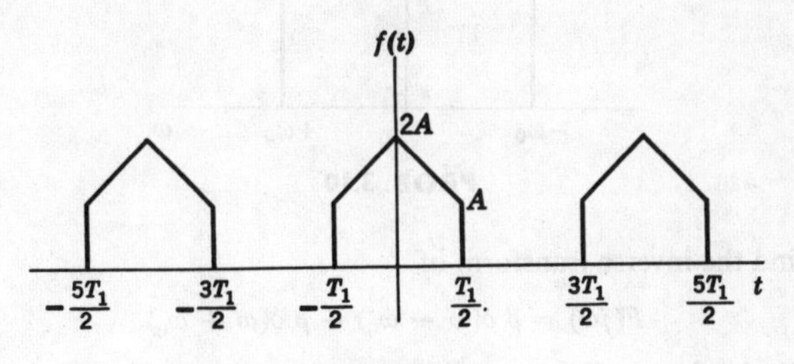

PROB. 3.6

3.7 Find the Fourier transform for the functions shown in the figure.

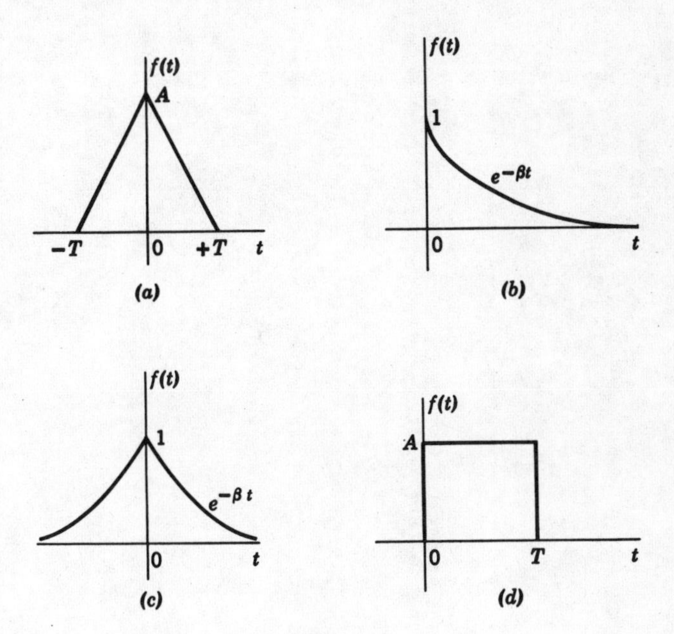

PROB. 3.7

3.8 Find the Fourier transform for

(*a*) $$f(t) = A\,\delta(t)$$

(*b*) $$f(t) = A \sin \omega_0 t$$

3.9 Prove that (*a*) if $f(t)$ is even, its Fourier transform $F(j\omega)$ is also an even function; (*b*) if $f(t)$ is odd, its Fourier transform is odd and pure imaginary.

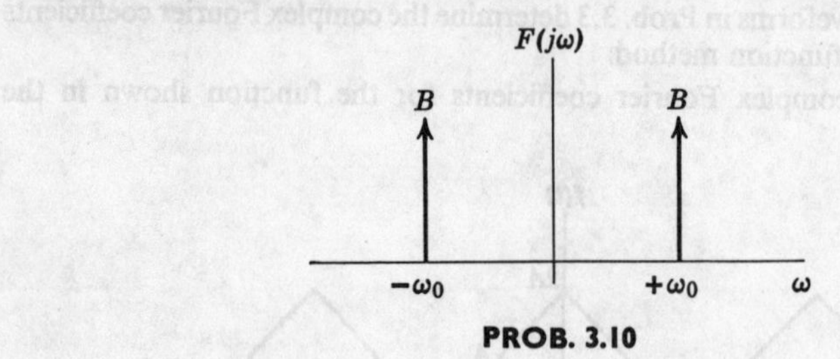

PROB. 3.10

3.10 Find the inverse transform of

$$F(j\omega) = \beta\,\delta(\omega - \omega_0) + \beta\,\delta(\omega + \omega_0)$$

as shown in the figure. What can you say about line spectra in the frequency domain?

chapter 4

Differential equations

4.1 INTRODUCTION

This chapter is devoted to a brief study of ordinary linear differential equations. We will concentrate on the mathematical aspects of differential equations and leave the physical applications for Chapter 5. The differential equations considered herein have the general form

$$F[x(t), x'(t), \ldots, x^{(n)}(t), t] = 0 \tag{4.1}$$

where t is the independent variable and $x(t)$ is a function dependent upon t. The superscripted terms $x^{(i)}(t)$ indicate the ith derivative of $x(t)$ with respect to t, namely,

$$x^{(i)}(t) = \frac{d^{(i)}\,x(t)}{dt^i} \tag{4.2}$$

The *solution* of $F = 0$ in Eq. 4.1 is $x(t)$ and must be obtained as an explicit function of t. When we substitute the explicit solution $x(t)$ into F, the equation must equal zero. If F in Eq. 4.1 is an ordinary linear differential equation, it is given by the general equation

$$a_n\, x^{(n)}(t) + a_{n-1}x^{(n-1)}(t) + \cdots + a_1\, x'(t) + a_0\, x(t) = f(t) \tag{4.3}$$

The *order* of the equation is n, the order of the highest derivative term. The term $f(t)$ on the right-hand side of the equation is the *forcing function* or *driver*, and is independent of $x(t)$. When $f(t)$ is identically zero, the equation is said to be *homogeneous*; otherwise, the equation is *nonhomogeneous*.

In this chapter we will restrict our study to *ordinary, linear* differential equations with *constant coefficients*. Let us now examine the meanings of these terms.

Ordinary. An ordinary differential equation is one in which there is only one independent variable (in our case, t). As a result there is no need for *partial* derivatives.

Constant coefficients. The coefficients $a_n, a_{n-1}, \ldots, a_2, a_1, a_0$ are constant, independent of the variable t.

Linear. A differential equation is linear if it contains only terms of the first degree in $x(t)$ and all its higher derivatives, as given by Eq. 4.3. For example, the equation

$$3x'(t) + 2x(t) = \sin t \tag{4.4}$$

is a linear differential equation. On the other hand,

$$3[x'(t)]^2 + 2x(t)\,x'(t) + 4x(t) = 5t \tag{4.5}$$

is nonlinear, because the terms $[x'(t)]^2$ and $x(t)\,x'(t)$ are nonlinear by the definition just given.

An important implication of the linearity property is the *superposition property*. According to the superposition property, if $x_1(t)$ and $x_2(t)$ are solutions of a given differential equation for forcing functions $f_1(t)$ and $f_2(t)$, respectively, then, if the forcing function were any linear combination of $f_1(t)$ and $f_2(t)$ such as

$$f(t) = a\,f_1(t) + b\,f_2(t) \tag{4.6}$$

the solution would be

$$x(t) = a\,x_1(t) + b\,x_2(t) \tag{4.7}$$

where a and b are arbitrary constants. It should be emphasized that the superposition property is extremely important and should be kept in mind in any discussion of linear differential equations.

4.2 HOMOGENEOUS LINEAR DIFFERENTIAL EQUATIONS

This section deals with some methods for the solution of homogeneous, linear differential equations with constant coefficients. First, let us find the solution to the equation

$$x'(t) - 2x(t) = 0 \tag{4.8}$$

Now, with a little prestidigitation, we *assume* the solution to be of the form

$$x(t) = Ce^{2t} \tag{4.9}$$

where C is any arbitrary constant. Let us check to see whether $x(t) = Ce^{2t}$ is truly a solution of Eq. 4.8. Substituting the assumed solution in Eq. 4.8, we obtain

$$2Ce^{2t} - 2Ce^{2t} = 0 \tag{4.10}$$

It can be shown, in general, that the solutions of homogeneous, linear differential equations consist of exponential terms of the form $C_i e^{p_i t}$. To obtain the solution of any differential equation, we substitute Ce^{pt} for $x(t)$ in the equation and determine those values of p for which the equation is zero. In other words, given the general equation

$$a_n x^{(n)}(t) + \cdots + a_1 x'(t) + a_0 x(t) = 0 \tag{4.11}$$

we let $x(t) = Ce^{pt}$, so that Eq. 4.11 becomes

$$Ce^{pt}(a_n p^n + a_{n-1}p^{n-1} + \cdots + a_1 p + a_0) = 0 \tag{4.12}$$

Since e^{pt} cannot be zero except at $p = -\infty$, the only nontrivial solutions for Eq. 4.12 occur when the polynomial

$$H(p) = a_n p^n + a_{n-1}p^{n-1} + \cdots + a_1 p + a_0 = 0 \tag{4.13}$$

Equation 4.13 is often referred to as the *characteristic equation*, and is denoted symbolically in this discussion as $H(p)$. The characteristic equation is zero only at its roots. Therefore, let us factor $H(p)$ to give

$$H(p) = a_n(p - p_0)(p - p_1)\cdots(p - p_{n-1}) \tag{4.14}$$

From Eq. 4.14, we note that $C_0 e^{p_0 t}, C_1 e^{p_1 t}, \ldots, C_{n-1}e^{p_{n-1}t}$ are all solutions of Eq. 4.11. By the *superposition principle*, the total solution is a linear combination of all the individual solutions. Therefore, the total solution of the differential equation is

$$x(t) = C_0 e^{p_0 t} + C_1 e^{p_1 t} + \cdots + C_{n-1}e^{p_{n-1}t} \tag{4.15}$$

where $C_0, C_1, \ldots, C_{n-1}$ are generally complex. The solution $x(t)$ in Eq. 4.15 is not unique unless the constants $C_0, C_1, \ldots, C_{n-1}$ are uniquely specified. In order to determine the constants C_i, we need n additional pieces of information about the equation. These pieces of information are usually specified in terms of values of $x(t)$ and its derivatives at $t = 0+$, and are therefore referred to as *initial conditions*. To obtain n coefficients, we must be given the values $x(0+), x'(0+), \ldots, x^{(n-1)}(0+)$. In a number of special cases, the values at $t = 0-$ are *not equal* to the values at $t = 0+$. If the initial specifications are given in terms of $x(0-)$, $x'(0-), \ldots, x^{(n-1)}(0-)$, we must determine the values at $t = 0+$ in order to solve for the constants C_i. This problem arises when the forcing function $f(t)$ is an impulse function or any of its derivatives. We will discuss this problem in detail in Section 4.4.

For example, in Eq. 4.9 if we are given that $x(0+) = 4$, then we obtain the constant from the equation

$$x(0+) = Ce^0 = C \tag{4.16}$$

so that $x(t)$ is uniquely determined to be

$$x(t) = 4e^{2t}$$

Example 4.1 Find the solution for

$$x''(t) + 5x'(t) + 4x(t) = 0 \tag{4.17}$$

given the initial conditions

$$x(0+) = 2 \qquad x'(0+) = -1$$

Solution. From the given equation, we first obtain the characteristic equation

$$H(p) = p^2 + 5p + 4 = 0 \tag{4.18}$$

which factors into

$$(p + 4)(p + 1) = 0 \tag{4.19}$$

The roots of the characteristic equation (referred to here as *characteristic values*) are $p = -1$; $p = -4$. Then $x(t)$ takes the form

$$x(t) = C_1e^{-t} + C_2e^{-4t} \tag{4.20}$$

From the initial condition $x(0+) = 2$, we obtain the equation

$$x(0+) = 2 = C_1 + C_2 \tag{4.21}$$

In order to solve for C_1 and C_2 explicitly, we need the additional initial condition $x'(0+) = -1$. Taking the derivative of $x(t)$ in Eq. 4.20, we have

$$x'(t) = -C_1e^{-t} - 4C_2e^{-4t} \tag{4.22}$$

At $t = 0+$, $x'(t)$ is

$$x'(0+) = -1 = -C_1 - 4C_2 \tag{4.23}$$

Solving Eqs. 4.21 and 4.23 simultaneously, we find that

$$C_1 = \tfrac{7}{3} \qquad C_2 = -\tfrac{1}{3}$$

Thus the final solution is

$$x(t) = \tfrac{7}{3}e^{-t} - \tfrac{1}{3}e^{-4t} \tag{4.24}$$

Next, we examine the case when the characteristic equation $H(p)$ has *multiple roots*. Specifically, let us consider the case where $H(p)$ has a root $p = p_0$ of multiplicity k as given by

$$H(p) = a_n(p - p_0)^k(p - p_1)\cdots(p - p_n) \tag{4.25}$$

It will be left to the reader to show that the solution must then contain k terms involving e^{p_0t} of the form

$$x(t) = C_{00}e^{p_0t} + C_{01}te^{p_0t} + C_{02}t^2e^{p_0t} + \cdots + C_{0k-1}t^{k-1}e^{p_0t} + C_1e^{p_1t} + C_2e^{p_2t} + \cdots + C_ne^{p_nt} \tag{4.26}$$

where the double-scripted terms in Eq. 4.26 denote the terms in the solution due to the multiple root, $(p - p_0)^k$.

Example 4.2 Solve the equation

$$x''(t) - 8x'(t) + 16x(t) = 0 \tag{4.27}$$

with

$$x(0+) = 2 \quad \text{and} \quad x'(0+) = 4$$

Solution. The characteristic equation is

$$H(p) = p^2 - 8p + 16 = (p - 4)^2 \tag{4.28}$$

Since $H(p)$ has a double root at $p = 4$, the solution must take the form

$$x(t) = C_1 e^{4t} + C_2 t e^{4t} \tag{4.29}$$

In order to determine C_1 and C_2, we evaluate $x(t)$ and $x'(t)$ at $t = 0+$ to give

$$\begin{aligned} x(0+) &= C_1 = 2 \\ x'(0+) &= 4C_1 + C_2 = 4 \end{aligned} \tag{4.30}$$

Thus the final solution is

$$x(t) = 2e^{4t} - 4te^{4t} \tag{4.31}$$

Another interesting case arises when $H(p)$ has complex conjugate roots. Consider the equation

$$H(p) = a_2(p - p_1)(p - p_2) \tag{4.32}$$

where p_1 and p_2 are complex conjugate roots, that is,

$$p_1, p_2 = \sigma \pm j\omega \tag{4.33}$$

The solution $x(t)$ then takes the form

$$x(t) = C_1 e^{(\sigma + j\omega)t} + C_2 e^{(\sigma - j\omega)t} \tag{4.34}$$

Expanding the term $e^{j\omega t}$ by Euler's equation, $x(t)$ can be expressed as

$$x(t) = C_1 e^{\sigma t}(\cos \omega t + j \sin \omega t) + C_2 e^{\sigma t}(\cos \omega t - j \sin \omega t) \tag{4.35}$$

which reduces to

$$x(t) = (C_1 + C_2)e^{\sigma t} \cos \omega t + j(C_1 - C_2)e^{\sigma t} \sin \omega t \tag{4.36}$$

Let us introduce two new constants, M_1 and M_2, so that $x(t)$ may be expressed in the more convenient form

$$x(t) = M_1 e^{\sigma t} \cos \omega t + M_2 e^{\sigma t} \sin \omega t \tag{4.37}$$

where M_1 and M_2 are related to the constants C_1 and C_2 by the equations

$$\begin{aligned} M_1 &= C_1 + C_2 \\ M_2 &= j(C_1 - C_2) \end{aligned} \tag{4.38}$$

The constants M_1 and M_2 are determined in the usual manner from initial conditions.

Another convenient form for the solution $x(t)$ can be obtained if we introduce still another pair of constants, M and ϕ, defined by the equations

$$M_1 = M \sin \phi$$
$$M_2 = M \cos \phi \tag{4.39}$$

With the constants M and ϕ we obtain another form of $x(t)$, namely,

$$x(t) = Me^{\sigma t} \sin (\omega t + \phi) \tag{4.40}$$

Example 4.3 Solve the equation

$$x''(t) + 2x'(t) + 5x(t) = 0 \tag{4.41}$$

with the initial conditions

$$x(0+) = 1 \qquad x'(0+) = 0$$

Solution. The characteristic equation $H(p)$ is

$$H(p) = p^2 + 2p + 5 = (p + 1 + j2)(p + 1 - j2) \tag{4.42}$$

so that, assuming the form of solution in Eq. 4.37, we have $\sigma = -1$ and $\omega = 2$. Then $x(t)$ is

$$x(t) = M_1e^{-t} \cos 2t + M_2e^{-t} \sin 2t \tag{4.43}$$

At $t = 0+$

$$x(0+) = 1 = M_1 \tag{4.44}$$

The derivative of $x(t)$ is

$$x'(t) = M_1(-e^{-t} \cos 2t - 2e^{-t} \sin 2t) + M_2(-e^{-t} \sin 2t + 2e^{-t} \cos 2t) \tag{4.45}$$

At $t = 0+$, we obtain the equation

$$x'(0+) = 0 = -M_1 + 2M_2 \tag{4.46}$$

Solving Eqs. 4.44 and 4.46 simultaneously, we find $M_1 = 1$ and $M_2 = +\frac{1}{2}$. Thus the final solution is

$$x(t) = e^{-t}(\cos 2t + \tfrac{1}{2} \sin 2t) \tag{4.47}$$

If we had used the form of $x(t)$ given in Eq. 4.40, we would have obtained the solution

$$x(t) = \sqrt{\tfrac{5}{4}}e^{-t} \sin [2t + \tan^{-1} (2)] \tag{4.48}$$

Now let us consider a differential equation that illustrates everything we have discussed concerning characteristic values.

Example 4.4 The differential equation is

$$x^{(5)}(t) + 9x^{(4)}(t) + 32x^{(3)}(t) + 58x''(t) + 56x'(t) + 24x(t) = 0 \tag{4.49}$$

The initial conditions are

$$x^{(4)}(0+) = 0 \qquad x^{(3)}(0+) = 1$$
$$x''(0+) = -1 \qquad x'(0+) = 0 \qquad x(0+) = 1$$

Solution. The characteristic equation is

$$H(p) = p^5 + 9p^4 + 32p^3 + 58p^2 + 56p + 24 = 0 \tag{4.50}$$

which factors into

$$H(p) = (p + 1 + j1)(p + 1 - j1)(p + 2)^2(p + 3) = 0 \tag{4.51}$$

From $H(p)$ we immediately write $x(t)$ as

$$x(t) = M_1e^{-t}\cos t + M_2e^{-t}\sin t + C_0e^{-2t} + C_1te^{-2t} + C_2e^{-3t} \tag{4.52}$$

Since there are five coefficients, we need a corresponding number of equations to evaluate the unknowns. These are

$$\begin{aligned}
x(0+) &= M_1 + C_0 + C_2 = 1 \\
x'(0+) &= -M_1 + M_2 - 2C_0 - 3C_2 + C_1 = 0 \\
x''(0+) &= -2M_2 + 4C_0 - 4C_1 + 9C_2 = -1 \\
x^{(3)}(0+) &= 2M_1 + 2M_2 - 8C_0 + 12C_1 - 27C_2 = 1 \\
x^{(4)}(0+) &= -4M_1 + 16C_0 + 81C_2 - 32C_1 = 0
\end{aligned} \tag{4.53}$$

Solving these five equations simultaneously, we obtain

$$M_1 = 0 \qquad M_2 = \tfrac{3}{2} \qquad C_0 = 1 \qquad C_1 = \tfrac{1}{2} \qquad C_2 = 0$$

so that the final solution is

$$x(t) = \tfrac{3}{2}e^{-t}\sin t + e^{-2t} + \tfrac{1}{2}te^{-2t} \tag{4.54}$$

We have seen that the solution of a homogeneous, differential equation may take different forms depending upon the roots of its characteristic equation. Table 4.1 should be useful in determining the particular form of solution.

TABLE 4.1

Roots of $H(p)$	Forms of Solution
1. Single real root, $p = p_0$	e^{p_0t}
2. Root of multiplicity, k, $(p - p_1)^k$	$C_0e^{p_1t} + C_1te^{p_1t} + \cdots + C_{k-1}t^{k-1}e^{p_1t}$
3. Complex roots at $p_{2,3} = \sigma \pm j\omega$	$M_1e^{\sigma t}\cos \omega t + M_2e^{\sigma t}\sin \omega t$ or $Me^{\sigma t}\sin(\omega t + \phi)$
4. Complex roots of multiplicity k at $p_{4,5} = \sigma \pm j\omega$	$M_0e^{\sigma t}\cos \omega t + M_1te^{\sigma t}\cos \omega t + \cdots$ $+ M_{k-1}t^{k-1}e^{\sigma t}\cos \omega t + N_0e^{\sigma t}\sin \omega t$ $+ N_1te^{\sigma t}\sin \omega t + \cdots$ $+ N_{k-1}t^{k-1}e^{\sigma t}\sin \omega t$

4.3 NONHOMOGENEOUS EQUATIONS

As we mentioned in the introduction to this chapter, a nonhomogeneous differential equation is one in which the forcing function $f(t)$ is not identically zero for all t. In this section, we will discuss methods for obtaining the solution $x(t)$ of an equation with constant coefficients

$$a_n x^{(n)}(t) + a_{n-1} x^{(n-1)}(t) + \cdots + a_0 x(t) = f(t) \tag{4.55}$$

Let $x_p(t)$ be a particular solution for Eq. 4.55, and let $x_c(t)$ be the solution of the homogeneous equation obtained by letting $f(t) = 0$ in Eq. 4.55. It is readily seen that

$$x(t) = x_p(t) + x_c(t) \tag{4.56}$$

is also a solution of Eq. 4.55. According to the *uniqueness theorem*, the solution $x(t)$ in Eq. 4.56 is the unique solution for the nonhomogeneous differential equation *if* it satisfies the specified initial conditions at $t = 0+$.[1] In Eq. 4.56, $x_p(t)$ is the *particular integral*; $x_c(t)$ is the *complementary function*; and $x(t)$ is the *total solution*.

Since we already know how to find the complementary function $x_c(t)$, we now have to find the particular integral $x_p(t)$. In solving for $x_p(t)$, a very reliable rule of thumb is that $x_p(t)$ usually takes the *same form* as the forcing function if $f(t)$ can be expressed as a sum of exponential functions. Specifically, $x_p(t)$ assumes the form of $f(t)$ plus all its derivatives. For example, if $f(t) = \alpha \sin \omega t$, then $x_p(t)$ takes the form

$$x_p(t) = A \sin \omega t + B \cos \omega t$$

The only unknowns that must be determined are the coefficients A and B of the terms in $x_p(t)$. The method for obtaining $x_p(t)$ is appropriately called the *method of undetermined coefficients* or unknown coefficients.

In illustrating the method of unknown coefficients, let us take $f(t)$ to be

$$f(t) = \alpha e^{\beta t} \tag{4.57}$$

where α and β are arbitrary constants. We then assume $x_p(t)$ to have a similar form, that is,

$$x_p(t) = Ae^{\beta t} \tag{4.58}$$

and A is the unknown coefficient. To determine A, we simply substitute the assumed solution $x_p(t)$ into the differential equation. Thus,

$$Ae^{\beta t}(a_n\beta^n + a_{n-1}\beta^{n-1} + \cdots + a_1\beta + a_0) = \alpha e^{\beta t} \tag{4.59}$$

[1] See, for example, C. R. Wylie, *Advanced Engineering Mathematics* (2nd ed.), McGraw-Hill Book Company, New York, 1960, pp. 83–84.

We see that the polynomial within the parentheses is the characteristic equation $H(p)$ with $p = \beta$. Consequently, the unknown coefficient is obtained as

$$A = \frac{\alpha}{H(\beta)} \tag{4.60}$$

provided that $H(\beta) \neq 0$.

Example 4.5 Determine the solution of the equation

$$x''(t) + 3x'(t) + 2x(t) = 4e^t \tag{4.61}$$

with the initial conditions, $x(0+) = 1$, $x'(0+) = -1$.

Solution. The characteristic equation is

$$H(p) = p^2 + 3p + 2 = (p + 2)(p + 1)$$

so that the complementary function is

$$x_c(t) = C_1e^{-t} + C_2e^{-2t}$$

For the forcing function $f(t) = 4e^t$, the constants in Eq. 4.60 are $\alpha = 4$, $\beta = 1$.

Then
$$A = \frac{4}{H(1)} = \frac{2}{3}$$

Thus we obtain
$$x_p(t) = \tfrac{2}{3}e^t$$

The total solution is

$$x(t) = x_c(t) + x_p(t) = C_1e^{-t} + C_2e^{-2t} + \tfrac{2}{3}e^t \tag{4.62}$$

To evaluate the constants C_1 and C_2, we substitute the given initial conditions, namely,

$$\begin{aligned} x(0+) &= 1 = C_1 + C_2 + \tfrac{2}{3} \\ x'(0+) &= -1 = -C_1 - 2C_2 + \tfrac{2}{3} \end{aligned} \tag{4.63}$$

Solving Eq. 4.63, we find that $C_1 = -1$, $C_2 = \frac{4}{3}$. Consequently,

$$x(t) = -e^{-t} + \tfrac{4}{3}e^{-2t} + \tfrac{2}{3}e^t \tag{4.64}$$

It should be pointed out that we solve for the constants C_1 and C_2 from the initial conditions for the *total solution.* This is because initial conditions are not given for $x_c(t)$ or $x_p(t)$, but for the total solution.

Next, let us consider an example of a constant forcing function $f(t) = \alpha$. We may use Eq. 4.60 if we resort to the artifice

$$f(t) = \alpha = \alpha e^{t0} \tag{4.65}$$

that is, $\beta = 0$. For the differential equation in Example 4.5 with $f(t) = 4$, we see that

$$x_p(t) = A = \frac{4}{H(0)} = 2 \tag{4.66}$$

and

$$x(t) = C_1 e^{-t} + C_2 e^{-2t} + 2 \tag{4.67}$$

When the forcing function is a sine or cosine function, we can still consider the forcing function to be of exponential form and make use of the method of undetermined coefficients and Eq. 4.60. Suppose

$$f(t) = \alpha e^{j\omega t} = \alpha(\cos \omega t + j \sin \omega t) \tag{4.68}$$

then the particular integral $x_{p1}(t)$ can be written as

$$x_{p1}(t) = \text{Re } x_{p1}(t) + j \text{ Im } x_{p1}(t) \tag{4.69}$$

From the superposition principle, we can show that

if $\quad f(t) = \alpha \cos \omega t \quad$ then $\quad x_p(t) = \text{Re } x_{p1}(t)$

if $\quad f(t) = \alpha \sin \omega t \quad$ then $\quad x_p(t) = \text{Im } x_{p1}(t)$

Consequently, whether the excitation is a cosine function $\alpha \cos \omega t$ or a sine function $\alpha \sin \omega t$, we can use an exponential driver $f(t) = \alpha e^{j\omega t}$; then we take the real or imaginary part of the resulting particular integral.

Example 4.6 Find the particular integral for the equation

$$x''(t) + 5x'(t) + 4x(t) = 2 \sin 3t \tag{4.70}$$

Solution. First, let us take the excitation to be

$$f_1(t) = 2e^{j3t} \tag{4.71}$$

so that the particular integral $x_{p1}(t)$ takes the form

$$x_{p1}(t) = Ae^{j3t} \tag{4.72}$$

From the characteristic equation

$$H(p) = p^2 + 5p + 4$$

we determine the coefficient A to be

$$A = \frac{2}{H(j3)} = \frac{2}{-5 + j15} = \frac{2}{5\sqrt{10}} e^{j[\tan^{-1}(3) - \pi]} \tag{4.73}$$

Then $x_{p1}(t)$ is

$$x_{p1}(t) = \frac{2}{5\sqrt{10}} e^{j[\tan^{-1}(3) + 3t - \pi]} \tag{4.74}$$

and the particular integral $x_p(t)$ for the original driver $f(t) = 2 \sin 3t$ is

$$x_p(t) = \text{Im}\, x_{p1}(t) = \frac{2}{5\sqrt{10}} \sin\,[3t + \tan^{-1}(3) - \pi] \tag{4.75}$$

There are certain limitations to the applicability of the method of undetermined coefficients. If $f(t)$ were, for example, a Bessel function $J_0(t)$, we could not assume $x_p(t)$ to be a Bessel function of the same form (if it is a Bessel function at all). However, we may apply the method to forcing functions of the following types:

1. $f(t) = A$; constant.
2. $f(t) = A(t^n + b_{n-1}t^{n-1} + \cdots + b_1 t + b_0)$; n, integer.
3. $f(t) = e^{pt}$; p real or complex.
4. Any function formed by multiplying terms of type 1, 2, or 3.

For the purposes of linear network analysis, the method is more than adequate.

Suppose the forcing function were

$$f(t) = At^k e^{pt} \qquad p = \sigma + j\omega$$

The particular integral can be written as

$$x_p(t) = (A_k t^k + A_{k-1}t^{k-1} + \cdots + A_1 t + A_0)e^{pt} \tag{4.76}$$

where the coefficients $A_k, A_{k-1}, \ldots, A_1, A_0$ are to be determined.

4.4 STEP AND IMPULSE RESPONSE

In this section we will discuss solutions of differential equations with step or impulse forcing functions. In physical applications these solutions are called, respectively, *step responses* and *impulse responses.* As physical quantities, the step and impulse responses of a linear system are highly significant measures of system performance. In Chapter 7 it will be shown that a precise mathematical description of a linear system is given by its impulse response. Moreover, a reliable measure of the transient behavior of the system is given by its step and impulse response. In this section, we will be concerned with the mathematical problem of solving for the impulse and step response, given a linear differential equation with initial conditions at $t = 0-$.

From Chapter 2, recall that the definition of the unit step function was

$$\begin{aligned} u(t) &= 1 \qquad t \geq 0 \\ &= 0 \qquad t < 0 \end{aligned}$$

and the unit impulse was shown to have the properties:

$$\begin{aligned} \delta(t) &= \infty \qquad t = 0 \\ &= 0 \qquad t \neq 0 \end{aligned}$$

and
$$\int_{0-}^{0+} \delta(t)\, dt = 1$$

In addition, we have the relationship

$$\delta(t) = \frac{d\,u(t)}{dt}$$

As the definitions of $\delta(t)$ and $u(t)$ indicate, both functions have discontinuities at $t = 0$. In dealing with initial conditions for step and impulse drivers, we must then recognize that the solution $x(t)$ *and its derivatives* $x'(t)$, $x''(t)$, *etc., may not be continuous at* $t = 0$. In other words, it may be that

$$\begin{aligned} x^{(n)}(0-) &\neq x^{(n)}(0+) \\ x^{(n-1)}(0-) &\neq x^{(n-1)}(0+) \\ &\cdots \\ x(0-) &\neq x(0+) \end{aligned}$$

In many physical problems, the initial conditions are given at $t = 0-$. However, to evaluate the unknown constants of the total solution, we must have the initial conditions at $t = 0+$. Our task, then, is to determine the conditions at $t = 0+$, given the initial conditions at $t = 0-$. The method discussed here is borrowed from electromagnetic theory and is often referred to as "integrating through a Green's function."[2]

Consider the differential equation with an impulse forcing function

$$a_n\, x^{(n)}(t) + a_{n-1}\, x^{(n-1)}(t) + \cdots + a_0\, x(t) = A\, \delta(t) \tag{4.77}$$

To insure that the right-hand side of Eq. 4.77 will equal the left-hand side, one of the terms $x^{(n)}(t), x^{(n-1)}(t), \ldots, x(t)$ must contain an impulse. The question is, "Which term contains the impulse?" A close examination shows that the highest derivative term $x^{(n)}(t)$ *must* contain the impulse, because if $x^{(n-1)}(t)$ contained the impulse, $x^{(n)}(t)$ would contain a *doublet* $C\,\delta'(t)$. This argument holds, similarly, for all lower derivative terms of $x(t)$. If the term $x^{(n)}(t)$ contains the impulse, then $x^{(n-1)}(t)$ would contain a step and $x^{(n-2)}(t)$, a ramp. *We conclude therefore that, for an impulse forcing function, the two highest derivative terms are discontinuous at* $t = 0$.

[2] The Green's function is another name for impulse response; see, for example, Morse and Feshbach, *Methods of Theoretical Physics,* McGraw-Hill Book Company, New York, 1952, Chapter 7.

For a step forcing function, only the highest derivative term is discontinuous at $t = 0$.

Since initial conditions are usually given at $t = 0-$, our task is to determine the values $x^{(n)}(0+)$ and $x^{(n-1)}(0+)$ for an impulse forcing function. Referring to Eq. 4.77, let us integrate the equation between $t = 0-$ and $t = 0+$, namely,

$$a_n \int_{0-}^{0+} x^{(n)}(t)\,dt + a_{n-1} \int_{0-}^{0+} x^{(n-1)}(t)\,dt + \cdots + a_0 \int_{0-}^{0+} x(t)\,dt = A \int_{0-}^{0+} \delta(t)\,dt \tag{4.78}$$

After integrating, we obtain

$$a_n[x^{(n-1)}(0+) - x^{(n-1)}(0-)] + a_{n-1}[x^{(n-2)}(0+) - x^{(n-2)}(0-)] + \cdots = A \tag{4.79}$$

We know that all derivative terms below $(n - 1)$ are continuous at $t = 0$. Consequently, Eq. 4.79 simplifies to

$$a_n[x^{(n-1)}(0+) - x^{(n-1)}(0-)] = A \tag{4.80}$$

so that

$$x^{(n-1)}(0+) = \frac{A}{a_n} + x^{(n-1)}(0-) \tag{4.81}$$

We must next determine $x^{(n)}(0+)$. At $t = 0+$, the differential equation in Eq. 4.77 is

$$a_n\, x^{(n)}(0+) + a_{n-1}\, x^{(n-1)}(0+) + \cdots + a_0\, x(0+) = 0 \tag{4.82}$$

Since all derivative terms below $(n - 1)$ are continuous, and since we have already solved for $x^{(n-1)}(0+)$, we find that

$$x^{(n)}(0+) = -\frac{1}{a_n}[a_{n-1}x^{(n-1)}(0+) + \cdots + a_1x'(0+) + a_0x(0+)] \tag{4.83}$$

For a step forcing function $A\,u(t)$, all derivative terms except $x^{(n)}(t)$, are continuous at $t = 0$. To determine $x^{(n)}(0+)$, we derive in a manner similar to Eq. 4.83, the expression

$$x^{(n)}(0+) = \frac{A}{a_n} - \frac{1}{a_n}[a_{n-1}x^{(n-1)}(0+) + \cdots + a_0x(0+)] \tag{4.84}$$

The process of determining initial conditions when the forcing function is an impulse or one of its higher derivatives can be simplified by the visual process shown in Eqs. 4.85 and 4.86. Above each derivative term we draw its associated highest-order singularity. Note that we need only go as low as a step in this visual aid.

$$a_n x^{(n)}(t) + a_{n-1} x^{(n-1)}(t) + a_{n-2} x^{(n-2)}(t) + \cdots + a_0 x(t) = \delta'(t) \qquad (4.85)$$

$$a_n x^{(n)}(t) + a_{n-1} x^{(n-1)}(t) + a_{n-2} x^{(n-2)}(t) + \cdots + a_0 x(t) = \delta(t) \qquad (4.86)$$

It should be noted that if a derivative term contains a certain singularity—for example, a doublet—it also contains all lower derivative terms. For example, in the equation

$$x''(t) \quad + \quad 3x'(t) \quad + \quad 2x(t) \quad = \quad 4\delta'(t) \qquad (4.87)$$

we assume the following forms for the derivative terms at $t = 0$:

$$\begin{aligned} x''(t) &= A\,\delta'(t) + B\,\delta(t) + C\,u(t) \\ x'(t) &= A\,\delta(t) + B\,u(t) \\ x(t) &= A\,u(t) \end{aligned} \qquad (4.88)$$

Substituting Eq. 4.88 into Eq. 4.87, we obtain

$$A\,\delta'(t) + B\,\delta(t) + C\,u(t) + 3A\,\delta(t) + 3B\,u(t) + 2A\,u(t) = 4\delta'(t) \qquad (4.89)$$

Or in a more convenient form, we have

$$A\,\delta'(t) + (B + 3A)\,\delta(t) + (C + 3B + 2A)\,u(t) = 4\delta'(t) \qquad (4.90)$$

Equating like coefficients on both sides of the Eq. 3.90 gives

$$\begin{aligned} A &= 4 \\ B + 3A &= 0 \\ C + 3B + 2A &= 0 \end{aligned} \qquad (4.91)$$

from which we obtain $B = -12$ and $C = 28$. Therefore, at $t = 0$, it is true that

$$
\begin{aligned}
x''(t) &= 4\delta'(t) - 12\delta(t) + 28u(t) \\
x'(t) &= 4\delta(t) - 12u(t) \qquad (4.92) \\
x(t) &= 4u(t)
\end{aligned}
$$

The $u(t)$ terms in Eq. 4.92 gives rise to the discontinuities in the initial conditions at $t = 0$. We are given the initial conditions at $t = 0-$. Once we evaluate the A, B, C coefficients in Eq. 4.88, we can obtain the initial conditions at $t = 0+$ by referring to the coefficients of the step terms. For example, if

$$
\begin{aligned}
x(0-) &= -2 \\
x'(0-) &= -1 \\
x''(0-) &= 7
\end{aligned}
$$

Then from Eq. 4.92 we obtain

$$
\begin{aligned}
x(0+) &= -2 + 4 = 2 \\
x'(0+) &= -1 - 12 = -13 \qquad (4.93) \\
x''(0+) &= 7 + 28 = 35
\end{aligned}
$$

The total solution of Eq. 4.87 is obtained as though it were a homogeneous equation, since $\delta'(t) = 0$ for $t \neq 0$. The only influence the doublet driver has is to produce discontinuities in the initial conditions at $t = 0$. Having evaluated the initial conditions at $t = 0+$, we can obtain the total solution with ease. Thus,

$$x(t) = C_1 e^{-t} + C_2 e^{-2t} \qquad (4.94)$$

From Eq. 4.93 we readily obtain

$$x(t) = (-9e^{-t} + 11e^{-2t})\, u(t) \qquad (4.95)$$

The total solution of a differential equation with a step or impulse forcing function is obtained in an equally straightforward manner. For a step forcing function, only the highest derivative term has a discontinuity at $t = 0$. Since we do not need the initial condition of the highest derivative term for our solution, we proceed as if we were solving a standard nonhomogeneous equation with a constant forcing function. For an impulse driver, once we determine the initial conditions at $t = 0+$, the equation is solved in the same manner as a homogeneous equation.

Example 4.7 Find the step and impulse response for the equation

$$2x''(t) + 4x'(t) + 10x(t) = f(t)$$

where $f(t) = \delta(t)$ and $f(t) = u(t)$, respectively. The initial conditions at $t = 0-$ are

$$x(0-) = x'(0-) = x''(0-) = 0$$

Solution. Let us first find the impulse response. We note that the x'' term contains an impulse and a step; the x' term contains a step; the x term contains a ramp, and is therefore continuous at $t = 0$. Thus $x(0+) = x(0-) = 0$. To obtain $x'(0+)$, we use Eq. 4.81

$$x'(0+) = \frac{K}{a_2} + x'(0-) = \frac{1}{2} + 0 = \frac{1}{2} \tag{4.96}$$

Note that we actually need only $x(0+)$ and $x'(0+)$ to evaluate the constants for the second-order differential equation. Next, we proceed to the complementary function $x_c(t)$. The characteristic equation is

$$H(p) = 2(p^2 + 2p + 5) = 2(p + 1 + j2)(p + 1 - j2) \tag{4.97}$$

Since $H(p)$ has a pair of complex conjugate roots, we use a standard form for $x_c(t)$:

$$x_c(t) = Me^{-t} \sin (2t + \phi) \tag{4.98}$$

Substituting the initial conditions at $t = 0+$, we obtain

$$\begin{aligned} x(0+) &= 0 = M \sin \phi \\ x'(0+) &= \tfrac{1}{2} = 2M \cos \phi - M \sin \phi \end{aligned} \tag{4.99}$$

from which we find $\phi = 0$ and $M = \frac{1}{4}$. Thus the impulse response, which we denote here as $x_\delta(t)$, is

$$x_\delta(t) = \tfrac{1}{4}e^{-t} \sin 2t \, u(t) \tag{4.100}$$

Next we must solve for the step response $x_u(t)$. For convenience, let us write the complementary function as

$$x_c(t) = e^{-t}(A_1 \sin 2t + A_2 \cos 2t) \tag{4.101}$$

The particular integral is evaluated by considering the forcing function a constant $f(t) = 1$ so that

$$x_p(t) = \frac{1}{H(0)} = \frac{1}{10} \tag{4.102}$$

The total solution is then

$$x(t) = (A_1 \sin 2t + A_2 \cos 2t)e^{-t} + \tfrac{1}{10} \tag{4.103}$$

Since $x'(t)$ and $x(t)$ must be continuous for a step forcing function,

$$\begin{aligned} x(0+) &= x(0-) = 0 \\ x'(0+) &= x'(0-) = 0 \end{aligned} \tag{4.104}$$

Substituting these initial conditions into $x(t)$ and $x'(t)$, we find that $A_1 = -0.05$ $A_2 = -0.1$. Therefore, the step response is

$$x_u(t) = 0.1[1 - e^{-t}(0.5 \sin 2t + \cos 2t)]\, u(t) \qquad (4.105)$$

$$x_\rho(t) \xrightarrow{d/dt} x_u(t) \xrightarrow{d/dt} x_\delta(t)$$

$$x_\delta(t) \xrightarrow{\int_{0-}^{t}} x_u(t) \xrightarrow{\int_{0-}^{t}} x_\rho(t)$$

FIG. 4.1.

Note that the impulse response and the step response are related by the equation

$$x_\delta(t) = \frac{d}{dt} x_u(t) \qquad (4.106)$$

We can demonstrate Eq. 4.106 by the following procedure. Let us substitute $x_u(t)$ into the original equation

$$2\frac{d^2}{dt^2} x_u(t) + 4\frac{d}{dt} x_u(t) + 10x_u(t) = u(t) \qquad (4.107)$$

Differentiating both sides, we have

$$2\frac{d^2}{dt^2}\left[\frac{d}{dt} x_u(t)\right] + 4\frac{d}{dt}\left[\frac{d}{dt} x_u(t)\right] + 10\left[\frac{d}{dt} x_u(t)\right] = \delta(t) \qquad (4.108)$$

from which Eq. 4.106 follows.

Generalizing, we see that, if we have the step response for a differential equation, we can obtain the impulse response by differentiating the step response. We can also obtain the response to a ramp function $f(t) = A\,\rho(t)$ (where A is the height of the step) by integrating the step response. The relationships discussed here are summarized in Fig. 4.1.

4.5 INTEGRODIFFERENTIAL EQUATIONS

In this section, we will consider an integrodifferential equation of the form

$$a_n x^n(t) + a_{n-1} x^{n-1}(t) + \cdots + a_0 x(t) + a_{-1}\int_0^t x(\tau)\, d\tau = f(t) \qquad (4.109)$$

where the coefficients $\{a_n, a_{n-1}, \ldots, a_{-1}\}$ are constants. In solving an equation of the form of Eq. 4.109 we use two very similar methods. The first method is to differentiate both sides of Eq. 4.109 to give

$$a_n x^{(n+1)}(t) + a_{n-1} x^{(n)}(t) + \cdots + a_0 x'(t) + a_{-1} x(t) = f'(t) \qquad (4.110)$$

The second method consists of a change of variables. We let $y'(t) = x(t)$; Eq. 4.109 then becomes

$$a_n y^{(n+1)}(t) + a_{n-1} y^{(n)}(t) + \cdots + a_0 y'(t) + a_{-1} y(t) = f(t) \qquad (4.111)$$

Note that from Eq. 4.110 we obtain $x(t)$ directly. From Eq. 4.111, we obtain $y(t)$, which we must then differentiate to obtain $x(t)$. An important point to keep in mind is that we might have to derive some additional initial conditions in order to have a sufficient number to evaluate the unknown constants.

Example 4.8 Solve the integrodifferential equation

$$x'(t) + 3x(t) + 2\int_{0-}^{t} x(\tau)\,d\tau = 5u(t) \tag{4.112}$$

The initial condition is $x(0-) = 1$.

Solution. Since the characteristic equation of Eq. 4.112 is of second degree, we need an additional initial condition $x'(0+)$. We obtain $x'(0+)$ from the given equation at $t = 0+$:

$$x'(0+) + 3x(0+) + 2\int_{0-}^{0+} x(\tau)\,d\tau = 5 \tag{4.113}$$

Since $x(t)$ is continuous at $t = 0$,

$$\int_{0-}^{0+} x(\tau)\,d\tau = 0 \tag{4.114}$$

and

$$x(0+) = x(0-) = 1 \tag{4.115}$$

Therefore,

$$x'(0+) = 5 - 3x(0+) = 2 \tag{4.116}$$

METHOD 1. Differentiating both sides of Eq. 4.112, we obtain

$$x''(t) + 3x'(t) + 2x(t) = 5\delta(t) \tag{4.117}$$

The complementary function is then

$$x_c(t) = C_1e^{-t} + C_2e^{-2t} \tag{4.118}$$

Using the initial conditions for $x(0+)$ and $x'(0+)$, we obtain the total solution

$$x(t) = 4e^{-t} - 3e^{-2t} \tag{4.119}$$

METHOD 2. Letting $y'(t) = x(t)$, the original differential equation then becomes

$$y''(t) + 3y'(t) + 2y(t) = 5u(t) \tag{4.120}$$

We know that

$$\begin{aligned} y'(0+) &= x(0+) = 1 \\ y''(0+) &= x'(0+) = 2 \end{aligned} \tag{4.121}$$

From Eq. 4.120, at $t = 0+$, we obtain

$$y(0+) = \tfrac{1}{2}[5 - y''(0+) - 3y'(0+)] = 0 \tag{4.122}$$

Without going into details, the total solution can be determined as

$$y(t) = -4e^{-t} + \tfrac{3}{2}e^{-2t} + \tfrac{5}{2} \tag{4.123}$$

Differentiating $y(t)$, we obtain

$$x(t) = y'(t) = 4e^{-t} - 3e^{-2t} \tag{4.124}$$

4.6 SIMULTANEOUS DIFFERENTIAL EQUATIONS

Up to this point, we have considered only differential equations with a single dependent variable $x(t)$. In this section, we will discuss equations with more than one dependent variable. We shall limit our discussion to equations with two unknowns, $x(t)$ and $y(t)$. The methods described here, however, are applicable to any number of unknowns. Consider first the system of homogeneous equations

$$\begin{aligned} \alpha_1 x'(t) + \alpha_0 x(t) + \beta_1 y'(t) + \beta_0 y(t) &= 0 \\ \gamma_1 x'(t) + \gamma_0 x(t) + \delta_1 y'(t) + \delta_0 y(t) &= 0 \end{aligned} \tag{4.125}$$

where $\alpha_i, \beta_i, \gamma_i, \delta_i$ are arbitrary constants. The complementary function is obtained by assuming that

$$x(t) = C_1 e^{pt}; \qquad y(t) = C_2 e^{pt}$$

so that the characteristic equation is given by the determinant

$$H(p) = \begin{vmatrix} (\alpha_1 p + \alpha_0) & (\beta_1 p + \beta_0) \\ (\gamma_1 p + \gamma_0) & (\delta_1 p + \delta_0) \end{vmatrix} \tag{4.126}$$

The roots of $H(p)$ are found by setting the determinant equal to zero, that is,

$$(\alpha_1 p + \alpha_0)(\delta_1 p + \delta_0) - (\beta_1 p + \beta_0)(\gamma_1 p + \gamma_0) = 0 \tag{4.127}$$

It is seen that a nontrivial solution of $H(p) = 0$ exists only if

$$(\alpha_1 p + \alpha_0)(\delta_1 p + \delta_0) \neq (\beta_1 p + \beta_0)(\gamma_1 p + \gamma_0) \tag{4.128}$$

Assuming that the preceding condition holds, we see that $H(p)$ is a second-degree polynomial in p and can be expressed in factored form as

$$H(p) = C(p - p_0)(p - p_1) \tag{4.129}$$

where C is a constant multiplier. The complementary functions are

$$\begin{aligned} x(t) &= K_1 e^{p_0 t} + K_2 e^{p_1 t} \\ y(t) &= K_3 e^{p_0 t} + K_4 e^{p_1 t} \end{aligned} \tag{4.130}$$

and the constants K_1, K_2, K_3, K_4 are determined from initial conditions. As in the case of a single unknown, if $H(p)$ has a pair of double roots; i.e., if $p_0 = p_1$, then

$$\begin{aligned} x(t) &= (K_1 + K_2 t)e^{p_0 t} \\ y(t) &= (K_3 + K_4 t)e^{p_0 t} \end{aligned} \tag{4.131}$$

If $H(p)$ has a pair of conjugate roots,

$$\left.\begin{matrix} p_1 \\ p_1{}^* \end{matrix}\right\} = \sigma \pm j\omega$$

then

$$\begin{aligned} x(t) &= M_1 e^{\sigma t} \sin(\omega t + \phi_1) \\ y(t) &= M_2 e^{\sigma t} \sin(\omega t + \phi_2) \end{aligned} \tag{4.132}$$

Example 4.9. Consider the system of equations

$$\begin{aligned} 2x'(t) + 4x(t) + y'(t) - y(t) &= 0 \\ x'(t) + 2x(t) + y'(t) + y(t) &= 0 \end{aligned} \tag{4.133}$$

with the initial conditions

$$\begin{aligned} x'(0+) &= 2 \qquad y'(0+) = -3 \\ x(0+) &= 0 \qquad y(0+) = 1 \end{aligned} \tag{4.134}$$

Solution. The characteristic equation is

$$H(p) = \begin{vmatrix} 2p + 4 & p - 1 \\ p + 2 & p + 1 \end{vmatrix} = 0 \tag{4.135}$$

Evaluating the determinant, we find that

$$H(p) = p^2 + 5p + 6 = (p + 2)(p + 3) \tag{4.136}$$

so that

$$\begin{aligned} y(t) &= K_1 e^{-2t} + K_2 e^{-3t} \\ x(t) &= K_3 e^{-2t} + K_4 e^{-3t} \end{aligned} \tag{4.137}$$

With the initial conditions, $x'(0+) = 2$, $x(0+) = 0$, we obtain $K_3 = 2$, $K_4 = -2$. From the conditions $y'(0+) = -3$, $y(0+) = 1$, we obtain $K_1 = 0$, $K_2 = 1$. Thus the final solutions are

$$\begin{aligned} x(t) &= 2e^{-2t} - 2e^{-3t} \\ y(t) &= e^{-3t} \end{aligned} \tag{4.138}$$

Next, let us determine the solutions for a set of nonhomogeneous differential equations. We use the method of undetermined coefficients here. Consider first an exponential forcing function given by the set of equations

$$\begin{aligned} \alpha_1 x' + \alpha_0 x + \beta_1 y' + \beta_0 y &= Ne^{\theta t} \\ \gamma_1 x' + \gamma_0 x + \delta_1 y' + \delta_0 y &= 0 \end{aligned} \tag{4.139}$$

We first assume that

$$\begin{aligned} x_p(t) &= Ae^{\theta t} \\ y_p(t) &= Be^{\theta t} \end{aligned} \tag{4.140}$$

Then Eq. 4.139 becomes

$$\begin{aligned} (\alpha_1\theta + \alpha_0)A + (\beta_1\theta + \beta_0)B &= N \\ (\gamma_1\theta + \gamma_0)A + (\delta_1\theta + \delta_0)B &= 0 \end{aligned} \tag{4.141}$$

The determinant for the set of equations 4.141 is

$$H(\theta) = \Delta(\theta) = \begin{vmatrix} \alpha_1\theta + \alpha_0 & \beta_1\theta + \beta_0 \\ \gamma_1\theta + \gamma_0 & \delta_1\theta + \delta_0 \end{vmatrix} \tag{4.142}$$

where $H(\theta)$ is the characteristic equation with $p = \theta$. We now determine the undetermined coefficients A and B from $\Delta(\theta)$ and its cofactors, namely,

$$\begin{aligned} A &= \frac{N\,\Delta_{11}(\theta)}{\Delta(\theta)} \\ B &= \frac{N\,\Delta_{12}(\theta)}{\Delta(\theta)} \end{aligned} \tag{4.143}$$

where Δ_{ij} is the ijth cofactor of $\Delta(\theta)$.

Example 4.10. Solve the set of equations

$$\begin{aligned} 2x' + 4x + y' - y &= 3e^{4t} \\ x' + 2x + y' + y &= 0 \end{aligned} \tag{4.144}$$

given the conditions $x'(0+) = 1$, $x(0+) = 0$, $y'(0+) = 0$, $y(0+) = -1$.

Solution. The complementary functions $x_c(t)$ and $y_c(t)$, as well as the characteristic equation $H(p)$, were determined in Example 4.9. Now we must find A and B in the equations

$$\begin{aligned} x_p(t) &= Ae^{4t} \\ y_p(t) &= Be^{4t} \end{aligned} \tag{4.145}$$

The characteristic equation with $p = 4$ is

$$H(4) = \begin{vmatrix} 2(4) + 4 & (4) - 1 \\ (4) + 2 & (4) + 1 \end{vmatrix} = 42 \tag{4.146}$$

Then we obtain from Eq. 4.143 the constants

$$A = \tfrac{5}{14} \qquad B = -\tfrac{3}{7}$$

The incomplete solutions are

$$\begin{aligned} x(t) &= K_1e^{-2t} + K_2e^{-3t} + \tfrac{5}{14}e^{4t} \\ y(t) &= K_3e^{-2t} + K_4e^{-3t} - \tfrac{3}{7}e^{4t} \end{aligned} \tag{4.147}$$

Substituting for the initial conditions, we finally obtain

$$\begin{aligned} y(t) &= \tfrac{1}{7}(-4e^{-3t} - 3e^{4t}) \\ x(t) &= \tfrac{1}{14}(-6e^{-2t} + e^{-3t} + 5e^{4t}) \end{aligned} \tag{4.148}$$

Example 4.11. Solve the system of equations

$$\begin{aligned} 2x' + 4x + y' + 7y &= 5u(t) \\ x' + x + y' + 3y &= 5\delta(t) \end{aligned} \tag{4.149}$$

given the initial conditions

$$x(0-) = x'(0-) = y(0-) = y'(0-) = 0$$

Solution. First we find the characteristic equation

$$H(p) = \Delta(p) = \begin{vmatrix} 2p+4 & p+7 \\ p+1 & p+3 \end{vmatrix} \tag{4.150}$$

which simplifies to give

$$H(p) = (p^2 + 2p + 5) = (p + 1 + j2)(p + 1 - j2) \tag{4.151}$$

The complementary functions $x_c(t)$ and $y_c(t)$ are then

$$\begin{aligned} x_c(t) &= A_1e^{-t}\cos 2t + A_2e^{-t}\sin 2t \\ y_c(t) &= B_1e^{-t}\cos 2t + B_2e^{-t}\sin 2t \end{aligned} \tag{4.152}$$

The particular solutions are obtained for the set of equations with $t > 0$, namely,

$$\begin{aligned} 2x' + 4x + y' + 7y &= 5 \\ x' + x + y' + 3y &= 0 \end{aligned} \tag{4.153}$$

Using the method of undetermined coefficients, we assume that x_p and y_p are constants: $x_p = C_1$; $y_p = C_2$. Since the forcing function 5 can be regarded as an exponential term with zero exponent, that is, $5 = 5e^{0t}$, we can solve for C_1 and C_2 with the use of the characteristic equation $H(p)$ with $p = 0$. Thus,

$$H(0) = \Delta(0) = \begin{vmatrix} 4 & 7 \\ 1 & 3 \end{vmatrix} = 5 \tag{4.154}$$

and

$$C_1 = \frac{5\Delta_{11}(0)}{\Delta(0)} = \frac{5(3)}{5} = 3$$
$$C_2 = -\frac{(5)1}{5} = -1 \tag{4.155}$$

The general solution is then

$$\begin{aligned} x(t) &= A_1 e^{-t}\cos 2t + A_2 e^{-t}\sin 2t + 3 \\ y(t) &= B_1 e^{-t}\cos 2t + B_2 e^{-t}\sin 2t - 1 \end{aligned} \tag{4.156}$$

In order to find A_1, A_2, B_1, B_2, we need the values $x(0+)$, $x'(0+)$, $y(0+)$, $y'(0+)$. The values for $x(0+)$ and $y(0+)$ are first obtained by integrating the original differential equations between $t = 0-$ and $t = 0+$. Thus,

$$\begin{aligned} \int_{0-}^{0+} (2x' + 4x + y' + 7y)\,dt &= \int_{0-}^{0+} 5u(t)\,dt \\ \int_{0-}^{0+} (x' + x + y' + 3y)\,dt &= \int_{0-}^{0+} 5\delta(t)\,dt \end{aligned} \tag{4.157}$$

We know that only the highest derivative terms in both equations contain impulses at $t = 0$. Moreover, both $x(t)$ and $y(t)$ contain, at most, step discontinuities at $t = 0$. Therefore, in the integration

$$\begin{aligned} \int_{0-}^{0+} (4x + 7y)\,dt &= 0 \\ \int_{0-}^{0+} (x + 3y)\,dt &= 0 \end{aligned} \tag{4.158}$$

After integrating, we obtain

$$\begin{aligned} 2x(0+) + y(0+) &= 0 \\ x(0+) + y(0+) &= 5 \end{aligned} \tag{4.159}$$

Solving, we find

$$x(0+) = -5 \qquad y(0+) = 10 \tag{4.160}$$

To find $x'(0+)$ and $y'(0+)$, we substitute the values for $x(0+)$ and $y(0+)$ into the original equations at $t = 0+$. Thus,

$$\begin{aligned} 2x'(0+) - 20 + y'(0+) + 70 &= 5 \\ x'(0+) - 5 + y'(0+) + 30 &= 0 \end{aligned} \tag{4.161}$$

so that

$$\begin{aligned} x'(0+) &= -20 \\ y'(0+) &= -5 \end{aligned} \tag{4.162}$$

Substituting these values into Eq. 4.156, we eventually obtain the final solutions

$$\begin{aligned} x(t) &= (-8e^{-t}\cos 2t - 14e^{-t}\sin 2t + 3)\,u(t) \\ y(t) &= (11e^{-t}\cos 2t + 3e^{-t}\sin 2t - 1)\,u(t) \end{aligned} \tag{4.163}$$

Problems

4.1 Show that

$$x_1(t) = M_1 e^{-t} \cos 2t$$

and
$$x_2(t) = M_2 e^{-t} \sin 2t$$

are solutions for the equation

$$x''(t) + 2x'(t) + 5x(t) = 0$$

Show that $x_1 + x_2$ is also a solution.

4.2 Determine only the *form* of the solution for the equation

(*a*) $$x''(t) + 4x'(t) + 3x(t) = 0$$
(*b*) $$x''(t) + 8x'(t) + 5x(t) = 0$$
(*c*) $$x''(t) - 5x(t) = 0$$
(*d*) $$x''(t) + 5x(t) = 0$$
(*e*) $$x''(t) + 6x'(t) + 25x(t) = 0$$
(*f*) $$x''(t) + 6x'(t) + 9x(t) = 0$$

4.3 Given the initial conditions $x(0+) = 1$, $x'(0+) = -1$, determine the solutions for

(*a*) $$x''(t) + 6x'(t) + 25x(t) = 0$$
(*b*) $$x''(t) + 8x'(t) + 16x(t) = 0$$
(*c*) $$x''(t) + 4.81x'(t) + 5.76x(t) = 0$$

4.4 Find only the particular integrals for the equations

(*a*) $$x''(t) + 7x'(t) + 12x(t) = e^{-3t}$$
(*b*) $$x''(t) + 3x'(t) + 2x(t) = 2 \sin 3t$$
(*c*) $$x''(t) + 2x'(t) + 5x(t) = e^{-t} \sin 2t$$
(*d*) $$x''(t) + 2x'(t) + 5x(t) = e^{-5t}/2$$
(*e*) $$x''(t) + 5.0x'(t) + 6.25x(t) = 6$$
(*f*) $$x''(t) + 6x'(t) + 5x(t) = 2e^{-t} + 3e^{-3t}$$

4.5 Given the initial conditions $x(0+) = 1$, $x'(0+) = 0$, determine the solutions for

(*a*) $$x''(t) + 4x'(t) + 3x(t) = 5e^{-t} \sin 2t$$
(*b*) $$x''(t) + 6x'(t) + 25x(t) = 2 \cos t$$
(*c*) $$x''(t) + 8x'(t) + 16x(t) = 2$$

4.6 (*a*) A system is described by the differential equation

$$y''(t) + 3y'(t) + 2y(t) = \delta(t)$$

The initial conditions are $y(0-) = 1$; $y'(0-) = 2$. Find $y(0+)$ and $y'(0+)$.

(*b*) Given the differential equation

$$x^{(3)}(t) + 14x''(t) + 8x(t) = 6\delta(t)$$

with the intial conditions $x(0-) = 12$, $x'(0-) = 6$, and $x''(0-) = -7$, find $x^{(3)}(0+)$, $x''(0+)$, $x'(0+)$, and $x(0+)$.

4.7 Given the initial conditions $x'(0-) = x''(0-) = 0$, find the solutions for the equations

(*a*) $$x''(t) + 2x'(t) + 2x(t) = 3\delta(t)$$

(*b*) $$x''(t) + 7x'(t) + 12x(t) = 5u(t)$$

4.8 Given the initial condition $x(0-) = -2$, solve the integrodifferential equations

(*a*) $$x'(t) + 5x(t) + 4\int_0^t x(\tau)\,d\tau = 2\sin t$$

(*b*) $$x'(t) + 2x(t) + 2\int_0^t x(\tau)\,d\tau = \frac{3}{2}e^{-t}$$

(*c*) $$x'(t) + 6x(t) + 9\int_0^t x(\tau)\,d\tau = 2u(t)$$

4.9 Given the set of equations

$$x'(t) + x(t) + y(t) = u(t)$$
$$x(t) + y'(t) + 2y(t) = \delta(t)$$

with $x(0-) = x'(0-) = y(0-) = y'(0-) = 0$, find $x(0+)$, $x'(0+)$, $y(0+)$, and $y'(0+)$.

4.10 Solve the set of equations

$$2x'(t) + 3x(t) + y'(t) + 6y(t) = \delta(t)$$
$$x'(t) + x(t) + y'(t) + 6y(t) = u(t)$$

All initial conditions at $t = 0-$ are zero.

4.11 Solve the set of equations

$$2x'(t) + 2x(t) + y'(t) - y(t) = 2\delta(t)$$
$$x'(t) + x(t) + y'(t) + 2y(t) = 3e^{-t}\,u(t)$$

The initial conditions are

$$x(0-) = 1 \qquad y(0-) = 0$$
$$x'(0-) = -1 \qquad y'(0-) = 0$$

chapter 5

Network analysis: I

5.1 INTRODUCTION

In this chapter we will apply our knowledge of differential equations to the analysis of linear, passive, time-invariant networks. We will assume that the reader is already familiar with Kirchhoff's current and voltage laws, and with methods for writing mesh and node equations for a-c or d-c circuits.[1] We will, therefore, consider only briefly the problem of writing mesh and node equations when the independent variable is time t. The problems in this chapter have the following format: Given an excitation signal from an energy source and the network, a specified response that is a current or voltage in the network is to be determined. When relating these problems to the mathematics in Chapter 4, we shall see that, physically, the forcing function corresponds to the excitation; the network is described by the differential equation; and the unknown variable $x(t)$ is the response.

The problems encountered will be twofold. First, we must *write* the differential problems of the network using Kirchhoff's current and voltage laws. Next, we must *solve* these equations for a specified current or voltage in the network. Both problems are equally important. It is useless, for example, to solve a differential equation which is set up incorrectly, or whose initial conditions are incorrectly specified.

The usual type of problem presented in this chapter might generally be described as follows. A switch is closed at $t = 0$, which connects an energy (voltage or current) source to a network (Fig. 5.1). The analog of a switch closing at $t = 0$ is the energy source whose output is $e(t)\,u(t)$. Before the

[1] For a comprehensive treatment, see H. H. Skilling, *Electrical Engineering Circuits* (2nd Edition), John Wiley and Sons, New York, 1965.

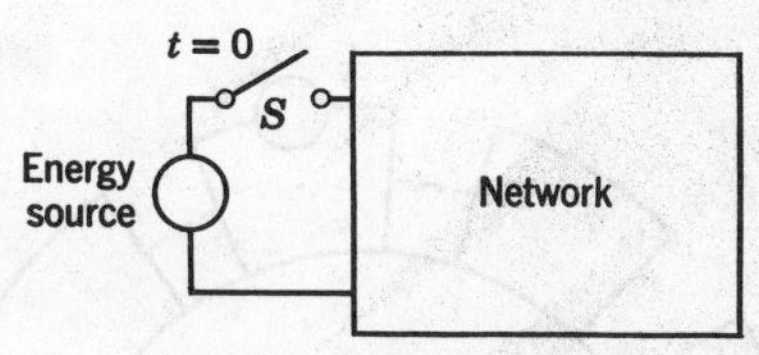

FIG. 5.1. Switching action.

switch is closed, the currents and voltages in the network have known values. These values at $t = 0-$ are the initial conditions. We must then determine the values of the currents and voltages just after the switch closes (at $t = 0+$) to solve the network equations. If the excitation is not an impulse function or any of its derivatives, the current and voltage variables are continuous at $t = 0$. For an impulse driver the values at $t = 0+$ can be determined from methods given in Chapter 4. Having obtained the initial conditions, we then go on to solve the network differential equations. Unless otherwise stated, all the solutions are valid *only* for $t \geq 0+$.

Since we are dealing only with linear circuits, it is essential that we bear in mind the all-important principle of *superposition*. According to the superposition principle, the current through any element in a linear circuit with n voltage and m current sources is equal to the algebraic sum of currents through the same element resulting from the sources taken one at a time, the other sources having been suppressed. Consider the linear network depicted in Fig. 5.2*a* with n voltage and m current sources.

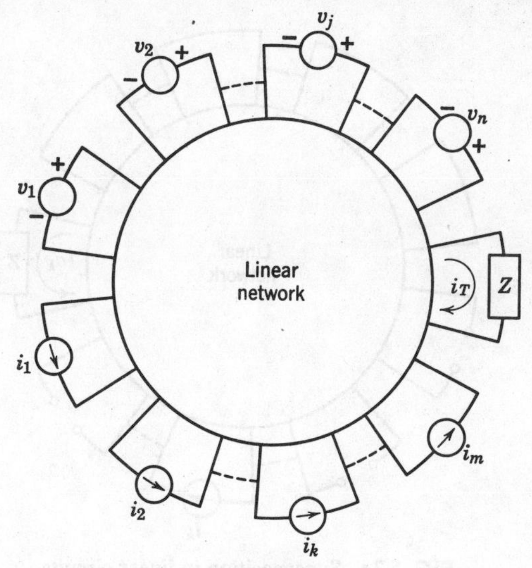

FIG. 5.2*a*

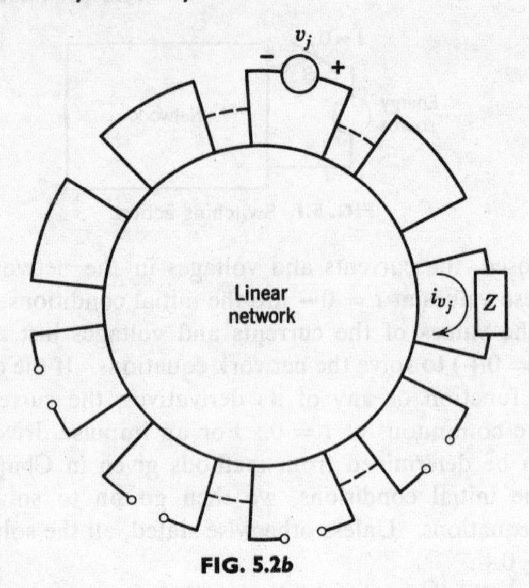

FIG. 5.2*b*

Suppose we are interested in the current $i_T(t)$ through a given element Z, as shown. Let us open-circuit all the current sources and short-circuit $n - 1$ voltage sources, leaving only $v_j(t)$ shown in Fig. 5.2*b*. By $i_{v_j}(t)$, we denote the current through Z due to the voltage source $v_j(t)$ alone. In similar fashion, by $i_{c_k}(t)$ we denote the current through Z due to the current source $i_k(t)$ alone, as depicted in Fig. 5.2*c*. By the superposition

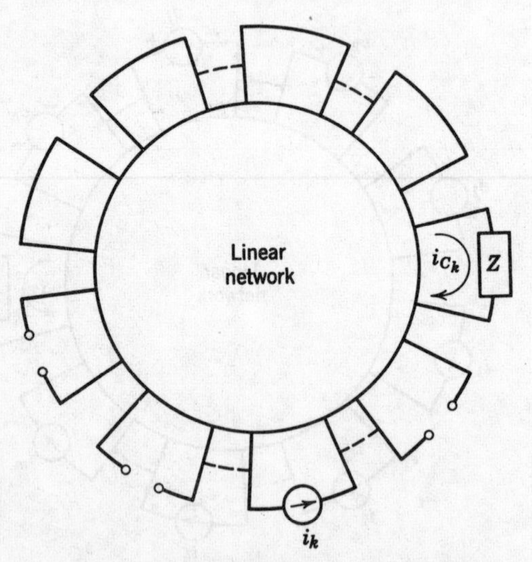

FIG. 5.2*c*. Superposition in linear circuits.

principle, the total current $i_T(t)$ due to all of the sources is equal to the algebraic sum

$$i_T = i_{v_1} + i_{v_2} + \cdots + i_{v_n} + i_{c_1} + i_{c_2} + \cdots + i_{c_m} \tag{5.1}$$

5.2 NETWORK ELEMENTS

In this section, we will discuss the voltage-current (v-i) relationships that exist for the basic network elements. Before we examine these relationships, it is important to assign first arbitrary reference polarities for the voltage across an element, and a reference direction of flow for the current through the element. For the purposes of our discussion, we assume that the positive polarity for voltage is at the tail of the current arrow, as shown by the resistor in Fig. 5.3. Now, let us review the voltage-current relationships for the resistor, the inductor, and the capacitor, which we first discussed in Chapter 1.

Resistor

The resistor shown in Fig. 5.3 defines a linear proportionality relationship between $v(t)$ and $i(t)$, namely,

$$\begin{aligned} v(t) &= R\,i(t) \\ i(t) &= G\,v(t) \qquad G = \frac{1}{R} \end{aligned} \tag{5.2}$$

where R is given in ohms and G in mhos.

Capacitor

For the capacitor shown in Fig. 5.4*a* the v-i relationships are

$$\begin{aligned} i(t) &= C\,\frac{d\,v(t)}{dt} \\ v(t) &= \frac{1}{C}\int_{0-}^{t} i(\tau)\,d\tau + v_C(0-) \end{aligned} \tag{5.3}$$

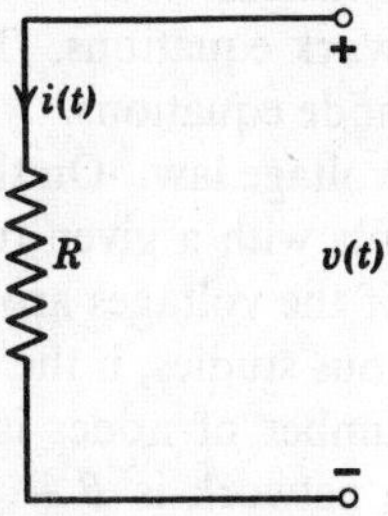

FIG. 5.3. Resistor.

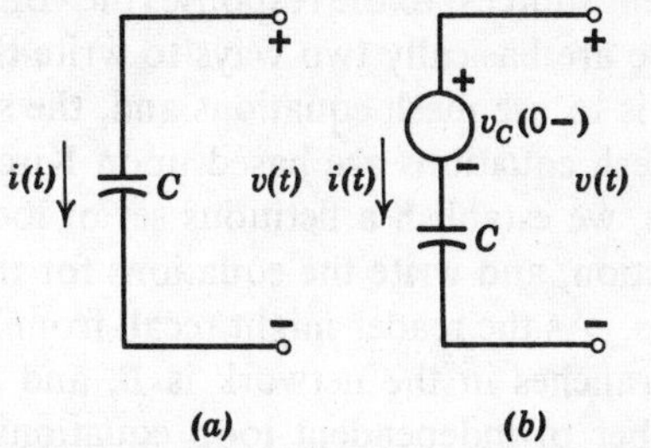

FIG. 5.4. (*a*) Capacitor. (*b*) Capacitor with initial voltage.

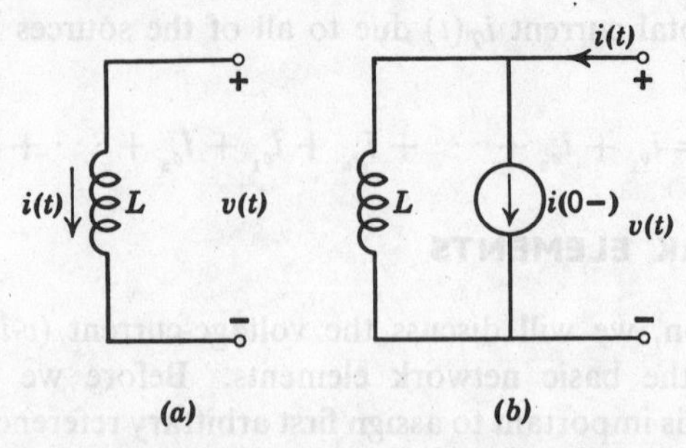

FIG. 5.5. (*a*) Inductor. (*b*) Inductor with initial current.

where C is given in farads. The initial value $v_C(0-)$ is the voltage across the capacitor just before the switching action. It can be regarded as an independent voltage source, as shown in Fig. 5.4*b*. We should point out also that $v_C(0-) = v_C(0+)$ for all excitations except impulses and derivatives of impulses.

Inductor

The inductor in Fig. 5.5*a* describes a dual relationship between voltage and current when compared to a capacitor. The *v-i* relationships are

$$v(t) = L\frac{di}{dt}$$

$$i(t) = \frac{1}{L}\int_{0-}^{t} v(\tau)\, d\tau + i_L(0-) \tag{5.4}$$

where L is given in henrys. The initial current $i_L(0-)$ can be regarded as an independent current source, as shown in Fig. 5.5*b*. As is true for the voltage across the capacitor, the current through the inductor is similarly continuous for all t, except in the case of impulse excitations.

When the network elements are interconnected, the resulting *i-v* equations are integrodifferential equations relating the excitation (voltage or current sources) to the response (the voltages and currents of the elements). There are basically two ways to write these network equations. The first way is to use mesh equations and, the second, node equations.

Mesh equations are based upon Kirchhoff's voltage law. On the mesh basis, we establish a fictitious set of loop currents with a given reference direction, and write the equations for the sum of the voltages around the loops. As the reader might recall from his previous studies, if the number of branches in the network is B, and if the number of nodes is N, the number of independent loop equations for the network is $B - N + 1$.[2]

[2] See Skilling, *op. cit.*

We must, in addition, choose the mesh currents such that at least one mesh current passes through every element in the network.

Example 5.1. In Fig. 5.6, a network is given with seven branches and five nodes. We therefore need $7 - 5 + 1 = 3$ independent mesh equations. The directions of the mesh currents i_1, i_2, i_3 are chosen as indicated. We also note that the

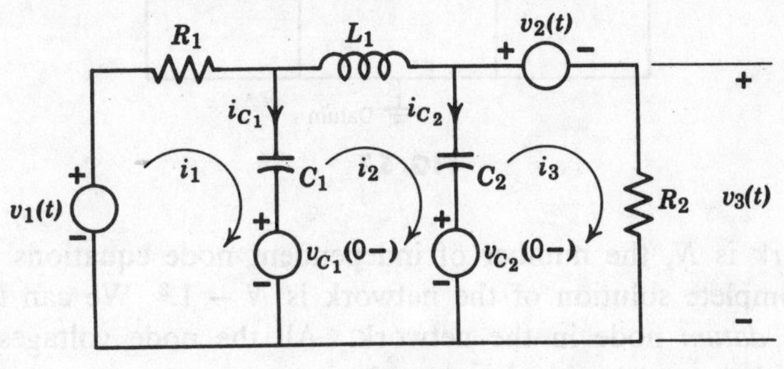

FIG. 5.6

capacitors in the circuit have associated initial voltages. These initial voltages are assigned reference polarities, as shown in the figure. Now we proceed to write the mesh equations.

Mesh i_1:

$$v_1(t) - v_{C_1}(0-) \doteq R_1 i_1(t) + \frac{1}{C_1}\int_{0-}^{t} i_1(\tau)\,d\tau - \frac{1}{C_1}\int_{0-}^{t} i_2(\tau)\,d\tau$$

Mesh i_2:

$$v_{C_1}(0-) - v_{C_2}(0-) = -\frac{1}{C_1}\int_{0-}^{t} i_1(\tau)\,d\tau + L_1\frac{di_2}{dt}$$

$$+ \left(\frac{1}{C_1} + \frac{1}{C_2}\right)\int_{0-}^{t} i_2(\tau)\,d\tau - \frac{1}{C_2}\int_{0-}^{t} i_3(\tau)\,d\tau$$

Mesh i_3:

$$-v_2(t) + v_{C_2}(0-) = \frac{-1}{C_2}\int_{0-}^{t} i_2(\tau)\,d\tau + \frac{1}{C_2}\int_{0-}^{t} i_3(\tau)\,d\tau + R_2\,i_3(t) \qquad (5.5)$$

After we find the three unknowns, i_1, i_2, and i_3, we can determine the branch currents and the voltages across the elements. For example, if we were required to find the branch currents i_{C_1} and i_{C_2} through the capacitors, we would use the following relationships

$$i_{C_1} = i_1 - i_2$$
$$i_{C_2} = i_2 - i_3 \qquad (5.6)$$

Alternatively, if the voltage $v_3(t)$ in Fig. 5.6 is our objective, we see that

$$v_3(t) = i_3(t)R_2 \qquad (5.7)$$

Network equations can also be written in terms of node equations, which are based upon Kirchhoff's current law. If the number of nodes in

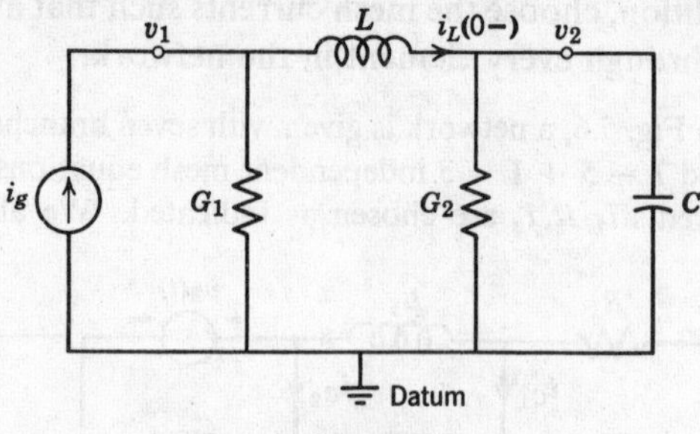

FIG. 5.7

the network is N, the number of independent node equations required for the complete solution of the network is $N - 1$.[3] We can therefore select one *datum* node in the network. All the node voltages will be positive with respect to this datum node.

Consider the network in Fig. 5.7. Let us write a set of node equations for the network with the datum node shown. Since the number of nodes in the network is $N = 3$, we need $N - 1 = 2$ independent node equations. These are written for nodes v_1 and v_2, as given below.

Node v_1:

$$i_g(t) - i_L(0-) = G_1 v_1(t) + \frac{1}{L}\int_{0-}^{t} v_1(\tau)\, d\tau - \frac{1}{L}\int_{0-}^{t} v_2(\tau)\, d\tau$$

Node v_2:

$$i_L(0-) = -\frac{1}{L}\int_{0-}^{t} v_1(\tau)\, d\tau + \frac{1}{L}\int_{0-}^{t} v_2(\tau)\, d\tau + C\frac{d\, v_2(t)}{dt} + G_2 v_2(t) \quad (5.8)$$

Further examples are given in the following sections.

5.3 INITIAL AND FINAL CONDITIONS

In this section, we consider some methods for obtaining initial conditions for circuit differential equations. We also examine ways to obtain particular integrals for networks with constant (d-c) or sinusoidal (a-c) excitations. In the solution of network differential equations, the complementary function is called the *transient solution* or *free response*. The particular integral is known as the *forced response*. In the case of constant or periodic excitations, the forced response at $t = \infty$ is the *steady-state* or *final* solution.

[3] See Skilling, *op. cit.*

There are two ways to obtain initial conditions at $t = 0+$ for a network: (*a*) through the differential equations describing the network, (*b*) through knowledge of the physical behavior of the *R*, *L*, and *C* elements in the network.

Initial conditions for a capacitor

For a capacitor, the voltage-current relationship at $t = 0+$ is

$$v_C(0+) = \int_{0-}^{0+} i(\tau)\, d\tau + v_C(0-) \tag{5.9}$$

If $i(t)$ does not contain impulses or derivatives of impulses, $v_C(0+) = v_C(0-)$. If q is the charge on the capacitor at $t = 0-$, the initial voltage is

$$v_C(0+) = v_C(0-) = \frac{q}{C} \tag{5.10}$$

When there is no initial charge on the capacitor, $v_C(0+) = 0$. We conclude that when there is no stored energy on a capacitor, its equivalent circuit at $t = 0+$ is a short circuit. This analogy is confirmed by examining the physical behavior of the capacitor. As a result of the *conservation of charge principle*, an instantaneous change in voltage across a capacitor implies instantaneous change in charge, which in turn means infinite current through the capacitor. Since we never encounter infinite current in physical situations, the voltage across a capacitor cannot change instantaneously. Therefore at $t = 0+$, we can replace the capacitor by a voltage source if an initial charge exists, or by a *short circuit* if there is no initial charge.

Example 5.2. Consider the *R-C* network in Fig. 5.8*a*. The switch is closed at $t = 0$, and we assume there is no initial charge in the capacitor. Let us find the initial conditions $i(0+)$ and $i'(0+)$ for the differential equation of the circuit

$$V\,u(t) = R\,i(t) + \frac{1}{C}\int_{0-}^{t} i(\tau)\, d\tau \tag{5.11}$$

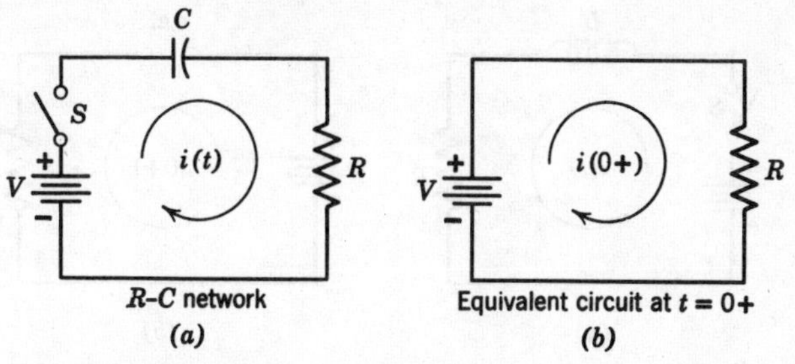

FIG. 5.8. (*a*) *R-C* network. (*b*) Equivalent circuit at $t = 0+$.

The equivalent circuit at $t = 0+$ is given in Fig. 5.8*b*, from which we obtain

$$i(0+) = \frac{V}{R} \tag{5.12}$$

To obtain $i'(0+)$, we must refer to the differential equation

$$V\,\delta(t) = R\,i'(t) + \frac{i(t)}{C} \tag{5.13}$$

At $t = 0+$ we have
$$0 = R\,i'(0+) + \frac{i(0+)}{C} \tag{5.14}$$

We then obtain
$$i'(0+) = -\frac{i(0+)}{RC} = -\frac{V}{R^2C} \tag{5.15}$$

The final condition, or steady-state solution, for the current in Fig. 5.8*a* is obtained from our knowledge of d-c circuits. We know that for a d-c excitation, a capacitor is an open circuit for d-c current. Thus the steady-state current is

$$i_p(t) = i(\infty) = 0$$

Initial conditions for an inductor

For an inductor, the voltage-current relationship at $t = 0+$ is

$$i_L(0+) = \frac{1}{L}\int_{0-}^{0+} v(\tau)\,d\tau + i_L(0-) \tag{5.16}$$

If $v(t)$ does not contain impulses, then $i_L(0+) = i_L(0-)$. If there is no initial current, $i_L(0+) = 0$, which corresponds to an *open circuit* at $t = 0+$. This analogy can also be obtained from the fact that the current through an inductor cannot change instantaneously due to the conservation of flux linkages.

Example 5.3 In Fig. 5.9*a*, the switch closes at $t = 0$. Let us find the initial conditions $i(0+)$ and $i'(0+)$ for the differential equation

$$V\,u(t) = L\frac{d\,i(t)}{dt} + R\,i(t) \tag{5.17}$$

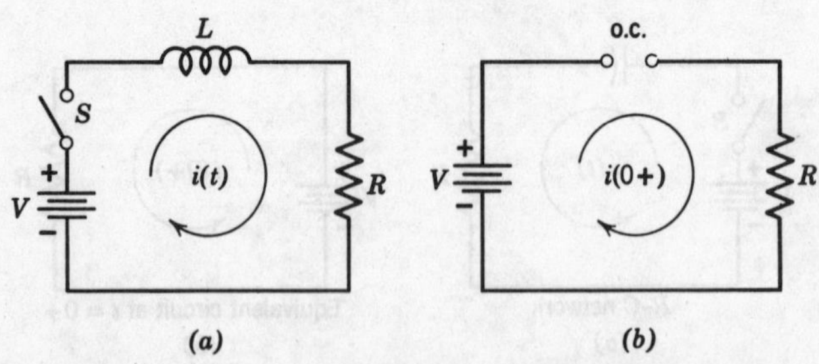

FIG. 5.9. (*a*) *R-L* network. (*b*) Equivalent circuit at $t = 0+$.

From the equivalent circuit at $t = 0+$, shown in Fig. 5.9*b*, we see that

$$i(0+) = 0 \tag{5.18}$$

We then refer to the differential equation to obtain $i'(0+)$.

$$V = L\,i'(0+) + R\,i(0+) \tag{5.19}$$

Thus

$$\begin{aligned} i'(0+) &= \frac{V}{L} - \frac{R}{L}\,i(0+) \\ &= \frac{V}{L} \end{aligned} \tag{5.20}$$

The steady-state solution for the circuit in Fig. 5.9*a* is obtained through the knowledge that for a d-c source, an indicator is a short circuit.

$$i_p(t) = i(\infty) = \frac{V}{R} \tag{5.21}$$

Example 5.4. In this example we consider the two-loop network of Fig. 5.10. As in Examples 5.2 and 5.3, we use the equivalent circuit models at $t = 0+$ and $t = \infty$ to obtain the initial conditions and steady-state solutions. At $t = 0$ the switch closes. The equivalent circuits at $t = 0+$ and $t = \infty$ are shown in Fig. 5.11*a* and *b*, respectively. The initial currents are

$$\begin{aligned} i_1(0+) &= \frac{V}{R_1} \\ i_2(0+) &= 0 \end{aligned} \tag{5.22}$$

The steady-state solutions are

$$i_1(\infty) = \frac{V}{R_1 + R_2} = i_2(\infty) \tag{5.23}$$

Final conditions for sinusoidal excitations

When the excitation is a pure sinusoid, the steady-state currents and voltages in the circuit are also sinusoids of the same frequency as the excitation. If the unknown is a voltage, for example, $v_1(t)$, the steady-state solution would take the form

$$v_{1p}(t) = |V(j\omega_0)| \sin\,[\omega_0 t - \phi(\omega_0)] \tag{5.24}$$

where ω_0 is the frequency of the excitation, and $|V(j\omega_0)|$ and $\phi(\omega_0)$ represent the magnitude and phase of $v_{1p}(t)$. A similar expression would hold if the unknown were a current.

To obtain the magnitude and phase, we follow standard procedures in a-c circuit analysis. For example, consider the *R-C* circuit in Fig. 5.12. The current generator is

$$i_g(t) = (I_0 \sin \omega_0 t)\, u(t) \tag{5.25}$$

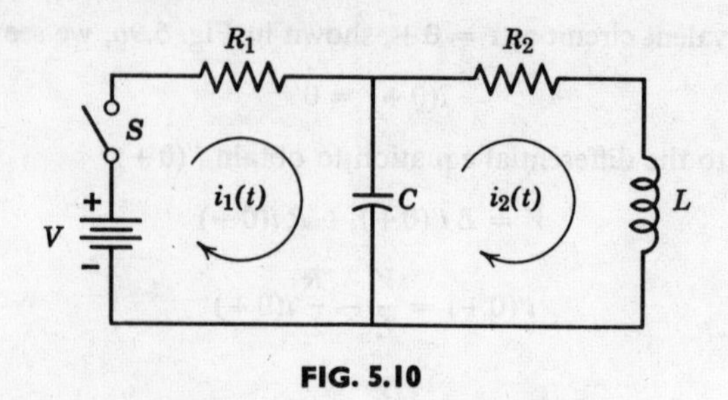

FIG. 5.10

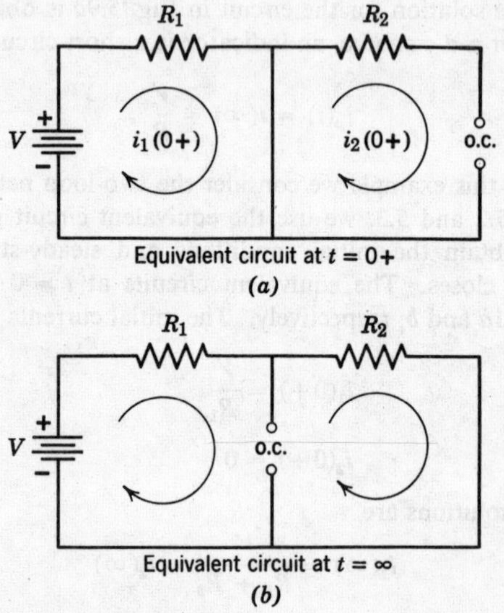

R_1 R_2

V $i_1(0+)$ $i_2(0+)$ o.c.

Equivalent circuit at $t = 0+$

(a)

R_1 R_2

V o.c.

Equivalent circuit at $t = \infty$

(b)

FIG. 5.11. *(a)* Equivalent circuit at $t = 0+$. *(b)* Equivalent circuit at $t = \infty$.

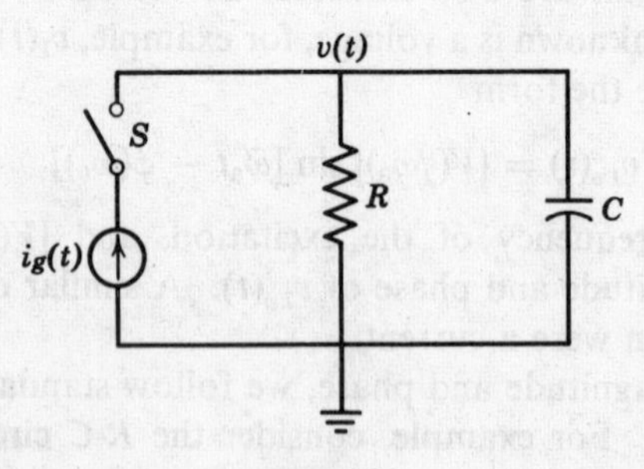

FIG. 5.12

If the steady-state voltage takes the form shown in Eq. 5.24,

$$\begin{aligned} |V(j\omega_0)| &= \frac{I_0}{|Y(j\omega_0)|} \\ &= \frac{I_0}{(G^2 + \omega_0{}^2C^2)^{1/2}} \end{aligned} \tag{5.26}$$

and

$$\phi(\omega_0) = \tan^{-1} \frac{\omega_0 C}{G} \tag{5.27}$$

so that

$$v_p(t) = \frac{I_0}{(G^2 + \omega_0{}^2C^2)^{1/2}} \sin \left(\omega_0 t - \tan^{-1} \frac{\omega_0 C}{G} \right) \tag{5.28}$$

We refer to this problem in Section 5.5.

5.4 STEP AND IMPULSE RESPONSE

As an introduction to the topic of solution of network differential equations, let us consider the important problem of obtaining the step and impulse responses for any voltage or current in the network. As we shall see in Chapter 7, the step and impulse responses are precise time-domain characterizations of the network. The problem of obtaining the step and impulse response is stated as follows: Given a network with zero initial energy, we are required to solve for a specified response (current or voltage) due to a given excitation function $u(t)$ or $\delta(t)$, which either can be a current or a voltage source. If the excitation is a step of voltage, the physical analogy is that of a switch—closing at time $t = 0$—which connects a 1-v battery to a circuit. The physical analogy of an impulse excitation is that of a very short (compared to the time constants of the circuit) pulse with large amplitude.

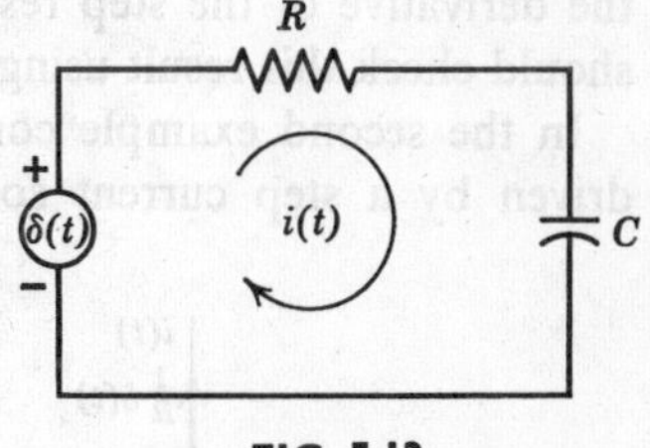

FIG. 5.13

The problems involved can best be illustrated by means of examples. Consider the series R-C circuit in Fig. 5.13. The differential equation of the circuit is

$$v(t) = \delta(t) = R\, i(t) + \frac{1}{C} \int_{0-}^{t} i(\tau)\, d\tau \tag{5.29}$$

We assume $v_C(0-) = 0$. Since Eq. 5.29 contains an integral, we substitute $x'(t)$ for $i(t)$ in the equation, giving us

$$\delta(t) = R\, x'(t) + \frac{1}{C}\, x(t) \tag{5.30}$$

Integrating both sides between 0− and 0+ gives

$$x(0+) = \frac{1}{R} \tag{5.31}$$

The characteristic equation is

$$H(p) = Rp + \frac{1}{C} \tag{5.32}$$

and with little effort we have

$$x(t) = \frac{1}{R}\, e^{-t/RC}\, u(t) \tag{5.33}$$

so that

$$i(t) = \frac{1}{R}\left[\delta(t) - \frac{1}{RC}\, e^{-t/RC}\, u(t)\right] \tag{5.34}$$

which is shown in Fig. 5.14*a*. We thus arrive at the current impulse response $i(t)$ as the result of an impulse voltage excitation. In the process we have also obtained the step response $x(t)$ in Fig. 5.14*b* since, by definition, the derivative of the step response is the impulse response. The reader should check this result using a step excitation of voltage.

In the second example consider the parallel *R-C* circuit in Fig. 5.12 driven by a step current source $i(t) = I_0\, u(t)$, where I_0 is a constant.

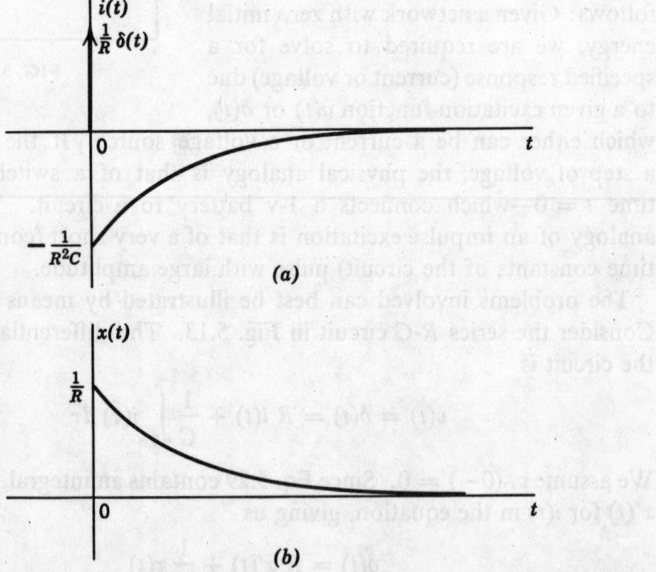

FIG. 5.14. (*a*) Impulse response of *R-C* circuit. (*b*) Step response of *R-C* circuit.

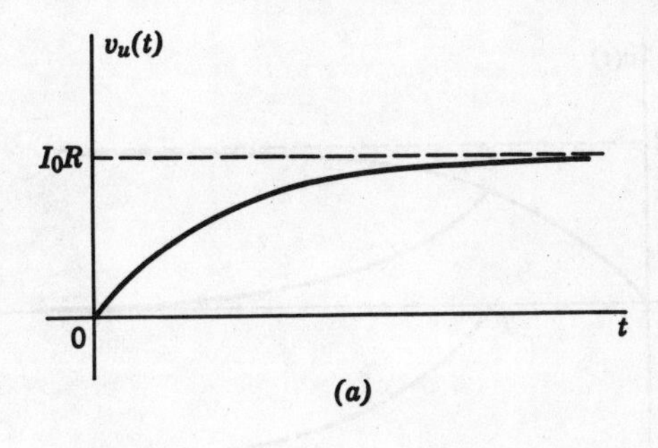

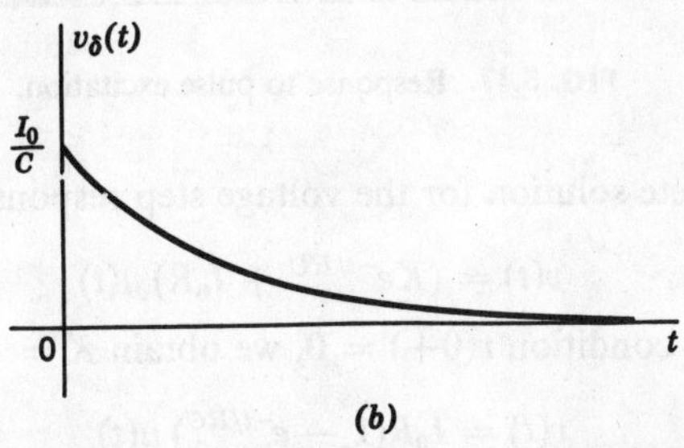

FIG. 5.15. (*a*) Step response of parallel *R-C* circuit. (*b*) Impulse response of parallel *R-C* circuit.

Assuming zero initial conditions, the differential equation is

$$I_0\, u(t) = Gv(t) + C\frac{d\, v(t)}{dt} \tag{5.35}$$

from which we obtain the characteristic equation

$$H(p) = Cp + G \tag{5.36}$$

The steady-state value of $v(t)$ is

$$v_p(t) = \frac{I_0}{H(0)} = \frac{I_0}{G} = I_0R \tag{5.37}$$

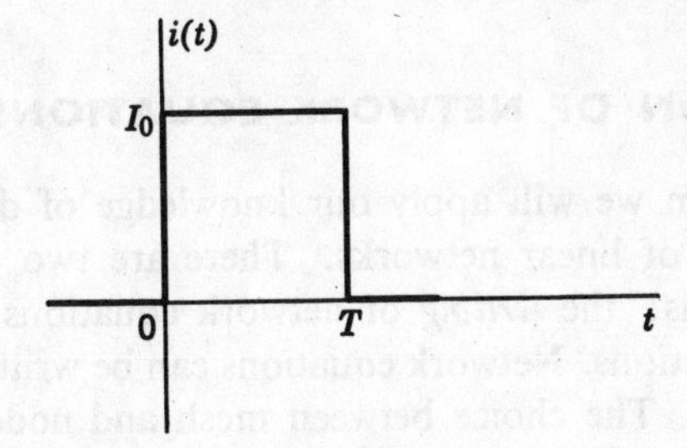

FIG. 5.16. Pulse excitation.

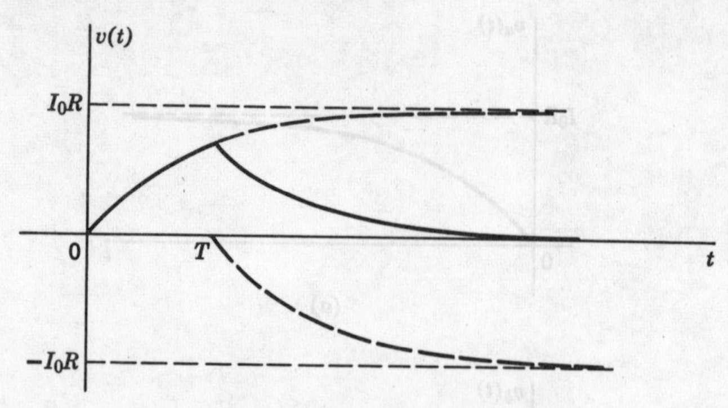

FIG. 5.17. Response to pulse excitation.

Thus the complete solution for the voltage step response is

$$v(t) = (Ke^{-t/RC} + I_0R)\,u(t) \tag{5.38}$$

From the initial condition $v(0+) = 0$, we obtain $K = -I_0R$ so that

$$v(t) = I_0R(1 - e^{-t/RC})\,u(t) \tag{5.39}$$

Differentiating Eq. 5.39 gives us the voltage impulse response

$$v(t) = \frac{I_0}{C}\,e^{-t/RC}\,u(t) \tag{5.40}$$

The step and impulse responses are plotted on Figs. 5.15*a* and *b*. Suppose the excitation in Fig. 5.12 were a pulse

$$i(t) = I_0[u(t) - u(t - T)] \tag{5.41}$$

shown in Fig. 5.16. Then, according to the superposition and time-invariance postulates of linear systems, the response would be

$$v(t) = I_0R[(1 - e^{-t/RC})\,u(t) - (1 - e^{-(t-T)/RC})\,u(t - T)] \tag{5.42}$$

which is shown in Fig. 5.17.

5.5 SOLUTION OF NETWORK EQUATIONS

In this section we will apply our knowledge of differential equations to the analysis of linear networks. There are two important points in network analysis: the *writing* of network equations and the *solution* of these same equations. Network equations can be written on a mesh, node, or mixed basis. The choice between mesh and node equations depends largely upon the unknown quantities for which we must solve. For instance, if the unknown quantity is a branch current, it is preferable to write mesh equations. On the other hand, if we wish to find a voltage across a certain element, then node equations are better. In many cases the choice is quite arbitrary. If, for example, we wish to find the voltage v across a resistor R, we can either find v directly by node equations or find the branch current through the resistor and then multiply by R.

Example 5.5. Find the current $i(t)$ for the network in Fig. 5.18, when the voltage source is $e(t) = 2e^{-0.5t}\,u(t)$ and $v_C(0-) = 0$.

Solution. The differential equation is

$$e(t) = Ri(t) + \frac{1}{C}\int_{0-}^{t} i(\tau)\,d\tau + v_C(0-) \tag{5.43}$$

Or, in terms of the numerical values, we have

$$2e^{-0.5t}\,u(t) = i(t) + 2\int_{0-}^{t} i(\tau)\,d\tau \tag{5.44}$$

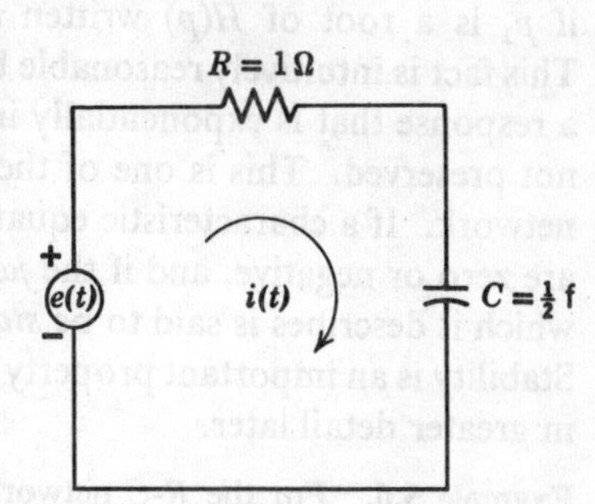

FIG. 5.18

Differentiating both sides of Eq. 5.44, we obtain

$$2\delta(t) - e^{-0.5t}\,u(t) = \frac{d\,i(t)}{dt} + 2i(t) \tag{5.45}$$

To obtain the initial condition $i(0+)$, we must integrate Eq. 5.45 between the limits $t = 0-$ and $t = 0+$ to give $i(0+) = 2$. From the characteristic equation

$$H(p) = p + 2 = 0 \tag{5.46}$$

we obtain the complementary function as

$$i_C(t) = Ke^{-2t} \tag{5.47}$$

If we assume the particular integral to be $i_p(t) = Ae^{-0.5t}$, then we obtain

$$A = -\frac{1}{H(-0.5)} = -\frac{2}{3} \tag{5.48}$$

The incomplete solution is

$$i(t) = Ke^{-2t} - \tfrac{2}{3}e^{-0.5t} \tag{5.49}$$

From the initial condition $i(0+) = 2$, we obtain the final solution,

$$i(t) = \left(\tfrac{8}{3}e^{-2t} - \tfrac{2}{3}e^{-0.5t}\right)u(t) \tag{5.50}$$

As we already noted in Section 5.3, in the solution of network differential equations, the complementary function is called the free response, whereas the forced response is a particular integral, and in the case of constant or periodic excitation, the forced response at $t = \infty$ is the steady-state solution. Note that the free response is a function of the network elements alone and is independent of excitation. On the other hand, the forced response depends on both the network and the excitation.

It is significant that, for networks which have only positive elements, the free response is made up of only damped exponential and/or sinusoids with constant peak amplitudes. In other words, the roots of the characteristic equation $H(p)$ all have negative or zero real parts. For example, if p_1 is a root of $H(p)$ written as $p_1 = \sigma \pm j\omega$, then $\mathrm{Re}\,(p_1) = \sigma \leq 0$. This fact is intuitively reasonable because, if a bounded excitation produces a response that is exponentially increasing, then conservation of energy is not preserved. This is one of the most important properties of a passive network. If a characteristic equation contains only roots whose real parts are zero or negative, and if the $j\omega$ axis roots are simple, then the network which it describes is said to be *stable*; otherwise, the network is *unstable*.[4] Stability is an important property of passive networks and will be discussed in greater detail later.

Example 5.6. For the *R-C* network in Fig. 5.19 with the excitation given by Eq. 5.25, find the voltage $v(t)$ across the capacitor; it is given that $v(0-) = v_C(0-) = 0$.

Solution. We have already obtained the particular integral in Eq. 5.28. Now let us find the complementary function. The differential equation on a node basis is

$$C\frac{dv}{dt} + Gv = I_0 \sin \omega t\, u(t) \tag{5.51}$$

from which we obtain the characteristic equation as

$$H(p) = Cp + G \tag{5.52}$$

so that

$$v_C(t) = Ke^{-Gt/C}\, u(t) \tag{5.53}$$

and the incomplete solution is

$$v(t) = Ke^{-Gt/C}u(t) + \frac{I_0}{(G^2 + \omega^2C^2)^{1/2}} \sin\left(\omega t - \tan^{-1}\frac{\omega C}{G}\right) u(t) \tag{5.54}$$

From the initial condition $v(0-) = 0$, we obtain

$$v\,(0+) = v(0-) = K - \frac{I_0}{(G^2 + \omega^2C^2)^{1/2}} \sin\left(\tan^{-1}\frac{\omega C}{G}\right) = 0 \tag{5.55}$$

From the argand diagram in Fig. 5.20 we see that,

$$\sin\left(\tan^{-1}\frac{\omega C}{G}\right) = \frac{\omega C}{(G^2 + \omega^2C^2)^{1/2}} \tag{5.56}$$

Consequently,

$$v(t) = \frac{I_0\, u(t)}{(G^2 + \omega^2C^2)^{1/2}}\left[\frac{\omega Ce^{-Gt/C}}{(G^2 + \omega^2C^2)^{1/2}} + \sin\left(\omega t - \tan^{-1}\frac{\omega C}{G}\right)\right] \tag{5.57}$$

[4] This is not a formal definition of stability, but it suffices for the moment.

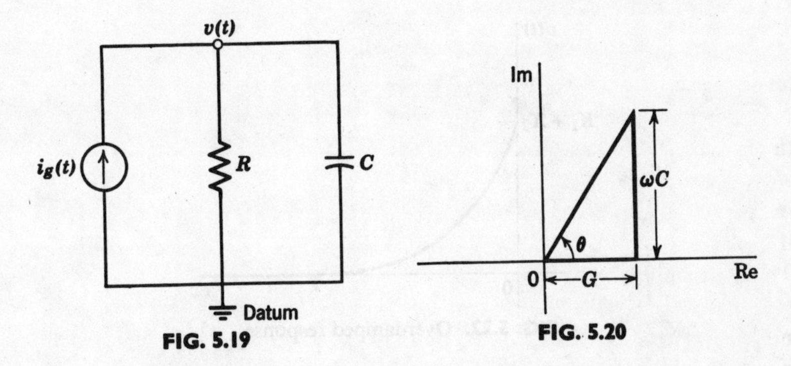

FIG. 5.19

FIG. 5.20

From the complementary function in Eq. 5.53, we see that the time constant of the circuit is $T = C/G = RC$.

Next, let us examine an example of the different kinds of free responses of a second-order network equation that depend on relative values of the network elements. Suppose we are given the network in Fig. 5.21; let us find the free response $v_C(t)$ for the differential equation

$$i_g(t) = C\frac{dv}{dt} + Gv + \frac{1}{L}\int_{0-}^{t} v(\tau)\,d\tau + i_L(0-) \tag{5.58}$$

Differentiating both sides of Eq. 5.58, we have

$$i'_g(t) = C\,v''(t) + G\,v'(t) + \frac{1}{L}\,v(t). \tag{5.59}$$

The characteristic equation is then

$$H(p) = Cp^2 + Gp + \frac{1}{L} = C\left(p^2 + \frac{G}{C}p + \frac{1}{LC}\right) \tag{5.60}$$

In factored form, $H(p)$ is

$$H(p) = C(p - p_1)(p - p_2) \tag{5.61}$$

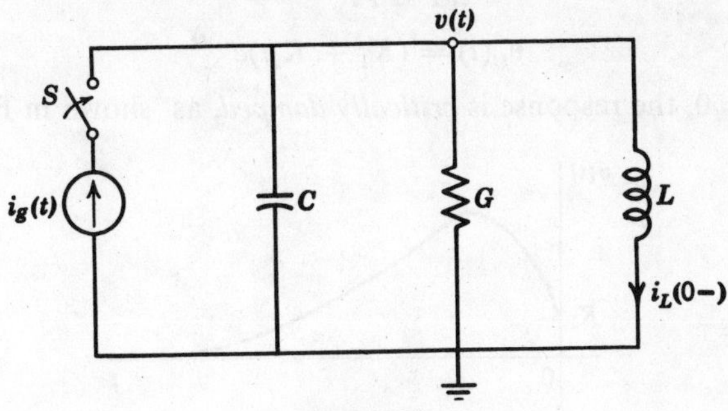

FIG. 5.21

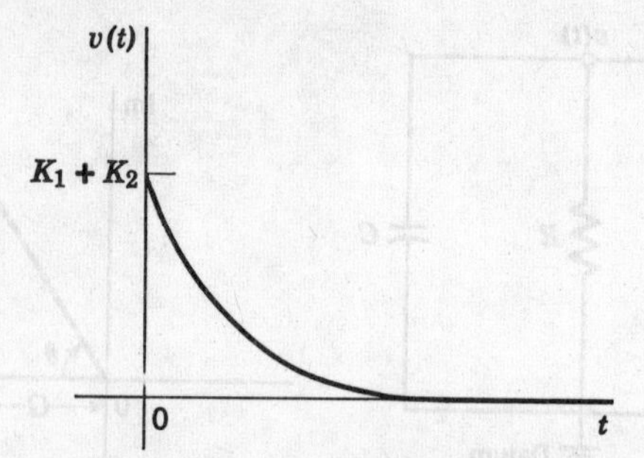

FIG. 5.22. Overdamped response.

where
$$\left.\begin{matrix} p_1 \\ p_2 \end{matrix}\right\} = -\frac{G}{2C} \pm \frac{1}{2}\left[\left(\frac{G}{C}\right)^2 - \frac{4}{LC}\right]^{1/2} \triangleq -A \pm B \tag{5.62}$$

There are three different kinds of responses depending upon whether B is real, zero, or imaginary.

CASE 1. B is real, that is,

$$\left(\frac{G}{C}\right)^2 > \frac{4}{LC}$$

then the free response is

$$v_C(t) = K_1 e^{-(A-B)t} + K_2 e^{-(A+B)t} \tag{5.63}$$

which is a sum of damped exponentials. In this case, the response is said to be *overdamped*. An example of an overdamped response is shown in Fig. 5.22.

CASE 2. $B = 0$, that is,

$$\left(\frac{G}{C}\right)^2 = \frac{4}{LC}$$

then
$$p_1 = p_2 = -A$$

so that
$$v_C(t) = (K_1 + K_2 t)e^{-At} \tag{5.64}$$

When $B = 0$, the response is *critically damped*, as shown in Fig. 5.23.

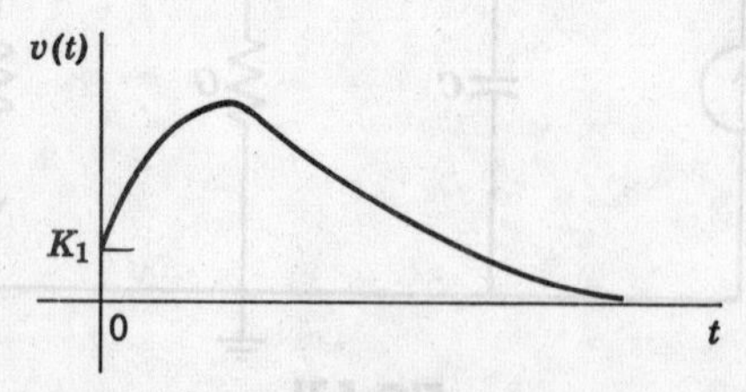

FIG. 5.23. Critically damped response.

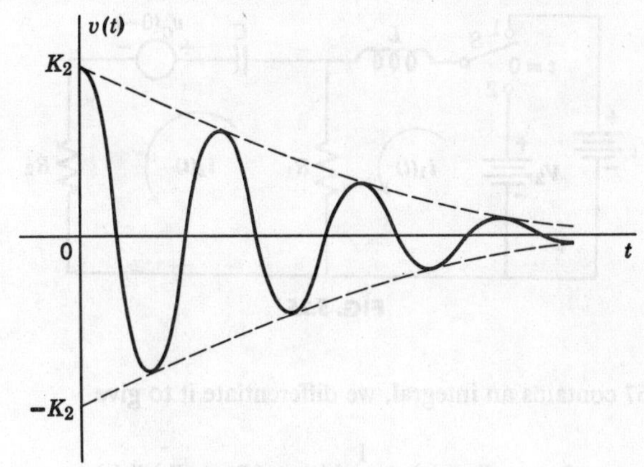

FIG. 5.24. Underdamped response.

CASE 3. B is imaginary, that is,

$$\left(\frac{G}{C}\right)^2 < \frac{4}{LC}$$

Letting $B = j\beta$, we have

$$v_C(t) = e^{-At}(K_1 \sin \beta t + K_2 \cos \beta t) \tag{5.65}$$

In this case, the response is said to be *underdamped*, and is shown by the damped oscillatory curve in Fig. 5.24.

Example 5.7. In this example we discuss the solution of a set of simultaneous network equations. As in the previous examples, we rely upon physical reasoning rather than formal mathematical operations to obtain the initial currents and voltages as well as the steady-state solutions. In the network of Fig. 5.25, the switch S is thrown from position 1 to position 2 at $t = 0$. It is known that prior to $t = 0$, the circuit had been in steady state. We make the idealized assumption that the switch closes instantaneously at $t = 0$. Our task is to find $i_1(t)$ and $i_2(t)$ after the switch position changes. The values of the batteries V_1 and V_2 are $V_1 = 2$ v, $V_2 = 3$ v; and the element values are given as

$$L = 1 \text{ h}, \qquad R_1 = 0.5\ \Omega$$
$$C = \tfrac{1}{3} \text{ f}, \qquad R_2 = 2.0\ \Omega$$

The mesh equations for $i_1(t)$ and $i_2(t)$ after $t = 0$ are

$$V_2 = L\, i'_1(t) + R_1\, i_1(t) - R_1\, i_2(t) \tag{5.66}$$

$$-v_C(0-) = -R_1\, i_1(t) + \frac{1}{C}\int_{0-}^{t} i_2(\tau)\, d\tau + (R_1 + R_2)\, i_2(t) \tag{5.67}$$

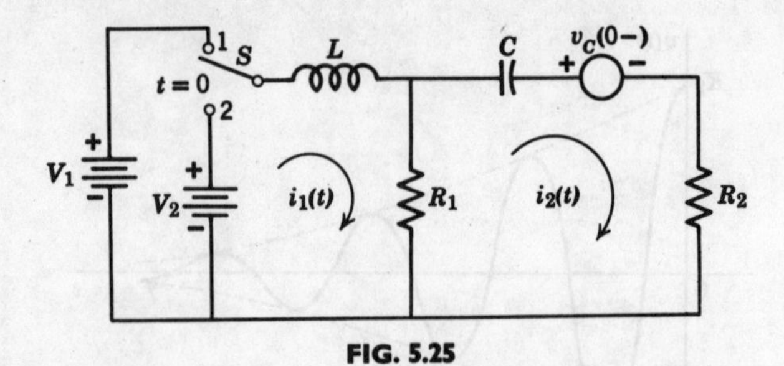

FIG. 5.25

Since Eq. 5.67 contains an integral, we differentiate it to give

$$0 = -R_1\, i'_1(t) + \frac{1}{C} i_2(t) + (R_1 + R_2)\, i'_2(t) \tag{5.68}$$

Using Eqs. 5.66 and 5.68 as our system of equations, we obtain the characteristic equation

$$H(p) = \begin{vmatrix} Lp + R_1 & -R_1 \\ -R_1 p & \dfrac{1}{C} + (R_1 + R_2)p \end{vmatrix}$$

$$= L(R_1 + R_2)p^2 + \left(\frac{L}{C} + R_1R_2\right)p + \frac{R_1}{C} \tag{5.69}$$

Substituting the element values into $H(p)$, we have

$$H(p) = 2.5p^2 + 4p + 1.5 = 2.5(p + 1)(p + 0.6) \tag{5.70}$$

The free responses are then

$$\begin{aligned} i_{1C}(t) &= (K_1 e^{-0.6t} + K_2 e^{-t})\, t(u) \\ i_{2C}(t) &= (K_3 e^{-0.6t} + K_4 e^{-t})\, u(t) \end{aligned} \tag{5.71}$$

The steady-state solutions for the mesh currents are obtained at $t = \infty$ by considering the circuit from a d-c viewpoint. The inductor is then a short circuit and the capacitor is an open circuit; thus we have

$$\begin{aligned} i_{1p}(t) &= \frac{V_2}{R_1} = 6 \text{ amp} \\ i_{2p}(t) &= 0 \end{aligned} \tag{5.72}$$

Now let us determine the initial currents and voltages, which, incidentally, have the same values at $t = 0-$ and $t = 0+$ because the voltage sources are not impulses. Before the switch is thrown at $t = 0$, the circuit with V_1 as the voltage source was at steady state. Consequently,

$$v_C(0-) = V_1 = 2 \text{ v}$$

$$i_1(0-) = \frac{V_1}{R_1} = 4 \text{ amp} \tag{5.73}$$

$$i_2(0-) = 0$$

We next find $i'_1(0+)$ from Eq. 5.66 at $t = 0+$.

$$V_2 = L\, i'_1(0+) + R_1\, i_1(0+) - R_1\, i_2(0+) \tag{5.74}$$

Substituting numerical values into Eq. 5.74, we find

$$i'_1(0+) = 1 \text{ amp/sec}$$

From Eq. 5.68 at $t = 0+$, we obtain similarly

$$i'_2(0+) = \frac{R_1}{R_1 + R_2}\, i'_1(0+) = 0.2 \text{ amp/sec} \tag{5.75}$$

With these initial values of $i_1(t)$ and $i_2(t)$, we can quickly arrive at the final solutions

$$i_1(t) = (0.5e^{-t} - 2.5e^{-0.6t} + 6)\, u(t)$$

$$i_2(t) = (-0.5e^{-t} + 0.5e^{-0.6t})\, u(t) \tag{5.76}$$

which are plotted in Fig. 5.26.

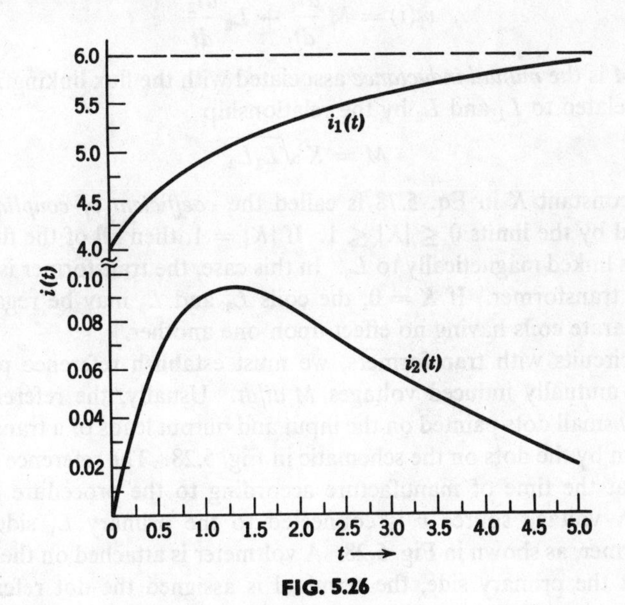

FIG. 5.26

5.6 ANALYSIS OF TRANSFORMERS

According to *Faraday's law of induction*, a current i_1 flowing in a coil L_1 may *induce* a current i_2 in a closed loop containing a second coil L_2. The sufficient conditions for inducing the current i_2 are: (*a*) part of the flux Φ_1 in the coil L_1 must be coupled magnetically to the coil L_2; (*b*) the flux Φ_1 must be changing with time.

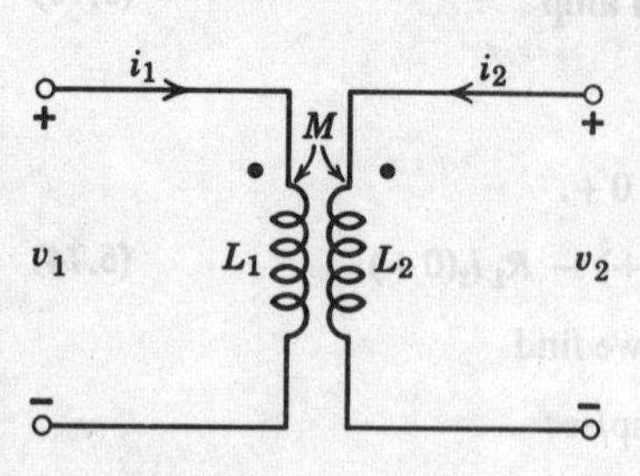

FIG. 5.27. Transformer.

In this section we will analyze circuits containing a device made up of two magnetically coupled coils known as a *transformer*. In Fig. 5.27, the schematic of a transformer is given. The L_1 side of the transformer is usually referred to as the *primary coil* and the L_2 side as the *secondary coil*. The only distinction between primary and secondary is that the energy source is generally at the primary side.

The transformer in Fig. 5.27 is described mathematically by the equations

$$
\begin{aligned}
v_1(t) &= L_1 \frac{di_1}{dt} + M \frac{di_2}{dt} \\
v_2(t) &= M \frac{di_1}{dt} + L_2 \frac{di_2}{dt}
\end{aligned}
\tag{5.77}
$$

where M is the *mutual inductance* associated with the flux linking L_1 to L_2, and is related to L_1 and L_2 by the relationship

$$M = K\sqrt{L_1 L_2} \tag{5.78}$$

The constant K in Eq. 5.78 is called the *coefficient of coupling*. It is bounded by the limits $0 \leq |K| \leq 1$. If $|K| = 1$, then all of the flux Φ_1 in coil L_1 is linked magnetically to L_2. In this case, the transformer is a *unity-coupled* transformer. If $K = 0$, the coils L_1 and L_2 may be regarded as two separate coils having no effect upon one another.

For circuits with transformers, we must establish reference polarities for the mutually induced voltages $M\, di/dt$. Usually, the references are given by small dots painted on the input and output leads of a transformer, as shown by the dots on the schematic in Fig. 5.28. The reference dots are placed at the time of manufacture according to the procedure outlined here. A voltage source v is connected to the primary L_1 side of the transformer, as shown in Fig. 5.28. A voltmeter is attached on the secondary. At the primary side, the terminal is assigned the dot reference to

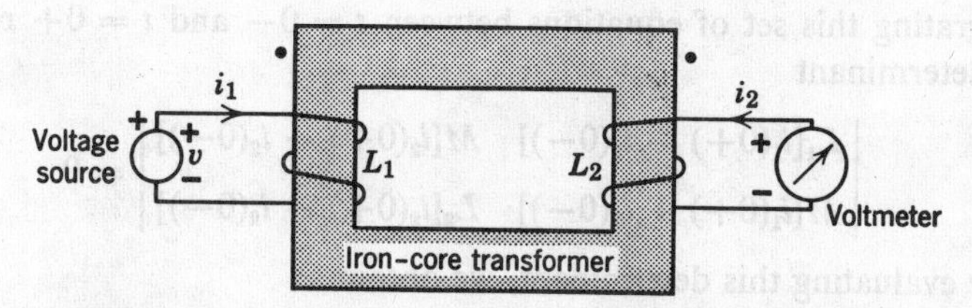

FIG. 5.28. An experiment to determine dot references.

which we connect the *positive* lead of the voltage source. The dot reference is placed on the secondary terminal at which the voltmeter indicates a positive voltage. In terms of the primary current i_1, the positive voltage at the secondary dot is due to the current i_1 flowing into the dot on the primary side. Since the positive voltage at the secondary dot corresponds to the current i_2 flowing into that dot, we can think of the dot references in the following way. If both currents are flowing into the dots or away from the dots, then the sign of the mutual voltage term $M\, di/dt$ is positive. When one current flows into a dot and the other away from the second dot, the sign of $M\, di/dt$ is negative.

If N_1 and N_2 are the number of turns of coils L_1 and L_2, then the *flux linkages* of L_1 and L_2 are given by $N_1\Phi_1$ and $N_2\Phi_2$, respectively. If both i_1 and i_2 flow into the dots, the sum of flux linkages of the transformer is

$$\sum \Phi \text{ linkages} = N_1\Phi_1 + N_2\Phi_2 \tag{5.79}$$

If, however, one of the currents, for example, i_1, flows into a dot, and the other i_2 flows out of the other dot, then

$$\sum \Phi \text{ linkages} = N_1\Phi_1 - N_2\Phi_2 \tag{5.80}$$

An important rule governing the behavior of a transformer is that the sum of flux linkages is continuous with time.

The differential equations for the transformer in Fig. 5.29 are

$$\begin{aligned} V\,u(t) &= L_1\, i'_1(t) + R_1\, i_1(t) + M\, i'_2(t) \\ 0 &= M\, i'_1(t) + R_2\, i_2(t) + L_2\, i'_2(t) \end{aligned} \tag{5.81}$$

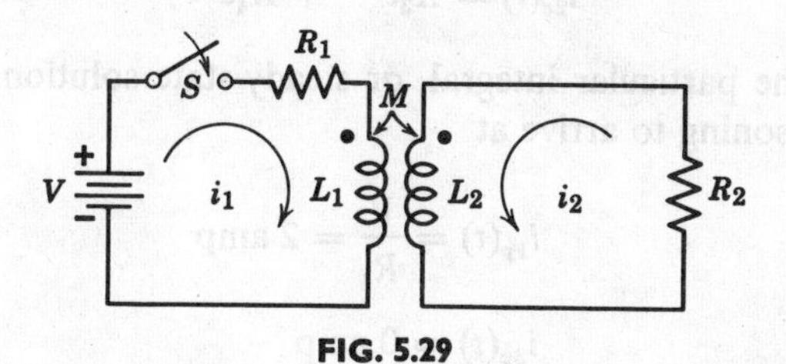

FIG. 5.29

Integrating this set of equations between $t = 0-$ and $t = 0+$ results in the determinant

$$\begin{vmatrix} L_1[i_1(0+) - i_1(0-)] & M[i_2(0+) - i_2(0-)] \\ M[i_1(0+) - i_1(0-)] & L_2[i_2(0+) - i_2(0-)] \end{vmatrix} = 0 \qquad (5.82)$$

By evaluating this determinant, we obtain

$$(L_1L_2 - M^2)[i_1(0+) - i_1(0-)][i_2(0+) - i_2(0-)] = 0 \qquad (5.83)$$

If $L_1L_2 > M^2$, that is, $K < 1$, then the currents must be continuous at $t = 0$ in order for the determinant in Eq. 5.82 to be equal to zero. Thus

$$i_1(0+) = i_1(0-), \qquad K < 1 \qquad (5.84)$$

and

$$i_2(0+) = i_2(0-), \qquad K < 1 \qquad (5.85)$$

Example 5.8. For the transformer circuit in Fig. 5.29, $L_1 = 1h$, $L_2 = 2h$, $R_1 = 3\Omega$, $R_2 = 8\Omega$, and $M = 1h$. The excitation is $V = 6u(t)$. Let us find $i_1(t)$ and $i_2(t)$, assuming that $i_1(0-) = i_2(0-) = 0$.

Solution. The differential equations for the circuit are

$$\begin{aligned} 6u(t) &= i'_1(t) + 3i_1(t) + i'_2(t) \\ 0 &= i'_1(t) + 2i'_2(t) + 8i_2(t) \end{aligned} \qquad (5.86)$$

The characteristic equation is given by the determinant

$$\begin{aligned} H(p) &= \begin{vmatrix} p+3 & p \\ p & 2p+8 \end{vmatrix} \\ &= (p+3)(2p+8) - p^2 \\ &= (p^2 + 14p + 24) \\ &= (p+2)(p+12) \end{aligned} \qquad (5.87)$$

Thus the complementary functions are

$$\begin{aligned} i_{1c}(t) &= K_1e^{-2t} + K_2e^{-12t} \\ i_{2c}(t) &= K_3e^{-2t} + K_4e^{-12t} \end{aligned} \qquad (5.88)$$

To obtain the particular integral, or steady-state solution, we rely upon physical reasoning to arrive at

$$\begin{aligned} i_{1p}(t) &= \frac{V}{R_1} = 2 \text{ amp} \\ i_{2p}(t) &= 0 \text{ amp} \end{aligned} \qquad (5.89)$$

Since the excitation does not contain an impulse (and since $L_1L_2 - M^2 \neq 0$, as we shall see later), we can assume that $i_1(0+)$ and $i_2(0+)$ are also zero. Then using Eq. 5.86 we can find $i'_1(0+)$ and $i'_2(0+)$.

$$\begin{aligned} 6 &= i'_1(0+) + i'_2(0+) \\ 0 &= i'_1(0+) + 2i'_2(0+) \end{aligned} \tag{5.90}$$

Solving, we find $i'_1(0+) = 12$ and $i'_2(0+) = 6$. With these initial conditions, we obtain the final solutions of $i_1(t)$ and $i_2(t)$

$$\begin{aligned} i_1(t) &= (2 - \tfrac{6}{5}e^{-2t} - \tfrac{4}{5}e^{-12t})\, u(t) \\ i_2(t) &= (-\tfrac{3}{5}e^{-2t} + \tfrac{3}{5}e^{-12t})\, u(t) \end{aligned} \tag{5.91}$$

Suppose, now, $L_1L_2 = M^2$, that is, $K = 1$, then $i_1(t)$ and $i_2(t)$ need not be continuous at $t = 0$. In fact, we will show that the currents are *discontinuous* at $t = 0$ for a unity-coupled transformer. Assuming that $K = 1$, consider the mesh equation for the secondary at $t = 0+$

$$R_2\, i_2(0+) = -M\, i'_1(0+) - L_2\, i'_2(0+) \tag{5.92}$$

$$= -\frac{M}{L_1}[L_1\, i'_1(0+) + M\, i'_2(0+)] \tag{5.93}$$

The mesh equation of the primary side then becomes

$$V = R_1\, i_1(0+) + [L_1\, i'_1(0+) + M\, i'_2(0+)] = R_1\, i_1(0+) - \frac{L_1}{M} R_2\, i_2(0+) \tag{5.94}$$

We need an additional equation to solve for $i_1(0+)$ and $i_2(0+)$. This is provided by the equation

$$L_1[i_1(0+) - i_1(0-)] + M[i_2(0+) - i_2(0-)] = 0 \tag{5.95}$$

which we obtained from Eq. 5.82. Since $i_1(0-) = i_2(0-) = 0$, we solve Eqs. 5.94 and 5.95 directly to give

$$\begin{aligned} i_1(0+) &= \frac{VL_2}{R_1L_2 + R_2L_1} \\ i_2(0+) &= \frac{-VM}{R_1L_2 + R_2L_1} \end{aligned} \tag{5.96}$$

Consider the following example. For the transformer in Fig. 5.29 the element values are

$$\begin{aligned} L_1 &= 4\text{ h}, & L_2 &= 1\text{ h} \\ R_1 &= 8\ \Omega, & R_2 &= 3\ \Omega \\ M &= 2\text{ h}, & V &= 10\text{ v} \end{aligned}$$

Assuming that the circuit is at steady state before the switch is closed at $t = 0$, let us find $i_1(t)$ and $i_2(t)$. The differential equations written on mesh basis are

$$10 = 4\frac{di_1}{dt} + 8i_1 + 2\frac{di_2}{dt}$$
$$0 = 2\frac{di_1}{dt} + \frac{di_2}{dt} + 3i_2 \tag{5.97}$$

The characteristic equation is

$$H(p) = \begin{vmatrix} 4(p+2) & 2p \\ 2p & (p+3) \end{vmatrix} = 0 \tag{5.98}$$

which yields $$H(p) = 20p + 24 = 20(p + \tfrac{6}{5}) \tag{5.99}$$

so that the complementary functions are

$$i_{1C}(t) = K_1 e^{-1.2t}$$
$$i_{2C}(t) = K_2 e^{-1.2t} \tag{5.100}$$

The particular integrals that we obtain by inspection are

$$i_{1p}(t) = \frac{V}{R_1} = \frac{10}{8} = \frac{5}{4}$$
$$i_{2p}(t) = 0 \tag{5.101}$$

The initial conditions are

$$i_1(0+) = \frac{VL_2}{R_1L_2 + R_2L_1} = \frac{10}{20} = 0.5$$
$$i_2(0+) = \frac{-VM}{R_1L_2 + R_2L_1} = -1.0 \tag{5.102}$$

We then find $K_1 = -0.75$ and $K_2 = -1.0$ so that

$$i_1(t) = (-0.75e^{-1.2t} + 1.25)\,u(t)$$
$$i_2(t) = -e^{-1.2t}\,u(t) \tag{5.103}$$

We see that as t approaches infinity $i_2(t) \to 0$, while $i_1(t)$ goes its steady-state value of 1.25.

Problems

5.1 Write the mesh equations for the network shown.

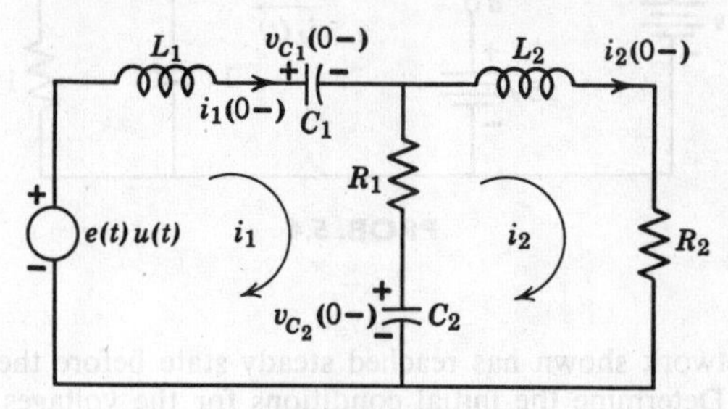

PROB. 5.1

5.2 Write a set of node equations to solve for the voltage $v_2(t)$ shown in the figure.

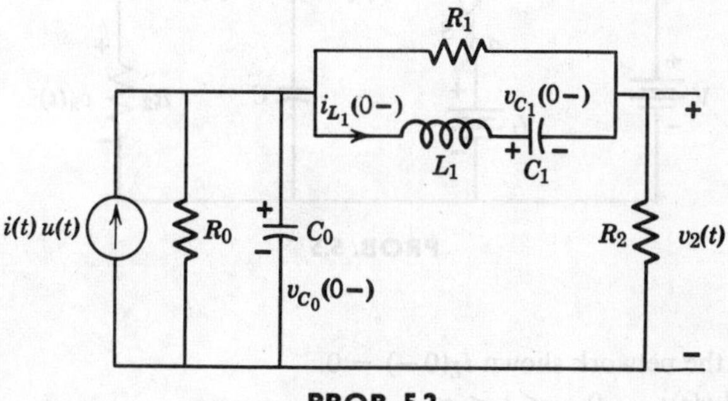

PROB. 5.2

5.3 The network shown has reached steady state before the switch S is opened at $t = 0$. Determine the initial conditions for the currents $i_1(t)$ and $i_2(t)$ and their derivatives.

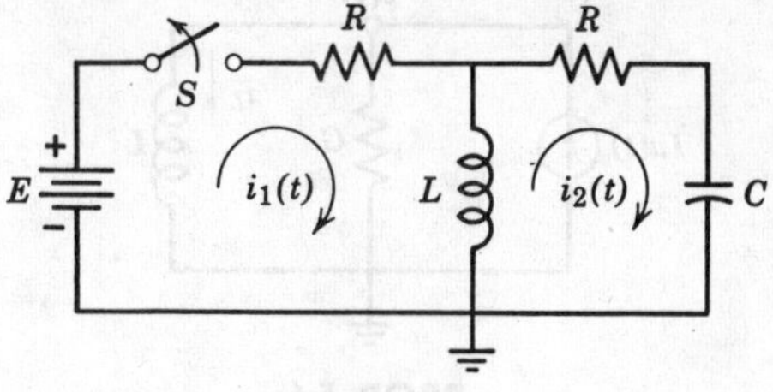

PROB. 5.3

5.4 The network shown has reached steady state before the switch moves from a to b. Determine the initial conditions for $i_L(t)$ and $v_C(t)$ and their first derivatives. Determine also the final values for $i_L(t)$ and $v_C(t)$.

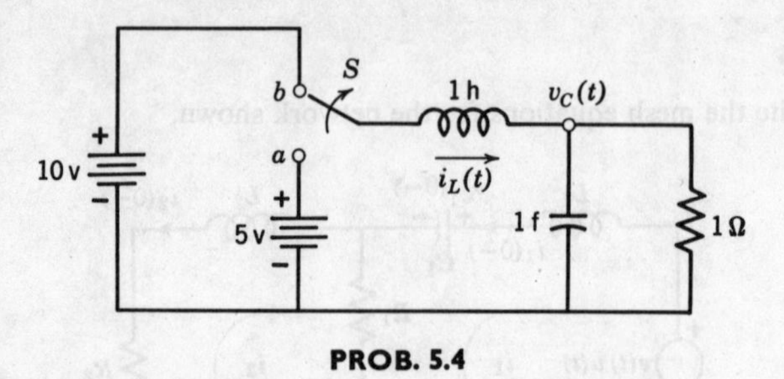

PROB. 5.4

5.5 The network shown has reached steady state before the switch moves from a to b. Determine the initial conditions for the voltages $v_1(t)$ and $v_2(t)$ and their first derivatives. Determine also the final values for $v_1(t)$ and $v_2(t)$.

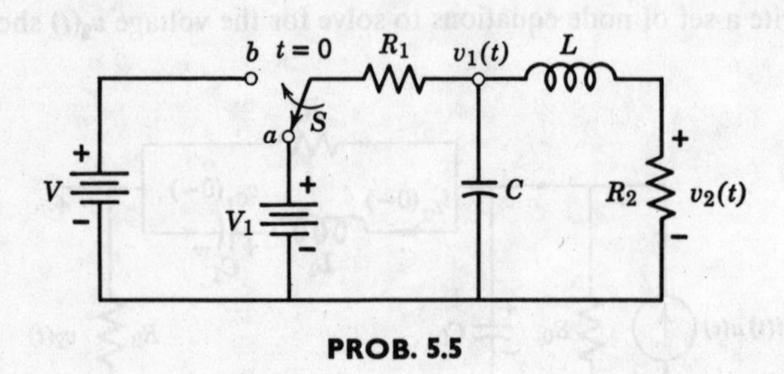

PROB. 5.5

5.6 For the network shown $i_L(0-) = 0$.

(*a*) Find $v(t)$; $0- < t < \infty$.

(*b*) Show that $v(t)$ approaches an impulse as $G \to 0$.

(*c*) Find the strength (area) of the impulse.

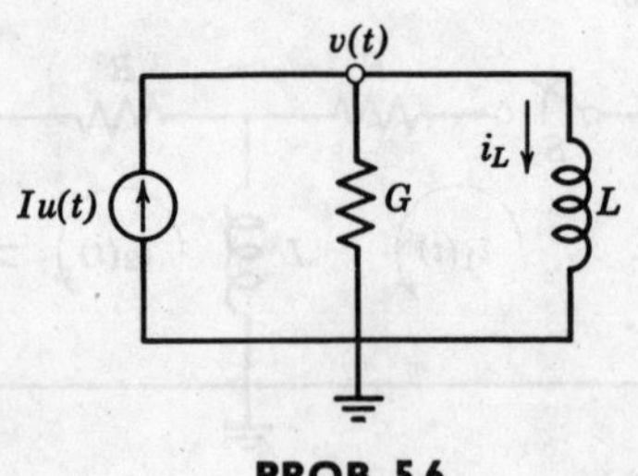

PROB. 5.6

5.7 For the network shown, $i(t) = \delta(t) - e^{-t}\,u(t)$ and $v(0-) = 4$. Find and sketch $i_C(t)$; $0- < t < \infty$.

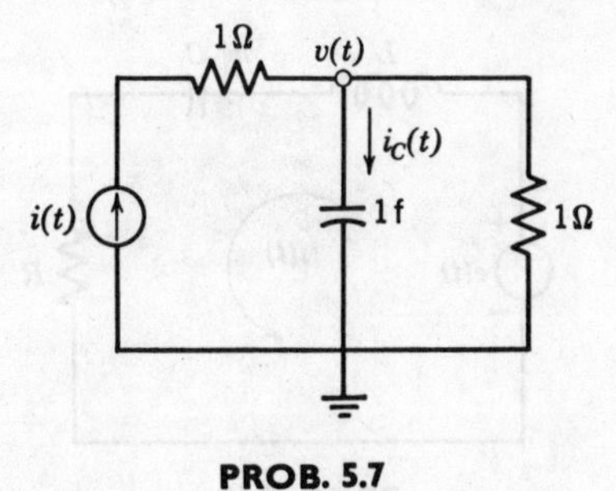

PROB. 5.7

5.8 For the network shown, before the switch moves from *a* to *b*, steady-state conditions prevailed. Find the current *i*(*t*).

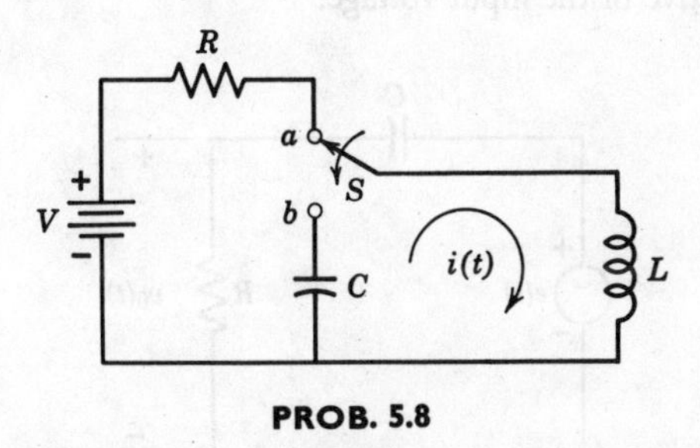

PROB. 5.8

5.9 For the circuit shown, switch *S* is opened at $t = 0$ after the circuit had been in steady state. Find *i*(*t*); $0- < t < \infty$.

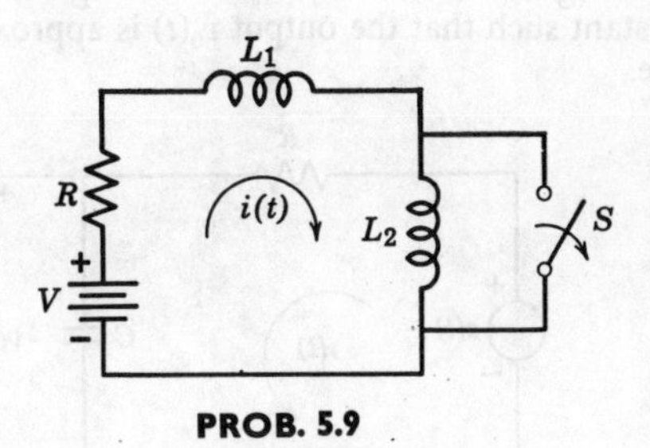

PROB. 5.9

5.10 Find the current *i*(*t*) in the network shown when the voltage source is a unit impulse. Discuss the three different kinds of impulse response waveforms possible depending upon the relative values of *R*, *L*, and *C*. All initial conditions are zero at $t = 0-$.

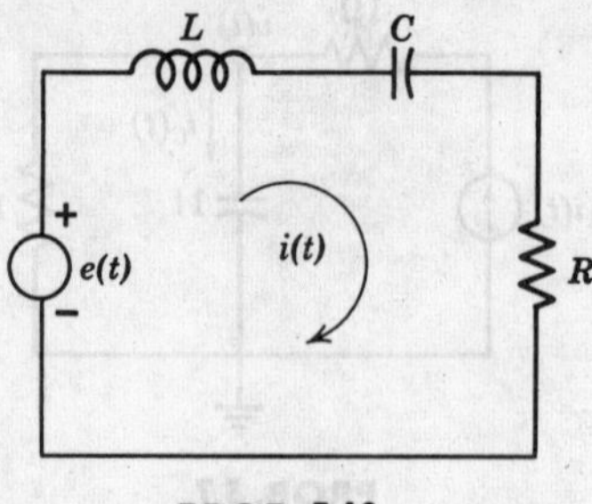

PROB. 5.10

5.11 An R-C differentiator circuit is shown in the figure. Find the requirements for the R-C time constant, such that the output voltage $v_0(t)$ is approximately the derivative of the input voltage.

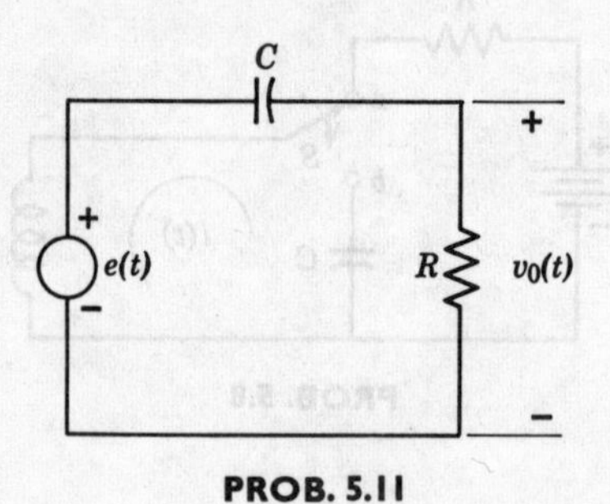

PROB. 5.11

5.12 An R-C integrator circuit is shown in the figure. Find the requirements for the time constant such that the output $v_0(t)$ is approximately the integral of the input voltage.

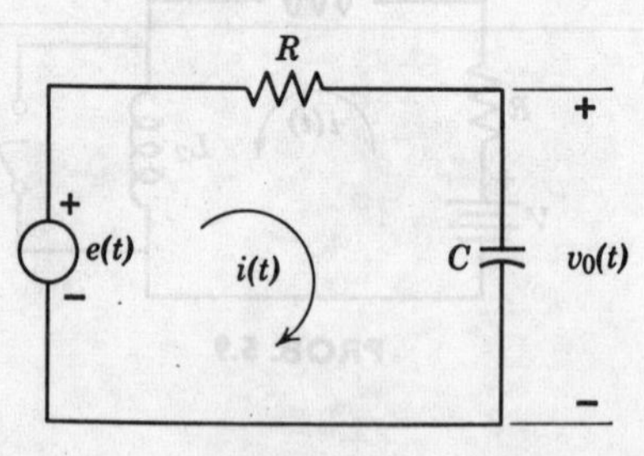

PROB. 5.12

5.13 Find the free response for $i(t)$ in the figure shown for (*a*) a current source; (*b*) a voltage source.

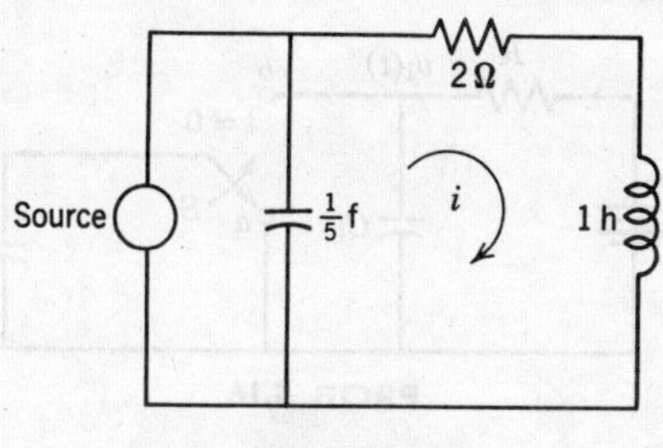

PROB. 5.13

5.14 At $t = 0$, the switch goes from position 1 to 2. Find $i(t)$, given that $e(t) = e^{-t} \sin 2t$. Assume the circuit had been in steady state for $t < 0$.

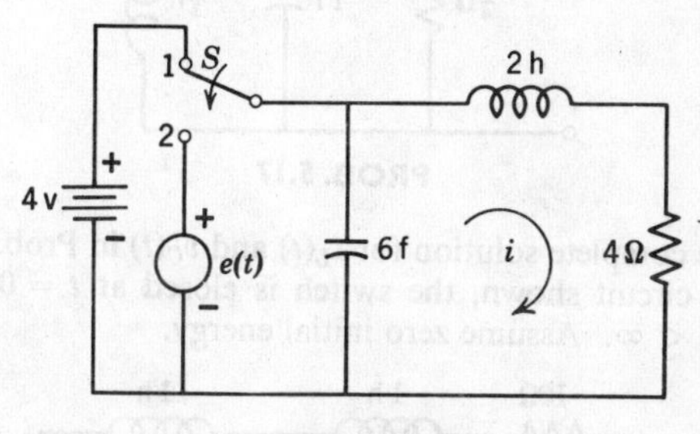

PROB. 5.14

5.15 For the circuit shown, switch S closes at $t = 0$. Solve for $i(t)$ when the initial conditions are zero and the voltage source is $e(t) = \sin(\omega t + \theta)$. What should θ be in terms of R, C, and ω so that the coefficient of the free response term is zero?

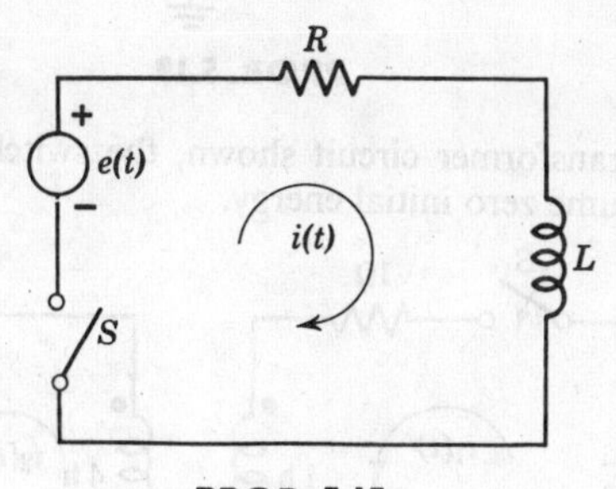

PROB. 5.15

5.16 For the circuit shown, the switch S moves from a to b at $t = 0$. Find and sketch $v_1(t)$ for $0- < t < \infty$. The circuit is in a steady state at $t = 0$.

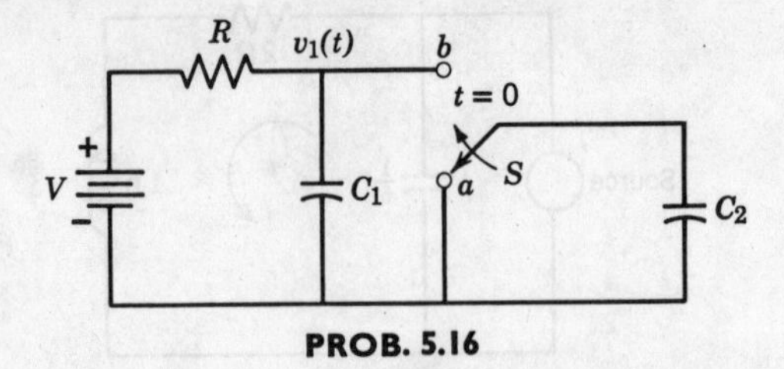

PROB. 5.16

5.17 For the circuit shown, $i(t) = 4e^{-2t}\,u(t)$. Find $v(t)$; $0- < t < \infty$.

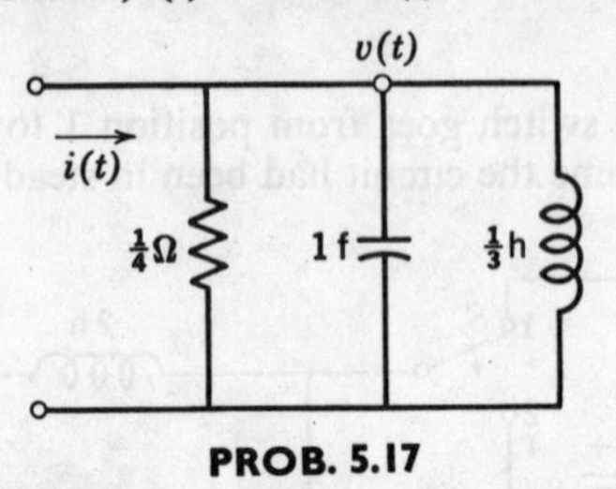

PROB. 5.17

5.18 Find the complete solution for $i_L(t)$ and $v_C(t)$ in Prob. 5.4.

5.19 For the circuit shown, the switch is closed at $t = 0$. Find $i_1(t)$ and $i_2(t)$ for $0- < t < \infty$. Assume zero initial energy.

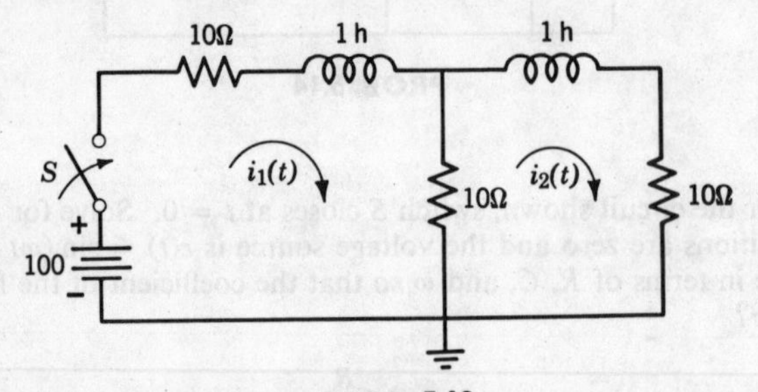

PROB. 5.19

5.20 For the transformer circuit shown, the switch closes at $t = 0$. Find $i_1(t)$ and $i_2(t)$. Assume zero initial energy.

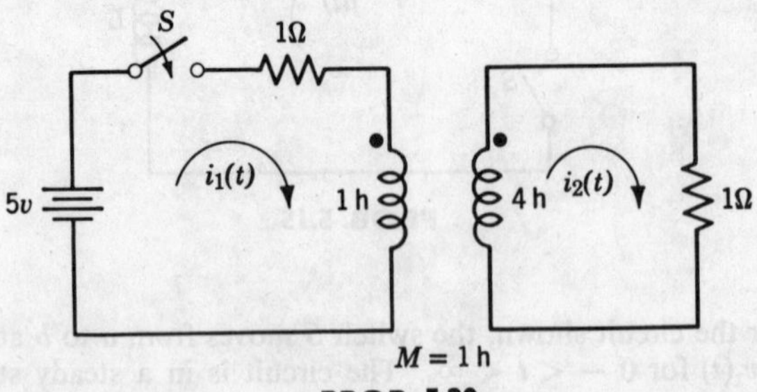

PROB. 5.20

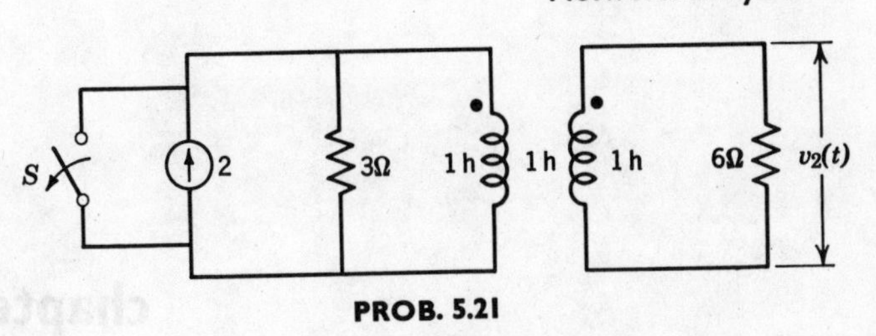

PROB. 5.21

5.21 For the transformer circuit shown, the switch S opens at $t = 0$. Find the voltage ($v_2 t$) for $0 - < t < \infty$. Assume zero initial energy.

chapter 6
The Laplace transform

6.1 THE PHILOSOPHY OF TRANSFORM METHODS

In Chapters 4 and 5, we discussed classical methods for solving differential equations. The solutions were obtained directly in the *time domain* since, in the process of solving the differential equation, we deal with functions of time at every step. In this chapter, we will use *Laplace* transforms to *transform* the differential equation to the *frequency domain*, where the independent variable is complex frequency s. It will be shown that differentiation and integration in the time domain are transformed into *algebraic* operations. Thus, the solution is obtained by simple algebraic operations in the frequency domain.

There is a striking analogy between the use of transform methods to solve differential equations and the use of logarithms for arithmetic operations. Suppose we are given two real numbers a and b. Let us find the product

$$C = a \times b \tag{6.1}$$

by means of logarithms. Since the logarithm of a product is the sum of the logarithms of the individual terms, we have

$$\log C = \log a \times b = \log a + \log b \tag{6.2}$$

so that

$$C = \log^{-1} (\log a + \log b) \tag{6.3}$$

If a and b were two six-digit numbers, the use of logarithms would probably facilitate the calculations, because logarithms transform multiplication into addition.

An analogous process is the use of transform methods to solve integro-differential equations. Consider the linear differential equation

$$y(x(t)) = f(t) \tag{6.4}$$

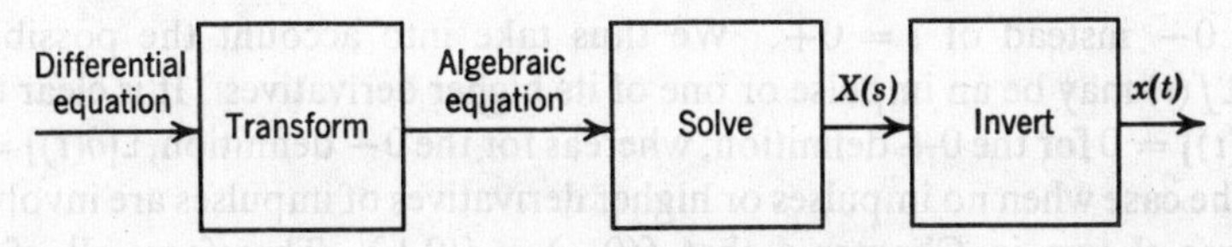

FIG. 6.1. Philosophy of transform methods.

where $f(t)$ is the forcing function, $x(t)$ is the unknown, and $y(x(t))$ is the differential equation. Let us denote the transformation process by $T(\cdot)$, and let s be the frequency variable. When we transform both sides of Eq. 6.4, we have

$$T[y(x(t))] = T[f(t)] \tag{6.5}$$

Since frequency domain functions are given by capital letters, let us write Eq. 6.5 as

$$Y(X(s), s) = F(s) \tag{6.6}$$

where $X(s) = T[x(t)]$, $F(s) = T[f(t)]$, and $Y(X(s), s)$ is an *algebraic* equation in s. The essence of the transformation process is that differential equations in time are changed into algebraic equations in frequency. We can then solve Eq. 6.6 algebraically to obtain $X(s)$. As a final step, we perform an inverse transformation to obtain

$$x(t) = T^{-1}[X(s)] \tag{6.7}$$

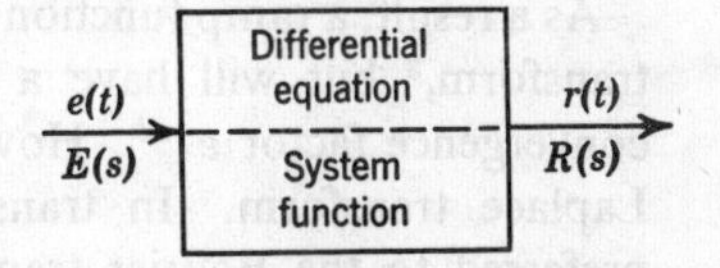

FIG. 6.2. Linear system.

In effecting the transition between the time and frequency domains, a table of transform pairs $\{x(t), X(s)\}$ can be very helpful. A diagram outlining the use of transform methods is given in Fig. 6.1. Figure 6.2 shows the relationship between excitation and response in both the time and frequency domains.

6.2 THE LAPLACE TRANSFORM

The Laplace transform of a function of time $f(t)$ is defined as

$$\mathfrak{L}[f(t)] = F(s) = \int_{0-}^{\infty} f(t)e^{-st}\, dt \tag{6.8}$$

where s is the complex frequency variable

$$s = \sigma + j\omega \tag{6.9}$$

This definition of the Laplace transform is different from the definition given in most standard texts,[1] in that the lower limit of integration is

[1] M. E. Van Valkenburg, *Network Analysis*, 2nd Ed. Prentice-Hall, Englewood Cliffs, New Jersey, 1964.

$t = 0-$ instead of $t = 0+$. We thus take into account the possibility that $f(t)$ may be an impulse or one of its higher derivatives. It is clear that $\mathcal{L}[\delta(t)] = 0$ for the $0+$ definition, whereas for the $0-$ definition, $\mathcal{L}[\delta(t)] = 1$. In the case when no impulses or higher derivatives of impulses are involved, it was shown in Chapter 4 that $f(0-) = f(0+)$. Therefore, all of the "strong results" resulting from a rigorous treatment of Laplace transforms[2] obtained by using $t = 0+$ as a lower limit also apply for the $0-$ definition.

In order for a function to possess a Laplace transform, it must obey the condition

$$\int_{0-}^{\infty} |f(t)|\, e^{-\sigma t}\, dt < \infty \tag{6.10}$$

for a real, positive σ. Note that, for a function to have a Fourier transform, it must obey the condition

$$\int_{-\infty}^{\infty} |f(t)|\, dt < \infty \tag{6.11}$$

As a result, a ramp function or a step function will not possess a Fourier transform,[3] but will have a Laplace transform because of the added convergence factor $e^{-\sigma t}$. However, the function e^{t^2} will not even have a Laplace transform. In transient problems, the Laplace transform is preferred to the Fourier transform, not only because a larger class of waveforms have Laplace transforms but also because the Laplace transform takes directly into account initial conditions at $t = 0-$ because of the lower limit of integration in the Laplace transform. In contrast, the Fourier transform has limits of integration $(-\infty, \infty)$, and, in order to take into account initial conditions due to a switch closing at $t = 0$, the forcing function must take a form as $f(t)\, u(t) + f(0-)\, \delta(t)$, where $f(0-)$ represents the initial condition.

The inverse transform $\mathcal{L}^{-1}[F(s)]$ is

$$f(t) = \frac{1}{2\pi j} \int_{\sigma_1 - j\infty}^{\sigma_1 + j\infty} F(s) e^{st}\, ds \tag{6.12}$$

where σ_1 is a real positive quantity that is greater than the σ convergence factor in Eq. 6.10. Note that the inverse transform, as defined, involves a

[2] D. V. Widder, *The Laplace Transform*, Princeton University Press, Princeton, 1941.

[3] These functions do not possess Fourier transforms in the strict sense, but many possess a *generalized* Fourier transform containing impulses in frequency; see M. J. Lighthill, *Fourier Analysis and Generalized Functions*, Cambridge University Press, New York, 1959.

complex integration known as a *contour integration*.[4] Since it is beyond the intended scope of this book to cover contour integration, we will use a partial fraction expansion procedure to obtain the inverse transform. To find inverse transforms by recognition, we must remember certain basic transform pairs and also use a table of Laplace transforms. Two of the most basic transform pairs are discussed here. Consider first the transform of a unit step function $u(t)$.

Example 6.1. $f(t) = u(t)$.

$$F(s) = \int_{0-}^{\infty} u(t)e^{-st}\,dt = -\left.\frac{e^{-st}}{s}\right|_{0-}^{\infty} = 0 - \left(-\frac{1}{s}\right) = \frac{1}{s} \tag{6.13}$$

Next, let us find the transform of an exponential function of time.

Example 6.2. $f(t) = e^{at}u(t)$.

$$F(s) = \int_{0-}^{\infty} e^{at}e^{-st}\,dt = -\left.\frac{e^{-(s-a)t}}{s-a}\right|_{0-}^{\infty} = \frac{1}{s-a} \tag{6.14}$$

With these two transform pairs, and with the use of the properties of Laplace transforms, which we discuss in Section 6.3, we can build up an extensive table of transform pairs.

6.3 PROPERTIES OF LAPLACE TRANSFORMS

In this section we will discuss a number of important properties of Laplace transforms. Using these properties we will build up a table of transforms. To facilitate this task, each property is illustrated by considering the transforms of important signal waveforms. First let us discuss the *linearity* property.

Linearity

The transform of a finite sum of time functions is the sum of the transforms of the individual functions, that is

$$\mathcal{L}\left[\sum_i f_i(t)\right] = \sum_i \mathcal{L}[f_i(t)] \tag{6.15}$$

This property follows readily from the definition of the Laplace transform.

Example 6.3. $f(t) = \sin \omega t$. Expanding $\sin \omega t$ by Euler's identity, we have

$$f(t) = \frac{1}{2j}(e^{j\omega t} - e^{-j\omega t}) \tag{6.16}$$

---

[4] For a lucid treatment, see S. Goldman, *Transformation Calculus and Electrical Transients*, Prentice-Hall, Englewood Cliffs, New Jersey, 1949, Chapter 7.

The Laplace transform of $f(t)$ is the sum of the transforms of the individual cisoidal $e^{\pm j\omega t}$ terms. Thus

$$\mathcal{L}[\sin \omega t] = \frac{1}{2j}\left(\frac{1}{s - j\omega} - \frac{1}{s + j\omega}\right) = \frac{\omega}{s^2 + \omega^2} \tag{6.17}$$

Real differentiation

Given that $\mathcal{L}[f(t)] = F(s)$, then

$$\mathcal{L}\left[\frac{df}{dt}\right] = s\,F(s) - f(0-) \tag{6.18}$$

where $f(0-)$ is the value of $f(t)$ at $t = 0-$.

Proof. By definition,

$$\mathcal{L}[f'(t)] = \int_{0-}^{\infty} e^{-st} f'(t)\,dt \tag{6.19}$$

Integrating Eq. 6.19 by parts, we have

$$\mathcal{L}[f'(t)] = e^{-st} f(t)\Big|_{0-}^{\infty} + s\int_{0-}^{\infty} f(t)e^{-st}\,dt \tag{6.20}$$

Since $e^{-st} \to 0$ as $t \to \infty$, and because the integral on the right-hand side is $\mathcal{L}[f(t)] = F(s)$, we have

$$\mathcal{L}[f'(t)] = s\,F(s) - f(0-) \tag{6.21}$$

Similarly, we can show for the nth derivative

$$\mathcal{L}\left[\frac{d^n f(t)}{dt^n}\right] = s^n F(s) - s^{n-1} f(0-) - s^{n-2} f'(0-) - \cdots - f^{(n-1)}(0-) \tag{6.22}$$

where $f^{(n-1)}$ is the $(n - 1)$st derivative of $f(t)$ at $t = 0-$. We thus see that differentiating by t in the time domain is equivalent to multiplying by s in the complex frequency domain. In addition, the initial conditions are taken into account by the terms $f^{(i)}(0-)$. It is this property that transforms differential equations in the time domain to algebraic equations in the frequency domain.

Example 6.4a. $f(t) = \sin \omega t$. Let us find

$$\mathcal{L}[\cos \omega t] = \mathcal{L}\left[\frac{1}{\omega}\frac{d}{dt}\sin \omega t\right] \tag{6.23}$$

By the real differentiation property, we have

$$\mathcal{L}\left[\frac{d}{dt}\sin \omega t\right] = s\left(\frac{\omega}{s^2 + \omega^2}\right) \tag{6.24}$$

so that
$$\mathcal{L}[\cos \omega t] = \frac{s}{\omega}\left(\frac{\omega}{s^2 + \omega^2}\right) = \frac{s}{s^2 + \omega^2} \tag{6.25}$$

In this example, note that $\sin \omega(0-) = 0$.

Example 6.4b. $f(t) = u(t)$. Let us find $\mathcal{L}[f'(t)]$, which is the transform of the unit impulse. We know that

$$\mathcal{L}[u(t)] = \frac{1}{s} \tag{6.26}$$

Then
$$\mathcal{L}[\delta(t)] = s\left(\frac{1}{s}\right) = 1 \tag{6.27}$$

since $u(0-) = 0$.

Real integration

If $\mathcal{L}[f(t)] = F(s)$, then the Laplace transform of the integral of $f(t)$ is $F(s)$ divided by s, that is,

$$\mathcal{L}\left[\int_{0-}^{t} f(\tau)\, d\tau\right] = \frac{F(s)}{s} \tag{6.28}$$

Proof. By definition,

$$\mathcal{L}\left[\int_{0-}^{t} f(\tau)\, d\tau\right] = \int_{0-}^{\infty} e^{-st}\left[\int_{0-}^{t} f(\tau)\, d\tau\right] dt \tag{6.29}$$

Integrating by parts, we obtain

$$\mathcal{L}\left[\int_{0-}^{t} f(\tau)\, d\tau\right] = -\frac{e^{-st}}{s}\int_{0-}^{t} f(\tau)\, d\tau\Bigg|_{0-}^{\infty} + \frac{1}{s}\int_{0-}^{\infty} e^{-st} f(t)\, dt \tag{6.30}$$

Since $e^{-st} \to 0$ as $t \to \infty$, and since

$$\int_{0-}^{t} f(\tau)\, d\tau\Bigg|_{t=0-} = 0 \tag{6.31}$$

we then have

$$\mathcal{L}\left[\int_{0-}^{t} f(\tau)\, d\tau\right] = \frac{F(s)}{s} \tag{6.32}$$

Example 6.5. Let us find the transform of the unit ramp function, $\rho(t) = t\, u(t)$.

We know that
$$\int_{-0}^{t} u(\tau)\, d\tau = \rho(t) \tag{6.33}$$

Since
$$\mathcal{L}[u(t)] = \frac{1}{s} \tag{6.34}$$

then
$$\mathcal{L}[\rho(t)] = \frac{\mathcal{L}[u(t)]}{s} = \frac{1}{s^2} \tag{6.35}$$

Differentiation by *s*

Differentiation by s in the complex frequency domain corresponds to multiplication by t in the domain, that is,

$$\mathcal{L}[tf(t)] = -\frac{d\,F(s)}{ds} \tag{6.36}$$

Proof. From the definition of the Laplace transform, we see that

$$\frac{d\,F(s)}{ds} = \int_{0-}^{\infty} f(t)\frac{d}{ds}\,e^{-st}\,dt = -\int_{0-}^{\infty} t\,f(t)e^{-st}\,dt = -\mathcal{L}[t\,f(t)] \tag{6.37}$$

Example 6.6. Given $f(t) = e^{-\alpha t}$, whose transform is

$$F(s) = \frac{1}{s+\alpha} \tag{6.38}$$

let us find $\mathcal{L}[te^{-\alpha t}]$. By the preceding theorem, we find that

$$\mathcal{L}[te^{-\alpha t}] = -\frac{d}{ds}\left(\frac{1}{s+\alpha}\right) = \frac{1}{(s+\alpha)^2} \tag{6.39}$$

Similarly, we can show that

$$\mathcal{L}[t^n e^{-\alpha t}] = \frac{n!}{(s+\alpha)^{n+1}} \tag{6.40}$$

where n is a positive integer.

Complex translation

By the complex translation property, if $F(s) = \mathcal{L}[f(t)]$, then

$$F(s-a) = \mathcal{L}[e^{at}f(t)] \tag{6.41}$$

where a is a complex number.

Proof. By definition,

$$\mathcal{L}[e^{at}f(t)] = \int_{0-}^{\infty} e^{at}f(t)e^{-st}\,dt = \int_{0-}^{\infty} e^{-(s-a)t}f(t)\,dt = F(s-a) \tag{6.42}$$

Example 6.7. Given $f(t) = \sin \omega t$, find $\mathcal{L}[e^{-at}\sin \omega t]$. Since

$$\mathcal{L}[\sin \omega t] = \frac{\omega}{s^2+\omega^2} \tag{6.43}$$

by the preceding theorem, we find that

$$\mathcal{L}[e^{-at}\sin \omega t] = \frac{\omega}{(s+a)^2+\omega^2} \tag{6.44}$$

Similarly, we can show that

$$\mathcal{L}[e^{-at}\cos \omega t] = \frac{s+a}{(s+a)^2+\omega^2} \tag{6.45}$$

Real translation (shifting theorem)

Here we consider the very important concept of the transform of a *shifted* or *delayed* function of time. If $\mathcal{L}[f(t)] = F(s)$, then the transform of the function delayed by time a is

$$\mathcal{L}[f(t-a)\,u(t-a)] = e^{-as}\,F(s) \tag{6.46}$$

Proof. By definition,

$$\mathcal{L}[f(t-a)\,u(t-a)] = \int_a^\infty e^{-st} f(t-a)\,dt \tag{6.47}$$

Introducing a new dummy variable $\tau = t - a$, we have

$$\mathcal{L}[f(\tau)\,u(\tau)] = \int_{0-}^\infty e^{-s(\tau+a)} f(\tau)\,d\tau = e^{-as}\int_{0-}^\infty f(\tau)e^{-s\tau}\,d\tau = e^{-as}\,F(s) \tag{6.48}$$

In Eq. 6.48 $f(\tau)\,u(\tau)$ is the shifted or delayed time function; therefore the theorem is proved.

It is important to recognize that the term e^{-as} is a time-delay operator. If we are given the function $e^{-as}\,G(s)$, and are required to find

$$\mathcal{L}^{-1}[e^{-as}\,G(s)] = g_1(t),$$

we can discard e^{-as} for the moment, find the inverse transform

$$\mathcal{L}^{-1}[G(s)] = g(t),$$

and then take into account the time delay by setting

$$g(t-a)\,u(t-a) = g_1(t) \tag{6.49}$$

Example 6.8a. Given the square pulse $f(t)$ in Fig. 6.3, let us first find its transform $F(s)$. Then let us determine the inverse transform of the square of $F(s)$, i.e., let us find

$$f_1(t) = \mathcal{L}^{-1}[F^2(s)] \tag{6.50}$$

Solution. The square pulse is given in terms of step functions as

$$f(t) = u(t) - u(t-a) \tag{6.51}$$

Its Laplace transform is then

$$F(s) = \frac{1}{s}(1 - e^{-as}) \tag{6.52}$$

Squaring $F(s)$, we obtain

$$F^2(s) = \frac{1}{s^2}(1 - 2e^{-as} + e^{-2as}) \tag{6.53}$$

To find the inverse transform of $F^2(s)$, we need only to determine the inverse transform

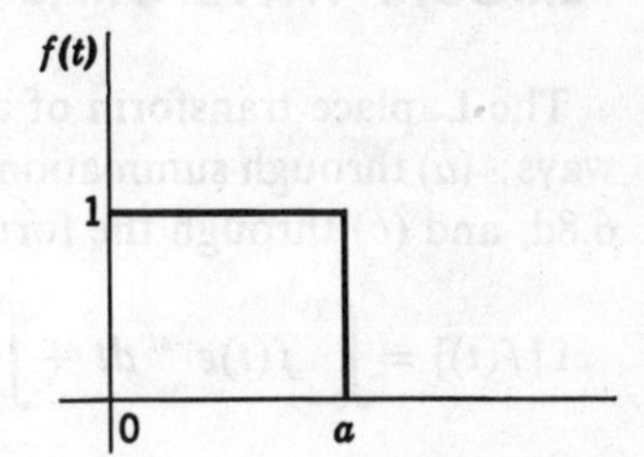

FIG. 6.3. Square pulse.

of the term with zero delay, which is

$$\mathcal{L}^{-1}\left[\frac{1}{s^2}\right] = t\,u(t) \tag{6.54}$$

Then $$\mathcal{L}^{-1}[F^2(s)] = t\,u(t) - 2(t-a)\,u(t-a) + (t-2a)\,u(t-2a) \tag{6.55}$$

so that the resulting waveform is shown in Fig. 6.4. From this example, we see that the square of a transformed function *does not* correspond to the square of its inverse transform.

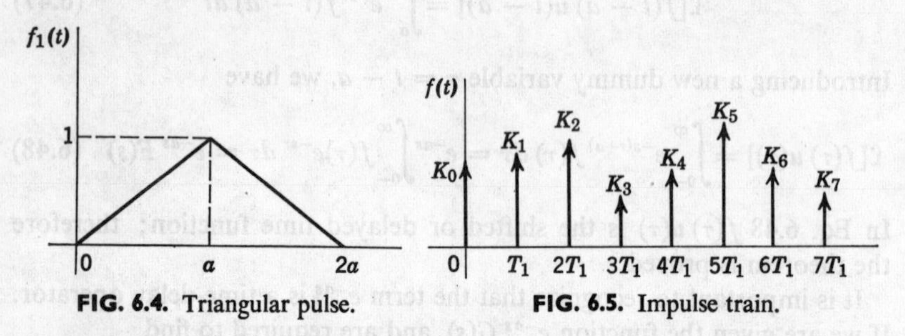

FIG. 6.4. Triangular pulse. **FIG. 6.5.** Impulse train.

Example 6.8b. In Fig. 6.5, the output of an ideal *sampler* is shown. It consists of a train of impulses

$$f(t) = K_0\,\delta(t) + K_1\,\delta(t - T_1) + \cdots + K_n\,\delta(t - nT_1) \tag{6.56}$$

The Laplace transform of this impulse train is

$$\mathcal{L}[f(t)] = K_0 + K_1 e^{-sT_1} + K_2 e^{-2sT_1} + \cdots + K_n e^{-nsT_1} \tag{6.57}$$

In dealing with sampled signals, the substitution $z = e^{sT_1}$ is often used. Then we can represent the transform of the impulse train as

$$\mathcal{L}[f(t)] = K_0 + \frac{K_1}{z} + \frac{K_2}{z^2} + \cdots + \frac{K_n}{z^n} \tag{6.58}$$

The transform in Eq. 6.58 is called the *z*-transform of $f(t)$. This transform is widely used in connection with sampled-data control systems.

PERIODIC WAVEFORMS

The Laplace transform of a periodic waveform can be obtained in two ways: (*a*) through summation of an infinite series as illustrated in Example 6.8d, and (*b*) through the formula derived below.

$$\mathcal{L}[f(t)] = \int_{0-}^{T} f(t)e^{-st}\,dt + \int_{T}^{2T} f(t)e^{-st}\,dt + \cdots + \int_{nT}^{(n+1)T} f(t)e^{-st}\,dt + \cdots \tag{6.59}$$

Since $f(t)$ is periodic, Eq. 6.59 reduces to

$$
\begin{aligned}
\mathcal{L}[f(t)] &= \int_{0-}^{T} f(t)e^{-st}\,dt + e^{-sT}\int_{0-}^{T} f(t)e^{-st}\,dt + \cdots \\
&\quad + e^{-snT}\int_{0-}^{T} f(t)e^{-st}\,dt + \cdots \\
&= (1 + e^{-sT} + e^{-s2T} + \cdots)\int_{0-}^{T} f(t)e^{-st}\,dt \\
&= \frac{1}{1 - e^{-sT}}\int_{0-}^{T} f(t)e^{-st}\,dt
\end{aligned}
\tag{6.60}
$$

Example 6.8c. Given the periodic pulse train in Fig. 6.6 let us use Eq. 6.60 to determine its Laplace transform.

$$
\begin{aligned}
F(s) &= \frac{1}{1 - e^{-sT}}\int_{0-}^{a} e^{-st}\,dt \\
&= \frac{-1}{s(1 - e^{-sT})}\, e^{-st}\Big|_{0-}^{a} \\
&= \frac{1}{s}\,\frac{1 - e^{-as}}{1 - e^{-sT}}
\end{aligned}
\tag{6.61}
$$

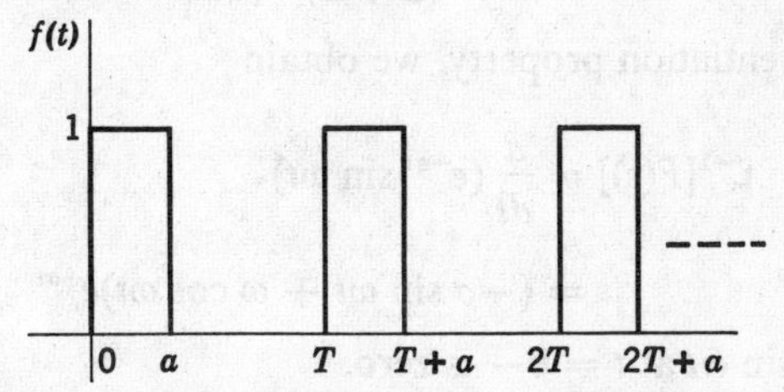

FIG. 6.6. Periodic pulse train.

Example 6.8d. In this example we calculate the Laplace transform of $f(t)$ in Fig. 6.6 using summation of an infinite series. The periodic pulse train can be represented as

$$
\begin{aligned}
f(t) = u(t) - u(t - a) + u(t - T) - u[t - (T + a)] \\
+ u(t - 2T) - u[t - (2T + a)] + \cdots
\end{aligned}
\tag{6.62}
$$

Its Laplace transform is

$$
\begin{aligned}
F(s) &= \frac{1}{s}(1 - e^{-as} + e^{-sT} - e^{-(T+a)s} + \cdots) \\
&= \frac{1}{s}[1 - e^{-as} + e^{-sT}(1 - e^{-as}) + e^{-2sT}(1 - e^{-as}) + \cdots]
\end{aligned}
\tag{6.63}
$$

which simplifies further to give

$$F(s) = \frac{1}{s}(1 - e^{-as})(1 + e^{-sT} + e^{-2sT} + \cdots) \tag{6.64}$$

We then see that $F(s)$ can be given in the closed form

$$F(s) = \frac{1}{s}\frac{1 - e^{-as}}{1 - e^{-sT}} \tag{6.65}$$

Other periodic pulse trains also can be given in closed form. The reader is referred to the problems at the end of this chapter.

At the end of the chapter is a table of Laplace transforms. Most of the entries are obtained through simple applications of the properties just discussed. It is important to keep those properties in mind, because many transform pairs that are not given in the table can be obtained by using these properties. For example, let us find the inverse transform of

$$F(s) = \frac{\omega s}{(s + \sigma)^2 + \omega^2} \tag{6.66}$$

Since the s in the numerator implies differentiation in the time domain, we can write

$$F(s) = s\frac{\omega}{(s + \sigma)^2 + \omega^2} \tag{6.67}$$

Using the differentiation property, we obtain

$$\begin{aligned} \mathcal{L}^{-1}[F(s)] &= \frac{d}{dt}(e^{-\sigma t}\sin \omega t) \\ &= (-\sigma \sin \omega t + \omega \cos \omega t)e^{-\sigma t} \end{aligned} \tag{6.68}$$

Note that $e^{-\sigma t} \sin \omega t$ at $t = 0-$ is zero.

6.4 USES OF LAPLACE TRANSFORMS

Evaluation of definite integrals

The Laplace transform is often useful in the evaluation of definite integrals. An obvious example occurs in the evaluation of

$$I = \int_0^\infty e^{-2t} \sin 5t \, dt \tag{6.69}$$

If we replace e^{-2t} by e^{-st}, the integral I then becomes the Laplace transform of $\sin 5t$, which is

$$\mathcal{L}[\sin 5t] = \frac{5}{s^2 + 25} \tag{6.70}$$

Replacing s by 2, we have

$$I = \frac{5}{2^2 + 25} = \frac{5}{29} \tag{6.71}$$

Perhaps a more subtle example is the evaluation of

$$I = \int_{-1}^{+1} t^2 e^{-2|t|}\, dt \tag{6.72}$$

First, we note that $t^2 e^{-2|t|}$ is an even function; therefore

$$I = 2\int_0^1 t^2 e^{-2t}\, dt \tag{6.73}$$

From the table of Laplace transforms we see that

$$\mathfrak{L}[t^2 e^{-2t}] = \frac{2}{(s+2)^3} \tag{6.74}$$

and the transform of

$$f(t) = \int_{0-}^{t} \tau^2 e^{-2\tau}\, d\tau \tag{6.75}$$

is

$$\mathfrak{L}[f(t)] = \frac{2}{s(s+2)^3} \tag{6.76}$$

Later we will see that the partial-fraction expansion of $\mathfrak{L}[f(t)]$ in Eq. 6.76 consists of three terms for the multiple root $(s+2)^3$, as given by

$$\mathfrak{L}[f(t)] = \frac{0.25}{s} - \frac{0.25}{s+2} - \frac{0.5}{(s+2)^2} - \frac{1}{(s+2)^3} \tag{6.77}$$

Taking the inverse transform of Eq. 6.77, we obtain

$$f(t) = 0.25(1 - e^{-2t} - 2te^{-2t} - 2t^2 e^{-2t})\, u(t) \tag{6.78}$$

Now observe that the definite integral in Eq. 6.73 is equal to $2f(t)$ at $t = 1$, that is,

$$t = 2f(1) = -2.5e^{-2} + 0.5 = 0.162 \tag{6.79}$$

Solution of integrodifferential equations

In Section 6.3 we said that the real differentiation and real integration properties of the Laplace transform change differential equations in time to algebraic equations in frequency. Let us consider some examples using Laplace transforms in solving differential equations.

Example 6.9. Let us solve the differential equation

$$x''(t) + 3x'(t) + 2x(t) = 4e^t \tag{6.80}$$

given the initial conditions $x(0-) = 1$, $x'(0-) = -1$.

Solution. We first proceed by taking the Laplace transform of the differential equation, which then becomes

$$[s^2 X(s) - s\,x(0-) - x'(0-)] + 3[s\,X(s) - x(0-)] + 2X(s) = \frac{4}{s-1} \quad (6.81)$$

Substituting the initial conditions into Eq. 6.81 and simplifying, we have

$$(s^2 + 3s + 2)\,X(s) = \frac{4}{s-1} + s + 2 \quad (6.82)$$

We then obtain $X(s)$ explicitly as

$$X(s) = \frac{s^2 + s + 2}{(s-1)(s+1)(s+2)} \quad (6.83)$$

To find the inverse transform $x(t) = \mathcal{L}^{-1}[X(s)]$, we expand $X(s)$ into partial fractions

$$X(s) = \frac{K_{-1}}{s-1} + \frac{K_1}{s+1} + \frac{K_2}{s+2} \quad (6.84)$$

Solving for K_{-1}, K_1, and K_2 algebraically, we obtain

$$K_{-1} = \tfrac{2}{3} \quad K_1 = -1 \quad K_2 = \tfrac{4}{3}$$

The final solution is the inverse transform of $X(s)$ or

$$x(t) = \tfrac{2}{3}e^{t} - e^{-t} + \tfrac{4}{3}e^{-2t} \quad (6.85)$$

In order to compare the Laplace transform method to the classical method of solving differential equations, the reader is referred to the example in Chapter 4, Eq. 4.64, where the differential equation in Eq. 6.80 is solved classically.

Example 6.10. Given the set of simultaneous differential equations

$$\begin{aligned} 2x'(t) + 4x(t) + y'(t) + 7y(t) &= 5u(t) \\ x'(t) + x(t) + y'(t) + 3y(t) &= 5\delta(t) \end{aligned} \quad (6.86)$$

with the initial conditions $x(0-) = y(0-) = 0$, let us find $x(t)$ and $y(t)$.

Solution. Transforming the set of equations, we obtain

$$\begin{aligned} 2(s+2)\,X(s) + (s+7)\,Y(s) &= \frac{5}{s} \\ (s+1)\,X(s) + (s+3)\,Y(s) &= 5 \end{aligned} \quad (6.87)$$

Solving for $X(s)$ and $Y(s)$ simultaneously, we have

$$\begin{aligned} X(s) &= \frac{(5/s)\Delta_{11} + 5\Delta_{21}}{\Delta} \\ Y(s) &= \frac{(5/s)\Delta_{12} + 5\Delta_{22}}{\Delta} \end{aligned} \quad (6.88)$$

where Δ is the determinant of the set of equations in Eq. 6.87, and Δ_{ij} is the ijth cofactor of Δ. More explicitly, $X(s)$ is

$$X(s) = \frac{-5s^2 - 30s + 15}{s(s^2 + 2s + 5)} \tag{6.89}$$

Expanding $X(s)$ in partial fractions, we have

$$X(s) = \frac{K_1}{s} + \frac{K_2 s + K_3}{s^2 + 2s + 5} \tag{6.90}$$

Multiplying both sides of Eq. 6.90 by s and letting $s = 0$, we find

$$K_1 = s\,X(s)\big|_{s=0} = 3.$$

K_2 and K_3 are then obtained from the equation

$$X(s) - \frac{3}{s} = \frac{-8s - 36}{s^2 + 2s + 5} \tag{6.91}$$

A further simplification occurs by completing the square of the denominator of $X(s)$, that is,

$$s^2 + 2s + 5 = (s + 1)^2 + 4 \tag{6.92}$$

As a result of Eq. 6.92, we can rewrite $X(s)$ as

$$X(s) = \frac{3}{s} - \frac{8(s + 1) + 14(2)}{(s + 1)^2 + (2)^2} \tag{6.93}$$

so that the inverse transform is

$$x(t) = (3 - 8e^{-t} \cos 2t - 14e^{-t} \sin 2t)\,u(t) \tag{6.94}$$

In similar fashion, we obtain $Y(s)$ as

$$Y(s) = -\frac{1}{s} + \frac{11s + 17}{s^2 + 2s + 5} = -\frac{1}{s} + \frac{11(s + 1) + 3(2)}{(s + 1)^2 + (2)^2} \tag{6.95}$$

The inverse transform is then seen to be

$$y(t) = (-1 + 11e^{-t} \cos 2t + 3e^{-t} \sin 2t)\,u(t) \tag{6.96}$$

This example is also solved by classical methods in Chapter 4, Eq. 4.163. We note one sharp point of contrast. While we had to find the initial conditions at $t = 0+$ in order to solve the differential equations directly in the time domain, the Laplace transform method works directly with the initial conditions at $t = 0-$. In addition, we obtain *both* the complementary function and the particular integral in a single operation when we use Laplace transforms. These are the reasons why the Laplace transform method is so effective in the solution of differential equations.

6.5 PARTIAL-FRACTION EXPANSIONS

As we have seen, the ease with which we use transform methods depends upon how quickly we are able to obtain the partial-fraction expansion of a given transform function. In this section we will elaborate on some simple and effective methods for partial-fraction expansions, and we discuss procedures for (*a*) simple roots, (*b*) complex conjugate roots, and (*c*) multiple roots.

It should be recalled that if the degree of the numerator is greater or equal to the degree of the denominator, we can divide the numerator by the denominator such that the remainder can be expanded more easily into partial fractions. Consider the following example:

$$F(s) = \frac{N(s)}{D(s)} = \frac{s^3 + 3s^2 + 3s + 2}{s^2 + 2s + 2} \tag{6.97}$$

Since the degree of $N(s)$ is greater than the degree of $D(s)$, we divide $D(s)$ into $N(s)$ to give

$$F(s) = s + 1 - \frac{s}{s^2 + 2s + 2} \tag{6.98}$$

Here we see the remainder term can be easily expanded into partial fractions. However, there is no real need at this point because the denominator $s^2 + 2s + 2$ can be written as

$$s^2 + 2s + 2 = (s + 1)^2 + 1 \tag{6.99}$$

We can then write $F(s)$ as

$$F(s) = s + 1 - \frac{(s + 1) - 1}{(s + 1)^2 + 1} \tag{6.100}$$

so that the inverse transform can be obtained directly from the transform tables, namely,

$$\mathcal{L}^{-1}[F(s)] = \delta'(t) + \delta(t) + e^{-t}(\sin t - \cos t) \tag{6.101}$$

From this example, we see that intuition and a knowledge of the transform table can often save considerable work. Consider some further examples in which intuition plays a dominant role.

Example 6.11. Find the partial-fraction expansion of

$$F(s) = \frac{2s + 3}{(s + 1)(s + 2)} \tag{6.102}$$

If we see that $F(s)$ can also be written as

$$F(s) = \frac{(s+1)+(s+2)}{(s+1)(s+2)} \tag{6.103}$$

then the partial-fraction expansion is trivially

$$F(s) = \frac{1}{s+1} + \frac{1}{s+2} \tag{6.104}$$

Example 6.12. Find the partial-fraction expansion of

$$F(s) = \frac{s+5}{(s+2)^2} \tag{6.105}$$

We see that $F(s)$ can be rewritten as

$$F(s) = \frac{(s+2)+3}{(s+2)^2} = \frac{1}{(s+2)} + \frac{3}{(s+2)^2} \tag{6.106}$$

Real roots

Now let us discuss some formal methods for partial-fraction expansions. First we examine a method for simple real roots. Consider the function

$$F(s) = \frac{N(s)}{(s-s_0)(s-s_1)(s-s_2)} \tag{6.107}$$

where s_0, s_1, and s_2 are distinct, real roots, and the degree of $N(s) < 3$. Expanding $F(s)$ we have

$$F(s) = \frac{K_0}{s-s_0} + \frac{K_1}{s-s_1} + \frac{K_2}{s-s_2} \tag{6.108}$$

Let us first obtain the constant K_0. We proceed by multiplying both sides of the equation by $(s - s_0)$ to give

$$(s-s_0)F(s) = K_0 + \frac{(s-s_0)K_1}{s-s_1} + \frac{(s-s_0)K_2}{s-s_2} \tag{6.109}$$

If we let $s = s_0$ in Eq. 6.109, we obtain

$$K_0 = (s-s_0)\, F(s)\big|_{s=s_0} \tag{6.110}$$

Similarly, we see that the other constants can be evaluated through the general relation

$$K_i = (s-s_i)\, F(s)\big|_{s=s_i} \tag{6.111}$$

Example 6.13. Let us find the partial-fraction expansion for

$$F(s) = \frac{s^2+2s-2}{s(s+2)(s-3)} = \frac{K_0}{s} + \frac{K_1}{s+2} + \frac{K_2}{s-3} \tag{6.112}$$

Using Eq. 6.111, we find

$$K_0 = sF(s)\big|_{s=0}$$

$$= \frac{s^2 + 2s - 2}{(s+2)(s-3)}\bigg|_{s=0} = \frac{1}{3}$$

$$K_1 = \frac{s^2 + 2s - 2}{s(s-3)}\bigg|_{s=-2} = -\frac{1}{5} \tag{6.113}$$

$$K_2 = \frac{s^2 + 2s - 2}{s(s+2)}\bigg|_{s=3} = \frac{13}{15}$$

Complex roots

Equation 6.111 is also applicable to a function with complex roots in its denominator. Suppose $F(s)$ is given by

$$F(s) = \frac{N(s)}{D_1(s)(s - \alpha - j\beta)(s - \alpha + j\beta)}$$

$$= \frac{K_1}{s - \alpha - j\beta} + \frac{K_2}{s - \alpha + j\beta} + \frac{N_1(s)}{D_1(s)} \tag{6.114}$$

where N_1/D_1 is the remainder term. Using Eq. 6.111, we have

$$K_1 = \frac{N(\alpha + j\beta)}{2j\beta\, D_1(\alpha + j\beta)}$$

$$K_2 = \frac{N(\alpha - j\beta)}{-2j\beta\, D_1(\alpha - j\beta)} \tag{6.115}$$

where we assume that $s = \alpha \pm j\beta$ are not zeros of $D_1(s)$.

It can be shown that the constants K_1 and K_2 associated with conjugate roots are themselves conjugate. Therefore, if we denote K_1 as

$$K_1 = A + jB \tag{6.116}$$

then

$$K_2 = A - jB = K_1{}^* \tag{6.117}$$

If we denote the inverse transform of the complex conjugate terms as $f_1(t)$, we see that

$$f_1(t) = \mathcal{L}^{-1}\left[\frac{K_1}{s - \alpha - j\beta} + \frac{K_1{}^*}{s - \alpha + j\beta}\right]$$

$$= e^{\alpha t}(K_1 e^{j\beta t} + K_1{}^* e^{-j\beta t})$$

$$= 2e^{\alpha t}(A \cos \beta t - B \sin \beta t) \tag{6.118}$$

A more convenient way to express the inverse transform $f_1(t)$ is to introduce the variables M and Φ defined by the equations

$$\begin{aligned} M \sin \Phi &= 2A \\ M \cos \Phi &= -2B \end{aligned} \tag{6.119}$$

where A and B are the real and imaginary parts of K_1 in Eq. 6.116. In terms of M and Φ, the inverse transform is

$$f_1(t) = Me^{\alpha t} \sin (\beta t + \Phi) \tag{6.120}$$

To obtain M and Φ from K_1, we note that

$$Me^{j\Phi} = M \cos \Phi + jM \sin \Phi = -2B + j2A = 2jK_1 \tag{6.121}$$

When related to the original function $F(s)$, we see from Eq. 6.115 that

$$Me^{j\Phi} = \frac{N(\alpha + j\beta)}{\beta D_1(\alpha + j\beta)} \tag{6.122}$$

Example 6.14. Let us find the inverse transform of

$$F(s) = \frac{s^2 + 3}{(s^2 + 2s + 5)(s + 2)} \tag{6.123}$$

For the simple root $s = -2$, the constant K is

$$K = (s + 2)\, F(s)\big|_{s=-2} = \tfrac{7}{5} \tag{6.124}$$

For the complex conjugate roots

$$s^2 + 2s + 5 = (s + 1 + j2)(s + 1 - j2) \tag{6.125}$$

We see that $\alpha = -1$, $\beta = 2$, thus

$$Me^{j\Phi} = \left.\frac{s^2 + 3}{2(s + 2)}\right|_{s=-1+j2} = \frac{2}{\sqrt{5}}\, e^{-j(\tan^{-1} 2 + \pi/2)} \tag{6.126}$$

The inverse transform is then

$$\begin{aligned} \mathcal{L}^{-1}[F(s)] &= \frac{7}{5} e^{-2t} + \frac{2}{\sqrt{5}} e^{-t} \sin \left(2t - \frac{\pi}{2} - \tan^{-1} 2 \right) \\ &= \frac{7}{5} e^{-2t} - \frac{2}{\sqrt{5}} e^{-t} \cos (2t - \tan^{-1} 2) \end{aligned} \tag{6.127}$$

Multiple roots

Next let us consider the case in which the partial fraction involves repeated or multiple roots. We will examine two methods. The first requires differentiation; the second does not.

Method A

Suppose we are given the function

$$F(s) = \frac{N(s)}{(s - s_0)^n \, D_1(s)} \tag{6.128}$$

with multiple roots of degree n at $s = s_0$. The partial fraction expansion of $F(s)$ is

$$F(s) = \frac{K_0}{(s - s_0)^n} + \frac{K_1}{(s - s_0)^{n-1}} + \frac{K_2}{(s - s_0)^{n-2}} + \cdots + \frac{K_{n-1}}{s - s_0} + \frac{N_1(s)}{D_1(s)} \tag{6.129}$$

where $N_1(s)/D_1(s)$ represents the remaining terms of the expansion. The problem is to obtain $K_0, K_1, \ldots, K_{n-1}$. For the K_0 term, we use the method cited earlier for simple roots, that is,

$$K_0 = (s - s_0)^n \, F(s) \,|_{s=s_0} \tag{6.130}$$

However, if we were to use the same formula to obtain the factors K_1, $K_2, \ldots, K_{n-1}$, we would invariably arrive at the indeterminate 0/0 condition. Instead, let us multiply both sides of Eq. 6.129 by $(s - s_0)^n$ and define

$$F_1(s) \equiv (s - s_0)^n \, F(s) \tag{6.131}$$

Thus

$$F_1(s) = K_0 + K_1(s - s_0) + \cdots + K_{n-1}(s - s_0)^{n-1} + R(s)(s - s_0)^n \tag{6.132}$$

where $R(s)$ indicates the remaining terms. If we differentiate Eq. 6.132 by s, we obtain

$$\frac{d}{ds} F_1(s) = K_1 + 2K_2(s - s_0) + \cdots + K_{n-1}(n - 1)(s - s_0)^{n-2} + \cdots \tag{6.133}$$

It is evident that

$$K_1 = \frac{d}{ds} F_1(s) \bigg|_{s=s_0} \tag{6.134}$$

On the same basis

$$K_2 = \frac{1}{2} \frac{d^2}{ds^2} F_1(s) \bigg|_{s=s_0} \tag{6.135}$$

and in general

$$K_j = \frac{1}{j!} \frac{d^j}{ds^j} F_1(s) \bigg|_{s=s_0}, \qquad j = 0, 1, 2, \ldots, n - 1 \tag{6.136}$$

Example 6.15. Consider the function

$$F(s) = \frac{s-2}{s(s+1)^3} \tag{6.137}$$

which we represent in expanded form as

$$F(s) = \frac{K_0}{(s+1)^3} + \frac{K_1}{(s+1)^2} + \frac{K_2}{s+1} + \frac{A}{s} \tag{6.138}$$

The constant A for the simple root at $s = 0$ is

$$A = s\,F(s)\big|_{s=0} = -2 \tag{6.139}$$

To obtain the constants for the multiple roots we first find $F_1(s)$.

$$F_1(s) = (s+1)^3\,F(s) = \frac{s-2}{s} \tag{6.140}$$

Using the general formula for the multiple root expansion, we obtain

$$K_0 = \frac{1}{0!}\frac{d^0}{ds^0}\left(\frac{s-2}{s}\right)\bigg|_{s=-1} = 3 \tag{6.141}$$

$$K_1 = \frac{1}{1!}\frac{d}{ds}\left(\frac{s-2}{s}\right)\bigg|_{s=-1} = \frac{2}{s^2}\bigg|_{s=-1} = 2 \tag{6.142}$$

$$K_2 = \frac{1}{2!}\frac{d}{ds}\left(\frac{2}{s^2}\right)\bigg|_{s=-1} = \left(-\frac{2}{s^3}\right)\bigg|_{s=-1} = 2 \tag{6.143}$$

so that
$$F(s) = \frac{3}{(s+1)^3} + \frac{2}{(s+1)^2} + \frac{2}{s+1} - \frac{2}{s} \tag{6.144}$$

Method B

The second method for arriving at the partial-fraction expansion for multiple roots requires no differentiation. It involves a modified power series expansion.[5] Let us consider $F(s)$ and $F_1(s)$, defined in Eqs. 6.128 and 6.131 in Method A. We define a new variable p such that $p = s - s_0$. Then we can write $F_1(s)$ as

$$F_1(p + s_0) = \frac{N(p + s_0)}{D_1(p + s_0)} \tag{6.145}$$

Dividing $N(p + s_0)$ by $D_1(p + s_0)$, with both polynomials written in *ascending* powers of p, we obtain

$$F_1(p + s_0) = K_0 + K_1 p + K_2 p^2 + \cdots + K_{n-1}p^{n-1} + \frac{K_n p^n}{D_1(p + s_0)} \tag{6.146}$$

[5] I am indebted to the late Professor Leonard O. Goldstone of the Polytechnic Institute of Brooklyn for showing me this method.

The original function $F(s)$ is related to $F_1(p + s_0)$ by the equation

$$F(p + s_0) = \frac{F_1(p + s_0)}{p^n} = \frac{K_0}{p^n} + \frac{K_1}{p^{n-1}} + \cdots + \frac{K_{n-1}}{p} + \frac{K_n}{D_1(p + s_0)} \tag{6.147}$$

Substituting $s - s_0 = p$ in Eq. 6.147, we obtain

$$F(s) = \frac{K_0}{(s - s_0)^n} + \frac{K_1}{(s - s_0)^{n-1}} + \cdots + \frac{K_{n-1}}{s - s_0} + \frac{K_n}{D_1(s)} \tag{6.148}$$

We have thus found the partial-fraction expansion for the multiple-root terms. The remaining terms $K_n/[D_1(s)]$ still must be expanded into partial fractions. Consider Example 6.16.

Example 6.16

$$F(s) = \frac{2}{(s + 1)^3(s + 2)} \tag{6.149}$$

Using the method just given, $F_1(s)$ is

$$F_1(s) = (s + 1)^3 F(s) = \frac{2}{s + 2} \tag{6.150}$$

Setting $p = s + 1$, we then have

$$F_1(p - 1) = \frac{2}{p + 1} \tag{6.151}$$

The expansion of $F_1(p - 1)$ into a series as given in Eq. 6.146 requires dividing the numerator 2 by the denominator $1 + p$, with both numerator and denominator arranged in ascending power of p. The division here is

$$\begin{array}{r} 2 - 2p + 2p^2 \\ 1 + p\overline{)\,2\qquad\qquad} \\ \underline{2 + 2p\qquad} \\ -2p\qquad \\ \underline{-2p - 2p^2} \\ 2p^2 \\ \underline{2p^2 + 2p^3} \\ -2p^3 \end{array}$$

Since the multiplicity of the root is $N = 3$, we terminate the division after we have three terms in the quotient. We then have

$$F_1(p - 1) = 2 - 2p + 2p^2 - \frac{2p^3}{p + 1} \tag{6.152}$$

The original function $F(p - 1)$ is

$$F(p - 1) = \frac{F_1(p - 1)}{p^3} = \frac{2}{p^3} - \frac{2}{p^2} + \frac{2}{p} - \frac{2}{p + 1} \tag{6.153}$$

Substituting $s + 1 = p$, we have

$$F(s) = \frac{2}{(s+1)^3} - \frac{2}{(s+1)^2} + \frac{2}{(s+1)} - \frac{2}{s+2} \tag{6.154}$$

Example 6.17. As a second example, consider the function

$$F(s) = \frac{s+2}{s^2(s+1)^2} \tag{6.155}$$

Since we have two sets of multiple roots here (at $s = 0$, and $s = -1$) we have a choice of expanding $F_1(s)$ about $s = 0$ or $s = -1$. Let us arbitrarily choose to expand about $s = 0$; since $p = s$ here, we do not have to make any substitutions. $F_1(s)$ is then

$$F_1(s) = \frac{s+2}{(s+1)^2} = \frac{2+s}{1+2s+s^2} \tag{6.156}$$

Expanding $F_1(s)$, we obtain

$$F_1(s) = 2 - 3s + \frac{s^2(3s+4)}{(s+1)^2} \tag{6.157}$$

$F(s)$ is then

$$F(s) = \frac{2}{s^2} - \frac{3}{s} + \frac{3s+4}{(s+1)^2} \tag{6.158}$$

We must now repeat this process for the term

$$\frac{3s+4}{(s+1)^2}$$

Fortunately we see that the term can be written as

$$\frac{3s+4}{(s+1)^2} = \frac{3(s+1)+1}{(s+1)^2} = \frac{3}{s+1} + \frac{1}{(s+1)^2} \tag{6.159}$$

The final answer is then

$$F(s) = \frac{2}{s^2} - \frac{3}{s} + \frac{3}{s+1} + \frac{1}{(s+1)^2} \tag{6.160}$$

6.6 POLES AND ZEROS

In this section we will discuss the many implications of a *pole-zero* description of a given rational function with real coefficients $F(s)$. We define the poles of $F(s)$ to be the roots of the denominator of $F(s)$. The zeros of $F(s)$ are defined as the roots of the numerator. In the complex s plane, a pole is denoted by a small cross, and a zero by a small circle. Thus, for the function

$$F(s) = \frac{s(s-1+j1)(s-1-j1)}{(s+1)^2(s+j2)(s-j2)} \tag{6.161}$$

the poles are at

$$s = -1 \quad \text{(double)}$$

$$s = -j2$$

$$s = +j2$$

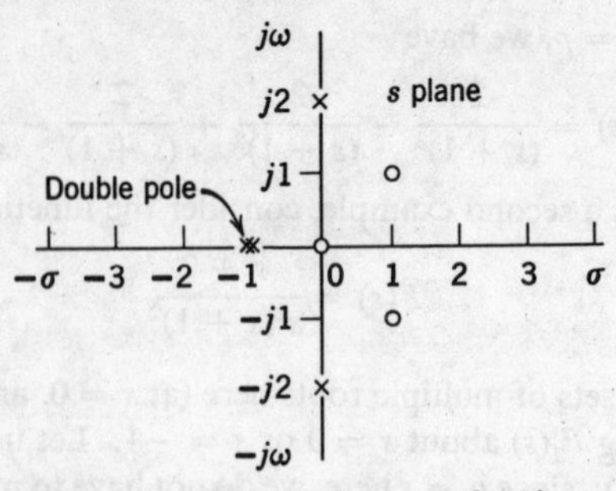

FIG. 6.7. Pole-zero diagram of $F(s)$.

and the zeros are at

$$s = 0$$

$$s = 1 + j1$$

$$s = 1 - j1$$

$$s = \infty$$

The poles and zeros of $F(s)$ are shown in Fig. 6.7.

Now let us consider some pole-zero diagrams corresponding to standard signals. For example, the unit step function is given in the complex frequency domain as

$$\mathcal{L}[u(t)] = \frac{1}{s} \tag{6.162}$$

and has a pole at the origin, as shown in Fig. 6.8*a*. The exponential signal $e^{-\sigma_0 t}$, where $\sigma_0 > 0$, has a transform

$$\mathcal{L}[e^{-\sigma_0 t}] = \frac{1}{s + \sigma_0} \tag{6.163}$$

which has a single pole at $s = -\sigma_0$, as indicated in Fig. 6.8*b*. The cosine function $\cos \omega_0 t$, whose transform is

$$\mathcal{L}[\cos \omega_0 t] = \frac{s}{s^2 + \omega_0^{\,2}} \tag{6.164}$$

has a zero at the origin and a pair of conjugate poles at $s = \pm j\omega_0$, as depicted in Fig. 6.8*c*. Figure 6.8*d* shows the pole-zero diagram corresponding to a damped cosine wave, whose transform is

$$\mathcal{L}[e^{-\sigma_0 t} \cos \omega_0 t] = \frac{s + \sigma_0}{(s + \sigma_0)^2 + \omega_0^{\,2}} \tag{6.165}$$

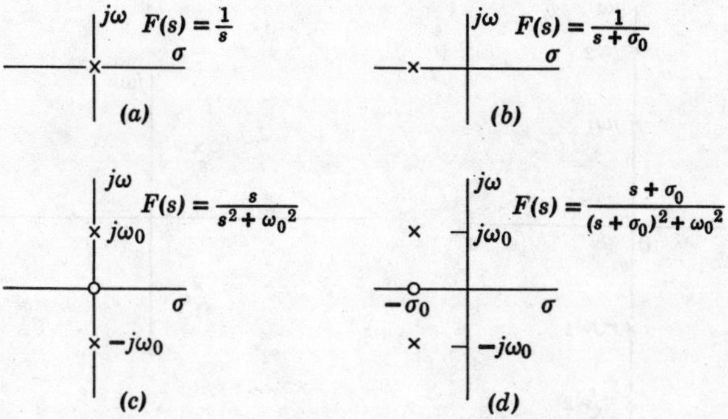

FIG. 6.8. Poles and zeros of various functions.

From these four pole-zero diagrams, we note that the poles corresponding to decaying exponential waves are on the $-\sigma$ axis and have zero imaginary parts. The poles and zeros corresponding to undamped sinusoids are on the $j\omega$ axis, and have zero real parts. Consequently, the poles and zeros for damped sinusoids must have real and imaginary parts that are both nonzero.

Now let us consider two exponential waves $f_1(t) = e^{-\sigma_1 t}$ and $f_2(t) = e^{-\sigma_2 t}$, where $\sigma_2 > \sigma_1 > 0$, so that $f_2(t)$ decays faster than $f_1(t)$, as shown in Fig. 6.9*a*. The transforms of the two functions are

$$F_1(s) = \frac{1}{s + \sigma_1}$$
$$F_2(s) = \frac{1}{s + \sigma_2} \tag{6.166}$$

as depicted by the pole-zero diagram in Fig. 6.9*b*. Note that the further the pole is from the origin on the $-\sigma$ axis, the more rapid the exponential decay. Now consider two sine waves, $\sin \omega_1 t$ and $\sin \omega_2 t$, where $\omega_2 > \omega_1 > 0$. Their corresponding poles are shown in Fig. 6.10. We note here

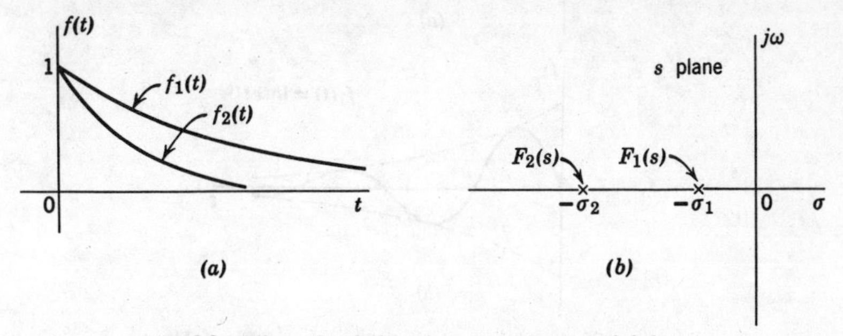

FIG. 6.9. Effect of pole location upon exponential decay.

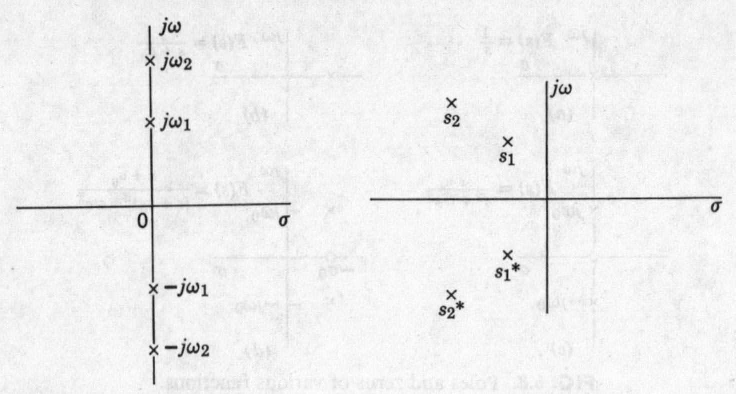

FIG. 6.10. Pole locations corresponding to $\sin \omega_2 t$ and $\sin \omega_1 t$.

FIG. 6.11

that the distance from the origin on the $j\omega$ axis represents frequency of oscillation; the greater the distance, the higher the frequency.

Using these rules of thumb, let us compare the time responses $f_1(t)$ and $f_2(t)$ corresponding to the pole pairs $\{s_1, s_1^*\}$ and $\{s_2, s_2^*\}$ shown in Fig. 6.11. We see that both pairs of poles corresponds to damped sinusoids. The damped sinusoid $f_1(t)$ has a smaller frequency of oscillation than $f_2(t)$ because the imaginary part of s_1 is less than the imaginary part of s_2. Also, $f_2(t)$ decays more rapidly than $f_1(t)$ because $\text{Re}\, s_1 > \text{Re}\, s_2$. The time responses $f_1(t)$ and $f_2(t)$ are shown in Fig. 6.12.

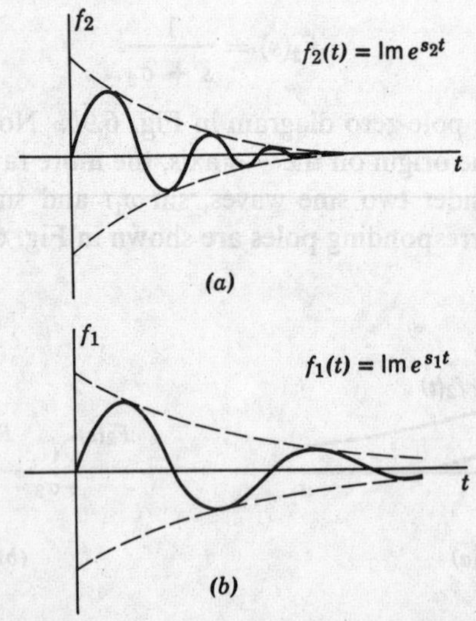

FIG. 6.12. Time responses for poles in Fig. 6.11.

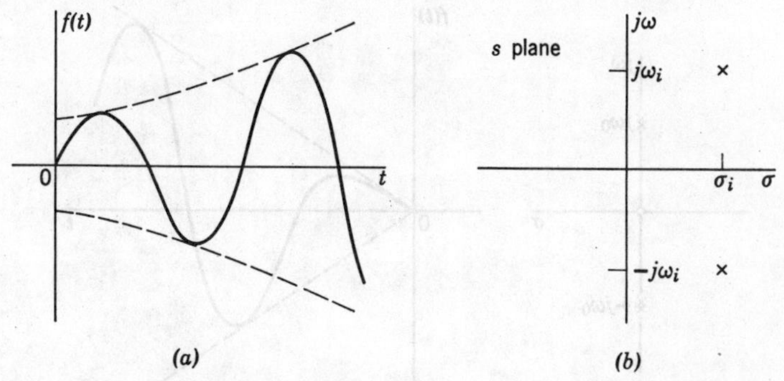

FIG. 6.13. Effect of right-half plane poles upon time response.

Let us examine more closely the effect of the positions of the poles in the complex frequency plane upon transient response. We denote a pole p_i as a complex number $p_i = \sigma_i + j\omega_i$. For a given function $F(s)$ with only first-order poles, consider the partial-fraction expansion

$$F(s) = \frac{K_0}{s - p_0} + \frac{K_1}{s - p_1} + \cdots + \frac{K_n}{s - p_n} \tag{6.167}$$

The inverse transform of $F(s)$ is

$$\begin{aligned} f(t) &= K_0 e^{p_0 t} + K_1 e^{p_1 t} + \cdots + K_n e^{p_n t} \\ &= K_0 e^{\sigma_0 t} e^{j\omega_0 t} + K_1 e^{\sigma_1 t} e^{j\omega_1 t} + \cdots + K_n e^{\sigma_n t} e^{j\omega_n t} \end{aligned} \tag{6.168}$$

In $f(t)$, we see that if the real part of a pole is positive, that is, $\sigma_i > 0$, then the corresponding term in the partial-fraction expansion

$$K_i e^{\sigma_i t} e^{j\omega_i t}$$

is an exponentially increasing sinusoid, as shown in Fig. 6.13*a*. We thus see that poles in the right half of the *s* plane (Fig. 6.13*b*) give rise to exponentially increasing transient responses. A system function that has poles in the right-half plane is, therefore, unstable. Another unstable situation arises if there is a pair of *double poles* on the $j\omega$ axis, such as for the function

$$F(s) = \frac{s}{(s^2 + \omega_0^2)^2} \tag{6.169}$$

whose pole-zero diagram is shown in Fig. 6.14*a*. The inverse transform of $F(s)$ is

$$f(t) = \frac{t}{2\omega_0} \sin \omega_0 t \tag{6.170}$$

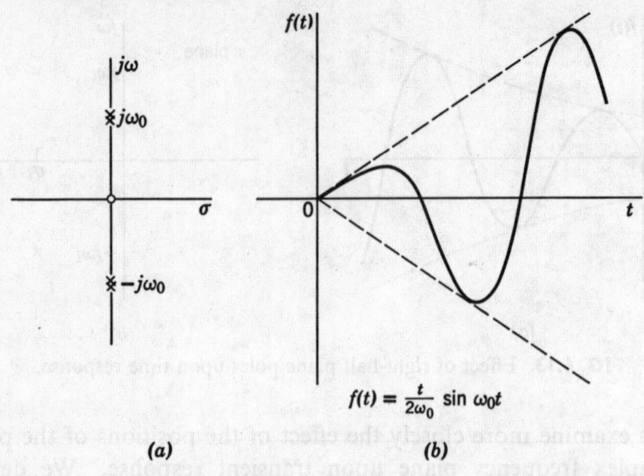

FIG. 6.14. Effect of double zeros on the $j\omega$ axis upon time response.

which is shown in Fig. 6.14*b*. It is apparent that a stable system function cannot also have multiple poles on the $j\omega$ axis.

Consider the system function $H(s) = N(s)/D(s)$. If we factor the numerator and denominator polynomials, we obtain

$$H(s) = \frac{H_0(s - z_0)(s - z_1) \cdots (s - z_n)}{(s - p_0)(s - p_1) \cdots (s - p_m)} \tag{6.171}$$

It is clear that $H(s)$ is completely specified in terms of its poles and zeros and an additional constant multiplier H_0. From the pole-zero plot of $H(s)$, we can obtain a substantial amount of information concerning the behavior of the system. As we have seen, we can determine whether the system is stable by checking the right-half plane for poles and the $j\omega$ axis for multiple poles. We obtain information concerning its transient behavior from the positions of its poles and zeros; and, as we will discuss in Section 8.1, the pole-zero diagram also gives us significant information concerning its steady-state ($j\omega$) amplitude and phase response. We thus see the importance of a pole-zero description.

Suppose we are given the poles and zeros of an excitation $E(s)$ and the system function $H(s)$. It is clear that the pole-zero diagram of the response function $R(s)$ is the superposition of the pole-zero diagrams of $H(s)$ and $E(s)$. Consider the system function

$$H(s) = \frac{H_0(s - z_0)}{(s - p_0)(s - p_1)} \tag{6.172}$$

and the excitation

$$E(s) = \frac{E_0(s - z_1)}{s - p_2} \tag{6.173}$$

Then the response function is

$$R(s) = \frac{H_0 E_0 (s - z_0)(s - z_1)}{(s - p_0)(s - p_1)(s - p_2)} \tag{6.174}$$

It is clear that $R(s)$ contains the poles and zeros of both $H(s)$ and $E(s)$, except in the case where a pole-zero cancellation occurs. As an example, let us take the system function

$$H(s) = \frac{2(s + 1)}{(s + 2 + j4)(s + 2 - j4)} \tag{6.175}$$

whose pole-zero diagram is shown in Fig. 6.15*a*. It is readily seen from the pole-zero diagrams of the excitation signals in Fig. 6.18*a* that the step response of the system has the pole-zero diagram indicated in Fig. 6.15*b*. The response to an excitation signal 3 cos 2*t* has the pole-zero representation of Fig. 6.15*c*. To specify completely the given function $F(s)$ on a pole-zero plot, we indicate the constant multiplier of the numerator on the plot itself.

Let us examine the significance of a pole or zero at the origin. We know that dividing a given function $H(s)$ by s corresponds to *integrating* the inverse transform $h(t) = \mathcal{L}^{-1}[H(s)]$. Since the division by s corresponds to a pole at the origin, we see that a pole at the origin implies an integration in the time domain. Because the inverse transform of $H(s)$ in Fig. 6.15*a* is the impulse response, placing a pole at the origin must correspond to the step response of the system. In similar manner, we deduce that a zero at the origin corresponds to a differentiation in the time domain. Suppose we consider the pole-zero diagram in Fig. 6.15*c* with the zero at the origin removed; then the resulting pole-zero plot is the response of the system to an excitation $E_0 \sin 2t$. Placing the zero at the origin must then give the response to an excitation $E_1 \cos 2t$, which, of

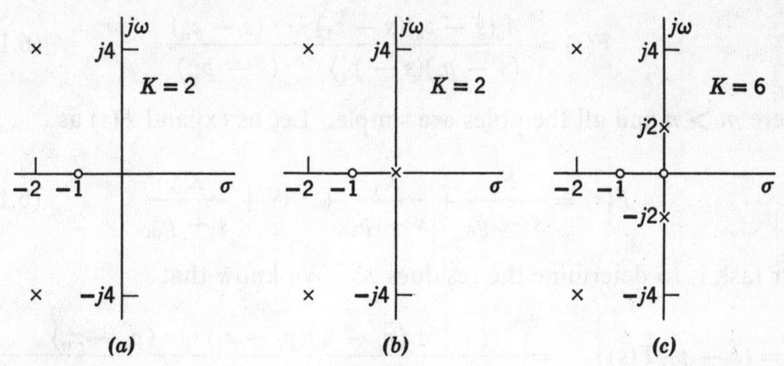

FIG. 6.15. (*a*) System function. (*b*) Response to unit step excitation. (*c*) Response to excitation $e(t) = 3 \cos 2t$.

course, is true. We therefore conclude that the system function of a differentiator must have a zero at the origin; that of an integrator must have a pole at the origin.

6.7 EVALUATION OF RESIDUES

Poles and zeros give a powerful graphical description of the behavior of a system. We have seen that the poles are the complex frequencies of the associated time responses. What role do the zeros play? To answer this question, consider the partial fraction expansion

$$F(s) = \sum_{i}^{N} \frac{K_i}{s - s_i} \tag{6.176}$$

where we shall assume that $F(s)$ has only simple poles and no poles at $s = \infty$.

The inverse transform is

$$f(t) = \sum_{i}^{N} K_i e^{s_i t} \tag{6.177}$$

It is clear that the time response $f(t)$ not only depends on the complex frequencies s_i but also on the constant multipliers K_i. These constants K_i are called *residues* when they are associated with first-order poles. We will show that the zeros as well as the poles play an important part in the determination of the residues K_i.

Earlier we discussed a number of different methods for obtaining the residues by partial-fraction expansion. Now we consider a graphical method whereby the residues are obtained directly from a pole-zero diagram. Suppose we are given

$$F(s) = \frac{A_0(s - z_0)(s - z_1) \cdots (s - z_n)}{(s - p_0)(s - p_1) \cdots (s - p_m)} \tag{6.178}$$

where $m > n$ and all the poles are simple. Let us expand $F(s)$ as

$$F(s) = \frac{K_0}{s - p_0} + \frac{K_1}{s - p_1} + \cdots + \frac{K_m}{s - p_m} \tag{6.179}$$

Our task is to determine the residues K_i. We know that

$$K_i = (s - p_i)\,F(s)\Big|_{s=p} = \frac{A_0(p_i - z_0)(p_i - z_1) \cdots (p_i - z_n)}{(p_i - p_0) \cdots (p_i - p_{i-1})(p_i - p_{i+1}) \cdots (p_i - p_m)} \tag{6.180}$$

When we interpret the Eq. 6.180 from a complex-plane viewpoint, we see that each one of the terms $(p_i - z_j)$ represents a vector drawn from a zero z_j to the pole in question, p_i. Similarly, the terms $(p_i - p_k)$, where $i \neq k$, represent vectors from the other poles to the pole p_i. In other words, the residue K_i of any pole p_i is equal to the ratio of the product of the vectors from the zeros to p_i, to the product of the vectors from the other poles to p_i. To illustrate this idea, let us consider the pole-zero plot of

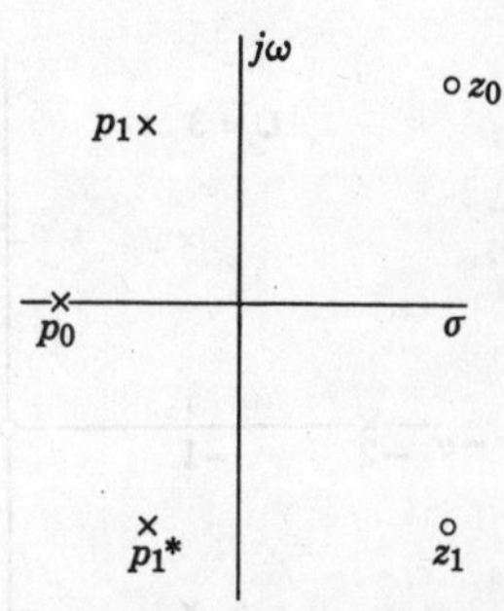

FIG. 6.16. Poles and zeros of $F(s)$.

$$F(s) = \frac{A_0(s - z_0)(s - z_1)}{(s - p_0)(s - p_1)(s - p_1^*)} \tag{6.181}$$

given in Fig. 6.16. The partial-fraction expansion of $F(s)$ is

$$F(s) = \frac{K_0}{s - p_0} + \frac{K_1}{s - p_1} + \frac{K_1^*}{s - p_1^*} \tag{6.182}$$

where the asterisk denotes *complex conjugate*. Let us find the residues K_0 and K_1 by means of the graphical method described. First we evaluate K_1 by drawing vectors from the poles and zeros to p_1, as shown in Fig. 6.17*a*. The residue K_1 is then

$$K_1 = \frac{A_0\mathbf{AB}}{\mathbf{CD}} \tag{6.183}$$

where symbols in boldface represent vectors. We know that the residue of the conjugate pole p_1^* is simply the conjugate of K_1 in Eq. 6.183. Next,

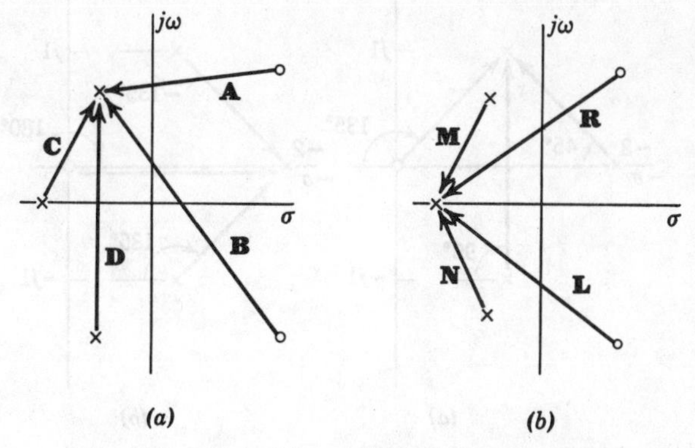

FIG. 6.17. Determining residues by vector method.

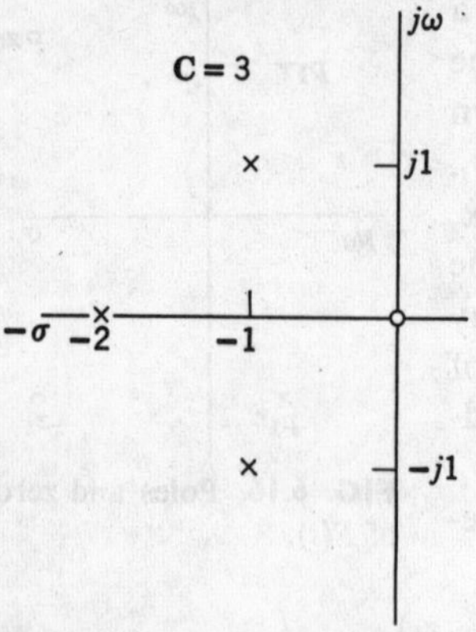

to evaluate K_0, we draw vectors from the poles and zeros to p_0, as indicated in Fig. 6.17*b*. We see that

$$K_0 = \frac{A_0\mathbf{RL}}{\mathbf{MN}} \tag{6.184}$$

With the use of a ruler and a protractor, we determine the lengths and the angles of the vectors so that the residues can be determined quickly and easily. Consider the following example. The pole-zero plot of

$$F(s) = \frac{3s}{(s+2)(s+1-j1)(s+1+j1)} \tag{6.185}$$

FIG. 6.18. Pole-zero diagram of $F(s)$.

is shown in Fig. 6.18. The partial-fraction expansion of $F(s)$ is

$$F(s) = \frac{K_1}{s+1-j1} + \frac{K_1^*}{s+1+j1} + \frac{K_2}{s+2} \tag{6.186}$$

First let us evaluate K_1. The phasors from the poles and zeros to the pole at $-1+j1$ are shown in Fig. 6.19*a*. We see that K_1 is

$$K_1 = 3 \times \frac{\sqrt{2}\,\angle 135^\circ}{\sqrt{2}\,\angle 45^\circ \times 2\angle 90^\circ} = \frac{3}{2}$$

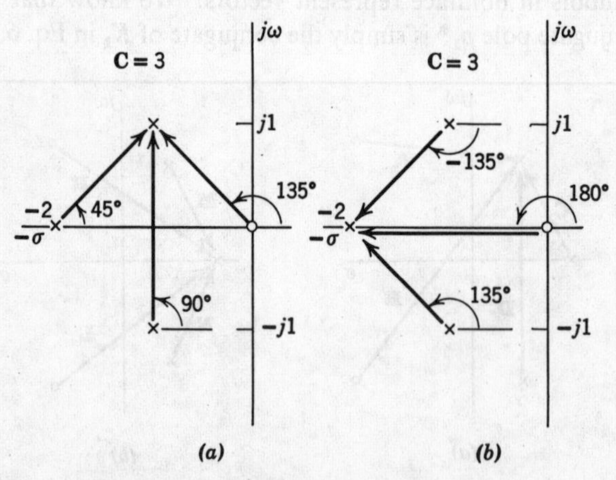

FIG. 6.19. Evaluation of residues of $F(s)$.

From Fig. 6.19*b* we find the value of the residue K_2 to be

$$K_2 = -\frac{3 \times 2}{\sqrt{2}\,\underline{/-135^\circ} \times \sqrt{2}\,\underline{/+135^\circ}} = -3$$

Therefore, the partial-fraction expansion of $F(s)$ is

$$F(s) = \frac{\frac{3}{2}}{s+1-j1} + \frac{\frac{3}{2}}{s+1+j1} - \frac{3}{s+2} \tag{6.187}$$

6.8 THE INITIAL AND FINAL VALUE THEOREMS

In this section we will discuss two very useful theorems of Laplace transforms. The first is the *initial value* theorem. It relates the initial value of $f(t)$ at $t = 0+$ to the limiting value of $s\,F(s)$ as s approaches infinity, that is,

$$\lim_{t \to 0+} f(t) = \lim_{s \to \infty} s\,F(s) \tag{6.188}$$

The only restriction is that $f(t)$ must be continuous or contain, at most, a step discontinuity at $t = 0$. In terms of the transform, $F(s) = \mathcal{L}[f(t)]$; this restriction implies that $F(s)$ must be a *proper fraction*, i.e., the degree of the denominator polynomial of $F(s)$ must be greater than the degree of the numerator of $F(s)$. Now consider the proof of the initial value theorem, which we give in two parts.

(*a*) The function $f(t)$ is continuous at $t = 0$, that is, $f(0-) = f(0+)$. From the relationship

$$\mathcal{L}[f'(t)] = \int_{0-}^{\infty} f'(t)e^{-st}\,dt = s\,F(s) - f(0-) \tag{6.189}$$

we obtain $$\lim_{s \to \infty} \mathcal{L}[f'(t)] = \lim_{s \to \infty} s\,F(s) - f(0-) = 0 \tag{6.190}$$

Therefore $$\lim_{s \to \infty} s\,F(s) = f(0-) = f(0+) \tag{6.191}$$

(*b*) The function $f(t)$ has a step discontinuity at $t = 0$. Let us represent $f(t)$ in terms of a continuous part $f_1(t)$ and a step discontinuity $D\,u(t)$, as shown in Fig. 6.20. We can then write $f(t)$ as

$$f(t) = f_1(t) + D\,u(t) \tag{6.192}$$

where $D = f(0+) - f(0-)$. The derivative of $f(t)$ is

$$f'(t) = f'_1(t) + D\,\delta(t) \tag{6.193}$$

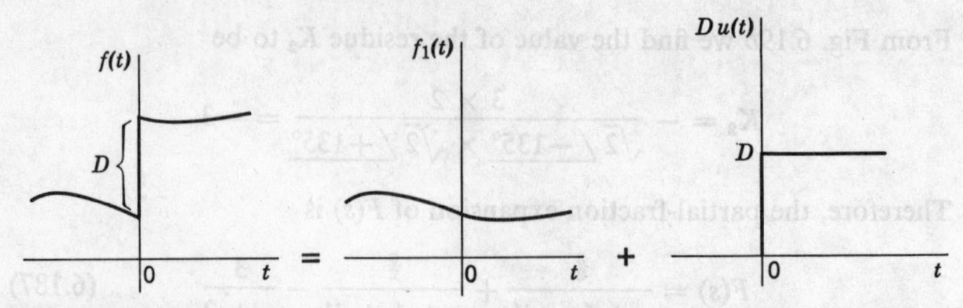

FIG. 6.20. Decomposition of a discontinuous function into a continuous function plus a step function.

Since $f_1(t)$ is continuous at $t = 0$, we know from part (a) that

$$\lim_{s\to\infty} s\,F_1(s) = f_1(0-) = f(0-) \tag{6.194}$$

Taking the Laplace transform of both sides of Eq. 6.193, we have

$$s\,F(s) - f(0-) = s\,F_1(s) - f_1(0-) + D \tag{6.195}$$

which simplifies to

$$s\,F(s) = s\,F_1(s) + D \tag{6.196}$$

Now, if we take the limit of Eq. 6.196 as $s \to \infty$ and let $f(0+) - f(0-) = D$, we have

$$\lim_{s\to\infty} s\,F(s) = \lim_{s\to\infty} s\,F_1(s) + f(0+) - f(0-) \tag{6.197}$$

By Eq. 6.194, we then obtain

$$\lim_{s\to\infty} s\,F(s) = f(0+) \tag{6.198}$$

Example 6.18. Given the function

$$F(s) = \frac{2(s+1)}{s^2+2s+5} \tag{6.199}$$

let us find $f(0+)$. Since the degree of the denominator is greater than the degree of the numerator of $F(s)$, the initial value theorem applies. Thus

$$\lim_{s\to\infty} s\,F(s) = \lim_{s\to\infty} \frac{2(s+1)s}{s^2+2s+5} = 2 \tag{6.200}$$

Since

$$\mathcal{L}^{-1}[F(s)] = 2e^{-t}\cos 2t \tag{6.201}$$

we see that $f(0+) = 2$.

Example 6.19. Now let us consider a case where the initial value theorem does not apply. Given the function

$$f(t) = \delta(t) + 3e^{-t} \tag{6.202}$$

we see that $f(0+) = 3$. The transform of $f(t)$ is

$$F(s) = 1 + \frac{3}{s+1} \tag{6.203}$$

so that
$$\lim_{s\to\infty} s\,F(s) = \lim_{s\to\infty} s\left(1 + \frac{3}{s+1} = \infty\right) \tag{6.204}$$

Next we consider the *final value* theorem, which states that

$$\lim_{t\to\infty} f(t) = \lim_{s\to 0} s\,F(s) \tag{6.205}$$

provided the poles of the denominator of $F(s)$ have negative or zero real parts, i.e., the poles of $F(s)$ must not be in the right half of the complex-frequency plane. The proof is quite simple. First

$$\int_{0-}^{\infty} f'(t)e^{-st}\,dt = s\,F(s) - f(0-) \tag{6.206}$$

Taking the limit as $s \to 0$ in the Eq. 6.206, we have

$$\int_{0-}^{\infty} f'(t)\,dt = \lim_{s\to 0} s\,F(s) - f(0-) \tag{6.207}$$

Evaluating the integral, we obtain

$$f(\infty) - f(0-) = \lim_{s\to 0} s\,F(s) - f(0-) \tag{6.208}$$

Consequently,
$$f(\infty) = \lim_{s\to 0} s\,F(s) \tag{6.209}$$

Example 6.20. Given the function

$$f(t) = 3u(t) + 2e^{-t} \tag{6.210}$$

which has the transform

$$F(s) = \frac{3}{s} + \frac{2}{s+1} = \frac{5s+3}{s(s+1)} \tag{6.211}$$

let us find the final value $f(\infty)$. Since the poles of $F(s)$ are at $s = 0$ and $s = -1$, we see the final value theorem applies. We find that

$$\lim_{s\to 0} s\,F(s) = 3 \tag{6.212}$$

which is the final value of $f(t)$ as seen from Eq. 6.210.

Example 6.21. Given
$$f(t) = 2e^{t} \tag{6.213}$$

we see that
$$\lim_{s\to\infty} f(t) = \infty \tag{6.214}$$

But from
$$s\,F(s) = \frac{2s}{s-1} \tag{6.215}$$

we have
$$\lim_{s\to 0} s\,F(s) = 0 \tag{6.216}$$

We see that the final value theorem does not apply in this case, because the pole $s = 1$ is in the right half of the complex-frequency plane.

TABLE 6.1

Laplace Transforms

	$f(t)$	$F(s)$
1.	$f(t)$	$F(s) = \int_{0-}^{\infty} f(t)e^{-st}\,dt$
2.	$a_1 f_1(t) + a_2 f_2(t)$	$a_1 F_1(s) + a_2 F_2(s)$
3.	$\frac{d}{dt} f(t)$	$s\,F(s) - f(0-)$
4.	$\frac{d^n}{dt^n} f(t)$	$s^n F(s) - \sum_{j=1}^{n} s^{n-j} f^{j-1}(0-)$
5.	$\int_{0-}^{t} f(\tau)\,d\tau$	$\frac{1}{s} F(s)$
6.	$\int_{0-}^{t}\int_{0-}^{t} f(\tau)\,d\tau\,d\sigma$	$\frac{1}{s^2} F(s)$
7.	$(-t)^n f(t)$	$\frac{d^n}{ds^n} F(s)$
8.	$f(t-a)\,u(t-a)$	$e^{-as} F(s)$
9.	$e^{at} f(t)$	$F(s-a)$
10.	$\delta(t)$	1
11.	$\frac{d^n}{dt^n}\,\delta(t)$	s^n
12.	$u(t)$	$\frac{1}{s}$
13.	t	$\frac{1}{s^2}$
14.	$\frac{t^n}{n!}$	$\frac{1}{s^{n+1}}$
15.	$e^{-\alpha t}$	$\frac{1}{s+\alpha}$
16.	$\frac{1}{\beta-\alpha}(e^{-\alpha t} - e^{-\beta t})$	$\frac{1}{(s+\alpha)(s+\beta)}$
17.	$\sin \omega t$	$\frac{\omega}{s^2+\omega^2}$

TABLE 6.1 (*cont.*)

18. $\cos \omega t$	$\dfrac{s}{s^2 + \omega^2}$
19. $\sinh at$	$\dfrac{a}{s^2 - a^2}$
20. $\cosh at$	$\dfrac{s}{s^2 - a^2}$
21. $e^{-\alpha t} \sin \omega t$	$\dfrac{\omega}{(s + \alpha^2) + \omega^2}$
22. $e^{-\alpha t} \cos \omega t$	$\dfrac{(s + \alpha)}{(s + \alpha)^2 + \omega^2}$
23. $\dfrac{e^{-\alpha t} t^n}{n!}$	$\dfrac{1}{(s + \alpha)^{n+1}}$
24. $\dfrac{t}{2\omega} \sin \omega t$	$\dfrac{s}{(s^2 + \omega^2)^2}$
25. $\dfrac{1}{\alpha^n} J_n(\alpha t);\ n = 0, 1, 2, 3, \ldots$ (Bessel function of first kind, nth order)	$\dfrac{1}{(s^2 + \alpha^2)^{1/2}[(s^2 + \alpha^2)^{1/2} - s]^{-n}}$
26. $(\pi t)^{-1/2}$	$s^{-1/2}$
27. t^k (k need not be an integer)	$\dfrac{\Gamma(k + 1)}{s^{k+1}}$

Problems

6.1 Find the Laplace transforms of

(*a*) $f(t) = \sin(\alpha t + \beta)$

(*b*) $f(t) = e^{-(t+\alpha)} \cos(\omega t + \beta)$

(*c*) $f(t) = (t^3 + 1)e^{-\alpha t}$

(*d*) $f(t) = K^t$; K is a real constant >1.

(*e*) $f(t) = (t^2 + ae^{-t})e^{-bt}$

(*f*) $f(t) = t\,\delta'(t)$

(*g*) $f(t) = u[t(t^2 - 4)]$

6.2 Find the Laplace transforms of

(*a*) $f(t) = \cos(t - \pi/4)\, u(t - \pi/4)$

(*b*) $f(t) = \cos(t - \pi/4)\, u(t)$

6.3 Find the Laplace transforms for the waveforms shown.

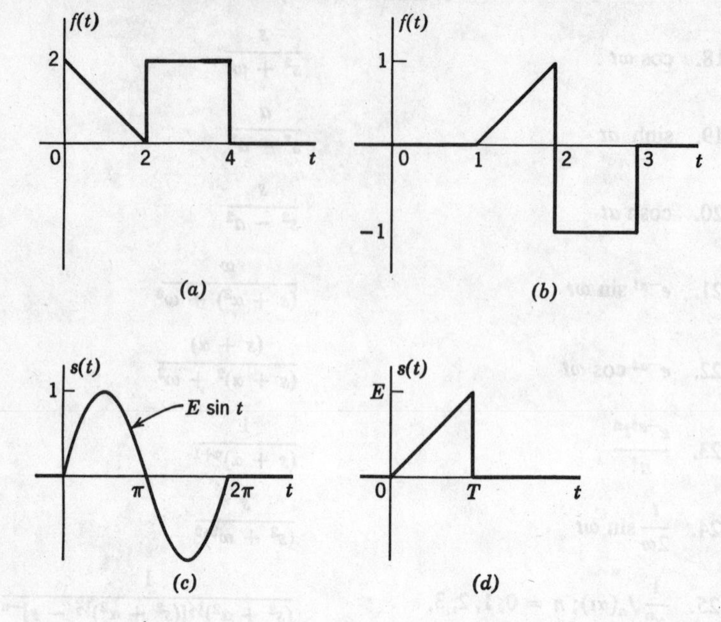

PROB. 6.3

6.4 Find the Laplace transforms for the derivatives of the waveforms in Prob. 6.3.

6.5 Find the inverse transforms for

(a) $$F(s) = \frac{2s + 9}{(s + 3)(s + 4)}$$

(b) $$F(s) = \frac{5s - 12}{s^2 + 4s + 13}$$

(c) $$F(s) = \frac{4s + 13}{s^2 + 4s - 5}$$

Note that all the inverse transforms may be obtained without resorting to *normal* procedures for partial-fraction expansions.

6.6 Find the Laplace transform of the waveforms shown. Use (*a*) the infinite series method and (*b*) the Laplace transform formula for periodic waveforms. Both answers should be in closed form and agree.

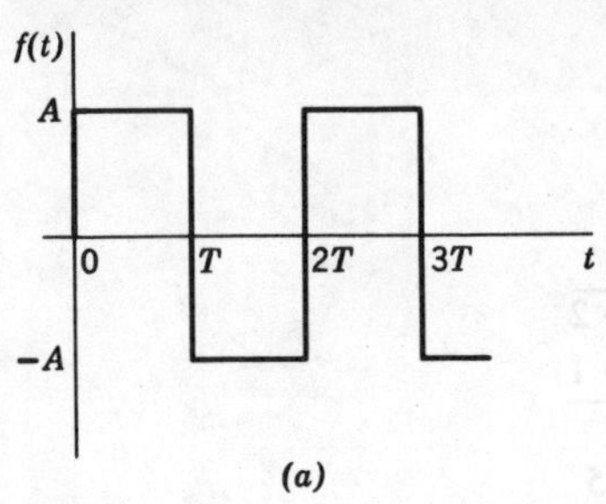

(*a*)

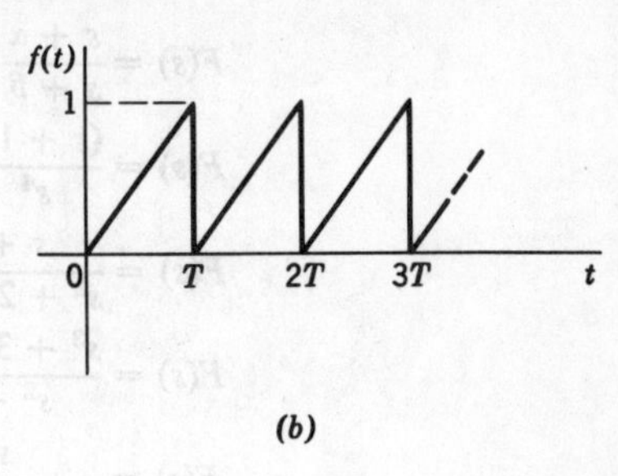

(*b*)

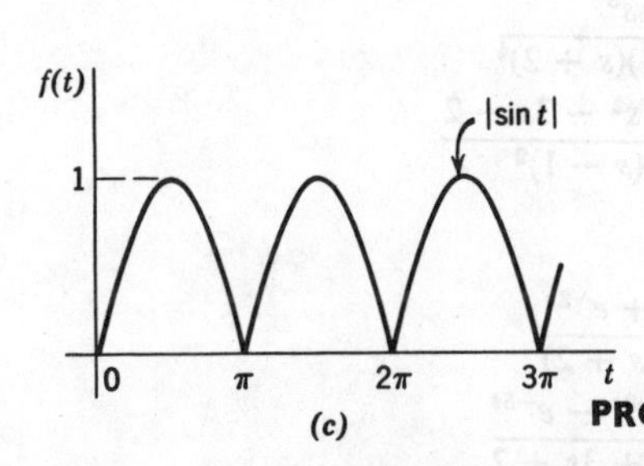

(*c*)

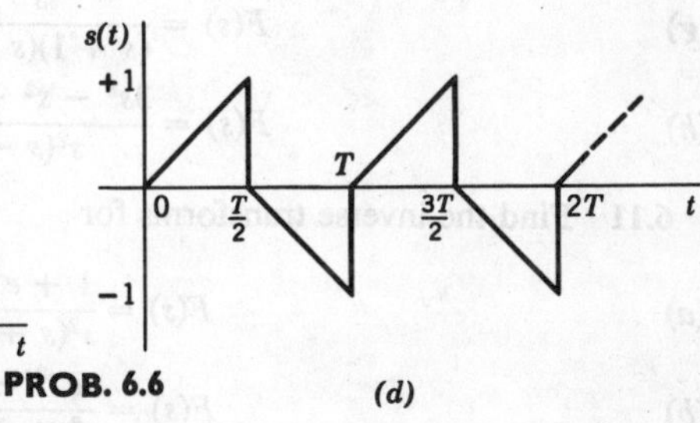

(*d*)

PROB. 6.6

6.7 Evaluate the definite integrals

(*a*) $$\int_0^{\pi} t \sin 2t \, dt$$

(*b*) $$\int_0^{\pi/2} t^2 \cos 3t \, dt$$

6.8 Solve the following differential equations using Laplace transforms

(*a*) $$x''(t) + 6x'(t) + 9x(t) = \cos 2t$$

(*b*) $$x''(t) + 3x'(t) + 2x(t) = \delta(t)$$

(*c*) $$x''(t) + 2x'(t) + 5x(t) = u(t)$$

(*d*) $$x''(t) + 5x'(t) + 4x(t) = (e^{-t} + e^{-4t})\, u(t)$$

(*e*) $$x''(t) + 2x'(t) = u(t) + e^{-t}\, u(t)$$

It is given that $x(0-) = x'(0-) = 0$ for all the equations.

6.9 Using Laplace transforms, solve the following sets of simultaneous equations:

(*a*)
$$2x'(t) + 4x(t) + y'(t) + 2y(t) = \delta(t)$$
$$x'(t) + 3x(t) + y'(t) + 2y(t) = 0$$

(*b*)
$$2x'(t) + x(t) + y'(t) + 4y(t) = 2e^{-t}$$
$$x'(t) + y'(t) + 8y(t) = \delta(t)$$

The initial conditions are all zero at $t = 0-$.

6.10 Find the inverse transforms for

(*a*) $$F(s) = \frac{s}{(s^2 + 9)(s + 3)}$$

(*b*) $$F(s) = \frac{s + \alpha}{s + \beta}$$

(*c*) $$F(s) = \frac{(s + 1)^3}{s^4}$$

(*d*) $$F(s) = \frac{s + 1}{s^2 + 2s + 2}$$

(*e*) $$F(s) = \frac{s^3 + 3s + 1}{s^2 + s}$$

(*f*) $$F(s) = \frac{s + 5}{(s + 1)^2(s + 2)^2}$$

(*g*) $$F(s) = \frac{s^3}{(s + 1)(s + 2)^4}$$

(*h*) $$F(s) = \frac{3s^3 - s^2 - 3s + 2}{s^2(s - 1)^2}$$

6.11 Find the inverse transforms for

(*a*) $$F(s) = \frac{1 + e^{-2s}}{s^2(s + 2)}$$

(*b*) $$F(s) = \frac{se^{-3s} - e^{-5s}}{s^2 + 3s + 2}$$

(*c*) $$F(s) = \frac{se^{-\alpha s}}{s + \beta}$$

6.12 Given the pole-zero diagrams shown in (*a*) and (*b*) of the figure, write the rational functions $F_1(s)$ and $F_2(s)$ as quotients of polynomials. What can you say about the relationship between the poles in the right-half plane and the signs of the coefficients of the denominator polynomials?

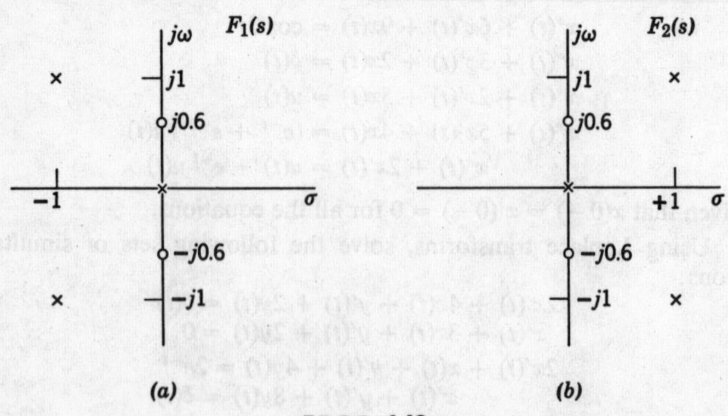

PROB. 6.12

6.13 For the pole-zero plots in Prob. 6.12, find the residues of the poles by the vector method.

6.14 For the pole-zero plots shown in the figure, find the residues of the poles.

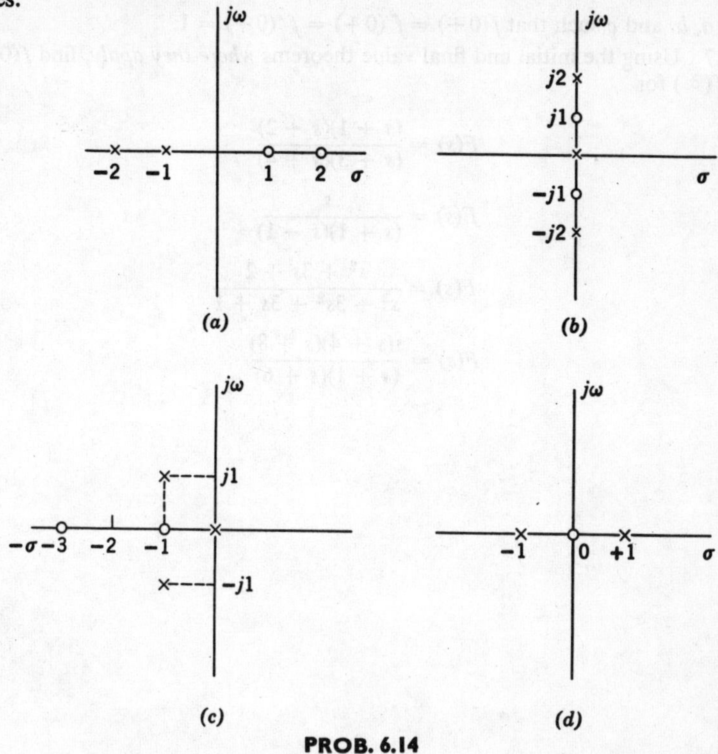

PROB. 6.14

6.15 Two response transforms, $R_1(s)$ and $R_2(s)$, have the pole-zero plots shown in (*a*) and (*b*) of the figures, respectively. In addition, it is known that $r_1(0+) = 4$ and $r_2(t) = 4$ for very large t. Find $r_1(t)$ and $r_2(t)$.

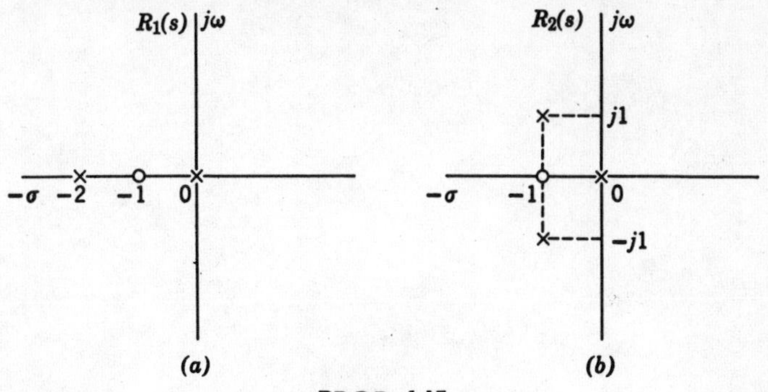

PROB. 6.15

6.16 It is given that

$$F(s) = \frac{as^2 + bs + c}{s^3 + 2s^2 + s + 1}$$

Find a, b, and c such that $f(0+) = f'(0+) = f''(0+) = 1$.

6.17 Using the initial and final value theorems *where they apply*, find $f(0+)$ and $f(\infty)$ for

(*a*)
$$F(s) = \frac{(s+1)(s+2)}{(s+3)(s+4)}$$

(*b*)
$$F(s) = \frac{s}{(s+1)(s-1)}$$

(*c*)
$$F(s) = \frac{s^2 + 3s + 2}{s^3 + 3s^2 + 3s + 1}$$

(*d*)
$$F(s) = \frac{s(s+4)(s+8)}{(s+1)(s+6)}$$

chapter 7

Transform methods in network analysis

7.1 THE TRANSFORMED CIRCUIT

In Chapter 5 we discussed the voltage-current relationships of network elements in the time domain. These basic relationships may also be represented in the complex-frequency domain. Ideal energy sources, for example, which were given in time domain as $v(t)$ and $i(t)$, may now be represented by their transforms $V(s) = \mathcal{L}[v(t)]$ and $I(s) = \mathcal{L}[i(t)]$. The resistor, defined by the v-i relationship

$$v(t) = R\,i(t) \tag{7.1}$$

is defined in the frequency domain by the transform of Eq. 7.1, or

$$V(s) = R\,I(s) \tag{7.2}$$

For an inductor, the defining v-i relationships are

$$\begin{aligned} v(t) &= L\frac{di}{dt} \\ i(t) &= \frac{1}{L}\int_{0-}^{t} v(\tau)\,d\tau + i(0-) \end{aligned} \tag{7.3}$$

Transforming both equations, we obtain

$$\begin{aligned} V(s) &= sLI(s) - L\,i(0-) \\ I(s) &= \frac{1}{sL}V(s) + \frac{i(0-)}{s} \end{aligned} \tag{7.4}$$

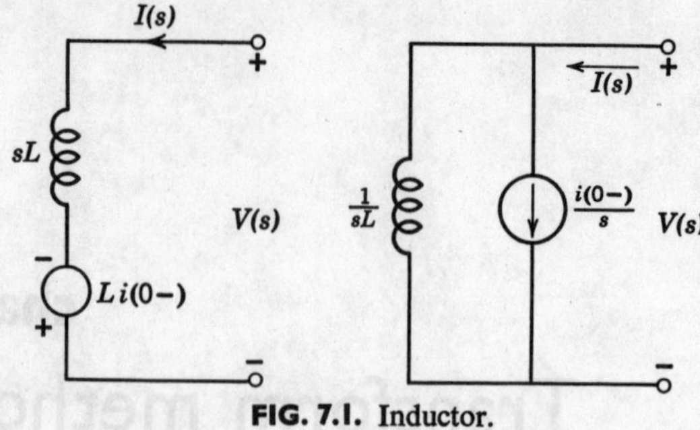

FIG. 7.1. Inductor.

The transformed circuit representation for an inductor is depicted in Fig. 7.1. For a capacitor, the defining equations are

$$v(t) = \frac{1}{C}\int_{0-}^{t} i(\tau)\, d\tau + v(0-)$$
$$i(t) = C\frac{dv}{dt} \tag{7.5}$$

The frequency domain counterparts of these equations are then

$$V(s) = \frac{1}{sC}I(s) + \frac{v(0-)}{s}$$
$$I(s) = sC\,V(s) - C\,v(0-) \tag{7.6}$$

as depicted in Fig. 7.2.

From this analysis we see that in the complex-frequency representation the network elements can be represented as *impedances* and *admittances* in series or parallel with energy sources. For example, from Eq. 7.4, we see that the complex-frequency impedance representation of an inductor is sL, and its associated admittance is $1/sL$. Similarly, the impedance of a capacitor is $1/sC$ and its admittance is sC. This fact is very useful in circuit analysis. Working from a transformed circuit diagram, we can write mesh and node equations on an impedance or admittance basis directly.

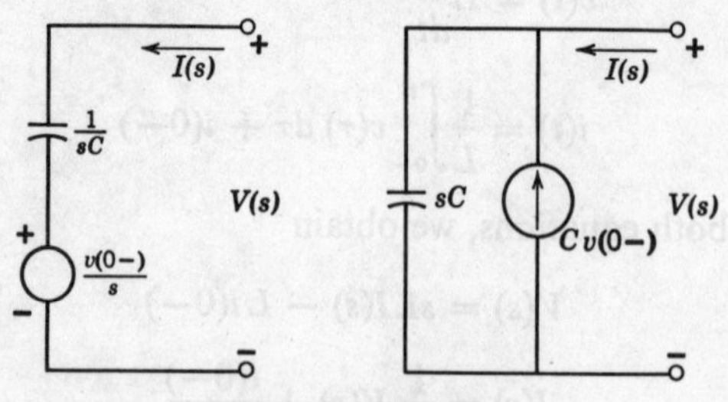

FIG. 7.2. Capacitor.

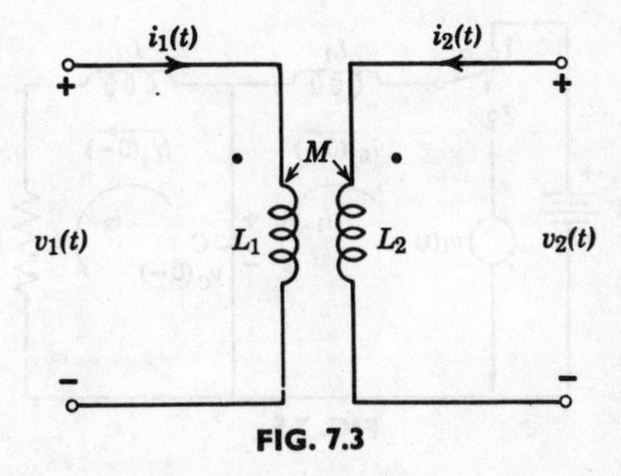

FIG. 7.3

The process of solving network differential equations with the use of transform methods has been given in Chapter 6. To analyze the circuit on a transform basis, the only additional step required is to represent all the network elements in terms of complex impedances or admittances with associated initial energy sources.

Consider the example of the transformer in Fig. 7.3. If we write the defining equations of the transformer directly in the time domain, we have

$$
\begin{aligned}
v_1(t) &= L_1 \frac{di_1}{dt} + M \frac{di_2}{dt} \\
v_2(t) &= M \frac{di_1}{dt} + L_2 \frac{di_2}{dt}
\end{aligned}
\tag{7.7}
$$

Transforming this set of equations, we obtain

$$
\begin{aligned}
V_1(s) &= sL_1\, I_1(s) - L_1\, i_1(0-) + sM\, I_2(s) - M\, i_2(0-) \\
V_2(s) &= sM\, I_1(s) - M\, i_1(0-) + sL_2\, I_2(s) - L_2\, i_2(0-)
\end{aligned}
\tag{7.8}
$$

This set of transform equations could also have been obtained by representing the circuit in Fig. 7.3 by its transformed equivalent given in Fig. 7.4. In general, the use of transformed equivalent circuits is considered an easier way to solve the problem.

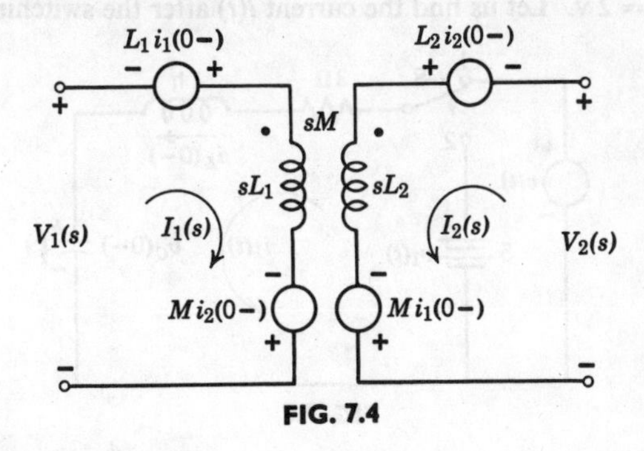

FIG. 7.4

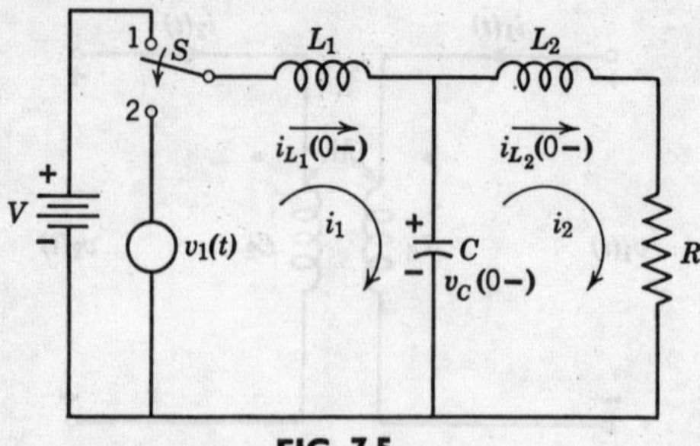

FIG. 7.5

Example 7.1. In Fig. 7.5, the switch is thrown from position 1 to 2 at $t = 0$. Assuming there is no coupling between L_1 and L_2, let us write the mesh equations from the transformed equivalent circuit in Fig. 7.6. The mesh equations are

$$
\begin{aligned}
V_1(s) + L_1\, i_{L_1}(0-) - \frac{v_C(0-)}{s} &= \left(sL_1 + \frac{1}{sC}\right) I_1(s) - \frac{1}{sC} I_2(s) \\
\frac{v_C(0-)}{s} + L_2\, i_{L_2}(0-) &= -\frac{1}{sC} I_1(s) + \left(\frac{1}{sC} + sL_2 + R\right) I_2(s)
\end{aligned}
\tag{7.9}
$$

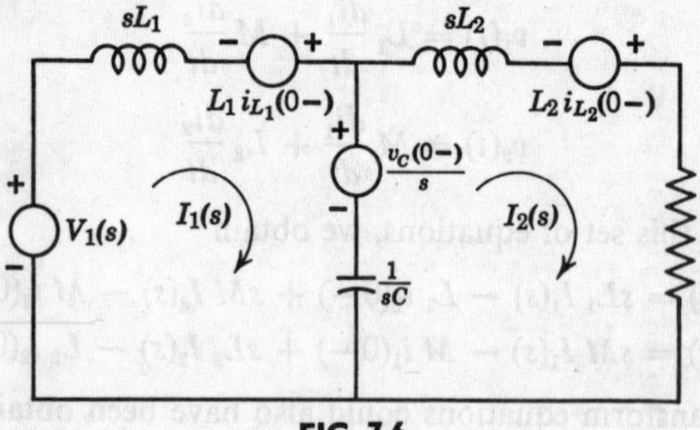

FIG. 7.6

Example 7.2. In Fig. 7.7, the switch is thrown from position 1 to 2 at time $t = 0$. Just before the switch is thrown, the initial conditions are $i_L(0-) = 2$ amp, $v_C(0-) = 2$ v. Let us find the current $i(t)$ after the switching action.

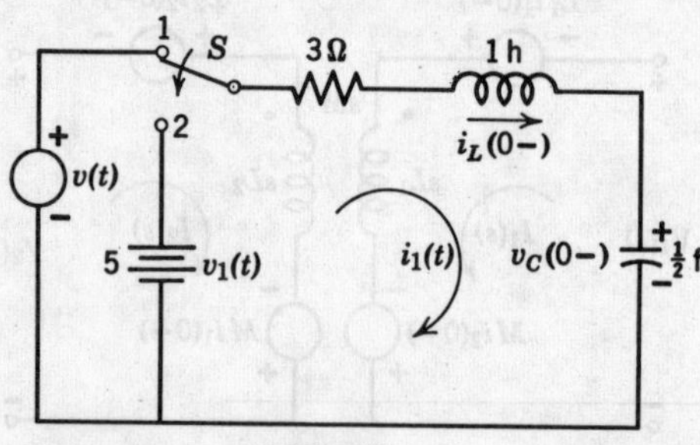

FIG. 7.7

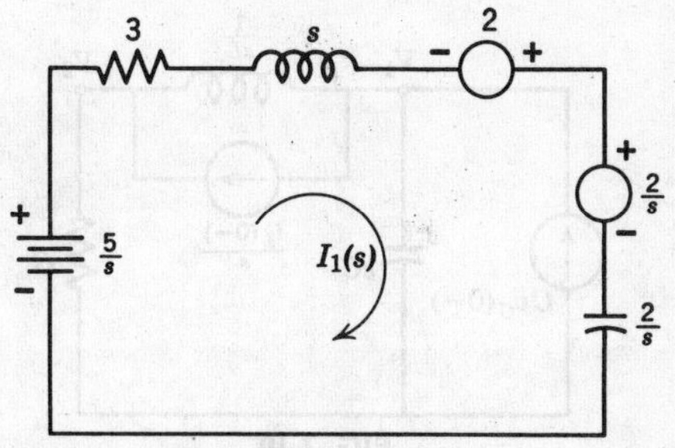

FIG. 7.8

Since the switch is closed at $t = 0$, we can regard the 5-v battery as an equivalent transformed source $5/s$. The circuit is now redrawn in Fig. 7.8 as a transformed circuit. The mesh equation for the circuit in Fig. 7.8 is

$$\frac{5}{s} - \frac{2}{s} + 2 = \left(3 + s + \frac{2}{s}\right) I(s) \tag{7.10}$$

Solving for $I(s)$, we have

$$I(s) = \frac{2s + 3}{(s + 1)(s + 2)} = \frac{1}{s + 1} + \frac{1}{s + 2} \tag{7.11}$$

Therefore,

$$i(t) = \mathcal{L}^{-1}[I(s)] = e^{-t} + e^{-2t} \tag{7.12}$$

Example 7.3. Consider the network in Fig. 7.9. At $t = 0$, the switch is opened. Let us find the node voltages $v_1(t)$ and $v_2(t)$ for the circuit. It is given that

$$L = \tfrac{1}{2}\text{ h} \qquad C = 1\text{ f}$$
$$G = 1\text{ mho} \qquad V = 1\text{ v}$$

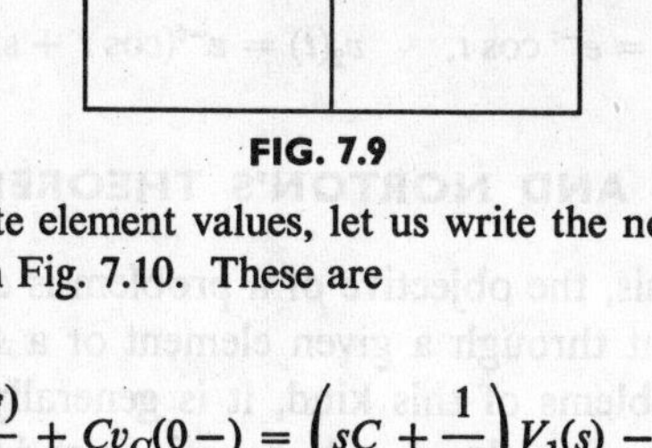

FIG. 7.9

Before we substitute element values, let us write the node equations for the transformed circuit in Fig. 7.10. These are

Node V_1:

$$-\frac{i_L(0-)}{s} + Cv_C(0-) = \left(sC + \frac{1}{sL}\right) V_1(s) - \frac{1}{sL} V_2(s)$$

Node V_2: (7.13)

$$\frac{i_L(0-)}{s} = -\frac{1}{sL} V_1(s) + \left(\frac{1}{sL} + G\right) V_2(s)$$

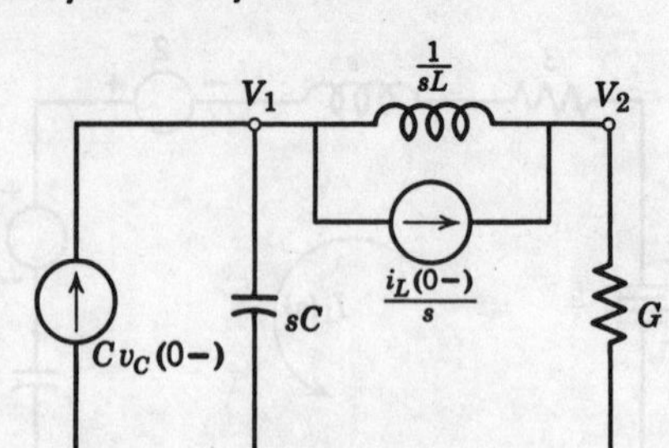

FIG. 7.10

If we assume that prior to the switch opening the circuit had been in steady state, then we have $v_C(0-) = 1$ v, $i_L(0-) = 1$ amp. Substituting numerical values into the set of node equations, we have

$$\begin{aligned} 1 - \frac{1}{s} &= \left(s + \frac{2}{s}\right) V_1(s) - \frac{2}{s} V_2(s) \\ \frac{1}{s} &= -\frac{2}{s} V_1(s) + \left(\frac{2}{s} + 1\right) V_2(s) \end{aligned} \tag{7.14}$$

Simplifying these equations, we obtain

$$\begin{aligned} s - 1 &= (s^2 + 2) V_1(s) - 2V_2(s) \\ 1 &= -2V_1(s) + (s + 2) V_2(s) \end{aligned} \tag{7.15}$$

Solving these equations simultaneously, we have

$$\begin{aligned} V_1(s) &= \frac{s + 1}{s^2 + 2s + 2} = \frac{s + 1}{(s + 1)^2 + 1} \\ V_2(s) &= \frac{s + 2}{s^2 + 2s + 2} = \frac{s + 2}{(s + 1)^2 + 1} \end{aligned} \tag{7.16}$$

so that the inverse transforms are

$$v_1(t) = e^{-t} \cos t, \qquad v_2(t) = e^{-t}(\cos t + \sin t) \tag{7.17}$$

7.2 THÉVENIN'S AND NORTON'S THEOREMS

In network analysis, the objective of a problem is often to determine a *single* branch current through a given element or a *single* voltage across an element. In problems of this kind, it is generally not practicable to write a complete set of mesh or node equations and to solve a system of equations for this one current or voltage. It is then convenient to use two very important theorems on equivalent circuits, known as *Thévenin's and Norton's theorems.*

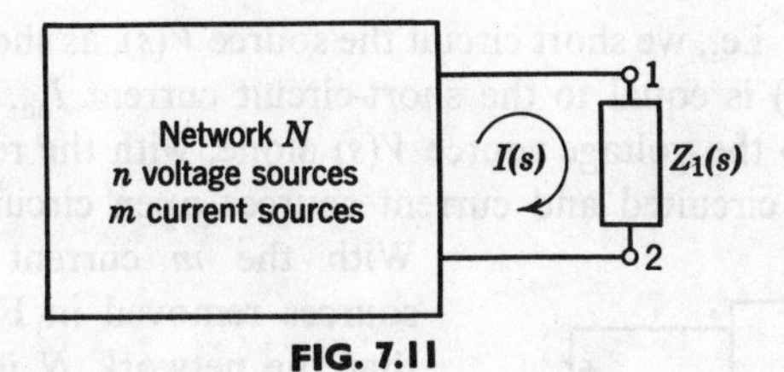

FIG. 7.11

Thévenin's theorem

From the standpoint of determining the current $I(s)$ through an element of impedance $Z_1(s)$, shown in Fig. 7.11, the rest of the network N can be replaced by an equivalent impedance $Z_e(s)$ in series with an equivalent voltage source $V_e(s)$, as depicted in Fig. 7.12. The equivalent impedance $Z_e(s)$ is the impedance "looking into" N from the terminals of $Z_1(s)$ when all voltage sources in N are short circuited and all current sources are open circuited. The equivalent voltage source $V_e(s)$ is the voltage which appears between the terminals 1 and 2 in Fig. 7.11, when the element $Z_1(s)$ is removed or open circuited. The only requirement for Thévenin's theorem is that the elements in Z_1 must not be magnetically coupled to any element in N.

The proof follows. The network in Fig. 7.11 contains n voltage and m current sources. We are to find the current $I(s)$ through an element that is not magnetically coupled to the rest of the circuit, and whose impedance is $Z_1(s)$. According to the compensation theorem,[1] we can replace $Z_1(s)$ by a voltage source $V(s)$, as shown in Fig. 7.13. Then by the superposition principle, we can think of the current $I(s)$ as the sum of two separate parts

$$I(s) = I_1(s) + I_2(s) \tag{7.18}$$

Let the current $I_1(s)$ be the current due to the n voltage and m current

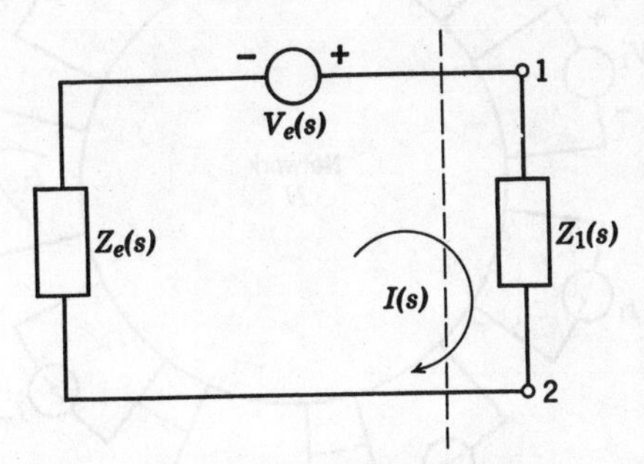

FIG. 7.12. Thévenin's equivalent circuit.

[1] See H. H. Skilling, *Electrical Engineering Circuits*, 2nd Ed. John Wiley and Sons, New York, 1965.

sources alone; i.e., we short circuit the source $V(s)$, as shown in Fig. 7.14*a*. Therefore $I_2(s)$ is equal to the short-circuit current I_{sc}. Let $I_2(s)$ be the current due to the voltage source $V(s)$ alone, with the rest of the voltage sources short circuited and current sources open circuited (Fig. 7.14*b*). With the m current and n voltage sources removed in Fig. 7.14*b*, we see that the network N is passive so that $I_2(s)$ is related to the source $V(s)$ by the relation

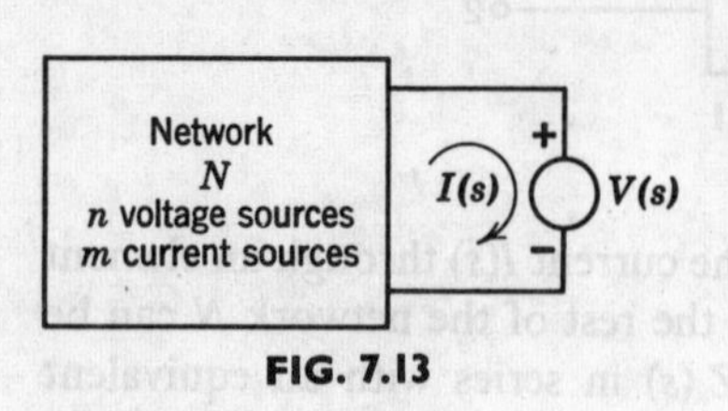

FIG. 7.13

$$I_2(s) = -\frac{V(s)}{Z_{in}(s)} \tag{7.19}$$

where $Z_{in}(s)$ is the input impedance of the circuit at the terminals of the source $V(s)$. We can now write $I(s)$ in Eq. 7.18 as

$$I(s) = I_{sc}(s) - \frac{V(s)}{Z_{in}(s)} \tag{7.20}$$

Since Eq. 7.20 must be satisfied in all cases, consider the particular case when we open circuit the branch containing $Z_1(s)$. Then $I(s) = 0$ and $V(s)$ is the open-circuit voltage $V_{oc}(s)$. From Eq. 7.20 we have

$$I_{sc}(s) = \frac{V_{oc}(s)}{Z_{in}(s)} \tag{7.21}$$

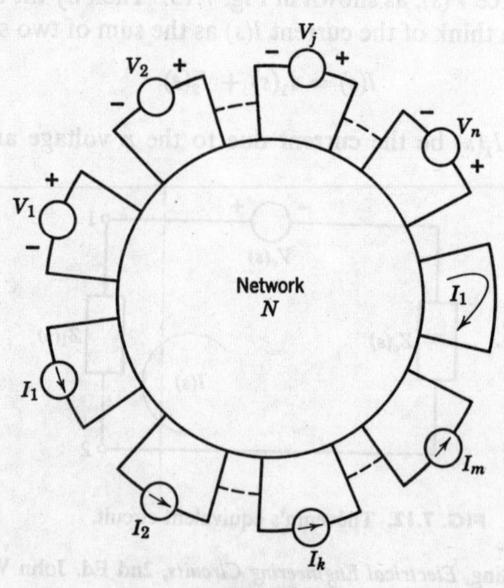

FIG. 7.14a

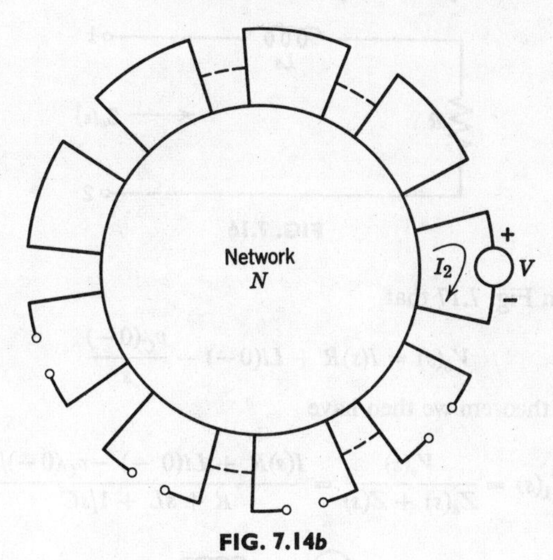

FIG. 7.14b

so that we can rewrite Eq. 7.20 as

$$Z_{\text{in}}(s)I(s) - V_{\text{oc}}(s) = -V(s) \tag{7.22}$$

In order to obtain the current $I(s)$ through $Z_1(s)$, the rest of the network N can be replaced by an equivalent source $V_e(s) = V_{\text{oc}}(s)$ in series, with an equivalent impedance $Z_e(s) = Z_{\text{in}}(s)$, as shown in Fig. 7.12.

Example 7.4. Let us determine by Thévenin's theorem the current $I_1(s)$ flowing through the capacitor in the network shown in Fig. 7.15. First, let us obtain

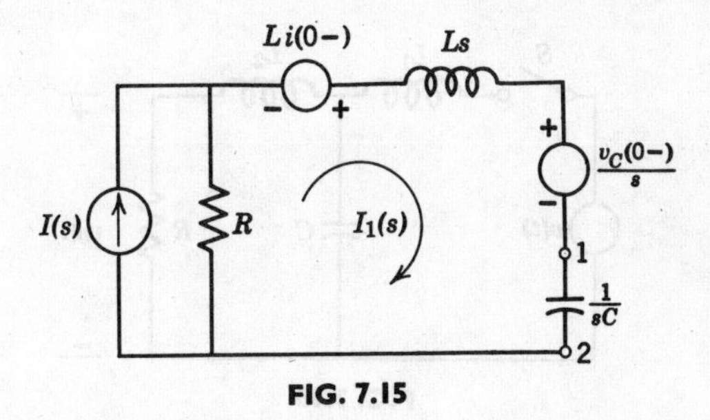

FIG. 7.15

$Z_e(s)$ by opening all current sources and short circuiting all voltage sources. Then, we have in the network in Fig. 7.16, where

$$Z_e(s) = R + sL \tag{7.23}$$

Next we find $V_e(s)$ by removing the capacitor so that the open-circuit voltage between the terminals 1 and 2 is $V_e(s)$, as shown in Fig. 7.17. We readily

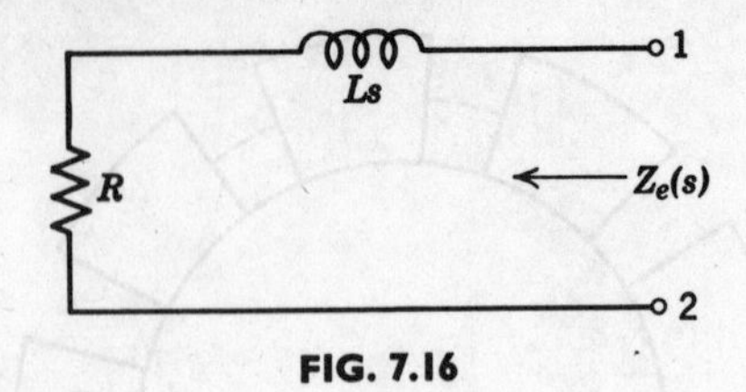

FIG. 7.16

determine from Fig. 7.17 that

$$V_e(s) = I(s)R + Li(0-) - \frac{v_C(0-)}{s} \tag{7.24}$$

By Thévenin's theorem we then have

$$I_1(s) = \frac{V_e(s)}{Z_e(s) + Z(s)} = \frac{I(s)R + Li(0\,-) - v_C(0-)/s}{R + sL + 1/sC} \tag{7.25}$$

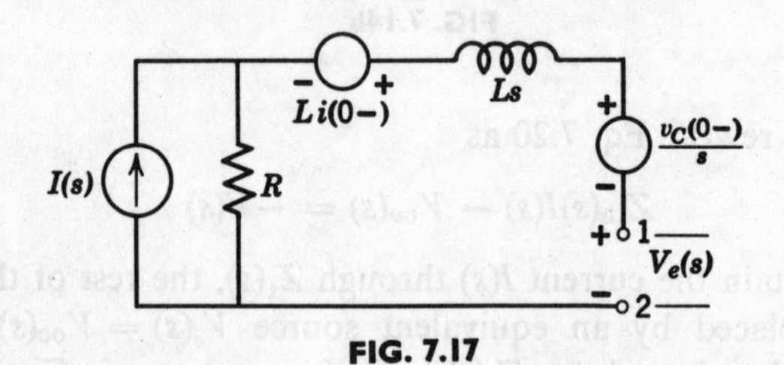

FIG. 7.17

Example 7.5. For the network in Fig. 7.18, let us determine the voltage $v_0(t)$ across the resistor by Thévenin's theorem. The switch closes at $t = 0$, and we assume that all initial conditions are zero at $t = 0$.

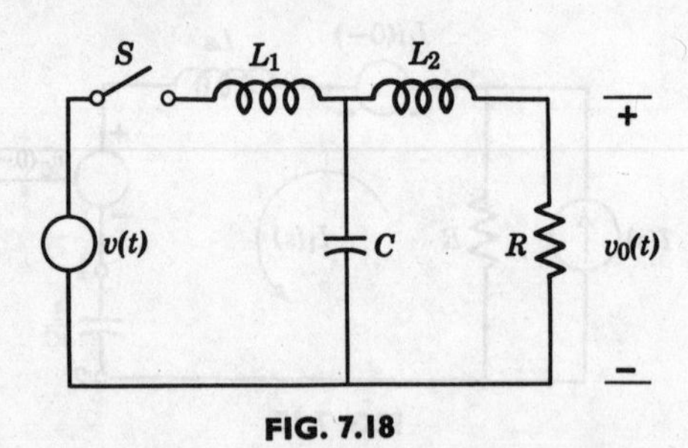

FIG. 7.18

First let us redraw the circuit in terms of its transformed representation, which is given in Fig. 7.19. We can almost determine by inspection that the Thévenin equivalent voltage source of the network to the left of N in Fig. 7.19 is

$$V_e(s) = \frac{V(s)(1/sC)}{L_1 s + 1/sC} \tag{7.26}$$

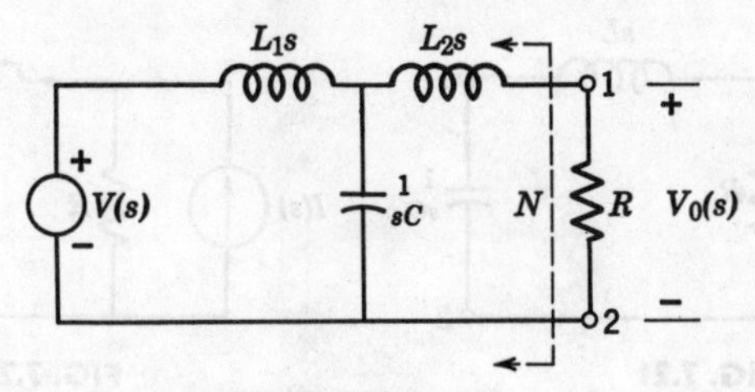

FIG. 7.19

and the input impedance to the left of N is

$$Z_e(s) = sL_2 + \frac{sL_1(1/sC)}{sL_1 + 1/sC} \tag{7.27}$$

We know that
$$V_0(s) = \frac{RV_e(s)}{Z_e(s) + R} \tag{7.28}$$

Therefore
$$V_0(s) = \frac{RV(s)(1/sC)}{(R + sL_2)(sL_1 + 1/sC) + L_1/C} \tag{7.29}$$

Finally
$$v_0(t) = \mathcal{L}^{-1}[V_0(s)] \tag{7.30}$$

Norton's theorem

When it is required to find the voltage across an element whose admittance is $Y_1(s)$, the rest of the network can be represented as an equivalent admittance $Y_e(s)$ in parallel with an equivalent current source $I_e(s)$, as shown in Fig. 7.20. The admittance $Y_e(s)$ is the reciprocal of the Thévenin impedance. The current $I_e(s)$ is that current which flows through a short circuit across $Y_1(s)$. From Fig. 7.20,

$$V_1(s) = \frac{I_e(s)}{Y_e(s) + Y_1(s)} \tag{7.31}$$

The element whose admittance is Y_1 must not be magnetically coupled to any element in the rest of the network.

Consider the network in Fig. 7.21. Let us find the voltage across the capacitor by Norton's theorem. First, the short-circuit current source $I_e(s)$ is found by placing a short circuit across the terminals 1 and 2 of the

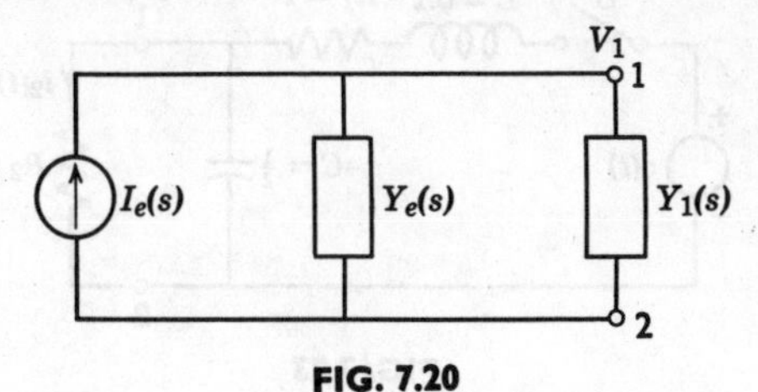

FIG. 7.20

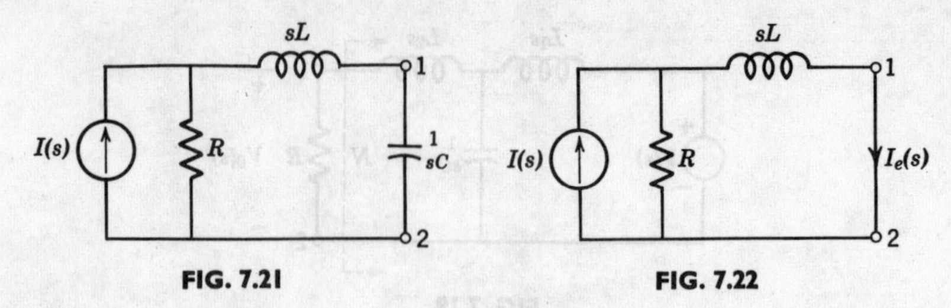

FIG. 7.21 FIG. 7.22

capacitor, as shown in Fig. 7.22. From Fig. 7.22 $I_e(s)$ is

$$I_e(s) = \frac{I(s)R}{sL + R} \tag{7.32}$$

The admittance $Y_e(s)$ is the reciprocal of the Thévenin impedance, or

$$Y_e(s) = \frac{1}{sL + R} \tag{7.33}$$

Then the voltage across the capacitor can be given as

$$V(s) = \frac{I_e(s)}{Y_e(s) + Y_C(s)} = \frac{I(s)R}{(sL + R)sC + 1} \tag{7.34}$$

Example 7.6. In the network in Fig. 7.23, the switch closes at $t = 0$. It is given that $v(t) = 0.1e^{-5t}$ and all initial currents and voltages are zero. Let us find the current $i_2(t)$ by Norton's theorem.

The transformed circuit is given in Fig. 7.24. To find the Norton equivalent current source, we short circuit points 1 and 2 in the network shown. Then $I_e(s)$ is the current flowing in the short circuit, or

$$I_e(s) = \frac{V(s)}{R_1 + sL} = \frac{0.1}{L(s + R/L)(s + 5)} = \frac{1}{(s + 5)(s + 10)} \tag{7.35}$$

The equivalent admittance of the circuit as viewed from points 1 and 2 is

$$Y_e(s) = sC + \frac{1}{R_1 + sL} = \frac{s^2LC + sR_1C + 1}{R_1 + sL} = \frac{0.5s^2 + 5s + 10}{s + 10} \tag{7.36}$$

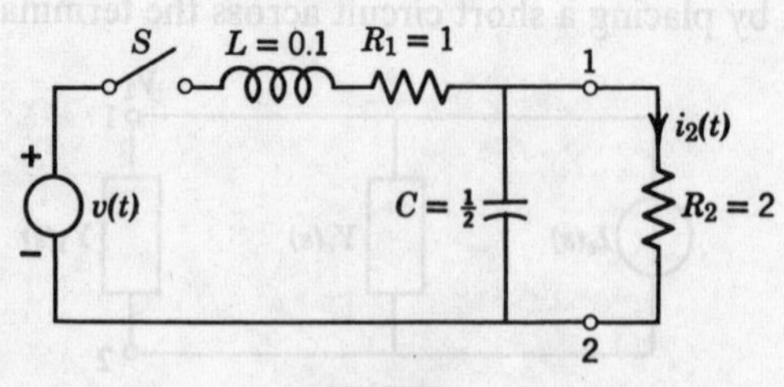

FIG. 7.23

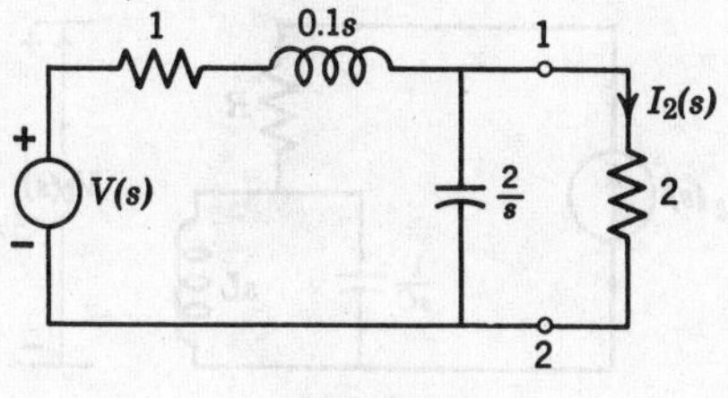

FIG. 7.24

$I_2(s)$ is then
$$I_2(s) = \frac{I_e(s)}{R_2[Y_e(s) + G_2]} \tag{7.37}$$
$$= \frac{1}{(s+5)^2(s+6)} \tag{7.38}$$

By inspection, we see that $I_2(s)$ can be written as
$$I_2(s) = \frac{(s+6)-(s+5)}{(s+5)^2(s+6)} = \frac{1}{(s+5)^2} - \frac{1}{(s+6)(s+5)} \tag{7.39}$$

Repeating this procedure, we then obtain
$$I_2(s) = \frac{1}{(s+5)^2} - \frac{1}{s+5} + \frac{1}{s+6} \tag{7.40}$$

Taking the inverse transform of $I_2(s)$, we finally obtain
$$i_2(t) = (te^{-5t} - e^{-5t} + e^{-6t})u(t) \tag{7.41}$$

7.3 THE SYSTEM FUNCTION

As we discussed earlier, a linear system is one in which the excitation $e(t)$ is related to the response $r(t)$ by a linear differential equation. When the Laplace transform is used in describing the system, the relation between the excitation $E(s)$ and the response $R(s)$ is an algebraic one. In particular, when we discuss initially inert systems, the excitation and response are related by the system function $H(s)$ as given the relation
$$R(s) = E(s)H(s) \tag{7.42}$$
We will discuss how a system function is obtained for a given network, and how this function can be used in determining the system response.

As mentioned in Chapter 1, the system function may assume many forms and may have special names such as *driving-point admittance*, *transfer*

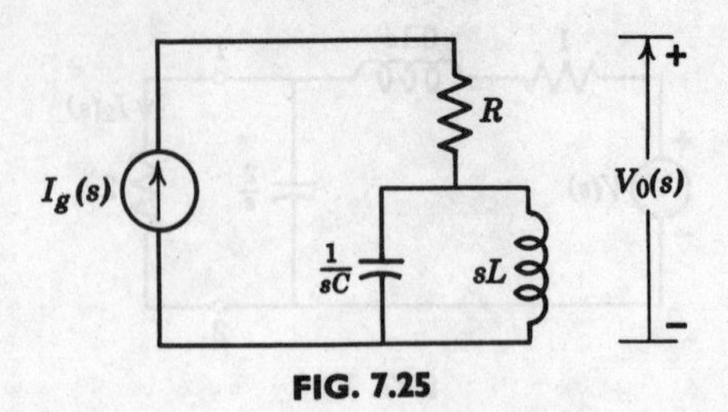

FIG. 7.25

impedance, voltage or *current-ratio transfer function.* This is because the form of the system function depends on whether the excitation is a voltage or current source, and whether the response is a specified current or voltage. We now discuss some specific forms of system functions.

Impedance

When the excitation is a current source and the response is a voltage, then the system function is an impedance. When both excitation and response are measured between the same pair of terminals, we have a driving-point impedance. An example of a driving-point impedance is given in Fig. 7.25, where

$$H(s) = \frac{V_0(s)}{I_g(s)} = R + \frac{(1/sC)sL}{sL + 1/sC} \tag{7.43}$$

Admittance

When the excitation is a voltage source and the response is a current, $H(s)$ is an admittance. In Fig. 7.26, the transfer admittance I_2/V_g is obtained from the network as

$$H(s) = \frac{I_2(s)}{V_g(s)} = \frac{1}{R_1 + sL + 1/sC_1} \tag{7.44}$$

Voltage-ratio transfer function

When the excitation is a voltage source and the response is also a voltage, then $H(s)$ is a voltage-ratio transfer function. In Fig. 7.27, the voltage-ratio transfer function $V_0(s)/V_g(s)$ is obtained as follows. We first find the

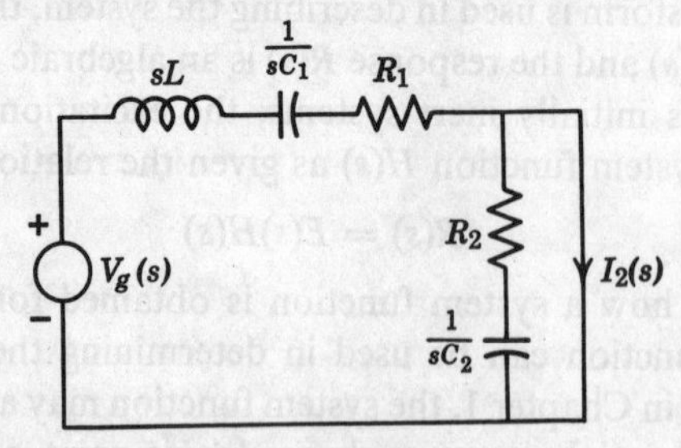

FIG. 7.26

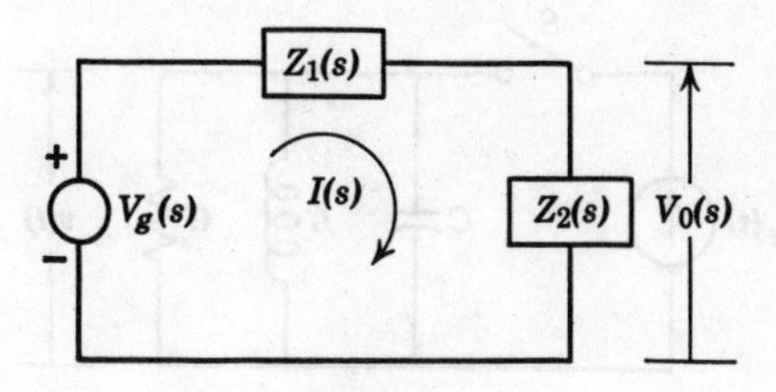

FIG. 7.27

current

$$I(s) = \frac{V_g(s)}{Z_1(s) + Z_2(s)} \tag{7.45}$$

Since

$$V_0(s) = Z_2(s)\, I(s) \tag{7.46}$$

then

$$\frac{V_0(s)}{V_g(s)} = \frac{Z_2(s)}{Z_1(s) + Z_2(s)} \tag{7.47}$$

Current-ratio transfer function

When the excitation is a current source and the response is another current in the network, then $H(s)$ is called a current-ratio transfer function. As an example, let us find the ratio I_0/I_g for the network given in Fig. 7.28. Referring to the depicted network, we know that

$$I_g(s) = I_1(s) + I_0(s), \qquad I_1(s)\frac{1}{sC} = I_0(s)(R + sL) \qquad (7.48, 7.49)$$

Eliminating the variable I_1, we find

$$I_g(s) = I_0(s)\left(1 + \frac{R + sL}{1/sC}\right) \tag{7.50}$$

so that the current-ratio transfer function is

$$\frac{I_0(s)}{I_g(s)} = \frac{1/sC}{R + sL + 1/sC} \tag{7.51}$$

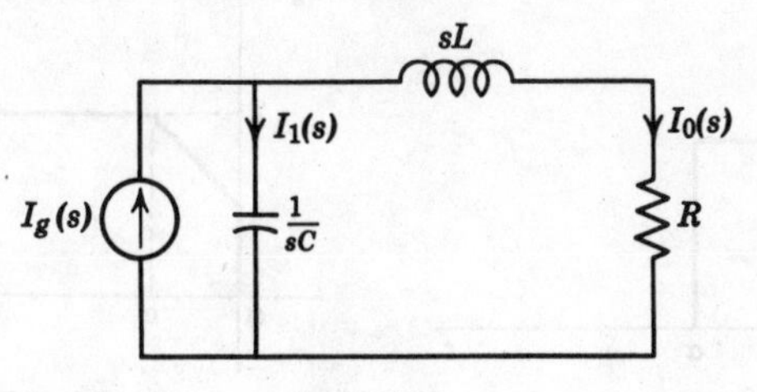

FIG. 7.28

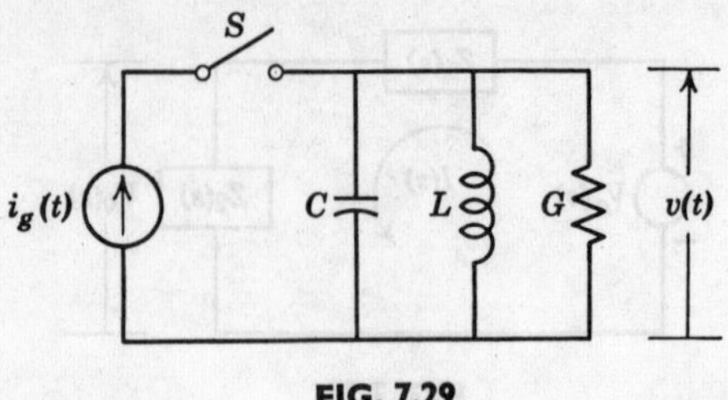

FIG. 7.29

From the preceding examples, we have seen that the system function is a function of the elements of the network alone, and is obtained from the network by a straightforward application of Kirchhoff's laws. Now let us obtain the response transform $R(s)$, given the excitation and the system function. Consider the network in Fig. 7.29, where the excitation is the current source $i_g(t)$ and the response is the voltage $v(t)$. We assume that the network is initially inert when the switch is closed at $t = 0$. Let us find the response $V(s)$ for the excitations:

1. $i_g(t) = (\sin \omega_0 t)\, u(t)$.
2. $i_g(t)$ is the square pulse in Fig. 7.30.
3. $i_g(t)$ has the waveform in Fig. 7.31.

First, we obtain the system function as

$$H(s) = \frac{1}{sC + 1/sL + G} = \frac{s}{C[s^2 + s(G/C) + 1/LC]} \tag{7.52}$$

1. $i_g(t) = (\sin \omega_0 t)\, u(t)$. The transform of $i_g(t)$ is

$$I_g(s) = \frac{\omega_0}{s^2 + \omega_0{}^2} \tag{7.53}$$

so that $$V(s) = I_g(s)H(s) = \frac{\omega_0}{s^2 + \omega_0{}^2} \cdot \frac{s}{C[s^2 + s(G/C) + 1/LC]} \tag{7.54}$$

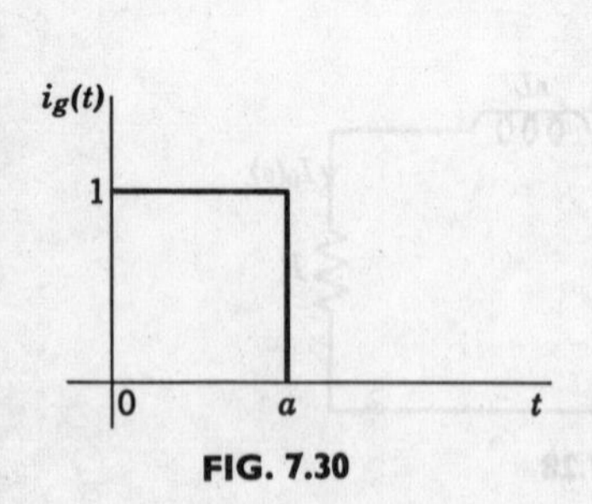

FIG. 7.30

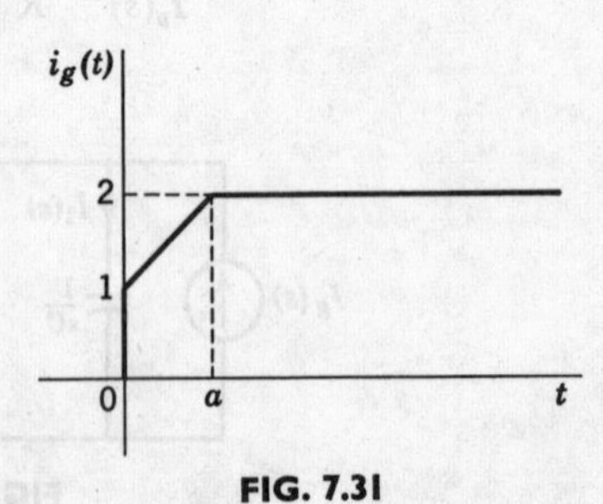

FIG. 7.31

2. For the square pulse in Fig. 7.30, $i_g(t)$ can be written as

$$i_g(t) = u(t) - u(t - a) \tag{7.55}$$

Its transform is

$$I_g(s) = \frac{1}{s}(1 - e^{-as}) \tag{7.56}$$

The response $V(s)$ is therefore given as

$$V(s) = \frac{1 - e^{-as}}{C[s^2 + s(G/C) + 1/LC]} \tag{7.57}$$

Note that in obtaining the inverse transform $\mathcal{L}^{-1}[V(s)]$, the factor e^{-as} must be regarded as only a delay factor in the time domain. Suppose we rewrite $V(s)$ in Eq. 7.57 as

$$V(s) = \frac{H(s)}{s}(1 - e^{-as}) \tag{7.58}$$

Then if we denote the inverse transform by $v_1(t)$,

$$v_1(t) = \mathcal{L}^{-1}\left[\frac{H(s)}{s}\right] \tag{7.59}$$

we obtain the time response

$$v(t) = v_1(t) - v_1(t - a)\, u(t - a) \tag{7.60}$$

Observe that $v_1(t)$ in Eq. 7.59 is the response of the system to a unit step excitation.

3. The waveform in Fig. 7.31 can be represented as

$$i_g(t) = u(t) + \frac{t}{a}\, u(t) - \frac{(t - a)}{a}\, u(t - a) \tag{7.61}$$

and its transform is

$$I_g(s) = \frac{1}{s}\left(1 + \frac{1}{as} - \frac{e^{-as}}{as}\right) \tag{7.62}$$

$V(s)$ is then

$$V(s) = \left(1 + \frac{1}{as} - \frac{e^{-as}}{as}\right)\frac{1}{C[s^2 + s(G/C) + 1/LC]} \tag{7.63}$$

If we denote by $v_2(t)$ the response of the system to a ramp excitation, we see that

$$\mathcal{L}^{-1}[V(s)] = v_1(t) + \frac{1}{a}\, v_2(t) - \frac{1}{a}\, v_2(t - a)\, u(t - a) \tag{7.64}$$

where $v_1(t)$ is the step response in Eq. 7.59.

Let us now discuss some further ramifications of the equation for the response $R(s) = H(s)\,E(s)$. Consider the partial-fraction expansion of $R(s)$:

$$R(s) = \sum_i \frac{A_i}{s - s_i} + \sum_j \frac{B_j}{s - s_j} \tag{7.65}$$

where s_i represent poles of $H(s)$, and s_j represent poles of $E(s)$. Taking the inverse transform of $R(s)$, we obtain

$$r(t) = \sum_i A_i e^{s_i t} + \sum_j B_j e^{s_j t} \tag{7.66}$$

The terms $A_i e^{s_i t}$ are associated with the system $H(s)$ and are called free response terms. The terms $B_j e^{s_j t}$ are due to the excitation and are known as forced response terms. The frequencies s_i are the *natural frequencies* of the system and s_j are the *forced frequencies.* It is seen from our discussion of system stability in Chapter 5 that the natural frequencies of a passive network have real parts which are zero or negative. In other words, if we denote s_i as $s_i = \sigma_i + j\omega_i$, then $\sigma_i \leq 0$.

Example 7.7. For the initially inert network in Fig. 7.32, the excitation is $v_g(t) = \frac{1}{2} \cos t\, u(t)$. Let us find the response $v_0(t)$ and determine the free and forced response parts of $v_0(t)$. The system function is

$$\frac{V_0(s)}{V_g(s)} = \frac{1/sC}{R + 1/sC} = \frac{2}{s + 2} \tag{7.67}$$

Since $V_g(s)$ is

$$V_g(s) = \frac{1}{2}\left(\frac{s}{s^2 + 1}\right) \tag{7.68}$$

the response is then

$$V_0(s) = V_g(s)\,H(s) = \frac{s}{(s_2 + 1)(s + 2)} = -\frac{0.4}{s + 2} + \frac{0.4s + 0.2}{s^2 + 1} \tag{7.69}$$

We next obtain the inverse transform

$$v_0(t) = -0.4e^{-2t} + 0.4 \cos t + 0.2 \sin t \tag{7.70}$$

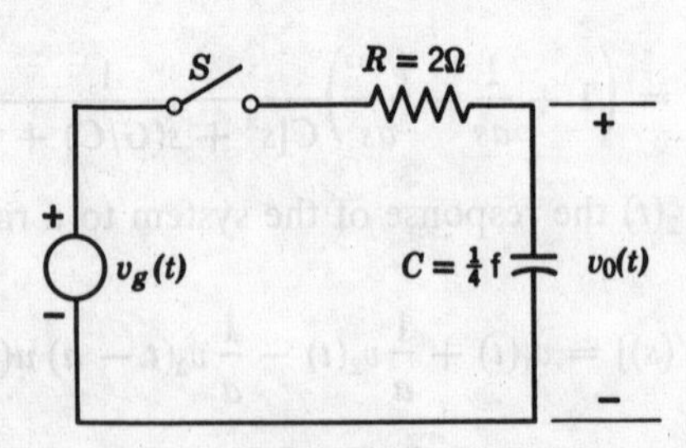

FIG. 7.32

It is apparent that the free response is $-0.4e^{-2t}$, and the forced response is $0.4 \cos t + 0.2 \sin t$.

As a final topic in our discussion, let us consider the basis of operation for the *R-C* differentiator and integrator shown in Figs. 7.33*a* and 7.33*b*. We use the Fourier transform in our analysis here, so that the system function is given as $H(j\omega)$, where ω is the ordinary radian frequency variable. Consider first the system function of the differentiator in Fig. 7.33*a*.

$$\frac{V_0(j\omega)}{V_g(j\omega)} = \frac{R}{R + 1/j\omega C} = \frac{j\omega RC}{j\omega RC + 1} \tag{7.71}$$

Let us impose the condition that

$$\omega RC \ll 1 \tag{7.72}$$

We have, approximately,

$$\frac{V_0(j\omega)}{V_0(j\omega)} \simeq j\omega RC \tag{7.73}$$

Then the response $V_0(j\omega)$ can be expressed as

$$V_0(j\omega) \simeq RC[j\omega\, V_g(j\omega)] \tag{7.74}$$

Taking the inverse transform of $V_0(j\omega)$, we obtain

$$v_0(t) \simeq RC \frac{d}{dt} v_g(t) \tag{7.75}$$

Note that the derivation of Eq. 7.75 depends upon the assumption that the *R-C* time constant is much less than unity. This is a necessary condition.

Next, for the *R-C* integrator in Fig. 7.33*b*, the voltage-ratio transfer function is

$$\frac{V_0(j\omega)}{V_g(j\omega)} = \frac{1/j\omega C}{1/j\omega C + R} = \frac{1}{j\omega CR + 1} \tag{7.76}$$

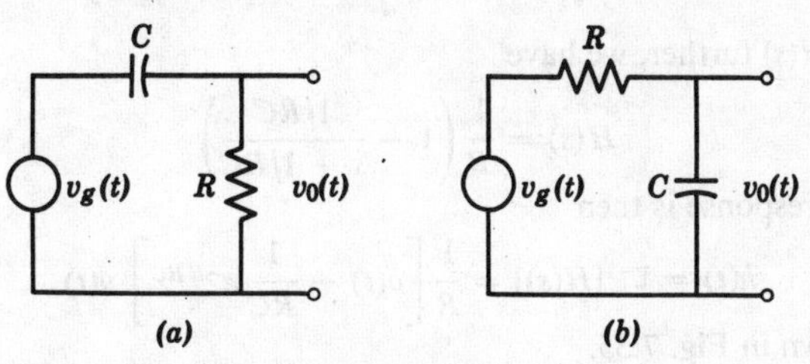

FIG. 7.33. (*a*) *R-C* differentiator. (*b*) *R-C* integrator.

If we assume that $\omega RC \gg 1$, then

$$V_0(j\omega) \simeq \frac{1}{j\omega RC} V_g(j\omega) \tag{7.77}$$

Under these conditions, the inverse transform is

$$v_0(t) \simeq \frac{1}{RC}\int_0^t v_g(\tau)\, d\tau \tag{7.78}$$

so that the *R-C* circuit in Fig. 7.33*b* is approximately an integrating circuit.

7.4 THE STEP AND IMPULSE RESPONSES

In this section we will show that the impulse response $h(t)$ and the system function $H(s)$ constitute a transform pair, so that we can obtain step and impulse responses directly from the system function.

We know, first of all, that the transform of a unit impulse $\delta(t)$ is unity, i.e., $\mathcal{L}[\delta(t)] = 1$. Suppose the system excitation were a unit impulse, then the response $R(s)$ would be

$$R(s) = E(s)\, H(s) = H(s) \tag{7.79}$$

We thus see that the impulse response $h(t)$ and the system function $H(s)$ constitute a transform pair, that is,

$$\begin{aligned} \mathcal{L}[h(t)] &= H(s) \\ \mathcal{L}^{-1}[H(s)] &= h(t) \end{aligned} \tag{7.80}$$

Since the system function is usually easy to obtain, it is apparent that we can find the impulse response of a system by taking the inverse transform of $H(s)$.

Example 7.8. Let us find the impulse response of the current $i(t)$ in the *R-C* circuit in Fig. 7.34. The system function is

$$H(s) = \frac{I(s)}{V_g(s)} = \frac{1}{R + 1/sC} = \frac{s}{R(s + 1/RC)} \tag{7.81}$$

Simplifying $H(s)$ further, we have

$$H(s) = \frac{1}{R}\left(1 - \frac{1/RC}{s + 1/RC}\right) \tag{7.82}$$

The impulse response is then

$$h(t) = \mathcal{L}^{-1}[H(s)] = \frac{1}{R}\left[\delta(t) - \frac{1}{RC} e^{-t/RC}\right] u(t) \tag{7.83}$$

which is shown in Fig. 7.35.

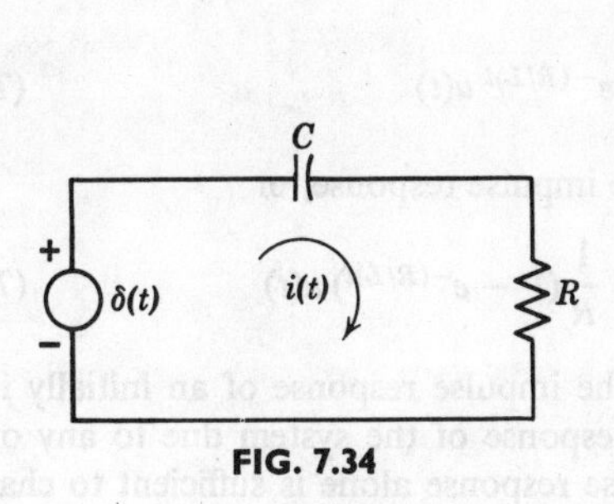

FIG. 7.34

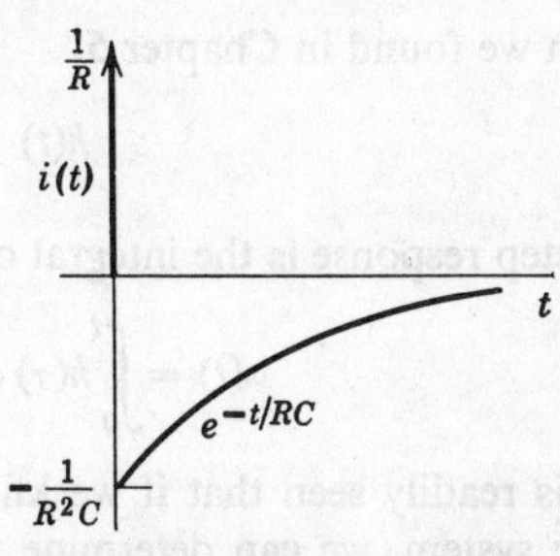

FIG. 7.35. Current impulse response of *R-C* network in Fig. 7.34.

Since the step response is the integral of the impulse response, we can use the integral property of Laplace transforms to obtain the step response as

$$\alpha(t) = \mathcal{L}^{-1}\left[\frac{H(s)}{s}\right] \tag{7.84}$$

where $\alpha(t)$ denotes the step response. Similarly, we obtain the unit ramp response from the equation

$$\gamma(t) = \mathcal{L}^{-1}\left[\frac{H(s)}{s^2}\right] \tag{7.85}$$

where $\gamma(t)$ denotes the ramp response. From this discussion, it is clear that a knowledge of the system function provides sufficient information to obtain all the transient response data that are needed to characterize the system.

Example 7.9. Let us find the current step response of the *R-L* circuit in Fig. 7.36. Since $I(s)$ is the response and $V_g(s)$ is the excitation, the system function is

$$H(s) = \frac{I(s)}{V_g(s)} = \frac{1}{R + sL} \tag{7.86}$$

Therefore $H(s)/s$ is

$$\frac{H(s)}{s} = \frac{1}{s(R + sL)} = \frac{1}{R}\left(\frac{1}{s} - \frac{1}{s + R/L}\right) \tag{7.87}$$

The step response $\alpha(t)$ is now obtained as

$$\alpha(t) = \frac{1}{R}(1 - e^{-(R/L)t})\, u(t) \tag{7.88}$$

To check this result, let us consider the impulse response of the *R-L* circuit, which we found in Chapter 5.

$$h(t) = \frac{1}{L} e^{-(R/L)t} u(t) \tag{7.89}$$

The step response is the integral of the impulse response, or

$$\alpha(t) = \int_0^t h(\tau)\, d\tau = \frac{1}{R}(1 - e^{-(R/L)t})\, u(t) \tag{7.90}$$

It is readily seen that if we know the impulse response of an initially inert linear system, we can determine the response of the system due to any other excitation. In other words, the impulse response alone is sufficient to characterize the system from the standpoint of excitation and response.

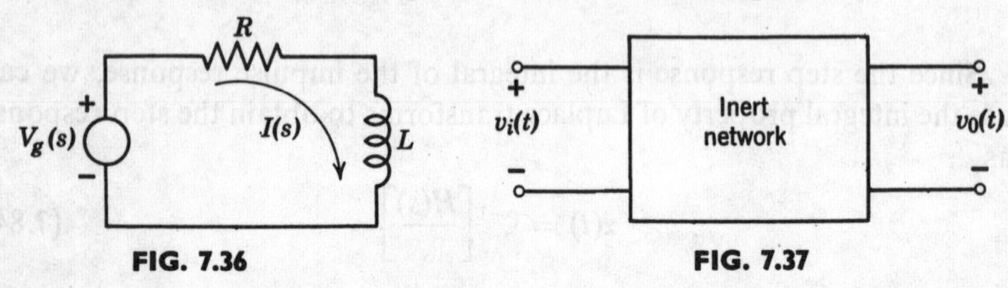

FIG. 7.36 FIG. 7.37

Example 7.10. In Fig. 7.37, the only information we possess about the system in the black box is: (1) it is an initially inert linear system; (2) when $v_i(t) = \delta(t)$, then

$$v_0(t) = (e^{-2t} + e^{-3t})\, u(t) \tag{7.91}$$

With this information, let us determine what the excitation $v_i(t)$ must be in order to produce a response $v_0(t) = te^{-2t} u(t)$. First, we determine the system function to be

$$H(s) = \frac{V_0(s)}{V_i(s)} = \frac{1}{s+2} + \frac{1}{s+3} = \frac{2s+5}{(s+2)(s+3)} \tag{7.92}$$

We next find the transform of $te^{-2t} u(t)$.

$$V_0(s) = \mathcal{L}[te^{-2t}] = \frac{1}{(s+2)^2} \tag{7.93}$$

The unknown excitation is then found from the equation

$$V_i(s) = \frac{V_0(s)}{H(s)} = \frac{(s+2)(s+3)}{2(s+2.5)(s+2)^2} \tag{7.94}$$

Simplifying and expanding $V_i(s)$ into partial fractions, we have

$$V_i(s) = \frac{(s+3)}{2(s+2.5)(s+2)} = \frac{1}{s+2} - \frac{0.5}{s+2.5} \tag{7.95}$$

The system excitation is then

$$v_i(t) = (e^{-2t} - 0.5e^{-2.5t})\, u(t) \tag{7.96}$$

7.5 THE CONVOLUTION INTEGRAL

In this section we will explore some further ramifications of the use of the impulse response $h(t)$ to determine the system response $r(t)$. Our discussion is based upon the important *convolution theorem* of Laplace (or Fourier) transforms. Given two functions $f_1(t)$ and $f_2(t)$, which are zero for $t < 0$, the convolution theorem states that if the transform of $f_1(t)$ is $F_1(s)$, and if the transform of $f_2(t)$ is $F_2(s)$, the transform of the *convolution* of $f_1(t)$ and $f_2(t)$ is the product of the individual transforms, $F_1(s)\,F_2(s)$, that is,

$$\mathcal{L}\left[\int_0^t f_1(t-\tau)f_2(\tau)\,d\tau\right] = F_1(s)\,F_2(s) \tag{7.97}$$

where the integral

$$\int_0^t f_1(t-\tau)\,f_2(\tau)\,d\tau$$

is the *convolution integral* or *folding integral*, and is denoted operationally as

$$\int_0^t f_1(t-\tau)f_2(\tau)\,d\tau = f_1(t)*f_2(t) \tag{7.98}$$

Proof. Let us prove that $\mathcal{L}[f_1*f_2] = F_1F_2$. We begin by writing

$$\mathcal{L}[f_1(t)*f_2(t)] = \int_0^\infty e^{-st}\left[\int_0^t f_1(t-\tau)f_2(\tau)\,d\tau\right]dt \tag{7.99}$$

From the definition of the shifted step function

$$\begin{aligned} u(t-\tau) &= 1 \qquad \tau \le t \\ &= 0 \qquad \tau > t \end{aligned} \tag{7.100}$$

we have the identity

$$\int_0^t f_1(t-\tau)f_2(\tau)\,d\tau = \int_0^\infty f_1(t-\tau)\,u(t-\tau)f_2(\tau)\,d\tau \tag{7.101}$$

Then Eq. 7.99 can be written as

$$\mathcal{L}[f_1(t)*f_2(t)] = \int_0^\infty e^{-st}\int_0^\infty f_1(t-\tau)\,u(t-\tau)f_2(\tau)\,d\tau\,dt \tag{7.102}$$

If we let $x = t - \tau$ so that

$$e^{-st} = e^{-s(x+\tau)} \tag{7.103}$$

then Eq. 7.102 becomes

$$\mathfrak{L}[f_1(t)*f_2(t)] = \int_0^\infty \int_0^\infty f_1(x)\, u(x) f_2(\tau) e^{-s\tau} e^{-sx}\, d\tau\, dx$$

$$= \int_0^\infty f_1(x)\, u(x) e^{-sx}\, dx \int_0^\infty f_2(\tau) e^{-s\tau}\, d\tau$$

$$= F_1(s)\, F_2(s) \tag{7.104}$$

The separation of the double integral in Eq. 7.104 into a product of two integrals is based upon a property of integrals known as the *separability property*.[2]

Example 7.11. Let us evaluate the convolution of the functions $f_1(t) = e^{-2t}\, u(t)$ and $f_2(t) = t\, u(t)$, and then compare the result with the inverse transform of $F_1(s)\, F_2(s)$, where

$$F_1(s) = \mathfrak{L}[f_1(t)] = \frac{1}{s+2}$$
$$F_2(s) = \frac{1}{s^2} \tag{7.105}$$

The convolution of $f_1(t)$ and $f_2(t)$ is obtained by first substituting the dummy variable $t - \tau$ for t in $f_1(t)$, so that

$$f_1(t - \tau) = e^{-2(t-\tau)}\, u(t - \tau) \tag{7.106}$$

Then $f_1(t)*f_2(t)$ is

$$\int_0^t f_1(t-\tau) f_2(\tau)\, d\tau = \int_0^t \tau e^{-2(t-\tau)}\, d\tau = e^{-2t} \int_0^t \tau e^{2\tau}\, d\tau \tag{7.107}$$

Integrating by parts, we obtain

$$f_1(t)*f_2(t) = \left(\frac{t}{2} - \frac{1}{4} + \frac{1}{4} e^{-2t}\right) u(t) \tag{7.108}$$

Next let us evaluate the inverse transform of $F_1(s)\, F_2(s)$. From Eq. 7.105, we have

$$F_1(s)\, F_2(s) = \frac{1}{s^2(s+2)} = \frac{\frac{1}{2}}{s^2} - \frac{\frac{1}{4}}{s} + \frac{\frac{1}{4}}{s+2} \tag{7.109}$$

so that

$$\mathfrak{L}^{-1}[F_1(s)\, F_2(s)] = \left(\frac{t}{2} - \frac{1}{4} + \frac{1}{4} e^{-2t}\right) u(t) \tag{7.110}$$

An important property of the convolution integral is expressed by the equation

$$\int_0^t f_1(t-\tau) f_2(\tau)\, d\tau = \int_0^t f_1(\tau) f_2(t-\tau)\, d\tau \tag{7.111}$$

---

[2] See, for example, P. Franklin, *A Treatise on Advanced Calculus*, John Wiley and Sons, New York, 1940.

This is readily seen from the relationships

$$\mathcal{L}[f_1(t) * f_2(t)] = F_1(s)\,F_2(s) \tag{7.112}$$

and

$$\mathcal{L}[f_2(t) * f_1(t)] = F_2(s)\,F_1(s) \tag{7.113}$$

To give the convolution integral a more intuitive meaning, let us examine the convolution or folding process from a graphical standpoint. Suppose we take the functions

$$\begin{aligned} f_1(\tau) &= u(\tau) \\ f_2(\tau) &= \tau\, u(\tau) \end{aligned} \tag{7.114}$$

as shown in Figs. 7.38*a* and 7.39*a*. In Fig. 7.38, the various steps for obtaining the integral

$$\int_0^t f_1(t-\tau) f_2(\tau)\, d\tau$$

are depicted. Part (*b*) of the figure shows $f_1(-\tau) = u(-\tau)$. The function

$$f_1(t-\tau) = u(t-\tau) \tag{7.115}$$

in part (*c*) merely advances $f_1(-\tau)$ by a variable amount t. Next, the product

$$f_1(t-\tau)\, f_2(\tau) = u(t-\tau)\tau\, u(\tau) \tag{7.116}$$

is shown in part (*d*). We see that the convolution integral is the area under the curve, as indicated by the cross-hatched area in part (*d*). Since the convolution integral has a variable upper limit, we must obtain the area under the curve of $f_1(t-\tau)\, f_2(\tau)$ for all t. With t considered a variable in Fig. 7.38*d*, the area under the curve is

$$f_1 * f_2 = f(t) = \frac{t^2}{2}\, u(t) \tag{7.117}$$

as plotted in part (*e*) of the figure.

In Fig. 7.39, we see that by folding $f_2(\tau)$ about a point t, we obtain the same result as in Fig. 7.38. The result in Eq. 7.117 can be checked by taking the inverse transform of $F_1(s)\,F_2(s)$, which is seen to be

$$\mathcal{L}^{-1}[F_1(s)\,F_2(s)] = \mathcal{L}^{-1}\left[\frac{1}{s^3}\right] = \frac{t^2}{2}\, u(t) \tag{7.118}$$

Let us next examine the role of the convolution integral in system analysis. From the familiar equation

$$R(s) = E(s)\, H(s) \tag{7.119}$$

we obtain the time response as

$$r(t) = \mathcal{L}^{-1}[E(s)\, H(s)] = \int_0^t e(\tau)\, h(t-\tau)\, d\tau \tag{7.120}$$

where $e(\tau)$ is the excitation and $h(\tau)$ is the impulse response of the system.

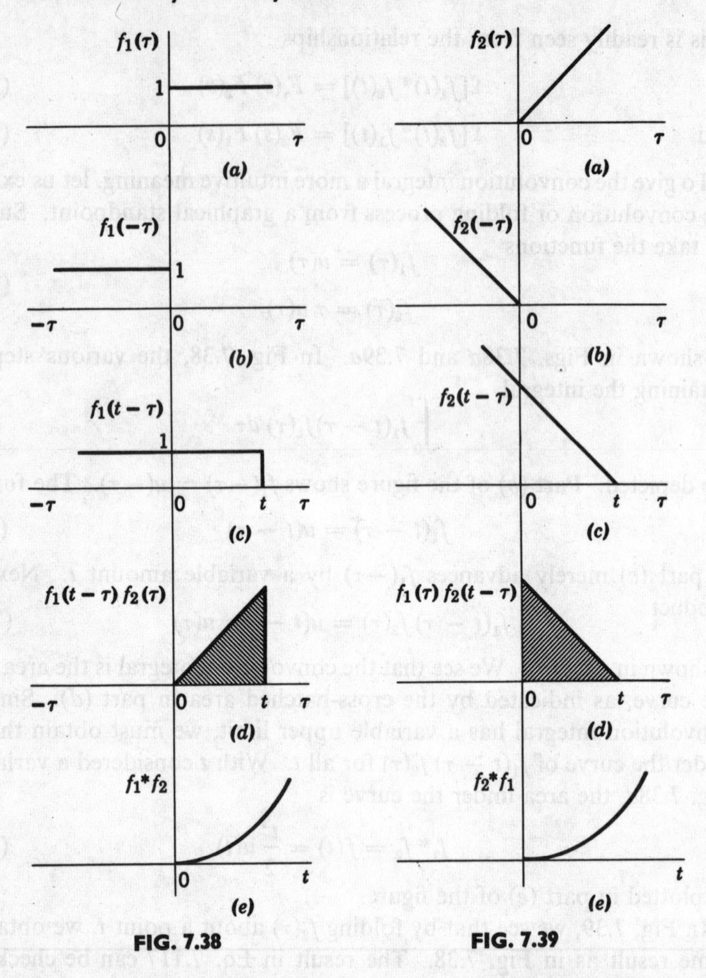

FIG. 7.38 **FIG. 7.39**

Using Eq. 7.120, we obtain the response of a system directly in the time domain. The only information we need about the system is its impulse response.

Example 7.12a. Let us find the response $i(t)$ of the R-L network in Fig. 7.40 due to the excitation

$$v(t) = 2e^{-t}\,u(t) \tag{7.121}$$

The impulse response for the current is

$$h(t) = \frac{1}{L}\,e^{-(R/L)t}\,u(t) \tag{7.122}$$

Therefore, for the R-L circuit under discussion

$$h(t) = e^{-2t}\,u(t) \tag{7.123}$$

Using the convolution integral, we obtain the response $i(t)$ as

$$i(t) = \int_0^t v(t-\tau)\,h(\tau)\,d\tau = 2\int_0^t e^{-(t-\tau)}e^{-2\tau}\,d\tau$$
$$= 2e^{-t}\int_0^t e^{-\tau}\,d\tau = 2(e^{-t} - e^{-2t})\,u(t) \tag{7.124}$$

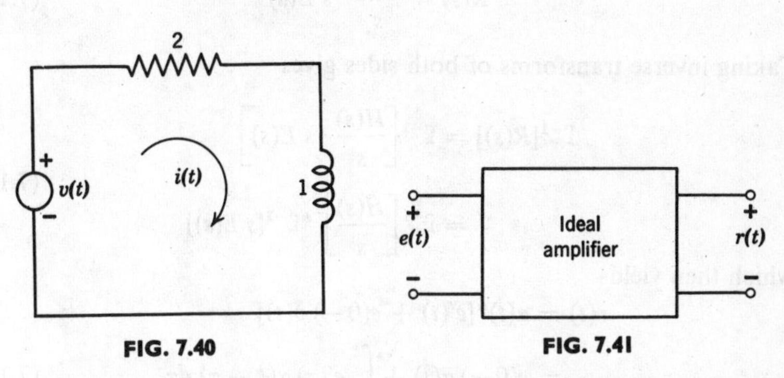

FIG. 7.40 FIG. 7.41

Example 7.12b. The ideal amplifier in Fig. 7.41 has a system function $H(s) = K$, where K is a constant. The impulse response of the ideal amplifier is then

$$h(t) = K\,\delta(t) \tag{7.125}$$

Let us show by means of the convolution integral that the response $r(t)$ is related to the excitation $e(t)$ by the equation

$$r(t) = K\,e(t) \tag{7.126}$$

Using the convolution integral, we have

$$r(t) = \int_0^t e(\tau)\,h(t-\tau)\,d\tau = K\int_0^t e(\tau)\,\delta(t-\tau)\,d\tau = K\,e(t) \tag{7.127}$$

Since t is a variable in the expression for $r(t)$, we see that the ideal amplifier in the time domain is an *impulse-scanning* device which scans the input $e(t)$ from $t = 0$ to $t = \infty$. Thus, the response of an ideal amplifier is a replica of the input $e(t)$ multiplied by the *gain* K of the amplifier.

7.6 THE DUHAMEL SUPERPOSITION INTEGRAL

In the Section 7.5, we discussed the role that the impulse response plays in determining the response of a system to an arbitrary excitation. In this section we will study the *Duhamel superposition integral*, which also describes an input-output relationship for a system. The superposition integral requires the step response $\alpha(t)$ to characterize the system behavior.

We plan to derive the superposition integral in two different ways.

The simplest is examined first. We begin with the excitation-response relationship

$$R(s) = E(s)\,H(s) \tag{7.128}$$

Multiplying and dividing by s gives

$$R(s) = \frac{H(s)}{s} \cdot s\,E(s) \tag{7.129}$$

Taking inverse transforms of both sides gives

$$\begin{aligned}\mathcal{L}^{-1}[R(s)] &= \mathcal{L}^{-1}\left[\frac{H(s)}{s} \cdot s\,E(s)\right] \\ &= \mathcal{L}^{-1}\left[\frac{H(s)}{s}\right] * \mathcal{L}^{-1}[s\,E(s)]\end{aligned} \tag{7.130}$$

which then yields

$$\begin{aligned}r(t) &= \alpha(t) * [e'(t) + e(0-)\,\delta(t)] \\ &= e(0-)\,\alpha(t) + \int_{0-}^{t} e'(\tau)\,\alpha(t-\tau)\,d\tau\end{aligned} \tag{7.131}$$

where $e'(t)$ is the derivative of $e(t)$, $e(0-)$ is the value of $e(t)$ at $t = 0-$, and $\alpha(t)$ is the step response of the system. Equation 7.131 is usually referred to as the *Duhamel superposition integral.*

Example 7.13. Let us find the current $i(t)$ in the R-C circuit in Fig. 7.42 when the voltage source is

$$v_g(t) = (A_1 + A_2 t)\,u(t) \tag{7.132}$$

as shown in Fig. 7.43. The system function of the R-C circuit is

$$H(s) = \frac{I(s)}{V_g(s)} = \frac{s}{R(s + 1/RC)} \tag{7.133}$$

Therefore the transform of the step response is

$$\frac{H(s)}{s} = \frac{1}{R(s + 1/RC)} \tag{7.134}$$

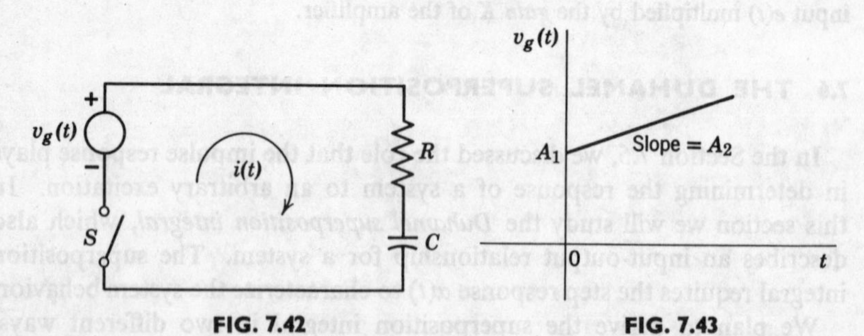

FIG. 7.42

FIG. 7.43

Taking the inverse transform of $H(s)/s$, we obtain the step response

$$\alpha(t) = \frac{1}{R} e^{-t/RC} u(t) \tag{7.135}$$

From the excitation in Eq. 7.132, we see that $e(0-) = 0$ and

$$v'_g(t) = A_1\,\delta(t) + A_2\,u(t) \tag{7.136}$$

The response is then

$$\begin{aligned} i(t) &= \int_{0-}^{t} v'_g(\tau)\,\alpha(t-\tau)\,d\tau \\ &= \int_{0-}^{t} A_1\,\delta(\tau)\,\alpha(t-\tau)\,d\tau + \int_{0-}^{t} A_2\,\alpha(t-\tau)\,d\tau \\ &= A_1\,\alpha(t) + \frac{A_2}{R}\int_{0-}^{t} e^{-(t-\tau)/RC}\,d\tau \\ &= \frac{A_1}{R} e^{-t/RC} u(t) + \frac{A_2}{R} e^{-t/RC}\int_{0-}^{t} e^{\tau/RC}\,d\tau \\ &= \left[\frac{A_1}{R} e^{-t/RC} + A_2 C(1 - e^{-t/RC})\right] u(t) \end{aligned} \tag{7.137}$$

As we see from the Example 7.13, $e(0-) = 0$, which is the case in many transient problems. Another method of deriving the Duhamel integral avoids the problem of discontinuities at the origin by assuming the lower limit of integration to be $t = 0+$. Consider the excitation $e(t)$ shown by the dotted curve in Fig. 7.44. Let us approximate $e(t)$ by a series of step functions, as indicated in the figure. We can write the staircase approximation of $e(t)$ as

$$\begin{aligned} e(t) = e(0+)\,u(t) &+ \Delta E_1\,u(t-\Delta\tau) \\ &+ \Delta E_2\,u(t-2\Delta\tau) + \cdots + \Delta E_n\,u(t-n\Delta\tau) \end{aligned} \tag{7.138}$$

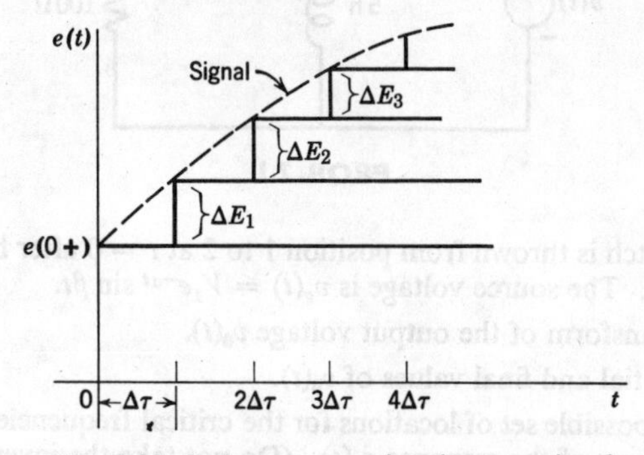

FIG. 7.44. Staircase approximation to a signal.

where ΔE_k is the height of the step increment at $t = k\Delta\tau$. Since we assume the system to be linear and time invariant, we know that if the response to a unit step is $\alpha(t)$, the response to a step $K_i\, u(t-\tau)$ is $K_i\, \alpha(t-\tau)$. Therefore we can write the response to the step approximation in Fig. 7.44 as

$$\begin{aligned} r(t) = e(0+)\,\alpha(t) &+ \Delta E_1\, \alpha(t-\Delta\tau) \\ &+ \Delta E_2\, \alpha(t-2\Delta\tau) + \cdots + \Delta E_n\, \alpha(t-n\Delta t) \end{aligned} \tag{7.139}$$

If $\Delta\tau$ is small, $r(t)$ can be given as

$$\begin{aligned} r(t) &= e(0+)\,\alpha(t) + \frac{\Delta E_1}{\Delta\tau}\alpha(t-\Delta\tau)\,\Delta\tau \\ &+ \frac{\Delta E_2}{\Delta\tau}\alpha(t-2\Delta\tau)\,\Delta\tau + \cdots + \frac{\Delta E_n}{\Delta\tau}\alpha(t-n\Delta\tau)\,\Delta\tau \end{aligned} \tag{7.140}$$

which, in the limit, becomes

$$\begin{aligned} r(t) &= e(0+)\,\alpha(t) + \lim_{\substack{\Delta\tau\to 0\\ n\to\infty}} \sum_{i=0+}^{n} \frac{\Delta E_i}{\Delta\tau}\alpha(t-i\Delta\tau)\,\Delta\tau \\ &= e(0+)\,\alpha(t) + \int_{0+}^{t} e'(\tau)\,\alpha(t-\tau)\,d\tau \end{aligned} \tag{7.141}$$

Problems

7.1 In the circuit shown $v(t) = 2u(t)$ and $i_L(0-) = 2$ amps. Find and sketch $i_2(t)$.

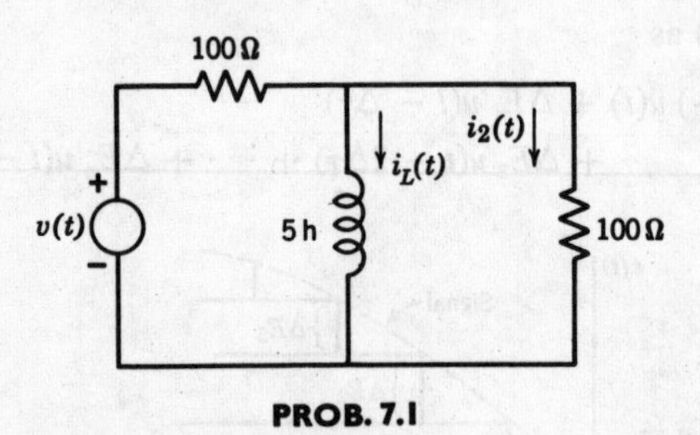

PROB. 7.1

7.2 The switch is thrown from position 1 to 2 at $t = 0$ after having been at 1 for a long time. The source voltage is $v_g(t) = V_1 e^{-\alpha t} \sin \beta t$.

(*a*) Find the transform of the output voltage $v_0(t)$.

(*b*) Find the initial and final values of $v_0(t)$.

(*c*) Sketch one possible set of locations for the critical frequencies in the s plane and write the *form* of the response $v_0(t)$. (Do not take the inverse transform.)

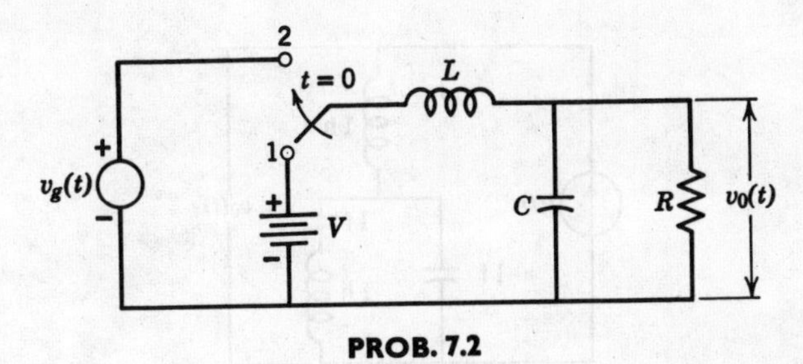

PROB. 7.2

7.3 In the circuit shown, all initial currents and voltages are zero. Find $i(t)$ for $t > 0$ using Thévenin's theorem.

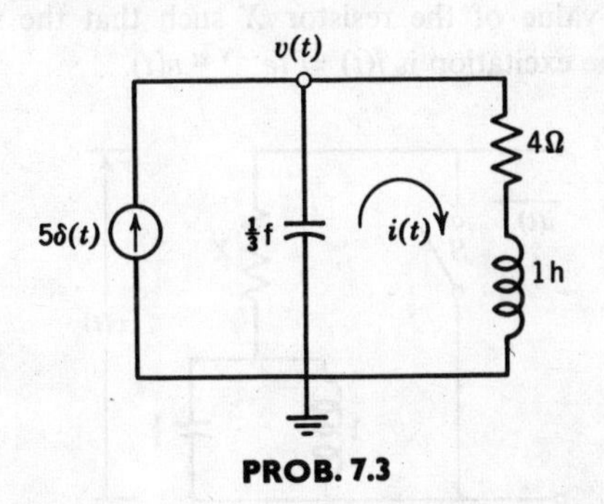

PROB. 7.3

7.4 In the circuit shown in (*a*), the excitation is the voltage source $e(t)$ described in (*b*). Determine the response $i(t)$ assuming zero initial conditions.

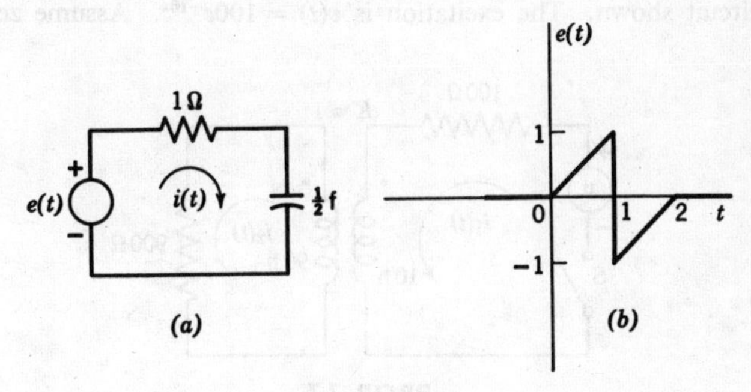

PROB. 7.4

7.5 Determine the expression for $v_0(t)$ when $i(t) = \delta(t)$, assuming zero initial conditions. Use transform methods.

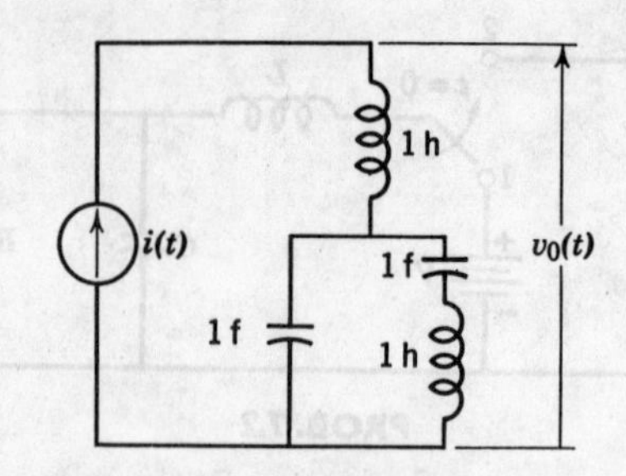

PROB. 7.5

7.6 The circuit shown has zero initial energy. At $t = 0$ the switch S is opened. Find the value of the resistor X such that the response is $v(t) = 0.5 \sin \sqrt{2}t\, u(t)$. The excitation is $i(t) = te^{-\sqrt{2}t}\, u(t)$.

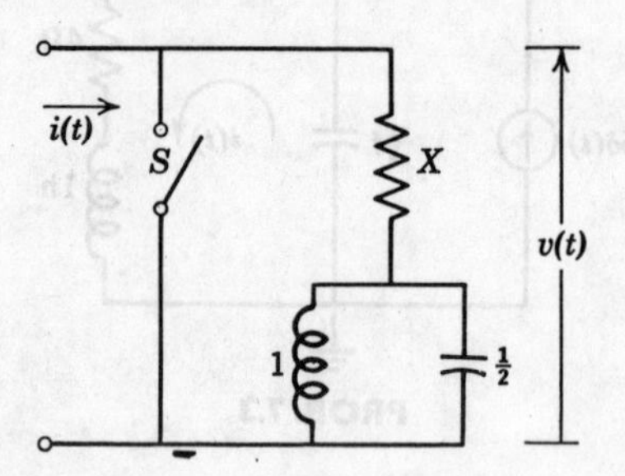

PROB. 7.6

7.7 Use transform methods to determine the expressions for $i_1(t)$ and $i_2(t)$ in the circuit shown. The excitation is $v(t) = 100e^{-10t}$. Assume zero initial energy.

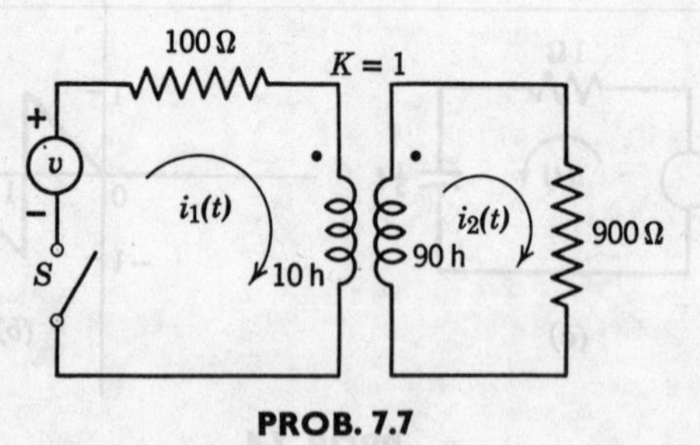

PROB. 7.7

7.8 For the circuit shown, the switch S is opened at $t = 0$. Use Thévenin's or Norton's theorem to determine the output voltage $v_2(t)$. Assume zero initial energy.

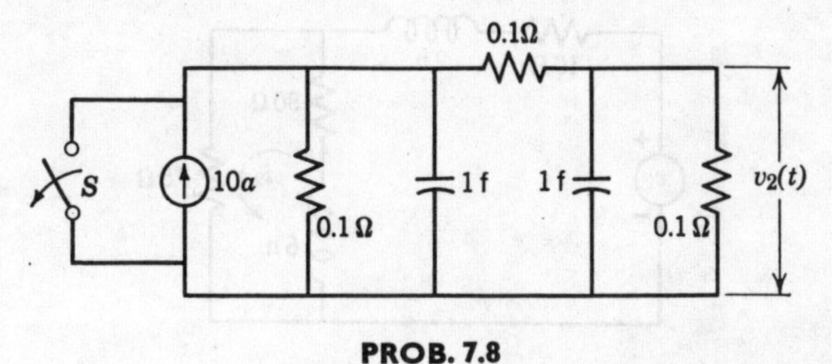

PROB. 7.8

7.9 For the transformer shown, find $i_1(t)$ and $i_2(t)$. It is given that $e(t) = 6u(t)$, and that prior to the switching action all initial energy was zero; also $M = 1h$.

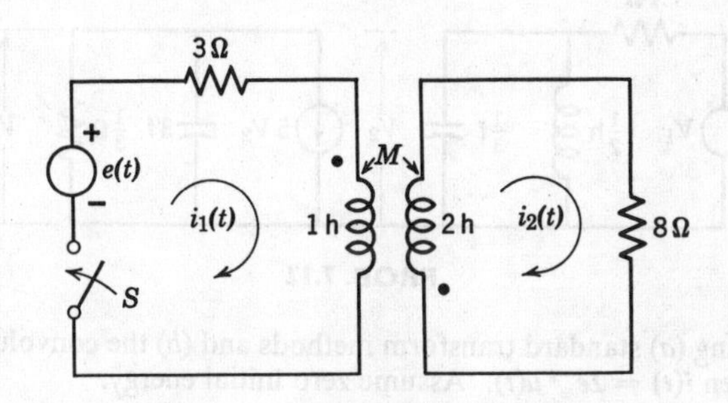

PROB. 7.9

7.10 For the circuit shown, find $i_2(t)$, given that the circuit had been in steady state prior to the switch closing at $t = 0$.

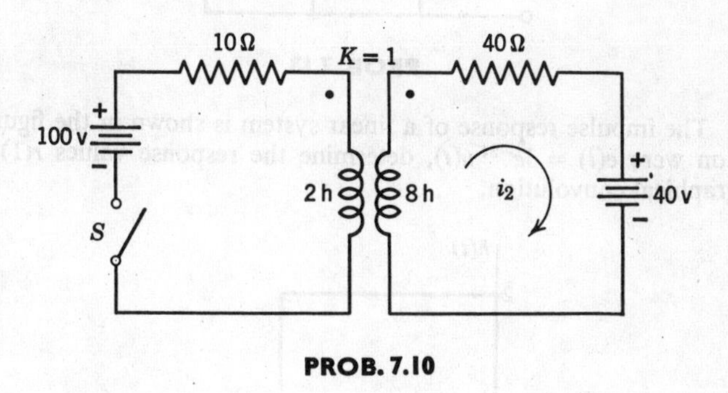

PROB. 7.10

7.11 Find $i_2(t)$ using Thévenin's theorem. The excitation is $e(t) = 100 \cos 20\, u(t)$. Assume zero initial energy.

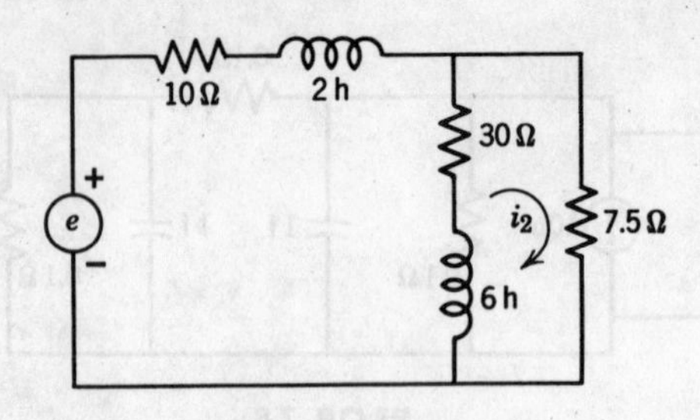

PROB. 7.11

7.12 Determine the transfer function $H(s) = V_3(s)/V_1(s)$. When $v_1(t) = u(t)$, find $v_3(t)$. Assume zero initial conditions.

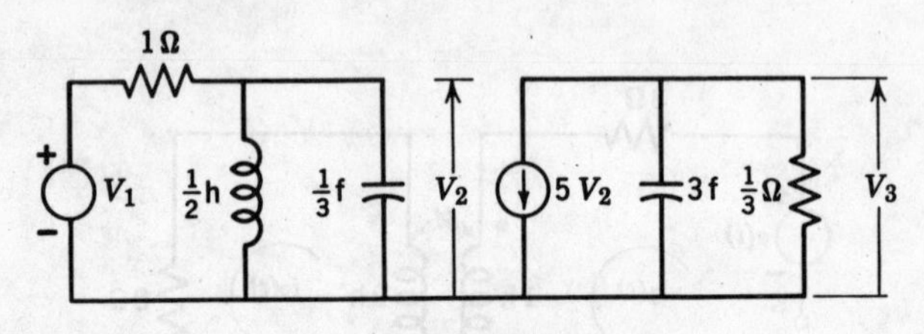

PROB. 7.12

7.13 Using (*a*) standard transform methods and (*b*) the convolution integral, find $v(t)$ when $i(t) = 2e^{-t}\,u(t)$. Assume zero initial energy.

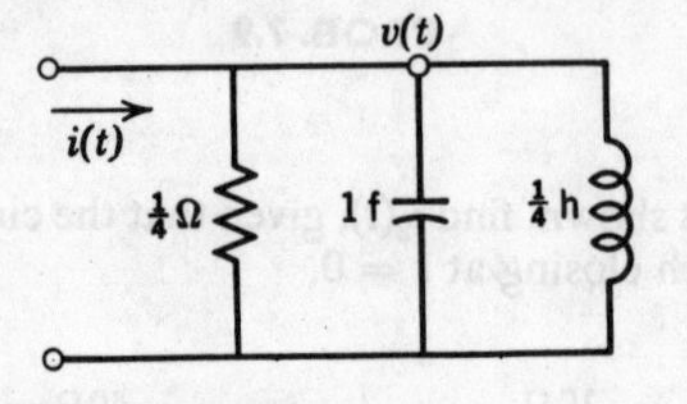

PROB. 7.13

7.14 The impulse response of a linear system is shown in the figure. If the excitation were $e(t) = 3e^{-2t}\,u(t)$, determine the response values $r(1)$ and $r(4)$ using graphical convolution.

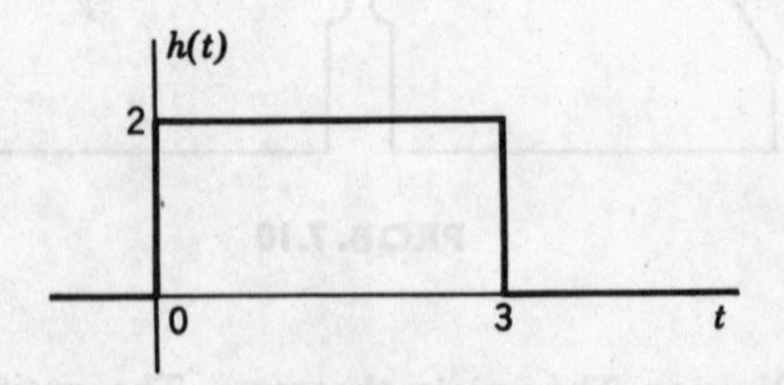

PROB. 7.14

7.15 The system function is given as $H(s) = 1/(s^2 + 9)^2$. If the excitation were $e(t) = 3\delta'(t)$, determine the response $r(t)$. (*Hint:* Use the convolution integral to break up the response transform $R(s)$.)

7.16 Solve the following integral equations for $x(t)$.

(*a*)
$$x(t) + \int_0^t (t - \tau)\, x(\tau)\, d\tau = 1$$

(*b*)
$$\sin t = \int_0^t x(\tau)\, e^{-(t-\tau)}\, d\tau$$

(*c*)
$$x(t) + \int_0^t x(t - \tau)\, e^{-\tau}\, d\tau = t$$

7.17 Using the convolution integral find the inverse transform of

(*a*)
$$F(s) = \frac{K}{(s + a)(s + b)}$$

(*b*)
$$F(s) = \frac{2(s + 2)}{(s^2 + 4)^2}$$

(*c*)
$$F(s) = \frac{s}{(s^2 + 1)^2}$$

7.18 By graphical means, determine the convolution of $f(t)$ shown in the figure with itself (i.e., determine $f(t)*f(t)$).

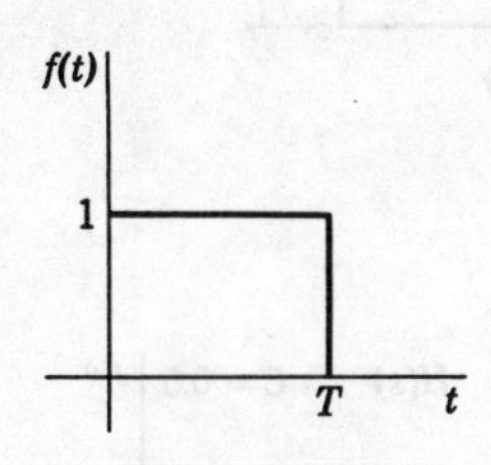

PROB. 7.18

7.19 Using (*a*) the convolution integral and (*b*) the Duhamel superposition integral, find $v(t)$ for $e(t) = 4e^{-3t}\, u(t)$. Assume zero initial conditions.

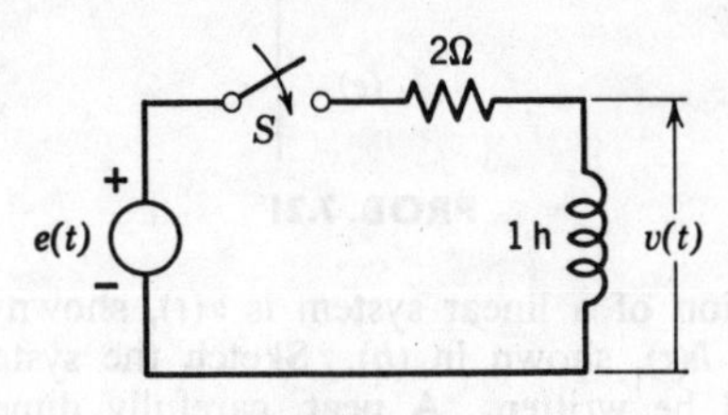

PROB. 7.19

7.20 Using (*a*) the convolution integral and (*b*) the Duhamel superposition integral, find $v(t)$ for $e(t) = 2e^{-3t}\,u(t)$. Assume zero initial conditions.

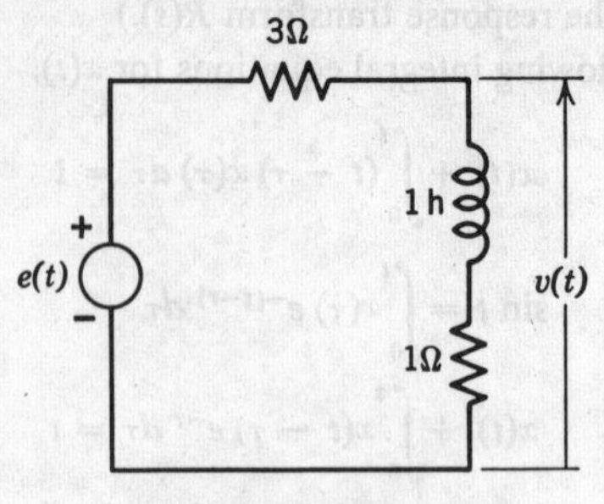

PROB. 7.20

7.21 For the circuit in (*a*), the system function $H(s) = V(s)/I(s)$ has the poles shown in (*c*). Find the element values for R and C. If the excitation $i(t)$ has the form shown in (*b*), use the convolution integral to find $v(t)$.

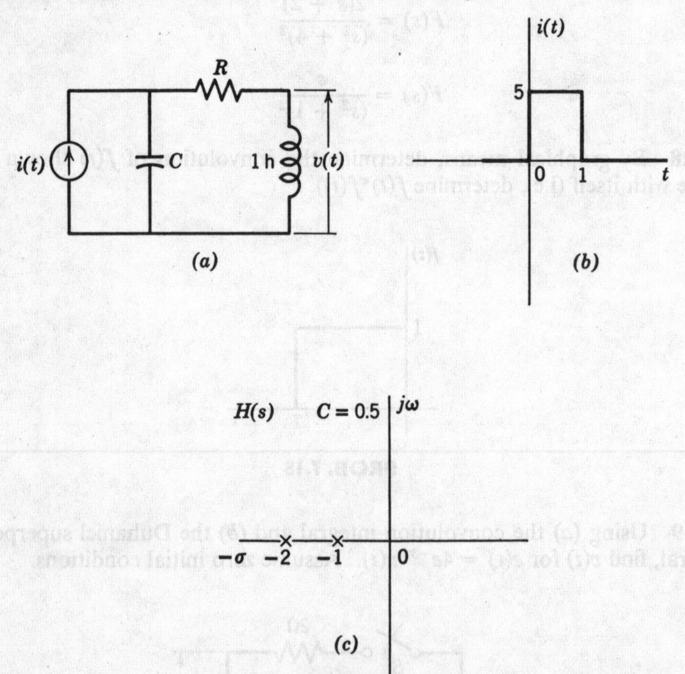

PROB. 7.21

7.22 The excitation of a linear system is $x(t)$, shown in (*a*). The system impulse response is $h(t)$, shown in (*b*). Sketch the system response to $x(t)$. (No equations need be written. A neat, carefully dimensioned sketch will suffice.)

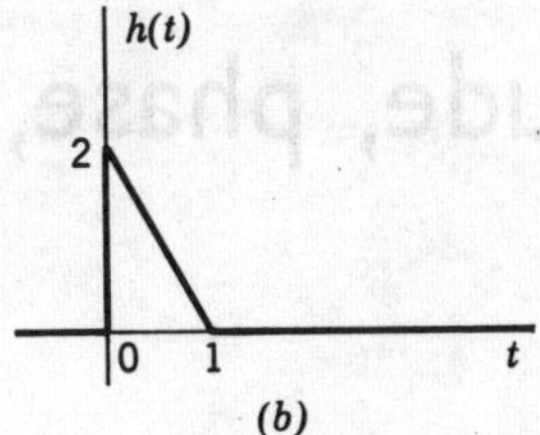

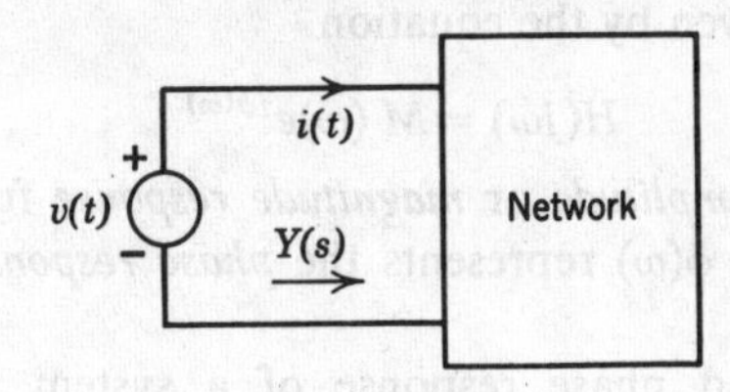

PROB. 7.22

7.23 A unit step of voltage is applied to the network and the resulting current is $i(t) = 0.01e^{-t} + 0.02$ amps.

(*a*) Determine the admittance $Y(s)$ for this network.

(*b*) Find a network that will yield this admittance function.

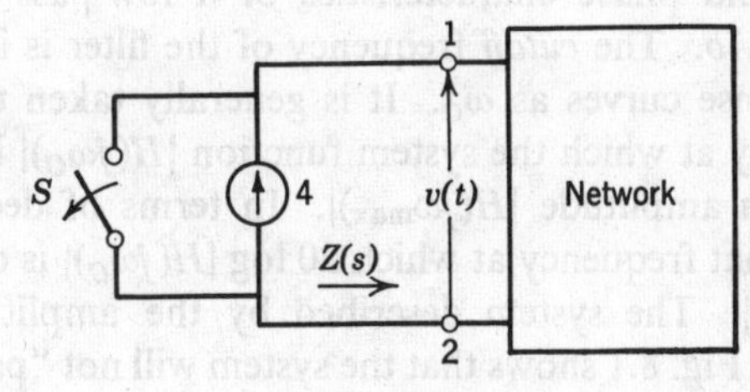

PROB. 7.23

1
S
4
v(t)
Network
Z(s)
2

PROB. 7.24

7.24 The current generator delivers a constant current of 4 amps. At $t = 0$ the switch S is opened and the resulting voltage across the terminals 1, 2 is $v(t) = 6e^{-4t} + 12$ v.

(*a*) Find $Z(s)$ looking into terminals 1, 2.

(*b*) Find a network realization for $Z(s)$.

chapter 8
Amplitude, phase, and delay

8.1 AMPLITUDE AND PHASE RESPONSE

In this section we will study the relationship between the poles and zeros of a system function and its steady-state sinusoidal response. In other words, we will investigate the effect of pole and zero positions upon the behavior of $H(s)$ along the $j\omega$ axis. The steady-state response of a system function is given by the equation

$$H(j\omega) = M(\omega)e^{j\phi(\omega)} \tag{8.1}$$

where $M(\omega)$ is the *amplitude* or *magnitude response* function, and is an even function in ω. $\phi(\omega)$ represents the *phase response*, and is an odd function of ω.

The amplitude and phase response of a system provides valuable information in the analysis and design of transmission circuits. Consider the amplitude and phase characteristics of a low-pass filter shown in Figs. 8.1*a* and 8.1*b*. The *cutoff* frequency of the filter is indicated on the amplitude response curves as ω_C. It is generally taken to be the "half-power" frequency at which the system function $|H(j\omega_C)|$ is equal to 0.707 of the maximum amplitude $|H(j\omega_{\max})|$. In terms of decibels, the half-power point is that frequency at which $20 \log |H(j\omega_C)|$ is down 3 db from $20 \log |H(j\omega_{\max})|$. The system described by the amplitude and phase characteristics in Fig. 8.1 shows that the system will not "pass" frequencies that are greater than ω_C. Suppose we consider a pulse train whose amplitude spectrum contains significant harmonics above ω_C. We know that the system will pass the harmonics below ω_C, but will block all harmonics above ω_C. Therefore the output pulse train will be distorted when compared to the original pulse train, because many higher harmonic terms will be missing. It will be shown in Chapter 13 that if the phase

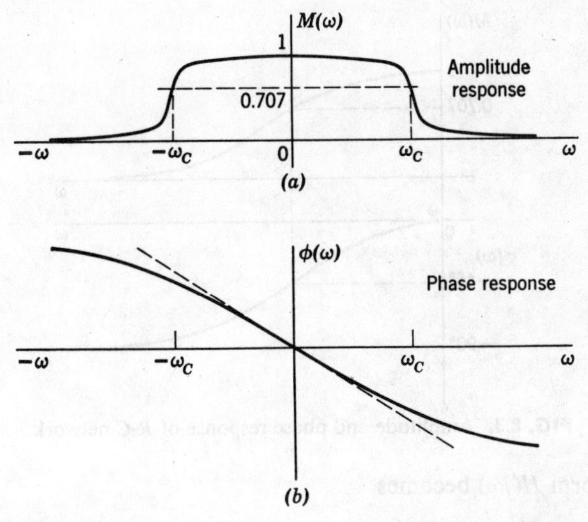

FIG. 8.1. Amplitude and phase response of low-pass filter.

response $\phi(\omega)$ is *linear*, then minimum pulse distortion will result. We see from the phase response $\phi(\omega)$ in Fig. 8.1 that the phase is approximately linear over the range $-\omega_C \leq \omega \leq +\omega_C$. If all the significant harmonic terms are less than ω_C, then the system will produce minimum *phase distortion*. With this example, we see the importance of an amplitude-phase description of a system. In the remaining part of this chapter, we will concentrate on methods to obtain amplitude and phase response curves, both analytically and graphically.

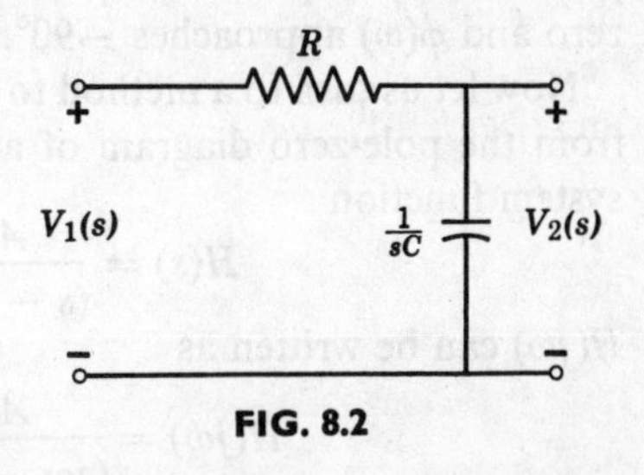

FIG. 8.2

To obtain amplitude and phase curves, we let $s = j\omega$ in the system function, and express $H(j\omega)$ in polar form. For example, for the amplitude and phase response of the voltage ratio V_2/V_1 of the *R-C* network shown in Fig. 8.2, the system function is

$$H(s) = \frac{V_2(s)}{V_1(s)} = \frac{1/RC}{s + 1/RC} \tag{8.2}$$

Letting $s = j\omega$, we see that $H(j\omega)$ is

$$H(j\omega) = \frac{1/RC}{j\omega + 1/RC} \tag{8.3}$$

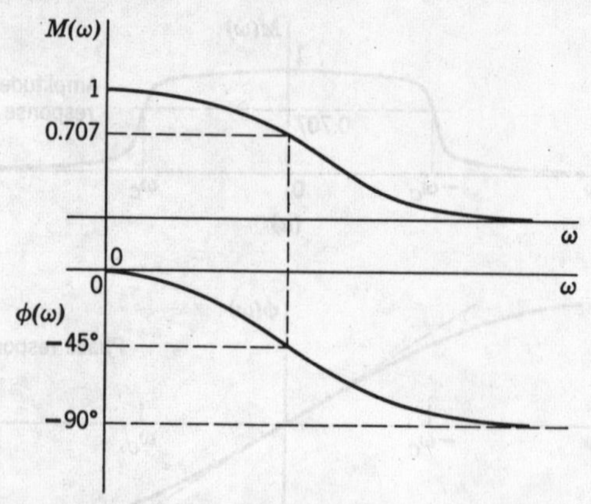

FIG. 8.3. Amplitude and phase response of *R-C* network.

In polar form $H(j\omega)$ becomes

$$H(j\omega) = \frac{1/RC}{(\omega^2 + 1/R^2C^2)^{1/2}} e^{-j\tan^{-1}\omega RC} = M(\omega)e^{j\phi(\omega)} \tag{8.4}$$

The amplitude and phase curves are plotted in Fig. 8.3. At the point $\omega = 0$, the amplitude is unity and the phase is zero degrees. As ω increases, the amplitude and phase decrease monotonically. When $\omega = 1/RC$, the amplitude is 0.707 and the phase is $-45°$. This point is the half-power point of the amplitude response. Finally as $\omega \to \infty$, $M(\omega)$ approaches zero and $\phi(\omega)$ approaches $-90°$.

Now let us turn to a method to obtain the amplitude and phase response from the pole-zero diagram of a system function. Suppose we have the system function

$$H(s) = \frac{A_0(s - z_0)(s - z_1)}{(s - p_0)(s - p_1)(s - p_2)} \tag{8.5}$$

$H(j\omega)$ can be written as

$$H(j\omega) = \frac{A_0(j\omega - z_0)(j\omega - z_1)}{(j\omega - p_0)(j\omega - p_1)(j\omega - p_2)} \tag{8.6}$$

Each one of the factors $j\omega - z_i$ or $j\omega - p_j$ corresponds to a vector from the zero z_i or pole p_j directed to any point $j\omega$ on the imaginary axis. Therefore, if we express the factors in polar forııı,

$$j\omega - z_i = \mathbf{N}_i e^{j\psi_i}, \qquad j\omega - p_j = \mathbf{M}_j e^{j\theta_j} \tag{8.7}$$

then $H(j\omega)$ can be given as

$$H(j\omega) = \frac{A_0\mathbf{N}_0\mathbf{N}_1}{\mathbf{M}_0\mathbf{M}_1\mathbf{M}_2} e^{j(\psi_0+\psi_1-\theta_0-\theta_1-\theta_2)} \tag{8.8}$$

as shown in Fig. 8.4, where we note that θ_1 is negative.

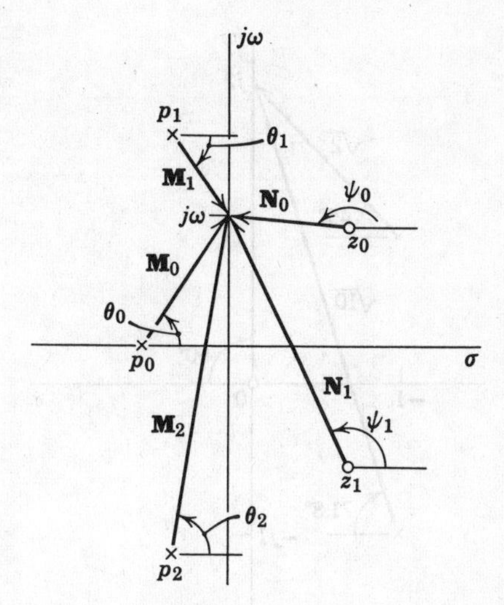

FIG. 8.4. Evaluation of amplitude and phase from pole-zero diagram.

In general, we can express the amplitude response $M(\omega)$ in terms of the following equation.

$$M(\omega) = \frac{\prod_{i=0}^{n} \text{vector magnitudes from the zeros to the point on the } j\omega \text{ axis}}{\prod_{j=0}^{m} \text{vector magnitudes from the poles to the point on the } j\omega \text{ axis}}$$

Similarly, the phase response is given as

$$\phi(\omega) = \sum_{i=0}^{n} \text{angles of the vectors from the zeros to the } j\omega \text{ axis}$$

$$- \sum_{j=0}^{m} \text{angles of the vectors from the poles to the } j\omega \text{ axis}$$

It is important to note that these relationships for amplitude and phase are point-by-point relationships only. In other words, we must draw vectors from the poles and zeros to every point on the $j\omega$ axis for which we wish to determine amplitude and phase. Consider the following example.

$$F(s) = \frac{4s}{s^2 + 2s + 2} = \frac{4s}{(s + 1 + j1)(s + 1 - j1)} \tag{8.9}$$

Let us find the amplitude and phase for $F(j2)$. From the poles and zeros of $F(s)$, we draw vectors to the point $\omega = 2$, as shown in Fig. 8.5. From

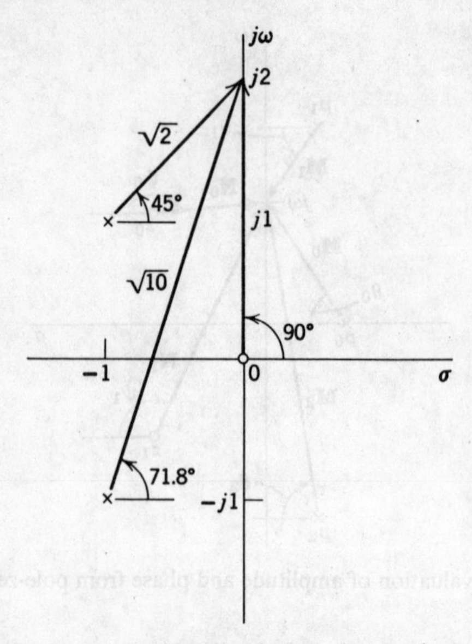

FIG. 8.5. Evaluation of amplitude and phase from pole-zero diagram.

the pole-zero diagram, it is clear that

$$M(j2) = 4\left(\frac{2}{\sqrt{2} \times \sqrt{10}}\right) = 1.78$$

and
$$\phi(j2) = 90° - 45° - 71.8° = -26.8°$$

With the values $M(j2)$ and $\phi(j2)$ and the amplitude and phase at three or four additional points, we have enough information for a rough estimate

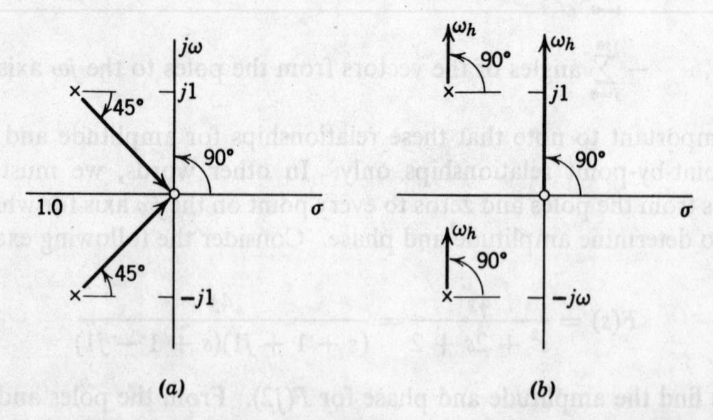

FIG. 8.6. Determining amplitude and phase at zero and very high frequencies.

of the amplitude and phase response. At $\omega = 0$, we see that the vector magnitude from the zero at the origin to $\omega = 0$, is of course, zero.

Consequently, $M(j0) = 0$. From Eq. 8.9 for $F(s)$, $F(j0)$ is

$$\lim_{\substack{\omega \to 0 \\ \omega > 0}} F(j\omega) = \frac{4(j0)}{(1 + j1)(1 - j1)} \tag{8.10}$$

From this equation, we see that the zero at the origin still contributes a 90° phase shift even though the vector magnitude is zero. From Fig. 8.6*a* we see that the net phase at $\omega = 0$ is $\phi(0) = 90° - 45° + 45° = 90°$. Next, at a very high frequency ω_h, where $\omega_h \gg 1$, all the vectors are approximately equal to $\omega_h e^{j90°}$, as seen in Fig. 8.6*b*. Then

$$M(\omega_h) \simeq \frac{4\omega_h}{\omega_h^{\ 2}} = \frac{4}{\omega_h}$$

and $\qquad \phi(\omega_h) \simeq 90° - 90° - 90° = -90°$

Extending this analysis for the frequencies listed in Table 8.1, we obtain values for amplitude and phase as given in the table. From this table we can sketch the amplitude and phase curves shown in Fig. 8.7.

Next let us examine the effect of poles and zeros on the $j\omega$ axis upon frequency response. Consider the function

$$F(s) = \frac{s^2 + 1.03}{s^2 + 1.23} = \frac{(s + j1.015)(s - j1.015)}{(s + j1.109)(s - j1.109)} \tag{8.11}$$

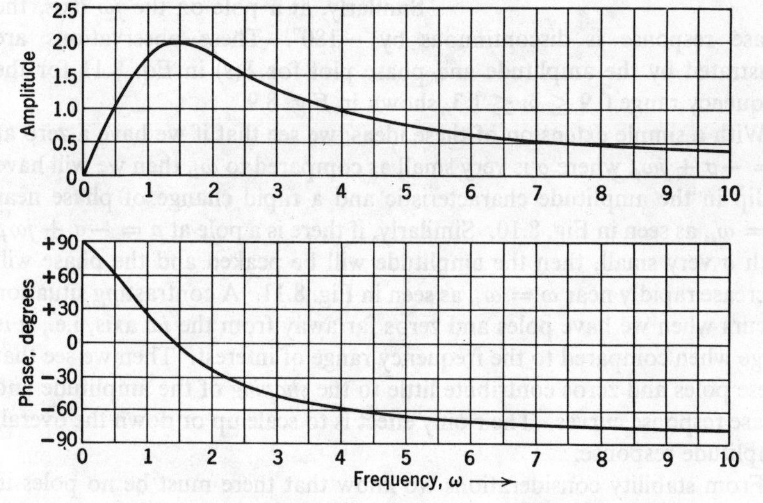

FIG. 8.7. Amplitude and phase response for $F(s)$ in Eq. 8.9.

TABLE 8.1

Frequency, ω	Amplitude	Phase, degrees
1.0	1.8	25.8
1.5	2.0	−5.3
3.0	1.3	−50.0
5.0	0.8	−66.0
10.0	0.4	−78.5

whose pole-zero diagram is shown in Fig. 8.8. At $\omega = 1.015$, the vector from zero to that frequency is of zero magnitude. Therefore at a zero on the $j\omega$ axis, the amplitude response is zero. At $\omega = 1.109$ the vector from the pole to that frequency is of zero magnitude. The amplitude response is therefore infinite at a pole as seen from Eq. 8.11. Next, consider the phase response. When $\omega < 1.015$, it is apparent from the pole-zero plot that the phase is zero. When $\omega > 1.015$ and $\omega < 1.109$, the vector from the zero at $\omega = 1.015$ is now pointing upward, while the vectors from the other poles and zeros are oriented in the same direction as for $\omega < 1.015$. We see that at a zero on the $j\omega$ axis, the phase response has a step discontinuity of $+180°$ for increasing frequency. Similarly, at a pole on the $j\omega$ axis, the phase response is discontinuous by $-180°$. These observations are illustrated by the amplitude and phase plot for $F(s)$ in Eq. 8.11 for the frequency range $0.9 \leq \omega \leq 1.3$, shown in Fig. 8.9.

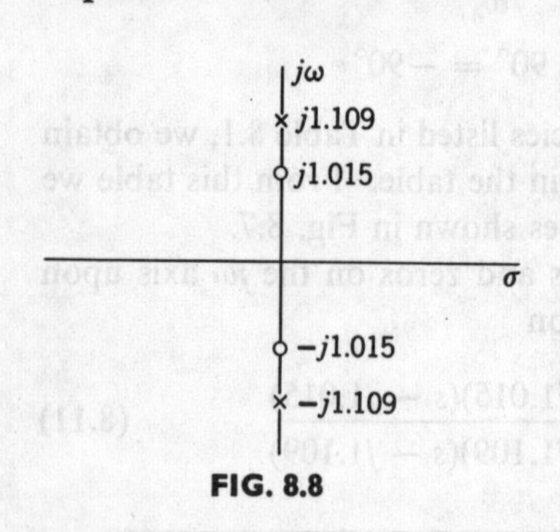

FIG. 8.8

With a simple extension of these ideas, we see that if we have a zero at $z = -\sigma \pm j\omega_i$, where σ is very small as compared to ω_i, then we will have a dip in the amplitude characteristic and a rapid change of phase near $\omega = \omega_i$, as seen in Fig. 8.10. Similarly, if there is a pole at $p = -\sigma \pm j\omega_j$, with σ very small, then the amplitude will be peaked and the phase will decrease rapidly near $\omega = \omega_j$, as seen in Fig. 8.11. A contrasting situation occurs when we have poles and zeros far away from the $j\omega$ axis, i.e., σ is large when compared to the frequency range of interest. Then we see that these poles and zeros contribute little to the *shaping* of the amplitude and phase response curves. Their only effect is to scale up or down the overall amplitude response.

From stability considerations we know that there must be no poles in the right half of the s plane. However, transfer functions may have zeros

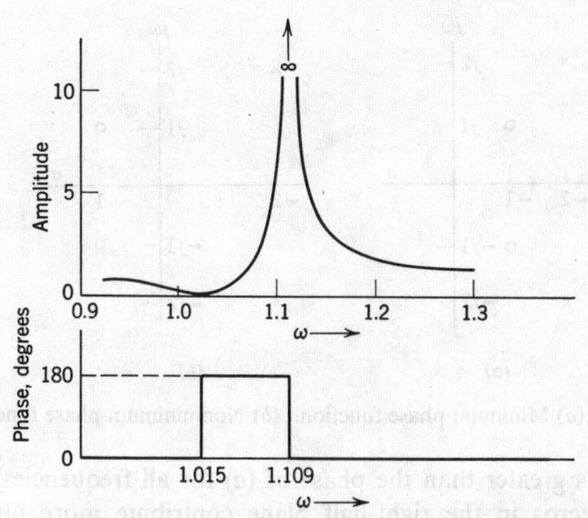

FIG. 8.9. Amplitude and phase for $F(s)$ in Fig. 8.8.

in the right-half plane. Consider the pole-zero diagrams in Figs. 8.12*a* and 8.12*b*. Both pole-zero configurations have the same poles; the only difference is that the zeros in (*a*) are in the left-half plane at $s = -1 \pm j1$, while the zeros in (*b*) are the mirror images of the zeros in (*a*), and are located at $s = +1 \pm j1$. Observe that the amplitude responses of the two configurations are the same because the lengths of the vectors correspond for both situations. We see that the absolute magnitude of the

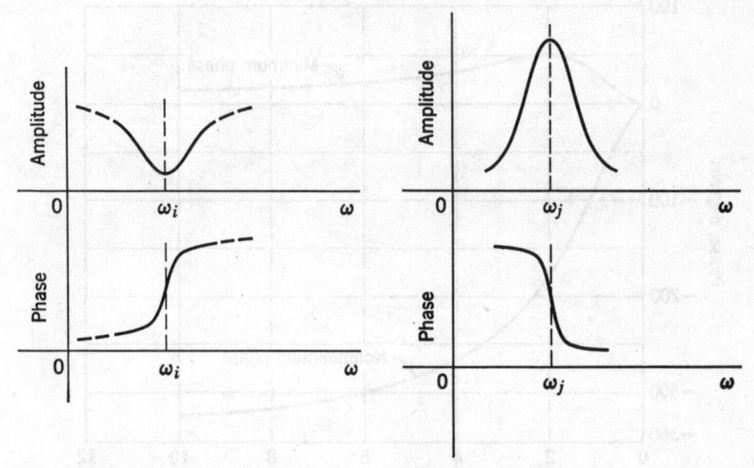

FIG. 8.10. Effect of zero very near the $j\omega$ axis.

FIG. 8.11. Effect of pole very near the $j\omega$ axis.

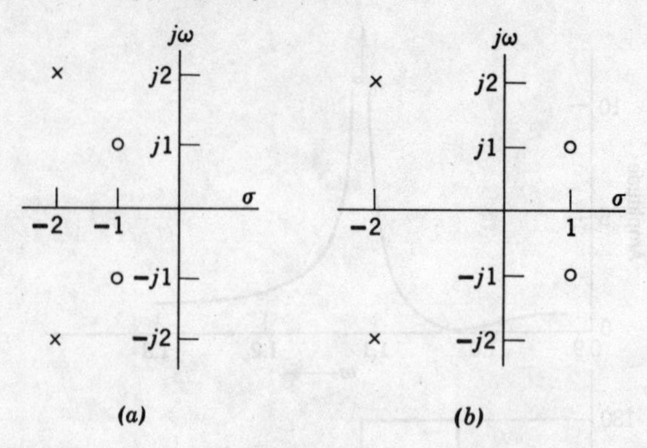

FIG. 8.12. (*a*) Minimum phase function. (*b*) Nonminimum phase function.

phase of (*b*) is greater than the phase of (*a*) for all frequencies. This is because the zeros in the right-half plane contribute more phase shift (on an absolute magnitude basis) than their counterparts in the left-half plane. From this reasoning, we have the following definitions. A system function with zeros in the left-half plane, or on the $j\omega$ axis only, is called a *minimum phase* function. If the function has one or more zeros in the right-half plane, it is a *nonminimum phase* function. In Fig. 8.13, we see the phase responses of the minimum and nonminimum phase functions in Figs. 8.12*a* and 8.12*b*.

Let us next consider the pole-zero diagram in Fig. 8.14. Observe that the zeros in the right-half plane are mirror images of the poles in the

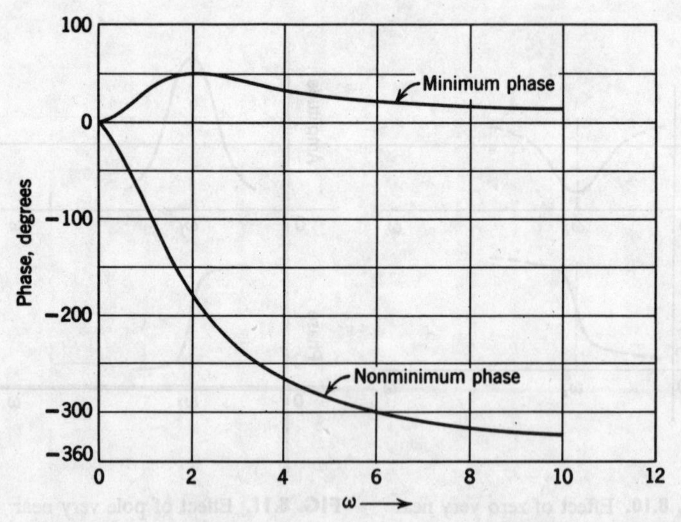

FIG. 8.13. Comparison of minimum and nonminimum phase functions.

left-half plane. Consequently, the vector drawn from a pole to any point ω_1 on the $j\omega$ axis is identical in magnitude with the vector drawn from its mirror image to ω_1. It is apparent that the amplitude response must be constant for all frequencies. The phase response, however, is anything but constant, as seen from the amplitude and phase response curves given in Fig. 8.15 for the pole-zero configuration in Fig. 8.14.

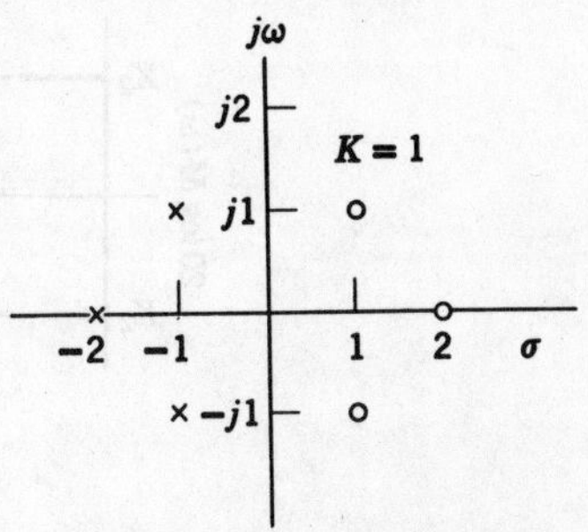

FIG. 8.14. All-pass function.

A system function whose poles are only in the left-half plane and whose zeros are mirror images of the poles about the $j\omega$ axis is called an *all-pass* function. The networks which have all-pass response characteristics are often used to correct for phase distortion in a transmission system.

8.2 BODE PLOTS

In this section we will turn our attention to semilogarithmic plots of amplitude and phase versus frequency. These plots are commonly known as *Bode plots*. Consider the system function

$$H(s) = \frac{N(s)}{D(s)} \tag{8.12}$$

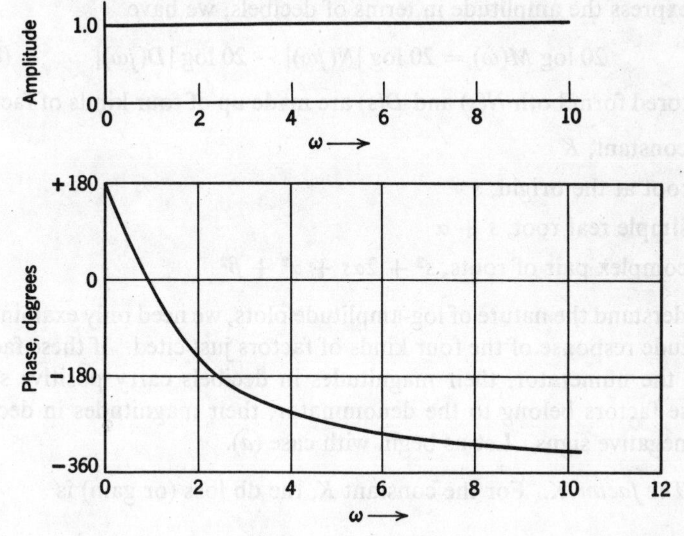

FIG. 8.15. Amplitude and phase of all-pass function in Fig. 8.14.

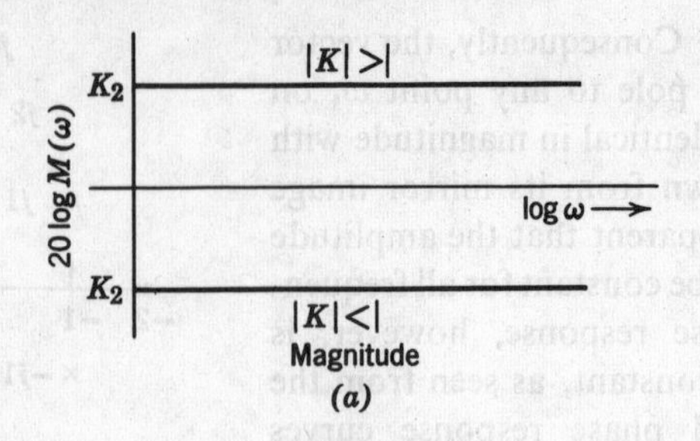

(a)

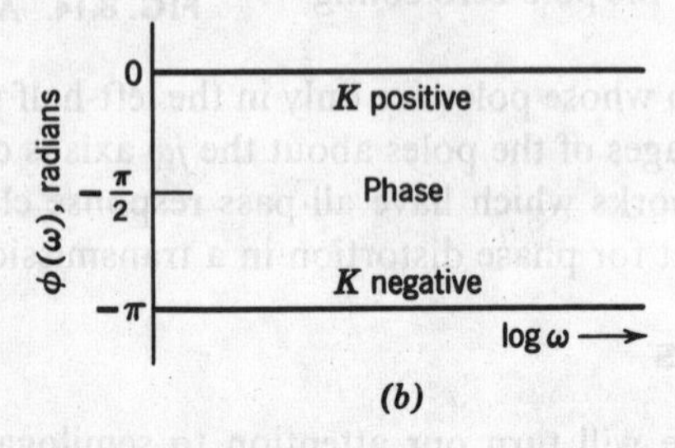

(b)

FIG. 8.16. Magnitude and phase of constant.

We know that the amplitude response is

$$M(\omega) = |H(j\omega)| = \frac{|N(j\omega)|}{|D(j\omega)|} \tag{8.13}$$

If we express the amplitude in terms of decibels, we have

$$20 \log M(\omega) = 20 \log |N(j\omega)| - 20 \log |D(j\omega)| \tag{8.14}$$

In factored form both $N(s)$ and $D(s)$ are made up of four kinds of factors:

(*a*) a constant, K

(*b*) a root at the origin, s

(*c*) a simple real root, $s + \alpha$

(*d*) a complex pair of roots, $s^2 + 2\alpha s + \alpha^2 + \beta^2$

To understand the nature of log-amplitude plots, we need only examine the amplitude response of the four kinds of factors just cited. If these factors are in the numerator, their magnitudes in decibels carry positive signs. If these factors belong to the denominator, their magnitudes in decibels carry negative signs. Let us begin with case (*a*).

(*a*) *The factor K.* For the constant K, the db loss (or gain) is

$$20 \log K = K_2 \tag{8.15}$$

The constant K_2 is either negative if $|K| < 1$, or positive if $|K| > 1$. The phase response is either zero or 180° depending on whether K is positive or negative. The Bode plots showing the magnitude and phase of a constant are given in Fig. 8.16.

(*b*) *The factor s*. The loss (gain) in decibels associated with a pole (zero) at the origin is $\pm 20 \log \omega$. Thus the plot of magnitude in decibels versus frequency in semilog coordinates is a straight line with slope of ± 20 db/decade or ± 6 db/octave. From the Bode plots in Fig. 8.17, we see that the zero loss point (in decibels) is at $\omega = 1$, and the phase is constant for all ω.

(*c*) *The factor* $s + \alpha$. For convenience, let us set $\alpha = 1$. Then the magnitude is

$$\pm 20 \log |j\omega + 1| = \pm 20 \log (\omega^2 + 1)^{1/2} \tag{8.16}$$

as shown in Fig. 8.18*a*. The phase is

$$\text{Arg}\,(j\omega + 1)^{\pm 1} = \pm \tan^{-1} \omega \tag{8.17}$$

as shown in Fig. 8.18*b*.

A straight-line approximation of the actual magnitude versus frequency curve can be obtained from examining the asymptotic behavior of the factor $j\omega + 1$. For $\omega \ll 1$, the low-frequency asymptote is

$$20 \log |j\omega + 1| \,\big|_{\omega \ll 1} \cong 20 \log 1 = 0 \text{ db} \tag{8.18}$$

For $\omega \gg 1$, the high-frequency asymptote is

$$20 \log |j\omega + 1| \,\big|_{\gg 1} \cong 20 \log \omega \text{ db} \tag{8.19}$$

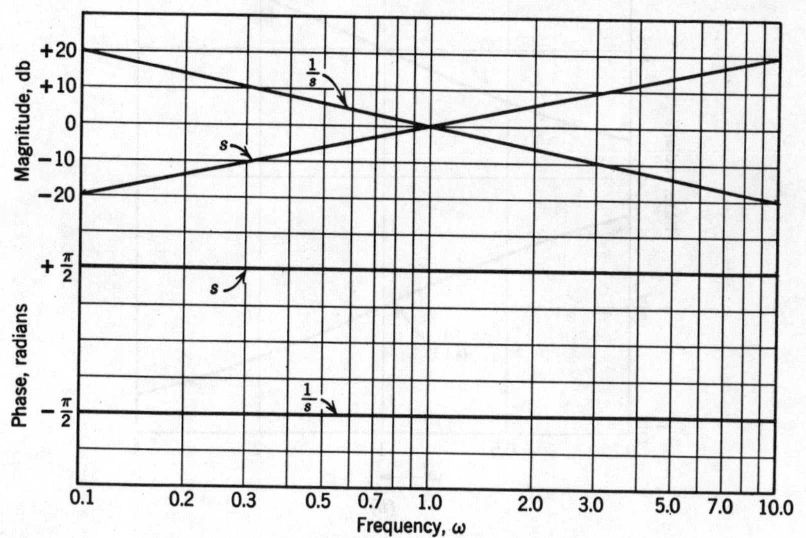

FIG. 8.17. Magnitude and phase of pole or zero at $s = 0$.

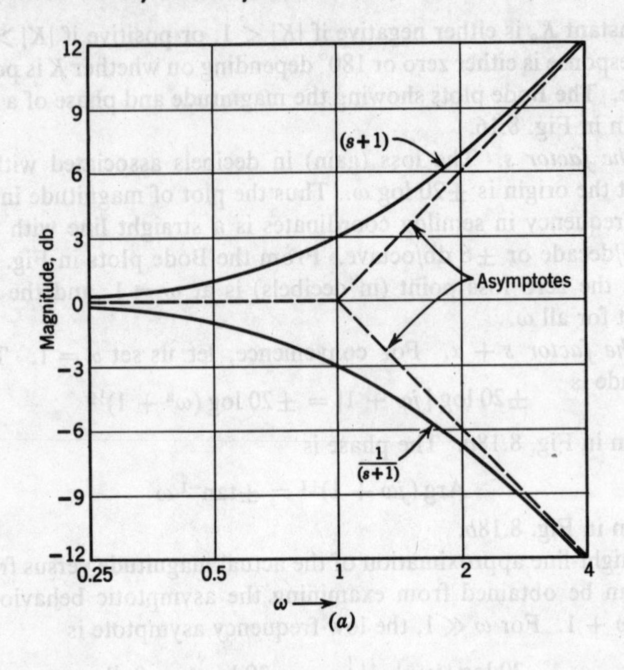

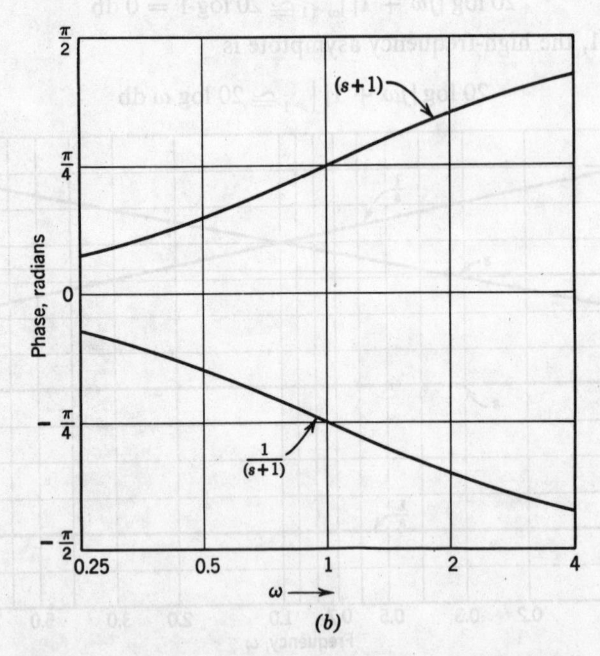

FIG. 8.18. Magnitude and phase of simple real pole or zero.

TABLE 8.2

Frequency		Actual Magnitude, db	Straight-Line Approximation, db	Error, db
$\omega = \frac{1}{4}$	2 octaves below	± 0.3	0	± 0.3
$\omega = \frac{1}{2}$	octave below	± 1	0	± 1
$\omega = 1$	break frequency	± 3	0	± 3
$\omega = 2$	octave above	± 7	± 6	± 1
$\omega = 4$	2 octaves above	± 12.3	± 12	± 0.3

which, as we saw in (*b*), has a slope of 20 db/decade or 6 db/octave. The low- and high-frequency asymptotes meet at $\omega = 1$, which we designate as the *break frequency* or *cutoff frequency*. The straight-line approximation is shown by the dashed lines in Fig. 8.18*a*. Table 8.2 shows the comparison between the actual magnitude versus the straight-line approximation. We see that the maximum error is at the break frequency $\omega = 1$, or in unnormalized form: $\omega = \alpha$.

For quick estimates of magnitude response, the straight-line approximation is an invaluable visual aid. An important example of the use of these straight-line approximations is in the design of linear control systems.

(*d*) *Complex conjugate roots*. For complex conjugate roots, it is convenient to adopt standard symbols so that we can use the universal curves that result therefrom. We describe the conjugate pole (zero) pair in terms of a magnitude ω_0 and an angle θ measured from the negative real axis, as shown in Fig. 8.19. Explicitly, the parameters that describe the pole (zero) positions are ω_0, which we call the *undamped frequency of oscillation*, and $\zeta = \cos\theta$, known as the *damping factor*. If the pole (zero) pair is given in terms of its real and imaginary parts,

$$p_{1,2} = -\alpha \pm j\beta \tag{8.20}$$

α and β are related to ζ and ω_0 by the following:

$$\begin{aligned} \alpha &= \omega_0 \cos\theta = \omega_0 \zeta \\ \beta &= \omega_0 \sin\theta = \omega_0 \sqrt{1-\zeta^2} \end{aligned} \tag{8.21}$$

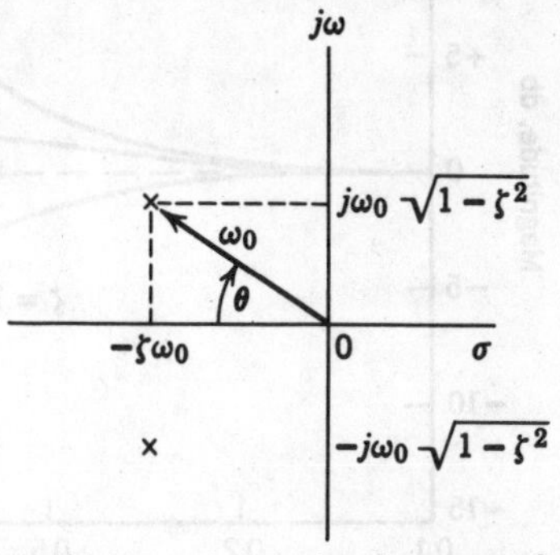

FIG. 8.19. Pole location in terms of ζ and ω_0.

Returning to the definition of the damping factor, $\zeta = \cos\theta$, we see that the closer the angle θ is to $\pi/2$, the smaller is the damping factor. When the angle θ is nearly zero degrees, the damping factor is nearly unity.

To examine the Bode plots for the conjugate pole (zero) pair, let us set $\omega_0 = 1$ for convenience. The magnitude is then

$$\pm 20 \log |1 - \omega^2 + j2\zeta\omega| = \pm 20 \log [(1 - \omega^2)^2 + 4\zeta^2\omega^2]^{1/2} \tag{8.22}$$

and the phase is

$$\phi(\omega) = \tan^{-1} \frac{2\zeta\omega}{1 - \omega^2} \tag{8.23}$$

If we examine the low- and high-frequency asymptotes of the magnitude, we see that the low-frequency asymptote is 0 decibels; the high-frequency asymptote (for $\omega \gg 1$) is $\pm 40 \log \omega$, which is a straight line of 40 db/decade or 12 db/octave slope. The damping factor ζ plays a significant part in the closeness of the straight-line approximation, however. In Fig. 8.20 the asymptotic approximation for a pair of conjugate poles ($\omega_0 = 1$) is indicated by the dashed line. Curves showing the magnitude for $\zeta = 0.1$, $\zeta = 0.6$, and $\zeta = 1.0$ are given by the solid lines. We see that only for $\zeta \simeq 0.6$ is the straight-line approximation a close one. Universal curves for magnitude and phase are plotted in Figs. 8.21 and 8.22 for the frequency normalized function

$$G(s) = \frac{1}{(s/\omega_0)^2 + 2\zeta(s/\omega_0) + 1} \tag{8.24}$$

We see that the phase response, as viewed from a semilog scale, is an odd function about $\omega/\omega_0 = 1$. The phase at $\omega = \omega_0$ is $-90°$ or $-\pi/2$ radians.

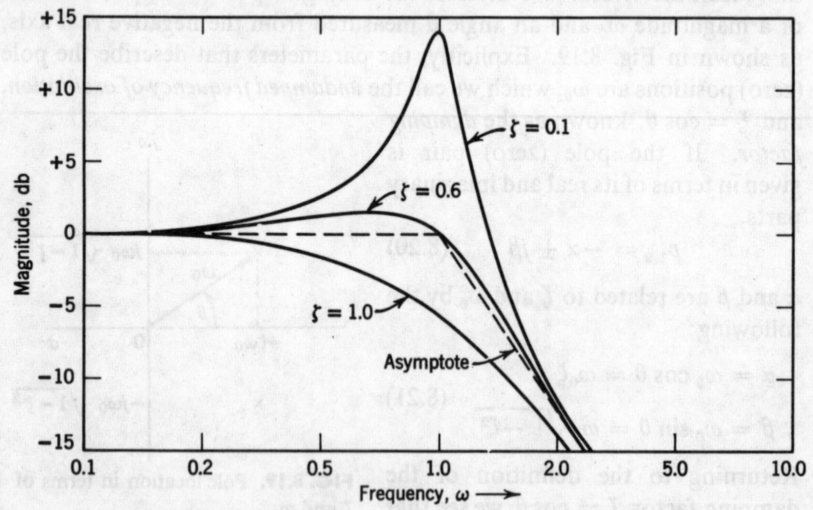

FIG. 8.20. Magnitude versus frequency for second-order pole.

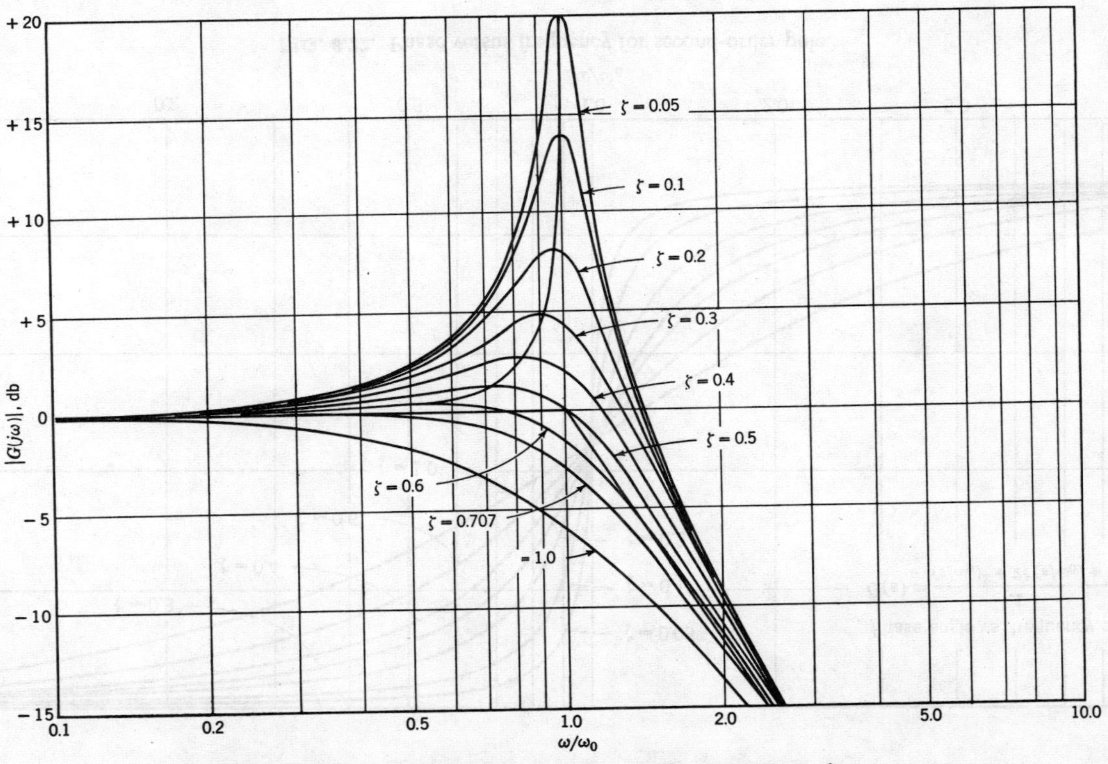

FIG. 8.21. Magnitude versus frequency for second-order pole.

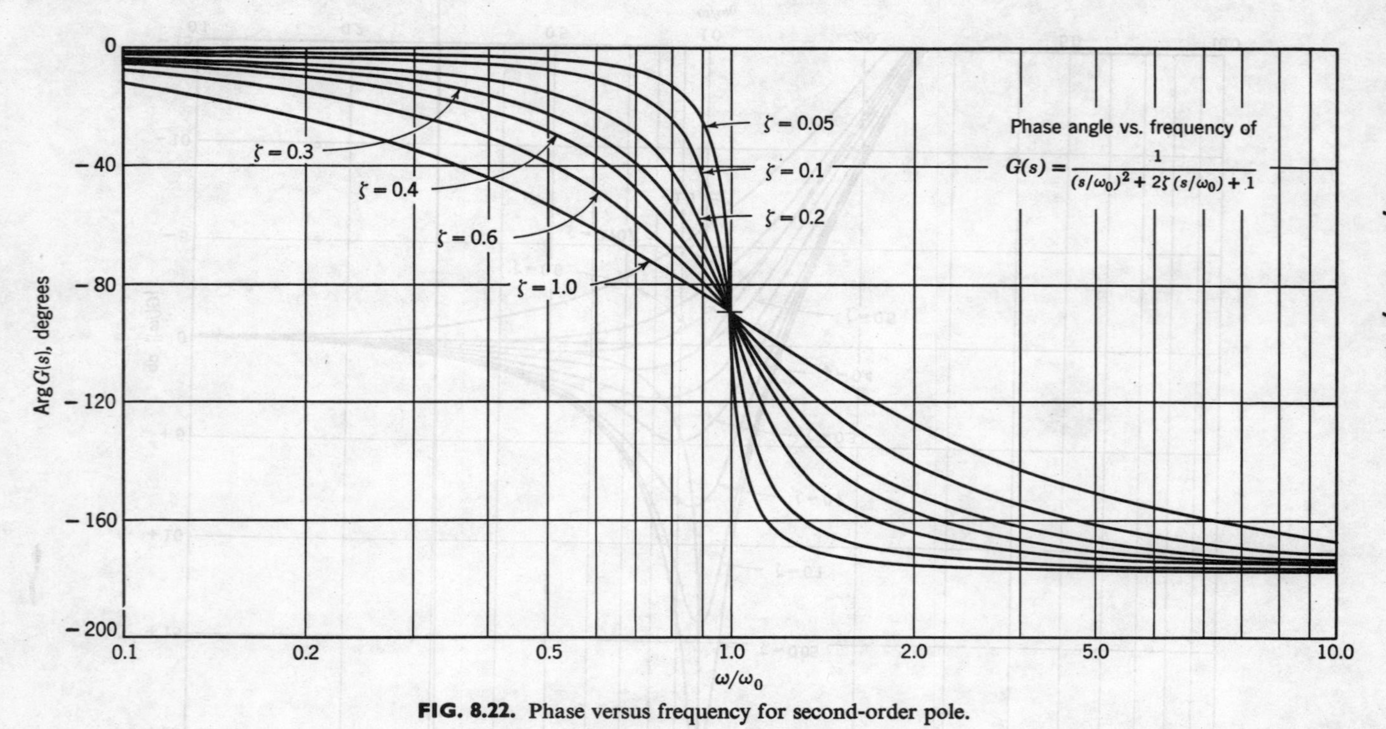

FIG. 8.22. Phase versus frequency for second-order pole.

For a conjugate pair of zeros, we need only reverse the signs on the scales of the magnitude and phase curves.

Example 8.1. Using Bode plot asymptotes, let us construct the magnitude versus frequency curve for

$$G(s) = \frac{0.1s}{\left(\frac{s}{50} + 1\right)\left(\frac{s^2}{16 \times 10^4} + \frac{s}{10^3} + 1\right)} \tag{8.25}$$

We see there are two first-order break frequencies at $\omega = 0$ and $\omega = 50$. In addition, there is a second-order break frequency at $\omega = 400$. With a quick calculation we find that $\zeta = 0.2$ for the second-order factor. The asymptotes are shown in Fig. 8.23. The magnitude and phase plots are given in Fig. 8.24 through a microfilm plot computer program.

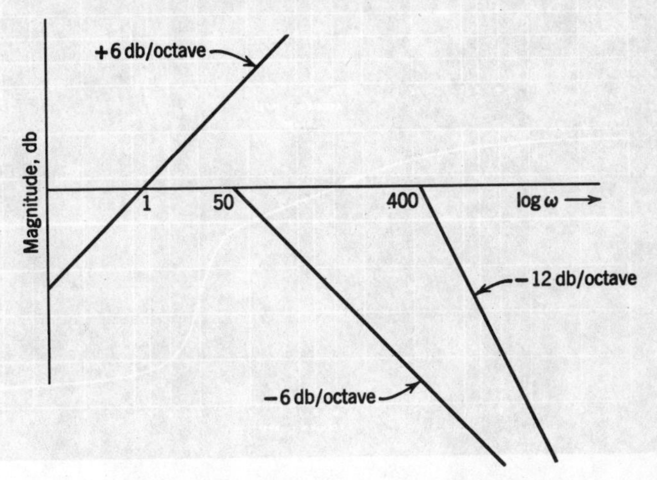

FIG. 8.23. Asymptotes for $G(s)$ in Eq. 8.25.

8.3 SINGLE-TUNED CIRCUITS

We will now study a class of circuits whose system functions can be described by a pair of conjugate poles. These circuits are called *single-tuned* circuits because they only need two reactive elements—an inductor and a capacitor. The undamped frequency of oscillation of the circuit is then $\omega_0 = (LC)^{-1/2}$. An example of a single-tuned circuit is the *R-L-C* circuit in Fig. 8.25, whose voltage-ratio transfer function is

$$H(s) = \frac{V_0(s)}{V_i(s)} = \frac{1/sC}{R + sL + 1/sC} = \frac{1/LC}{s^2 + (R/L)s + 1/LC} \tag{8.26}$$

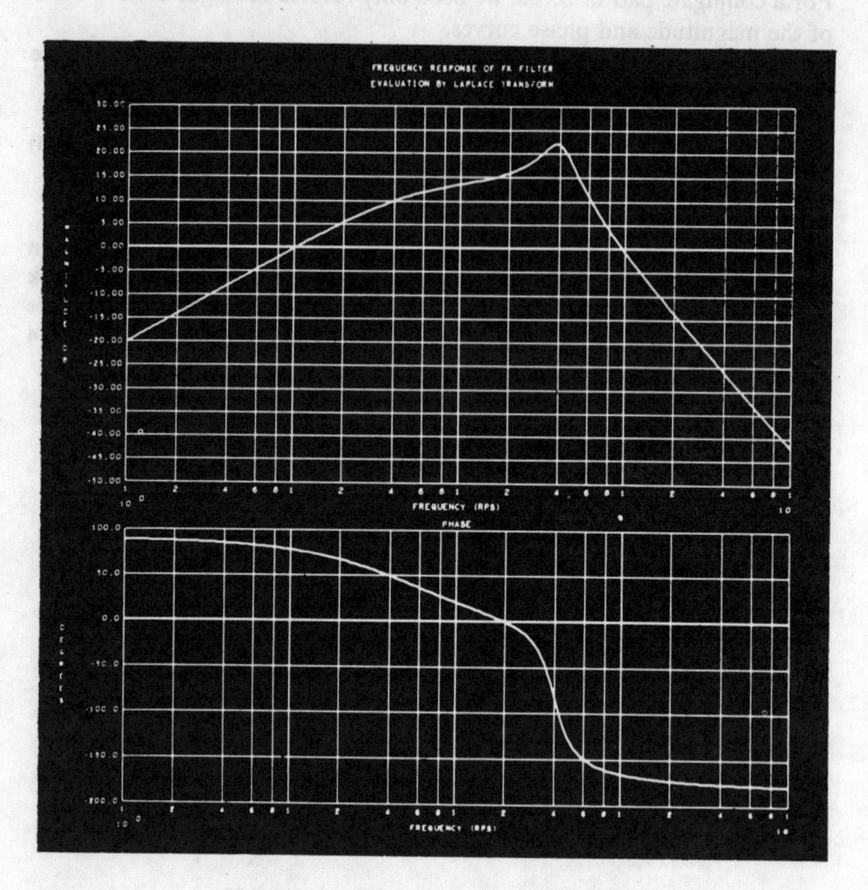

FIG. 8.24*a*. Magnitude of $G(s)$ in Eq. 8.25.
FIG. 8.24*b*. Phase of $G(s)$ in Eq. 8.25.

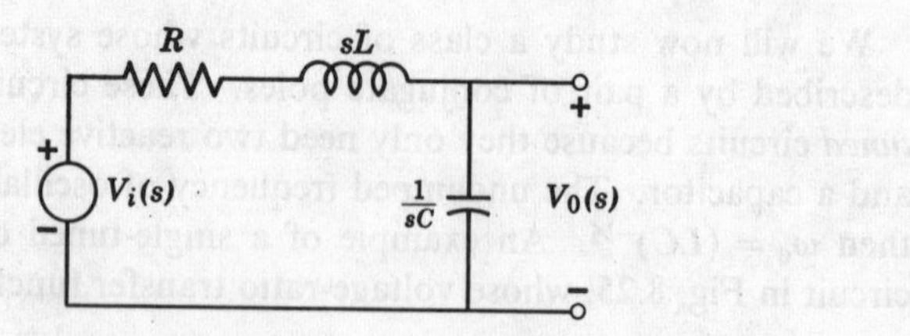

FIG. 8.25. Single-tuned circuit.

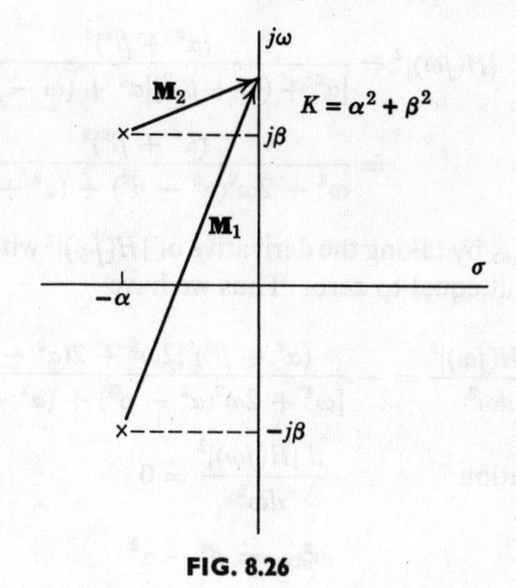

FIG. 8.26

The poles of $H(s)$ are

$$p_{1,2} = -\frac{R}{2L} \pm \frac{j}{2}\left(\frac{4}{LC} - \frac{R^2}{L^2}\right)^{1/2} = -\alpha \pm j\beta \tag{8.27}$$

where we assume that $(R^2/L^2) < (4/LC)$. In terms of α and β in Eq. 8.27, $H(s)$ is

$$H(s) = \frac{\alpha^2 + \beta^2}{(s + \alpha + j\beta)(s + \alpha - j\beta)} \tag{8.28}$$

From the pole-zero diagram of $H(s)$ shown in Fig. 8.26, we will determine the amplitude response $|H(j\omega)|$. Let us denote the vectors from the poles to the $j\omega$ axis as $|\mathbf{M}_1|$ and $|\mathbf{M}_2|$ as seen in Fig. 8.26. We can then write

$$|H(j\omega)| = \frac{K}{|\mathbf{M}_1|\,|\mathbf{M}_2|} \tag{8.29}$$

where $K = \alpha^2 + \beta^2$ and

$$\begin{aligned} |\mathbf{M}_1| &= [\alpha^2 + (\omega + \beta)^2]^{1/2} \\ |\mathbf{M}_2| &= [\alpha^2 + (\omega - \beta)^2]^{1/2} \end{aligned} \tag{8.30}$$

In characterizing the amplitude response, the point $\omega = \omega_{max}$, at which $|H(j\omega)|$ is maximum, is highly significant from both the analysis and design aspects. Since $|H(j\omega)|$ is always positive, the point at which $|H(j\omega)|^2$ is maximum corresponds exactly to the point at which $|H(j\omega)|$ is maximum. Since $|H(j\omega)|^2$ can be written as

$$|H(j\omega)|^2 = \frac{(\alpha^2+\beta^2)^2}{[\alpha^2+(\omega+\beta)^2][\alpha^2+(\omega-\beta)^2]}$$
$$= \frac{(\alpha^2+\beta^2)^2}{\omega^4+2\omega^2(\alpha^2-\beta^2)+(\alpha^2+\beta^2)^2} \tag{8.31}$$

we can find $\omega_{\max}$ by taking the derivative of $|H(j\omega)|^2$ with respect to ω^2 and setting the result equal to zero. Thus we have

$$\frac{d\,|H(j\omega)|^2}{d\omega^2} = -\frac{(\alpha^2+\beta^2)^2[2\omega^2+2(\alpha^2-\beta^2)]}{[\omega^4+2\omega^2(\alpha^2-\beta^2)+(\alpha^2+\beta^2)^2]^2} \tag{8.32}$$

From the equation
$$\frac{d\,|H(j\omega)|^2}{d\omega^2} = 0 \tag{8.33}$$

we determine
$$\omega_{\max}^2 = \beta^2 - \alpha^2 \tag{8.34}$$

Expressed in terms of the natural frequency of oscillation ω_0 and the damping factor ζ, $\omega_{\max}^2$ is

$$\omega_{\max}^2 = (\omega_0\sqrt{1-\zeta^2})^2 - (\zeta\omega_0)^2 = \omega_0^2(1-2\zeta^2) \tag{8.35}$$

Since $\omega_{\max}$ must always be real, the condition for $\omega_{\max}$ to exist, i.e., the condition for $|H(j\omega)|$ to possess a maximum, is given by the equation

$$2\zeta^2 \le 1 \tag{8.36}$$

so that $\zeta \le 0.707$. Since $\zeta = \cos\theta$, $\omega_{\max}$ does not exist for $\theta < 45°$. When $\theta = 45°$, we have the limiting case for which $\omega_{\max}$ exists. In this case, $\zeta = 0.707$ and the real and imaginary parts of the poles have the same magnitude, i.e., $\alpha = \beta$, or

$$\zeta\omega_0 = \omega_0\sqrt{1-\zeta^2} \tag{8.37}$$

We see from Eq. 8.34 that, when $\alpha = \beta$, then $\omega_{\max} = 0$. This is the lowest frequency at which $\omega_{\max}$ may be located. For $\zeta > 0.707$, or

$$\zeta\omega_0 > \omega_0\sqrt{1-\zeta^2} \tag{8.38}$$

$\omega_{\max}$ is imaginary; it therefore does not exist. To summarize, the key point in this analysis is that the imaginary part of the pole must be greater or equal to the real part of the pole in order for $\omega_{\max}$ to exist. Interpreted graphically, if we draw a circle in the s plane with the center at $-\alpha$ and the radius equal to β, then the circle must intersect the $j\omega$ axis in order for $\omega_{\max}$ to exist, as seen in Fig. 8.27. Moreover, the *point* at which the circle intersects the positive $j\omega$ axis is $\omega_{\max}$. This is readily seen from the triangle

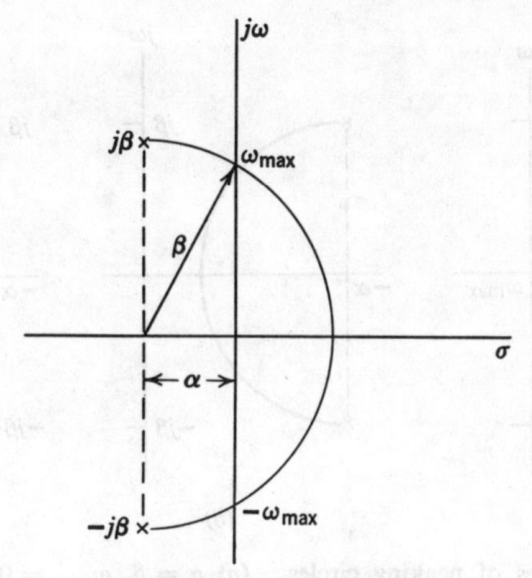

FIG. 8.27. Peaking circle.

with sides α, β, ω_{max} in Fig. 8.27. By the Pythagorean theorem, we find that

$$\omega_{max}^2 = \beta^2 - \alpha^2 \tag{8.39}$$

The circle described in Fig. 8.27 is called the *peaking circle*. When $\alpha = \beta$, the peaking circle intersects the $j\omega$ axis at $\omega = 0$, as seen in Fig. 8.28*a*. When $\alpha > \beta$, the circle does not intersect the $j\omega$ axis at all (Fig. 8.28*b*); therefore, ω_{max} cannot exist. When the imaginary part of the pole is much greater than the real part, i.e., when $\beta \gg \alpha$, then the circle intersects the $j\omega$ axis at approximately $\omega = \omega_0$, the natural frequency of oscillation of the circuit (Fig. 8.28*c*).

A figure of merit often used in describing the "peaking" of a tuned circuit is the circuit Q, which is defined in pole-zero notation as

$$Q \triangleq \frac{1}{2\zeta} = \frac{1}{2\cos\theta} \tag{8.40}$$

From this definition, we see that poles near the $j\omega$ axis (ζ small) represent high-Q systems, as given in Fig. 8.28*c*, and poles far removed from the $j\omega$ axis represent *low-Q* circuits (Fig. 8.28*a*). Although the Q of the circuit given by the pole-zero plot of Fig. 8.28*b* is theoretically defined, it has no practical significance because the circuit does not possess a maximum point in its amplitude.

By means of the peaking circle, we can also determine the half-power point, which is the frequency ω_C at which the amplitude response is $|H(j\omega_C)| = 0.707\,|H(j\omega_{max})|$.

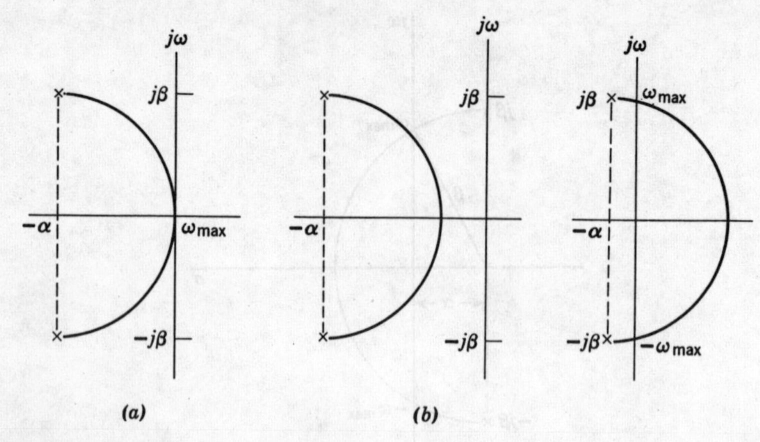

FIG. 8.28. Examples of peaking circles. (*a*) $\alpha = \beta$, $\omega_{\text{max}} = 0$. (*b*) $\alpha > \beta$, ω_{max} undetermined. (*c*) $\beta \gg \alpha$, $\omega_{\text{max}} \simeq \beta$.

We will now describe a method to obtain ω_C by geometrical construction. Consider the triangle in Fig. 8.29, whose vertices are the poles $\{p_1, p_1^*\}$ and a point ω_i on the $j\omega$ axis. The area of the triangle is

$$\text{Area}\ (\Delta p_1 p_1^* \omega_i) = \beta\alpha \tag{8.41}$$

In terms of the vectors $|\mathbf{M}_1|$ and $|\mathbf{M}_2|$ from the poles to ω_i, the area can also be expressed as

$$\text{Area}\ (\Delta p_1 p_1^* \omega_i) = \frac{|\mathbf{M}_1|\ |\mathbf{M}_2| \sin \psi}{2} \tag{8.42}$$

where ψ is the angle at ω_i, as seen in the figure. From Eqs. 8.41 and 8.42, we see that the product $|\mathbf{M}_1|\ |\mathbf{M}_2|$ is equal to

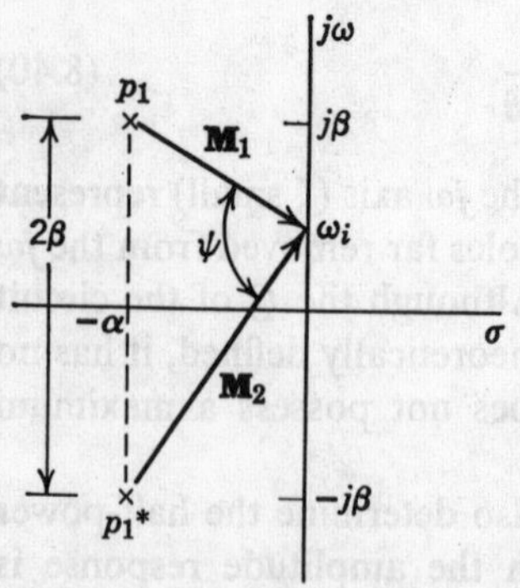

FIG. 8.29

$$|\mathbf{M}_1|\ |\mathbf{M}_2| = \frac{2\beta\alpha}{\sin \psi} \tag{8.43}$$

Since the amplitude response is

$$|H(j\omega_i)| = \frac{K}{|\mathbf{M}_1|\ |\mathbf{M}_2|} \tag{8.44}$$

where K is a constant, then

$$|H(j\omega_i)| = \frac{K \sin \psi}{2\beta\alpha} \tag{8.45}$$

For a given pole pair $\{p_1, p_1^*\}$ the parameters β, α, and K are prespecified. Therefore, we have derived $|H(j\omega)|$ in terms of a single variable parameter, the angle ψ. When the angle $\psi = \pi/2$ rad, then $\sin\psi = 1$, $\omega_i = \omega_{\max}$, and,

$$|H(j\omega_{\max})| = \frac{K}{2\beta\alpha} \tag{8.46}$$

When $\psi = \pi/4$ rad, then $\sin\psi = 0.707$ and

$$|H(j\omega_i)| = 0.707\,|H(j\omega_{\max})|$$

so that $\omega_i = \omega_C$. Let us consider now a geometric construction to obtain ω_C. Let us first draw the peaking circle as shown in Fig. 8.30. We will denote by A the point at which the peaking circle intersects the positive real axis. Now we draw a second circle with its center at A, and its radius equal to AB, the distance from A to either one of the poles, as seen in Fig. 8.30. The point where this second circle intersects the $j\omega$ axis is ω_C. The reason is that, at this point, the inscribed angle is $\psi_1 = \pi/4$ because it is equal to one-half the intercepted arc, which, by construction, is $\pi/2$.

When $\omega_{\max} = 0$, the half-power point ω_C is also called the *half-power bandwidth* of the tuned circuit. In Fig. 8.31*a* the half-power point is given when $\omega_{\max} = 0$. For a high-Q circuit, where $\omega_{\max} \simeq \omega_0$, the amplitude is highly peaked at $\omega = \omega_{\max}$, as shown in Fig. 8.31*b*. In this case, if $|H(j0)| < 0.707\,|H(j\omega_{\max})|$, there are two half-power points ω_{C_1} and ω_{C_2}

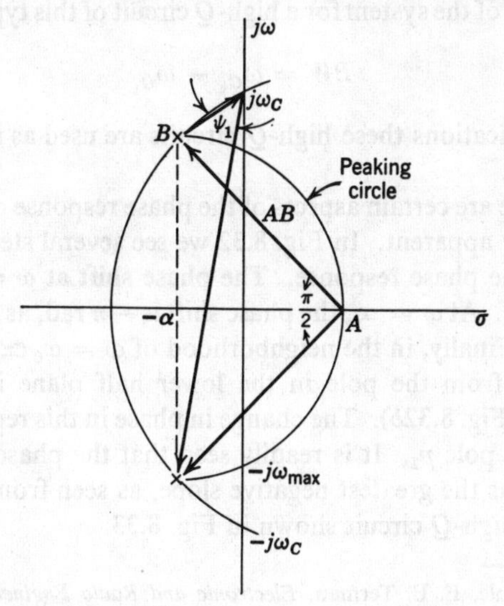

FIG. 8.30. Geometric construction to obtain half-power point.

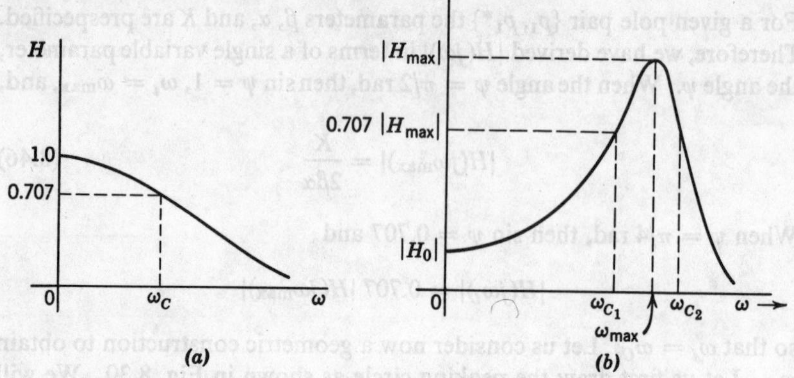

FIG. 8.31. (*a*) Low-Q circuit response. (*b*) High-Q circuit response.

about the point ω_{max}, as seen in Fig. 8.31*b*. By the construction process just described, we obtain the upper half-power point ω_{C_2}. It can be shown[1] that the point ω_{max} is the *geometric mean* of ω_{C_1} and ω_{C_2}, that is,

$$\omega_{C_1}\omega_{C_2} = \omega_{max}^2 \tag{8.47}$$

As a result, the lower half-power point is

$$\omega_{C_1} = \frac{\omega_{max}^2}{\omega_{C_2}} \tag{8.48}$$

The bandwidth of the system for a high-Q circuit of this type is described by

$$BW = \omega_{C_2} - \omega_{C_1} \tag{8.49}$$

In design applications these high-Q circuits are used as narrow bandpass filters.

Finally, there are certain aspects of the phase response of high-Q circuits that are readily apparent. In Fig. 8.32 we see several steps in the process of obtaining the phase response. The phase shift at $\omega = 0$ is 0, as seen from Fig. 8.32*a*. At $\omega = \infty$, the phase shift is $-\pi$ rad, as shown in part (*c*) of the figure. Finally, in the neighborhood of $\omega = \omega_0 \simeq \omega_{max}$, the phase shift resulting from the pole in the lower half plane is approximately $-\theta_2 = -\pi/2$ (Fig. 8.32*b*). The change in phase in this region is controlled in large by the pole p_1. It is readily seen that the phase response in the region of ω_0 has the greatest negative slope, as seen from a typical phase response of a high-Q circuit shown in Fig. 8.33.

[1] See for example, F. E. Terman, *Electronic and Radio Engineering*, McGraw-Hill Book Company, New York, 1953.

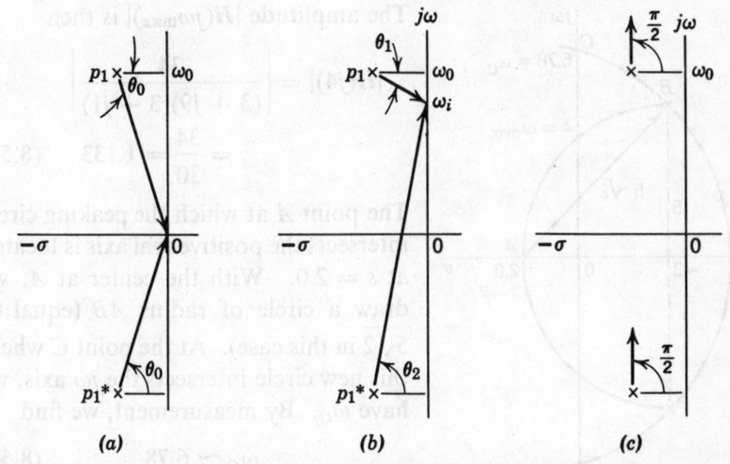

FIG. 8.32. Several steps in obtaining phase response for high-Q circuit. (a) $\omega_i = 0$. (b) $\omega_i \simeq \omega_0$. (c) $\omega_i = \alpha$.

Finally, as an example to illustrate our discussion of single-tuned circuits, let us find the amplitude response for the system function

$$H(s) = \frac{34}{s^2 + 6s + 34} \tag{8.50}$$

Now we determine the maximum and half-power points $\omega_{\max}$ and ω_C, and also the amplitudes $|H(j\omega_{\max})|$ and $|H(j\omega_C)|$. In factored form, $H(s)$ is

$$H(s) = \frac{34}{(s + 3 + j5)(s + 3 - j5)} \tag{8.51}$$

and the poles of $H(s)$ are shown in Fig. 8.34. We next draw the peaking circle with the center at $s = -3$ and the radius equal to 5. At the point where the circle intersects the $j\omega$ axis, we see that $\omega_{\max} = 4$. To check this result, the equation $\omega_{\max}^2 = \beta^2 - \alpha^2$ gives

$$\omega_{\max} = (5^2 - 3^2)^{1/2} = 4 \tag{8.52}$$

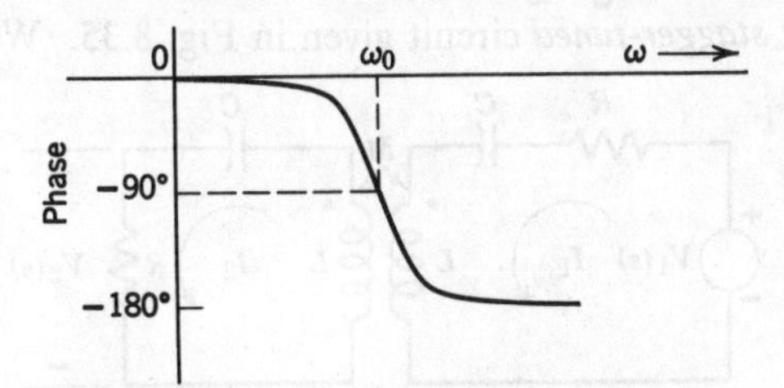

FIG. 8.33. Phase response of high-Q circuit.

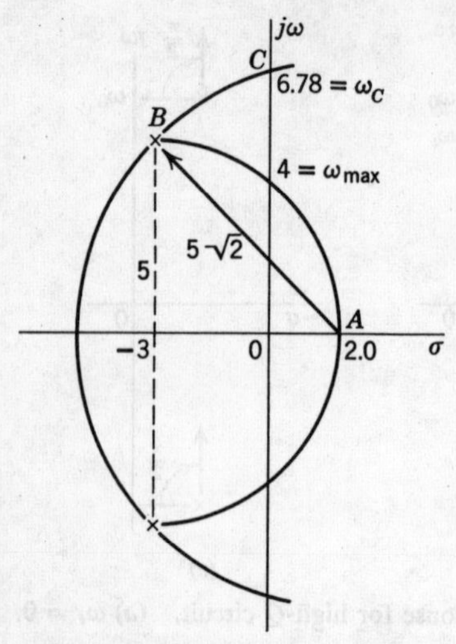

FIG. 8.34. Peaking circle construction example.

The amplitude $|H(j\omega_{max})|$ is then

$$|H(j4)| = \left|\frac{34}{(3+j9)(3-j1)}\right|$$
$$= \frac{34}{30} = 1.133 \qquad (8.53)$$

The point A at which the peaking circle intersects the positive real axis is located at $s = 2.0$. With the center at A, we draw a circle of radius AB (equal to $5\sqrt{2}$ in this case). At the point C where this new circle intersects the $j\omega$ axis, we have ω_C. By measurement, we find

$$\omega_C \simeq 6.78 \qquad (8.54)$$

Let us check this result. Referring to Fig. 8.34, we know that the line segment AB is of length $5\sqrt{2}$; it follows that AC is also $5\sqrt{2}$ units long. The line segment AO is of length

$$AO = 5 - 3 = 2 \text{ units} \qquad (8.55)$$

Then ω_C is given as

$$\omega_C = \sqrt{(AC)^2 - (AO)^2} = \sqrt{46} = 6.782 \qquad (8.56)$$

Finally, we obtain $|H(j\omega_C)|$ as

$$|H(j6.782)| = \frac{34}{\sqrt{(34-46)^2 + (6\sqrt{46})^2}} = 0.802 \qquad (8.57)$$

which is precisely 0.707 $|H(j\omega_{max})|$.

8.4 DOUBLE-TUNED CIRCUITS

In Section 8.3, we studied the frequency response for a pair of conjugate poles. Now we turn our attention to the amplitude response of two pairs of conjugate poles in a high-Q situation. The circuit we analyze here is the *double-tuned* or *stagger-tuned* circuit given in Fig. 8.35. We will consider

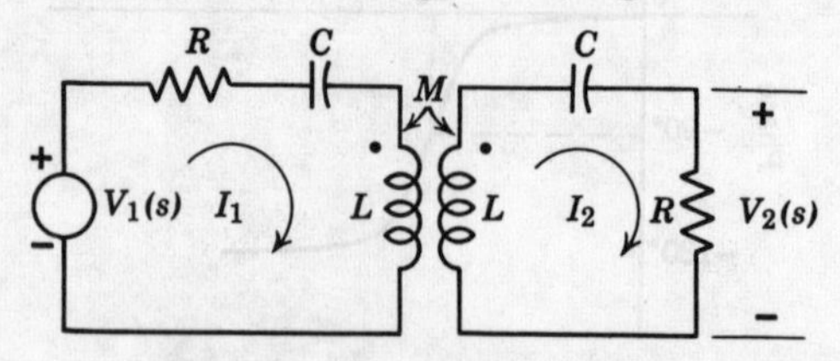

FIG. 8.35. Double-tuned circuit.

the special case when the R, L, and C elements in the primary circuit are equal in value to their counterparts in the secondary. Since the primary and secondary inductances are equal, the mutual inductance is

$$M = KL \tag{8.58}$$

In this analysis we assume the coefficient of coupling K to be a variable parameter. Let us determine the amplitude response for the voltage-ratio transfer function $V_2(s)/V_1(s)$. From the mesh equations

$$\begin{aligned} V_1(s) &= \left(R + sL + \frac{1}{sC}\right) I_1(s) - sM\, I_2(s) \\ 0 &= -sM\, I_1(s) + \left(R + sL + \frac{1}{sC}\right) I_2(s) \end{aligned} \tag{8.59}$$

we readily determine

$$H(s) = \frac{V_2(s)}{V_1(s)} = \frac{s^3RM/L^2}{[s^2 + (R/L)s + 1/LC]^2 - s^4K^2} \tag{8.60}$$

Using tuned-circuit notation, we set

$$\begin{aligned} 2\zeta\omega_0 &= \frac{R}{L} \\ \omega_0^{\,2} &= \frac{1}{LC} \end{aligned} \tag{8.61}$$

$H(s)$ can then be written as

$$H(s) = \frac{s^3RM/L^2}{(1 - K^2)\left(s^2 + \dfrac{2\zeta\omega_0}{1 + K}s + \dfrac{\omega_0^{\,2}}{1 + K}\right)\left(s^2 + \dfrac{2\zeta\omega_0}{1 - K}s + \dfrac{\omega_0^{\,2}}{1 - K}\right)} \tag{8.62}$$

If we set

$$A = \frac{RM}{L^2(1 - K^2)} = \frac{2\zeta\omega_0 K}{1 - K^2} \tag{8.63}$$

then we can write

$$H(s) = \frac{As^3}{(s - s_1)(s - s_1^*)(s - s_2)(s - s_2^*)} \tag{8.64}$$

where

$$\begin{aligned} \{s_1, s_1^*\} &= -\frac{\zeta\omega_0}{1 + K} \pm j\omega_0\left[\frac{1}{1 + K} - \frac{\zeta^2}{(1 + K)^2}\right]^{1/2} \\ \{s_2, s_2^*\} &= -\frac{\zeta\omega_0}{1 - K} \pm j\omega_0\left[\frac{1}{1 - K} - \frac{\zeta^2}{(1 - K)^2}\right]^{1/2} \end{aligned} \tag{8.65}$$

Let us restrict our analysis to a high-Q circuit so that $\zeta^2 \ll 1$. Furthermore, let us assume that the circuit is loosely coupled so that $K \ll 1$. Under these assumptions, we can approximate the pole locations by discarding the terms involving ζ^2 under the radicals in Eq. 8.65. Then the poles $\{s_1, s_1^*\}$ can be given approximately as

$$\{s_1, s_1^*\} \simeq -\zeta\omega_0 \pm j\omega_0\left(1 - \frac{K}{2}\right) \tag{8.66}$$

Similarly, $\{s_2, s_2^*\}$ can be given as

$$\{s_2, s_2^*\} \simeq -\zeta\omega_0 \pm j\omega_0\left(1 + \frac{K}{2}\right) \tag{8.67}$$

The pole-zero diagram of $H(s)$ is given in Fig. 8.36. The real part of the poles $-\zeta\omega_0$ is greatly enlarged in comparison to the imaginary parts for

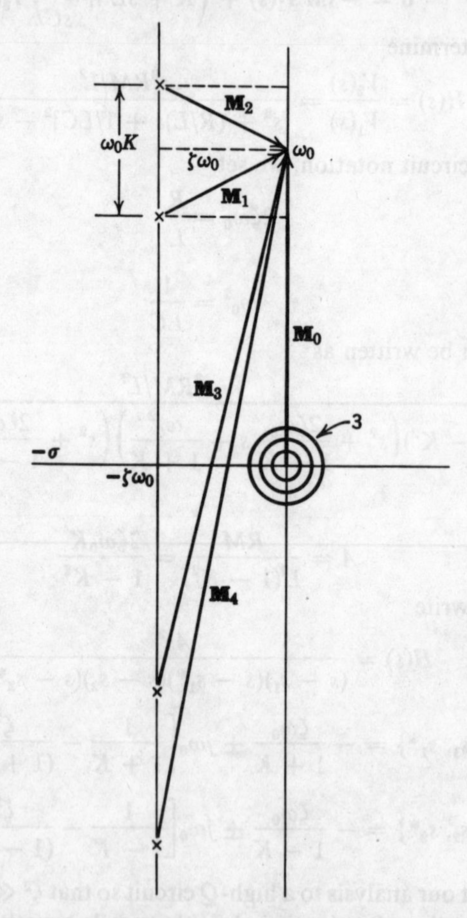

FIG. 8.36. Poles and zeros of a double-tuned circuit.

clarity purposes. Note that we have a triple zero at the origin. In terms of the vectors in Fig. 8.36, the amplitude response is

$$|H(j\omega)| = \frac{A\,|\mathbf{M}_0|^3}{|\mathbf{M}_1|\,|\mathbf{M}_2|\,|\mathbf{M}_3|\,|\mathbf{M}_4|} \tag{8.68}$$

Since the circuit is high-Q in the vicinity of $\omega = \omega_0$, we have

$$|\mathbf{M}_3| \simeq |\mathbf{M}_4| \simeq 2\,|\mathbf{M}_0| \simeq 2\omega_0 \tag{8.69}$$

so that in the neighborhood of ω_0

$$|H(j\omega)| \simeq \frac{A\omega_0}{4\,|\mathbf{M}_1|\,|\mathbf{M}_2|} \tag{8.70}$$

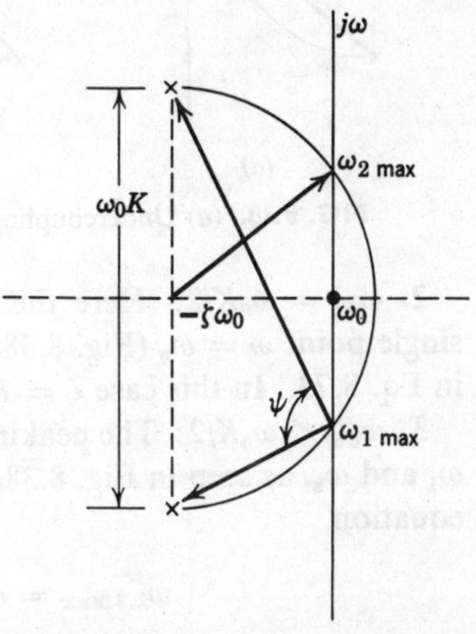

FIG. 8.37. Peaking circle for double-tuned circuit.

It is evident that the amplitude response of $|H(j\omega)|$ in the neighborhood of ω_0 depends only upon the pair of vectors $|\mathbf{M}_1|$ and $|\mathbf{M}_2|$. The double-tuned problem has thus been reduced to a single-tuned problem in the neighborhood of ω_0. Consequently, we can use all the results on the peaking circle that were derived in Section 8.3. Let us draw a peaking circle with the center at

$$s = -\zeta\omega_0 + j\omega_0 \tag{8.71}$$

and with a radius equal to $\omega_0 K/2$, as shown in Fig. 8.37. The inscribed angle ψ then determines the location of the maxima and half-power points of the response. Without going into the derivation, the amplitude response can be expressed as a function of ψ according to the equation

$$|H(j\omega)| = \frac{A\omega_0 \sin\psi}{4\omega_0\, K(\zeta\omega_0)} = \frac{\sin\psi}{2(1 - K^2)} \tag{8.72}$$

Referring to the peaking circle in Fig. 8.37, let us consider the following situations:

1. $\omega_0\zeta > \omega_0 K/2$: In this case, the peaking circle never intersects the $j\omega$ axis; ψ is always less than $\pi/2$ (Fig. 8.38*a*), and the amplitude response never attains the theoretical maximum

$$H_{\max} = \frac{1}{2(1 - K^2)} \tag{8.73}$$

as seen by the curve labeled (*a*) in Fig. 8.39. In this case, $K < 2\zeta$, and the circuit is said to be *undercoupled*.

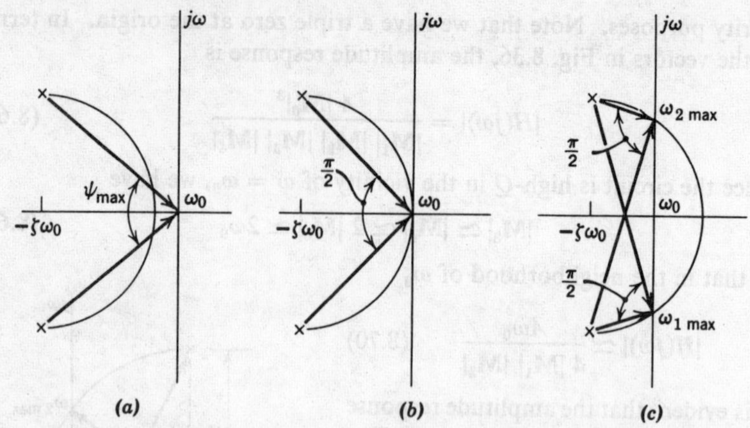

FIG. 8.38. (a) Undercoupling. (b) Critical coupling. (c) Overcoupling.

2. $\omega_0\zeta = \omega_0 K/2$: Here the peaking circle intersects the $j\omega$ axis at a single point $\omega = \omega_0$ (Fig. 8.38b). At ω_0, the amplitude is equal to H_{max} in Eq. 8.73. In this case $\zeta = K/2$, and we have *critical coupling*.

3. $\omega_0\zeta < \omega_0 K/2$: The peaking circle intersects the $j\omega$ axis at two points. ω_1 and ω_2, as seen in Fig. 8.38c. The intersecting points are given by the equation

$$\omega_{1,2\,\max} = \omega_0 \pm \omega_0\left[\left(\frac{K}{2}\right)^2 - \zeta^2\right]^{1/2} \tag{8.74}$$

Consequently, the amplitude response attains the theoretical maximum H_{max} at two points, as shown by curve (c) in Fig. 8.39. In this situation, the circuit is said to be *overcoupled*.

Note that in Fig. 8.39 the undercoupled and critically coupled curves have their maximum points at ω_0. The overcoupled curve, however, is

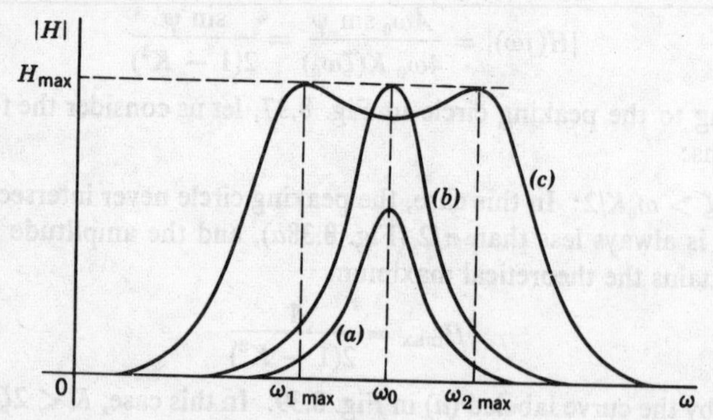

FIG. 8.39. (a) Undercoupled case. (b) Critically coupled case. (c) Overcoupled case.

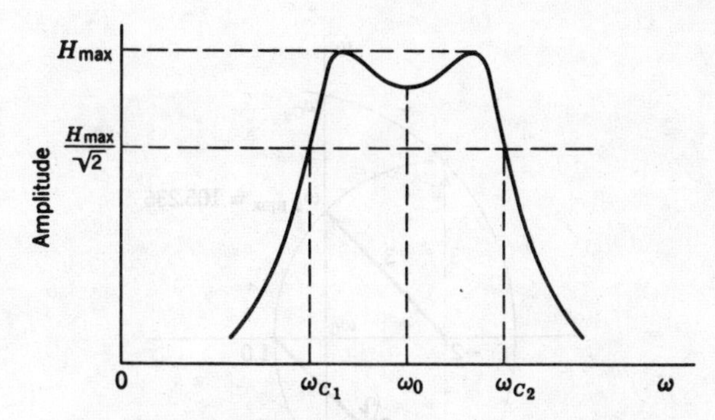

FIG. 8.40. Half-power points of overcoupled circuit.

maximum at ω_1 and ω_2. In the case of overcoupling and critical coupling, we can determine the half-power points by using the geometrical construction method given in Section 8.3. Observe that there are two half-power points ω_{C_2} and ω_{C_1}, as shown in the overcoupled curve in Fig. 8.40. The bandwidth of the circuit is then

$$BW = \omega_{C_2} - \omega_{C_1} \tag{8.75}$$

Example 8.2. The voltage-ratio transfer function of a double-tuned circuit is given as

$$H(s) = \frac{As^3}{(s+2+j100)(s+2-j100)(s+2+j106)(s+2-j106)} \tag{8.76}$$

From $H(s)$, let us determine the following: (*a*) the maximum points $\omega_{1\,\max}$ and $\omega_{2\,\max}$; (*b*) the 3 db bandwidth BW; (*c*) the damping factor ζ; (*d*) the coefficient of coupling K; (*e*) the gain constant A; and (*f*) the maximum of the amplitude response $H_{\max}$.

Solution. (*a*) The natural frequency of oscillation ω_0 is taken to be approximately halfway between the two poles, that is, $\omega_0 = 103$ radians. In the neighborhood of ω_0, we draw the poles $s = -2 + j100$ and $s = -2 + j106$, as shown in Fig. 8.41. From the peaking circle centered at the point $s = -2 + j\omega_0$, shown in Fig. 8.41, we obtain

$$\omega_{2\,\max} - \omega_0 = \sqrt{3^2 - 2^2} = 2.236 \text{ radians} \tag{8.77}$$

so that

$$\begin{aligned} \omega_{2\,\max} &= \omega_0 + 2.236 = 105.236 \text{ radians} \\ \omega_{1\,\max} &= \omega_0 - 2.236 = 100.764 \text{ radians} \end{aligned} \tag{8.78}$$

(*b*) Next we draw a circle centered at $s = 1 + j\omega_0$ with radius $3\sqrt{2}$. Where this circle intersects the $j\omega$ axis, we have ω_{C_2} so that

$$\omega_{C_2} - \omega_0 = \sqrt{(3\sqrt{2})^2 - 1} = 4.123 \text{ radians} \tag{8.79}$$

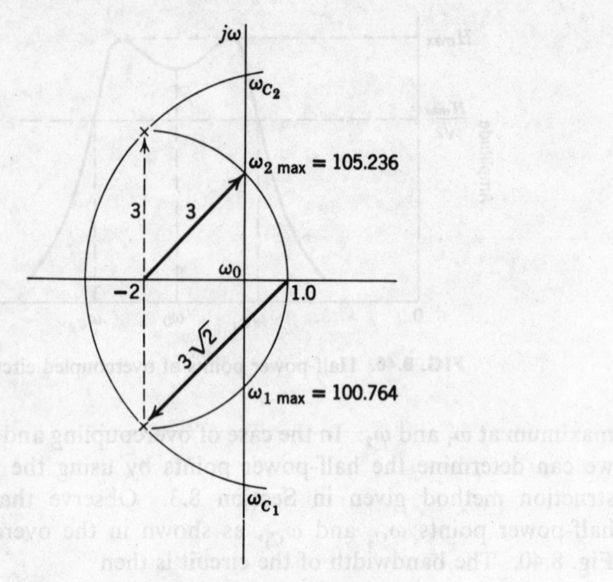

FIG. 8.41. Peaking circle for Example 8.2.

The 3 db bandwidth is then

$$BW = 2(\omega_{C_2} - \omega_0) = 8.246 \text{ radians} \tag{8.80}$$

(*c*) The damping factor ζ is obtained from the real part of the poles $\zeta\omega_0 = 2$, from which we obtain

$$\zeta = \frac{2}{103} = 0.0194 \tag{8.81}$$

(*d*) The coefficient of coupling K is obtained from the radius of the peaking circle, which is

$$\frac{\omega_0 K}{2} = 3 \tag{8.82}$$

We thus have

$$K = \frac{6}{\omega_0} = 0.0582 \tag{8.83}$$

(*e*) The gain constant A is equal to

$$A = \frac{2\zeta\omega_0 K}{1 - K^2} = \frac{2(2)(0.0582)}{1 - (0.0582)^2} = 0.2328 \tag{8.84}$$

(*f*) Finally, the maximum amplitude $H_{\max}$ is

$$H_{\max} = \frac{1}{2(1 - K^2)} = 0.5009 \tag{8.85}$$

8.5 ON POLES AND ZEROS AND TIME DELAY

What is time delay? How do we relate it to frequency response? We will attempt to answer these questions in this section. First consider the transfer function of pure delay

$$H(s) = e^{-sT} \tag{8.86}$$

For a system described by Eq. 8.86, any excitation $e(t)$ produces an identical response signal $e(t - T)$, which is delayed by time T with respect to the excitation. This is shown by the Laplace transform relationship,

$$R(s) = \mathcal{L}[e(t - T)] = e^{-sT}\,\mathcal{L}[e(t)] \tag{8.87}$$

Let us examine the amplitude and phase response of the pure delay. From the equation

$$H(j\omega) = e^{-j\omega T} \tag{8.88}$$

we obtain the amplitude response

$$|H(j\omega)| = 1 \tag{8.89}$$

and the phase response

$$\phi(\omega) = -\omega T \tag{8.90}$$

We see that the delay T is equal to minus the derivative of the phase response, that is,

$$T = -\frac{d\phi(\omega)}{d\omega} \tag{8.91}$$

The magnitude, phase and delay characteristics of $H(j\omega) = e^{-j\omega T}$ are given in Fig. 8.42*a*, *b*, and *c*.

If we define delay as in Eq. 8.91, we can readily deduce that for the response to be nearly identical to the excitation, the system amplitude response should be constant, and its phase response should be linear over the frequency range of interest. If the phase is not linear, we have what is known as *delay distortion*. To visualize delay distortion more clearly,

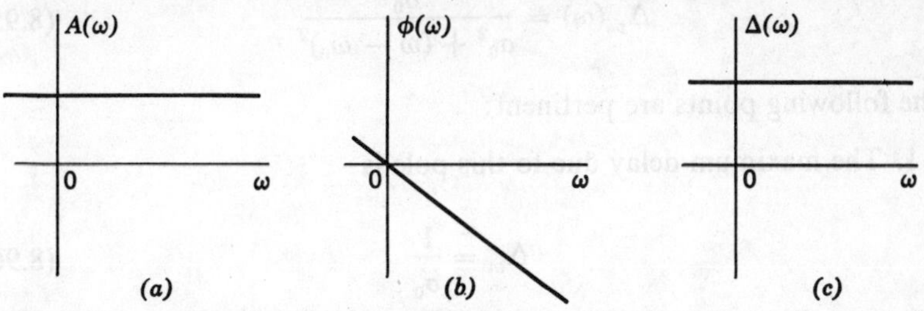

FIG. 8.42. Amplitude, phase, and delay of ideal delay function. (*a*) Amplitude. (*b*) Phase. (*c*) Delay.

we recall from Fourier analysis that any signal is made up of different frequency components. An ideal transmission system should delay each frequency component equally. If the frequency components are delayed by different amounts, the reconstruction of the output signal from its Fourier components would produce a signal of different shape as the input. For pulse applications, delay distortion is an essential design consideration.

Let us next examine how we relate delay, or *envelope delay* (as it is sometimes called) to the poles and zeros of a transfer function. For any transfer function

$$H(s) = \frac{\prod_{i=1}^{m} (s - z_i)}{\prod_{j=1}^{n} (s - p_j)} \tag{8.92}$$

with zeros at $z_i = -\sigma_i \pm j\omega_i$ and poles at $p_j = -\sigma_j \pm j\omega_j$, the phase for real frequencies is

$$\phi(\omega) = \sum_{i=1}^{m} \tan^{-1} \frac{\omega \pm \omega_i}{\sigma_i} - \sum_{j=1}^{n} \tan^{-1} \frac{\omega \pm \omega_j}{\sigma_j} \tag{8.93}$$

Envelope delay is

$$-\frac{d\phi(\omega)}{d\omega} = -\sum_{i=1}^{m} \frac{\sigma_i}{\sigma_i^2 + (\omega \pm \omega_i)^2} + \sum_{j=1}^{n} \frac{\sigma_j}{\sigma_j^2 + (\omega \pm \omega_j)^2} \tag{8.94}$$

We see that the shapes of the delay versus frequency characteristic are the same for all poles and zeros. The zeros contribute "negative" delay; the poles, positive delay. However, linear physical systems do not have transfer functions with zeros alone. The inductor $H(s) = Ls$ is the only exception. Its phase is $\phi(\omega) = \pi/2$; thus the delay is zero.

Now let us consider the delay due to *one* singularity, for example, a pole at $p_0 = -\sigma_0 + j\omega_0$. The delay due to the one pole is

$$\Delta_{p0}(\omega) = \frac{\sigma_0}{\sigma_0^2 + (\omega - \omega_0)^2} \tag{8.95}$$

The following points are pertinent:

1. The maximum delay due to this pole is

$$\Delta_m = \frac{1}{\sigma_0} \tag{8.96}$$

and occurs at $\omega = \omega_0$. The delay versus frequency curve is symmetric about $\omega = \omega_0$.

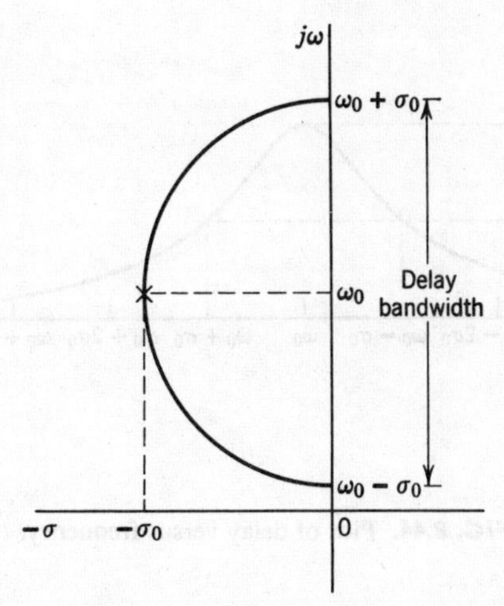

FIG. 8.43. Graphic construction to obtain delay bandwidth.

2. The frequency at which the delay is half the maximum, or $1/2\sigma_0$, is

$$\omega_{1/2} = \omega_0 \pm \sigma_0 \tag{8.97}$$

3. The "effective delay bandwidth" is then $\omega_0 - \sigma_0 < \omega < \omega_0 + \sigma_0$ or simply $2\sigma_0$. The upper and lower half-bandwidth points can be obtained graphically by drawing a circle with center at $\omega = \omega_0$ and radius σ_0. The intersections of the circle with the $j\omega$ axis are the half-bandwidth points, as shown in Fig. 8.43.

4. The product of the maximum delay and delay bandwidth is always 2. Thus, if we wish to obtain large delay by placing zeros or poles near the $j\omega$ axis, the effective delay bandwidth is then very narrow.

5. The delay of an all-pass function is twice the delay due to the poles alone.

The delay versus frequency curve for the pole is shown in Fig. 8.44. We see that the delay-bandwidth concept is only useful for a rough approximation, since the delay versus frequency characteristic only falls as $1/\omega$. To calculate the delay versus frequency characteristic for a transfer function with a number of poles and zeros, it is convenient to obtain the delay curves for the individual singularities and then to add the separate delays. It is not as desirable to obtain the total phase response and then differentiate numerically.

Finally, it should be pointed out that envelope delay only has meaning when the phase response goes through the origin. If it does not, there is a frequency-shift component in addition to the delay that is hard to account for analytically.

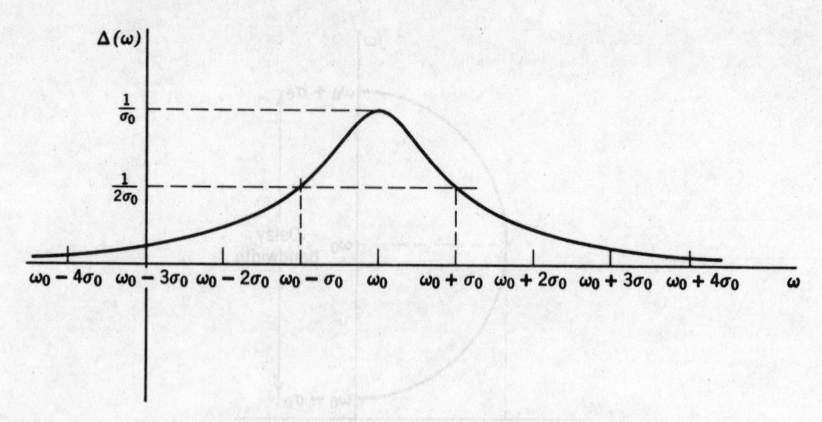

FIG. 8.44. Plot of delay versus frequency.

Problems

8.1 Find the poles and zeros of the impedances of the following networks and plot on a scaled s plane.

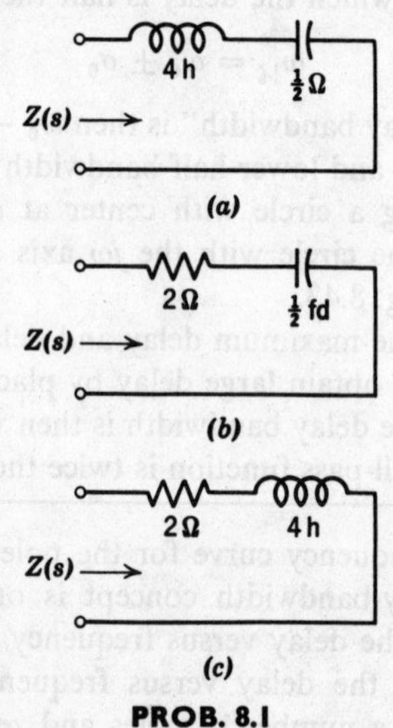

PROB. 8.1

8.2 The circuit shown in the figure is a *shunt peaking* circuit often used in video amplifiers.

(*a*) Show that the admittance $Y(s)$ is of the form

$$Y(s) = \frac{K(s - s_1)(s - s_2)}{(s - s_3)}$$

Express s_1, s_2, and s_3 in terms of R, L, and C.

(*b*) When $s_1 = -10 + j10^3$, $s_2 = -10 - j10^3$, and $Y(j0) = 10^{-2}$ mhos, find the values of R, L, and C and determine the numerical value of s_3.

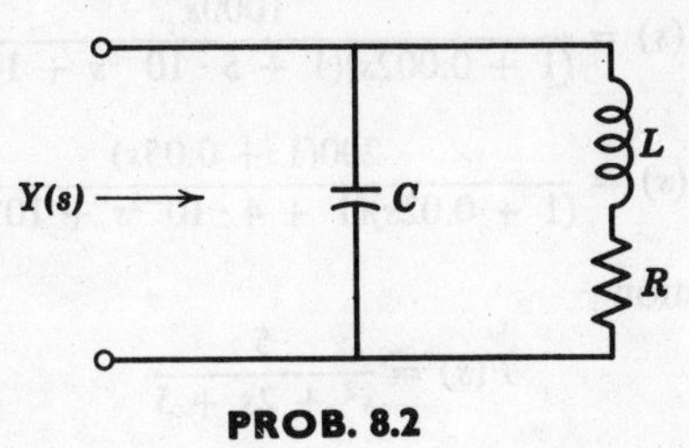

PROB. 8.2

8.3 Find the amplitude and phase response for the following functions and sketch.

(*a*) $F(s) = \dfrac{K}{s+K}$ (*b*) $F(s) = \dfrac{s}{s+K}$

(*c*) $F(s) = \dfrac{s}{s^2+\omega_0^2}$ (*d*) $F(s) = \dfrac{\omega_0}{s^2+\omega_0^2}$

Note that K and ω_0 are positive quantities.

8.4 Given the function

$$G(j\omega) = \frac{A(\omega) + jB(\omega)}{C(\omega) + jD(\omega)}$$

determine the amplitude and phase of $G(j\omega)$ in terms of A, B, C, D. Show that the amplitude function is even and the phase function is odd.

8.5 By means of the vector method, *sketch* the amplitude and phase response for

(*a*) $F(s) = \dfrac{s+0.5}{s(s+10)}$ (*c*) $F(s) = \dfrac{s-1}{s+1}$

(*b*) $F(s) = \dfrac{s}{s^2+2s+2}$ (*d*) $F(s) = \dfrac{s}{s^2-2s+2}$

(*e*) $F(s) = \dfrac{s+1}{s-1}$ (*g*) $F(s) = \dfrac{s^2-2s+5}{(s+2)(s+1)}$

(*f*) $F(s) = \dfrac{s^2+4}{(s+2)(s^2+9)}$ (*h*) $F(s) = \dfrac{s^2+2s+5}{(s+2)(s+1)}$

8.6 Plot on semilog paper the Bode plots of magnitude and phase for

(*a*)
$$F(s) = \frac{100(1+0.5s)}{s(s+2)}$$

(*b*)
$$F(s) = \frac{50(1+0.025s)(1+0.1s)}{(1+0.05s)(1+0.01s)}$$

8.7 Plot on semilog paper the Bode plots of magnitude and phase for

(*a*) $$F(s) = \frac{1000s}{(1 + 0.002s)(1 + 5 \cdot 10^{-5}s + 10^{-8}s^2)}$$

(*b*) $$F(s) = \frac{200(1 + 0.05s)}{(1 + 0.02s)(1 + 4 \cdot 10^{-2}s + 10^{-4}s^2)}$$

8.8 For the function

$$F(s) = \frac{5}{s^2 + 2s + 5}$$

determine $\omega_{\max}$, $|F(j\omega_{\max})|$, the half-power point ω_C, and $|F(j\omega_C)|$. Sketch the amplitude and phase response.

8.9 For the circuit shown, determine the current ratio I_L/I_g and find: (*a*) the point $\omega_{\max}$, where its amplitude is maximum; (*b*) the half-power point ω_C; (*c*) the point ω_u where $|I_L(\omega_u)/I_g(\omega_u)| = 1$. Use geometric construction.

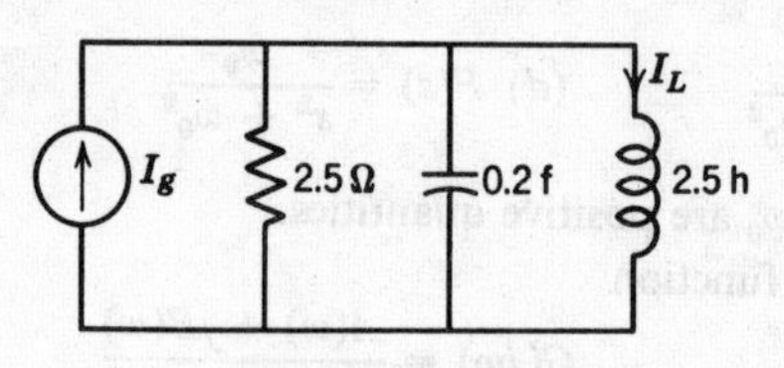

PROB. 8.9

8.10 A network function consists of two poles at $p_{1,2} = r_i e^{\pm j(\pi - \theta)} = -\sigma_i \pm j\omega_i$, as given in the figure. Show that the square of the amplitude response $M^2(\omega)$ is maximum at $\omega_m{}^2 = r_i{}^2 |\cos 2\theta|$.

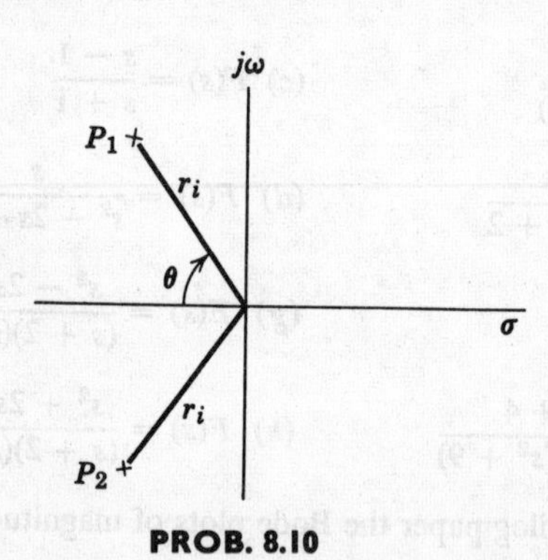

PROB. 8.10

8.11 In connection with Prob. 8.2 plot the poles and zeros of the impedance function $Z(s) = 1/Y(s)$. Find, approximately, the maximum point of the amplitude response. In addition, find the bandwidth at the half-power points and the circuit Q.

8.12 The pole configuration for a system function $H(s)$ is given in the figure. From the plot, calculate:

(*a*) The undamped frequency of oscillation ω_0

(*b*) The bandwidth and Q.

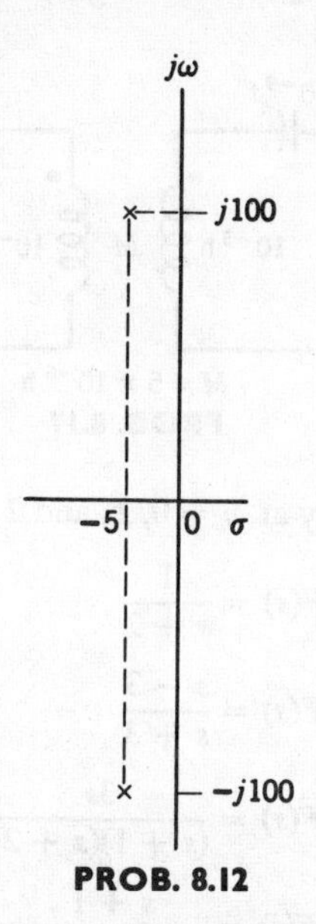

PROB. 8.12

8.13 In connection with Prob. 8.10 determine the ratio $M^2(\omega_{\max})/M^2(0)$.

8.14 Determine the amplitude and phase response for the admittance $Y(s)$ of the circuit shown. Is the peaking circle applicable here? What can you say about the shape of the amplitude response curve in a high-Q situation? Determine the bandwidth of the circuit and the circuit Q.

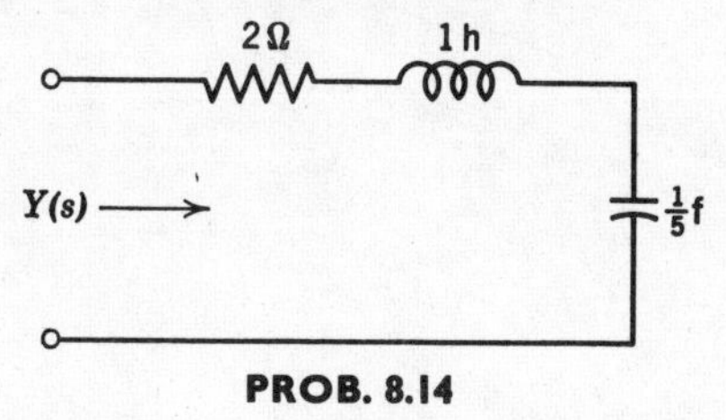

PROB. 8.14

8.15 For the overcoupled case of a double-tuned circuit, derive an expression for the peak-to-valley ratio that is, $M(\omega_{\max})/M(\omega_0)$, where $M(\cdot)$ denotes amplitude. Use the notation in Section 8.4. (*Hint:* see Prob. 8.13.)

8.16 For the voltage ratio of a double-tuned circuit

$$H(s) = \frac{As^2}{(s + 4 + j50)(s + 4 - j50)(s + 4 + j60)(s + 4 - j60)}$$

Use the peaking circle to determine the maximum and half-power points and the circuit Q. Find the gain constant A and the coefficient of coupling K.

8.17 For the double-tuned circuit shown, determine the maximum and half-power points and the circuit Q. Find the gain constant A and the coefficient of coupling K.

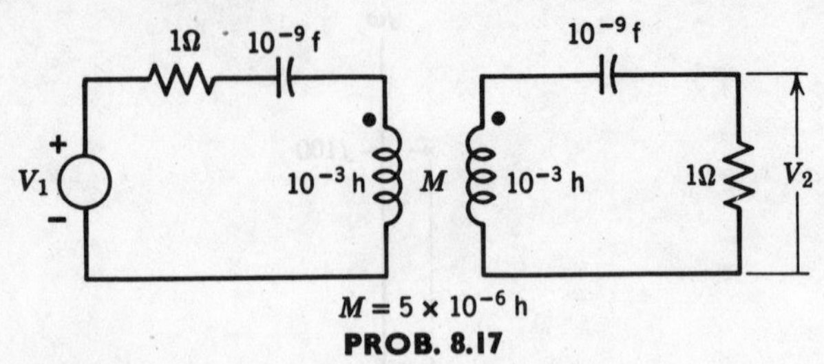

PROB. 8.17

8.18 Determine the delay at $\omega = 0$, 1, and 2 for

(a)
$$F(s) = \frac{1}{s+2}$$

(b)
$$F(s) = \frac{s-3}{s+3}$$

(c)
$$F(s) = \frac{3s}{(s+1)(s+2)}$$

(d)
$$F(s) = \frac{s+1}{s^2+2s+5}$$

(e)
$$F(s) = \frac{s(s+1)}{(s+2)(s^2+2s+2)}$$

chapter 9

Network analysis II

9.1 NETWORK FUNCTIONS

In electric network theory, the word *port* has a special meaning. A port may be regarded as a pair of terminals in which the current into one terminal equals the current out of the other. For the one-port network shown in Fig. 9.1, $I = I'$. A one-port network is completely specified when the voltage-current relationship at the terminals of the port is given. For example, if $V = 10$ v and $I = 2$ amp, then we know that the *input or driving-point* impedance of the one-port is

$$Z_{in} = \frac{V}{I} = 5\,\Omega \tag{9.1}$$

Whether the one-port is actually a single 5-Ω resistor, two 2.5-Ω resistors in series, or two 10-Ω resistors in parallel, is of little importance because the primary concern is the current-voltage relationship at the port. Consider the example in which $I = 2s + 3$ and $V = 1$; then the input admittance of the one-port is

$$Y_{in} = \frac{I}{V} = 2s + 3 \tag{9.2}$$

which corresponds to a 2-f capacitor in parallel with a $\frac{1}{3}$-Ω resistor in its simplest case (Fig. 9.2).

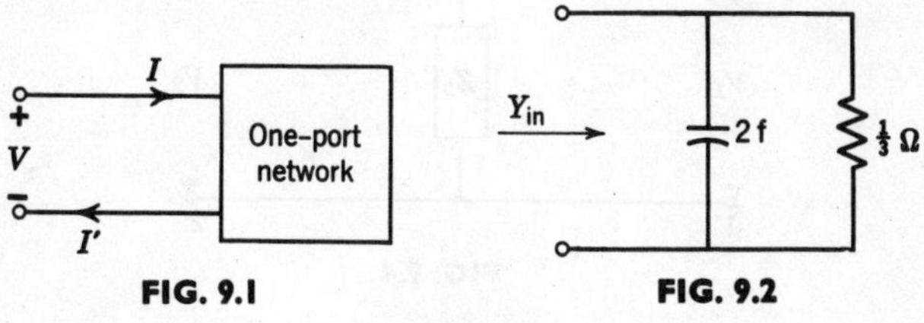

FIG. 9.1 FIG. 9.2

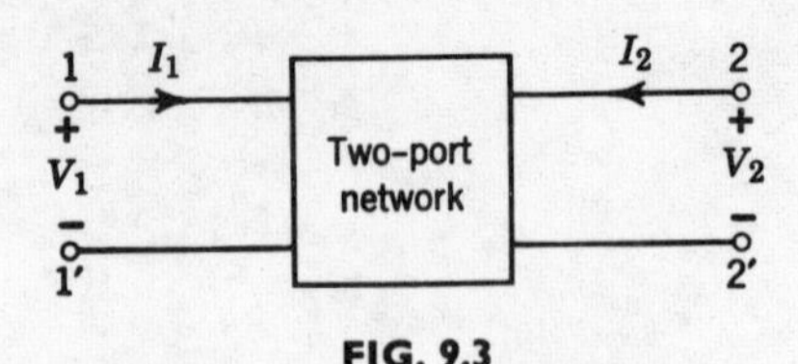

FIG. 9.3

Two-port parameters

A general *two-port* network, shown in Fig. 9.3, has two pairs of voltage-current relationships. The variables are V_1, V_2, I_1, I_2. Two of these are *dependent* variables; the other two are *independent* variables. The number of possible combinations generated by four variables taken two at a time is six. Thus there are six possible sets of equations describing a two-port network. We will discuss the four most useful descriptions here.

The z parameters

A particular set of equations that describe a two-port network are the z-parameter equations

$$
\begin{aligned}
V_1 &= z_{11}I_1 + z_{12}I_2 \\
V_2 &= z_{21}I_1 + z_{22}I_2
\end{aligned}
\tag{9.3}
$$

In these equations the variables V_1 and V_2 are dependent, and I_1, I_2 are independent. The individual z parameters are defined by

$$
\begin{aligned}
z_{11} &= \left.\frac{V_1}{I_1}\right|_{I_2=0} \qquad z_{12} = \left.\frac{V_1}{I_2}\right|_{I_1=0} \\
z_{21} &= \left.\frac{V_2}{I_1}\right|_{I_2=0} \qquad z_{22} = \left.\frac{V_2}{I_2}\right|_{I_1=0}
\end{aligned}
\tag{9.4}
$$

It is observed that all the z parameters have the dimensions of impedance. Moreover, the individual parameters are specified only when the current in one of the ports is zero. This corresponds to one of ports being *open circuited,* from which the z parameters also derive the name *open-circuit*

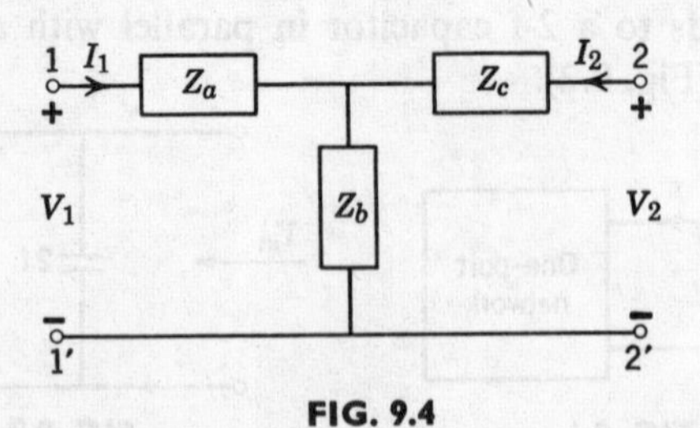

FIG. 9.4

parameters. Note that z_{11} relates the current and voltage in the 1–1′ port only; whereas z_{22} gives the current-voltage relationship for the 2–2′ port. Such parameters are called open-circuit driving-point impedances. On the other hand, the parameters z_{12} and z_{21} relate the voltage in one port to the current in the other. These are known as (open-circuit) transfer impedances.

As an example, let us find the open-circuit parameters for the T circuit in Fig. 9.4. We obtain the z parameters by inspection

$$
\begin{aligned}
z_{11} &= \left.\frac{V_1}{I_1}\right|_{I_2=0} = Z_a + Z_b \\
z_{22} &= \left.\frac{V_2}{I_2}\right|_{I_1=0} = Z_b + Z_c \\
z_{12} &= \left.\frac{V_1}{I_2}\right|_{I_1=0} = Z_b \\
z_{21} &= \left.\frac{V_2}{I_1}\right|_{I_2=0} = Z_b
\end{aligned}
\tag{9.5}
$$

Observe that $z_{12} = z_{21}$. When the open-circuit transfer impedances of a two-port network are equal, the network is *reciprocal*. It will be shown later that most passive time-invariant networks are reciprocal.[1]

Most two-port networks, whether passive or active, can be characterized by a set of open-circuited parameters. Usually, the network is sufficiently complicated so that we cannot obtain the z parameters by inspection, as we did for the T circuit in Fig. 9.4. The question is now, "How do we obtain the z parameters for *any* circuit in general?" The procedure is as follows. We write a set of node equations with the voltages at the ports V_1 and V_2, and other node voltages within the two-port $V_3, V_4, \ldots, V_k$ as the dependent variables. The independent variables are the currents I_1 and I_2, which we will take to be current sources. We then proceed to write a set of node equations.

$$
\begin{aligned}
I_1 &= n_{11}V_1 + n_{12}V_2 + n_{13}V_3 + \cdots + n_{1k}V_k \\
I_2 &= n_{21}V_1 + n_{22}V_2 + \cdots \qquad \cdots + n_{2k}V_k \\
0 &= n_{31}V_1 + \cdots \quad + \cdots \qquad + n_{3k}V_k \\
&\cdots \\
0 &= n_{k1}V_1 + \cdots \quad + \cdots \qquad + n_{kk}V_k
\end{aligned}
\tag{9.6}
$$

[1] One important exception is the *gyrator* discussed later in this chapter.

where n_{ij} represents the admittance between the ith and jth nodes, that is,

$$n_{ij} = G_{ij} + sC_{ij} + \frac{1}{sL_{ij}} \tag{9.7}$$

If the circuit is made up of R-L-C elements only, then it is clear that $n_{ij} = n_{ji}$. As a result, the ijth cofactor of the determinant of the node equations, Δ_{ij}, must be equal to the jith cofactor, Δ_{ji}, that is, $\Delta_{ij} = \Delta_{ji}$. This result leads directly to the reciprocity condition $z_{21} = z_{12}$, as we shall see.

Returning to the set of node equations in Eq. 9.6, let us solve for V_1 and V_2. We obtain

$$\begin{aligned} V_1 &= \frac{\Delta_{11}}{\Delta} I_1 + \frac{\Delta_{21}}{\Delta} I_2 \\ V_2 &= \frac{\Delta_{12}}{\Delta} I_1 + \frac{\Delta_{22}}{\Delta} I_2 \end{aligned} \tag{9.8}$$

In relating this last set of equations to the defining equations for the z parameters, it is clear that

$$\begin{aligned} z_{11} &= \frac{\Delta_{11}}{\Delta} \qquad z_{12} = \frac{\Delta_{21}}{\Delta} \\ z_{21} &= \frac{\Delta_{12}}{\Delta} \qquad z_{22} = \frac{\Delta_{22}}{\Delta} \end{aligned} \tag{9.9}$$

Since for a passive network $\Delta_{21} = \Delta_{12}$, it follows that $z_{21} = z_{12}$, the network is then reciprocal.

As an example, let us find the z parameters of the *Pi* circuit in Fig. 9.5. First, the node equations are

$$\begin{aligned} I_1 &= (Y_A + Y_C)V_1 - Y_C V_2 \\ I_2 &= -Y_C V_1 + (Y_B + Y_C)V_2 \end{aligned} \tag{9.10}$$

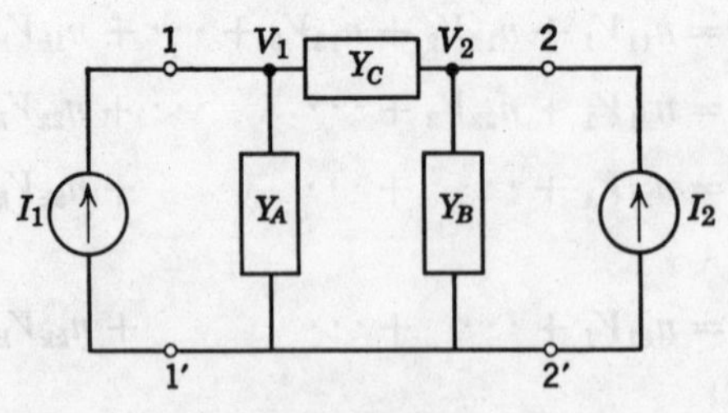

FIG. 9.5

The determinant for this set of equations is

$$\Delta Y = Y_A Y_B + Y_A Y_C + Y_B Y_C \tag{9.11}$$

In terms of ΔY, the open-circuit parameters for the *Pi* circuit are

$$z_{11} = \frac{Y_B + Y_C}{\Delta Y} \qquad z_{21} = \frac{Y_C}{\Delta Y}$$
$$z_{12} = \frac{Y_C}{\Delta Y} \qquad z_{22} = \frac{Y_A + Y_C}{\Delta Y} \tag{9.12}$$

Now let us perform a *delta-wye* transformation for the circuits in Figs. 9.4 and 9.5. In other words, let us find relationships between the immittances of the two circuits so that they both have the same z parameters. We readily obtain

$$z_{12} = Z_b = \frac{Y_C}{\Delta Y}$$
$$z_{22} = Z_b + Z_c = \frac{Y_A + Y_C}{\Delta Y} \tag{9.13}$$
$$z_{11} = Z_a + Z_b = \frac{Y_B + Y_C}{\Delta Y}$$

We then find

$$Z_a = \frac{Y_B}{\Delta Y}$$
$$Z_c = \frac{Y_A}{\Delta Y} \tag{9.14}$$

The y parameters

Suppose we were to write a set of mesh equations for the two port in Fig. 9.3. Then the voltages V_1 and V_2 would become independent sources, and the currents I_1 and I_2 would be just two of the dependent mesh currents. Consider the general set of mesh equations

$$\begin{aligned}
V_1 &= m_{11}I_1 + m_{12}I_2 + \cdots + m_{1k}I_k \\
V_2 &= m_{21}I_1 + m_{22}I_2 + \cdots + m_{2k}I_k \\
0 &= m_{31}I_1 + \cdots \quad \cdots + m_{3k}I_k \\
&\cdots \\
0 &= m_{k1}I_1 + \cdots \quad \cdots + m_{kk}I_k
\end{aligned} \tag{9.15}$$

where m_{ii} represents the sum of the impedances in the ith mesh and m_{ij} is the common impedance between mesh i and mesh j. We note here again that for an *R-L-C* network, $m_{ij} = m_{ji}$ for all i and j. Thus reciprocity holds.

Solving the set of mesh equations for I_1 and I_2, we obtain the following equations.

$$
\begin{aligned}
I_1 &= \frac{\Delta_{11}}{\Delta} V_1 + \frac{\Delta_{21}}{\Delta} V_2 \\
I_2 &= \frac{\Delta_{12}}{\Delta} V_1 + \frac{\Delta_{22}}{\Delta} V_2
\end{aligned}
\tag{9.16}
$$

The equations of 9.16 define the *short-circuit admittance parameters* as

$$
\begin{aligned}
I_1 &= y_{11}V_1 + y_{12}V_2 \\
I_2 &= y_{21}V_1 + y_{22}V_2
\end{aligned}
\tag{9.17}
$$

where $y_{ij} = \Delta_{ji}/\Delta$ for all i and j.

Let us find the y parameters for the *bridged*-T circuit given in Fig. 9.6. The mesh equations for the circuit are

$$
\begin{aligned}
V_1 &= \left(\frac{1}{s} + 1\right)I_1 + I_2 - \frac{1}{s} I_3 \\
V_2 &= I_1 + \left(\frac{1}{s} + 1\right)I_2 + \frac{1}{s} I_3 \\
0 &= -\frac{1}{s} I_1 + \frac{1}{s} I_2 + 2\left(\frac{1}{s} + 1\right)I_3
\end{aligned}
\tag{9.18}
$$

In straightforward fashion we obtain

$$
\begin{aligned}
\Delta &= \frac{2(2s + 1)}{s^2} \\
\Delta_{11} = \Delta_{22} &= \frac{2s^2 + 4s + 1}{s^2} \\
\Delta_{12} = \Delta_{21} &= -\frac{2s^2 + 2s + 1}{s^2}
\end{aligned}
\tag{9.19}
$$

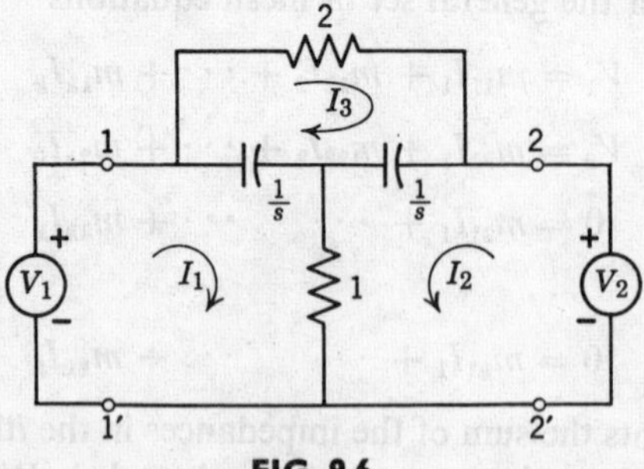

FIG. 9.6

The short-circuit parameters are then

$$y_{11} = y_{22} = \frac{2s^2 + 4s + 1}{2(2s + 1)}$$
$$y_{21} = y_{12} = -\frac{2s^2 + 2s + 1}{2(2s + 1)} \tag{9.20}$$

When $y_{11} = y_{22}$ or $z_{11} = z_{22}$, the network is *symmetrical*.[2]

Returning to Eq. 9.17, which defines the y parameters, we see that the y parameters are expressed explicitly as

$$y_{11} = \left.\frac{I_1}{V_1}\right|_{V_2=0}$$
$$y_{12} = \left.\frac{I_1}{V_2}\right|_{V_1=0}$$
$$y_{21} = \left.\frac{I_2}{V_1}\right|_{V_2=0} \tag{9.21}$$
$$y_{22} = \left.\frac{I_2}{V_2}\right|_{V_1=0}$$

The reason that the y parameters are also called ***short-circuit admittance*** parameters is now apparent. In obtaining y_{11} and y_{21}, the 2–2′ port must be short circuited, and when we find y_{22} and y_{12}, the 1–1′ port must be short circuited, as shown in Figs. 9.7*a* and 9.7*b*.

As a second example, let us obtain the y parameters of the *Pi* circuit in Fig. 9.5. To obtain y_{11} and y_{21}, we short circuit terminals 2–2′. We then have

$$y_{11} = Y_A + Y_C$$
$$y_{21} = -Y_C \tag{9.22}$$

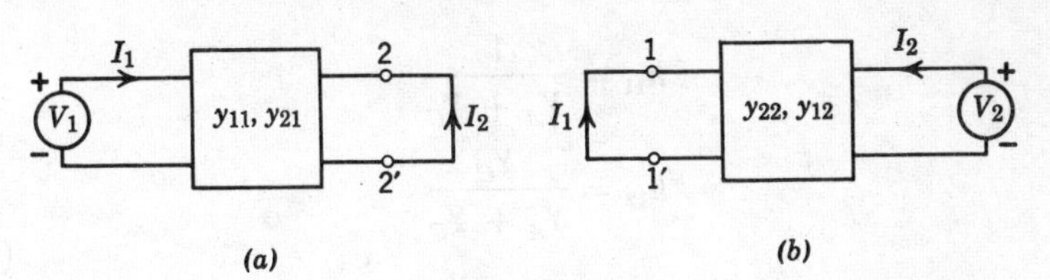

FIG. 9.7

[2] A symmetrical network is easily recognized because by interchanging the 1–1′ and 2–2′ port designations, the network remains unchanged.

We next short-circuit terminals 1–1′ to obtain

$$\begin{aligned} y_{22} &= Y_B + Y_C \\ y_{12} &= -Y_C \end{aligned} \tag{9.23}$$

The h parameters

A set of parameters that are extremely useful in describing transistor circuits are the h parameters given by the equations

$$\begin{aligned} V_1 &= h_{11}I_1 + h_{12}V_2 \\ I_2 &= h_{21}I_1 + h_{22}V_2 \end{aligned} \tag{9.24}$$

The individual parameters are defined by the relationships

$$\begin{aligned} h_{11} &= \frac{V_1}{I_1}\bigg|_{V_2=0} \qquad h_{12} = \frac{V_1}{V_2}\bigg|_{I_1=0} \\ h_{21} &= \frac{I_2}{I_1}\bigg|_{V_2=0} \qquad h_{22} = \frac{I_2}{V_2}\bigg|_{I_1=0} \end{aligned} \tag{9.25}$$

We see that h_{11} and h_{21} are short-circuit type parameters, and h_{12} and h_{22} are open-circuit type parameters. The parameter h_{11} can be interpreted as the input impedance at port 1 with port 2 short circuited. It is easily seen that h_{11} is merely the reciprocal of y_{11}.

$$h_{11} = \frac{1}{y_{11}} \tag{9.26}$$

The parameter h_{22} is an open-circuit admittance parameter and is related to z_{22} by

$$h_{22} = \frac{1}{z_{22}} \tag{9.27}$$

Both the remaining h parameters are transfer functions; h_{21} is a short-circuit current ratio, and h_{12} is an open-circuit voltage ratio. Their relationships to the z and y parameters is discussed later in this chapter.

For the *Pi* circuit in Fig. 9.5, the h parameters are

$$\begin{aligned} h_{11} &= \frac{1}{Y_A + Y_C} \\ h_{12} &= \frac{Y_C}{Y_A + Y_C} \\ h_{21} &= -\frac{Y_C}{Y_A + Y_C} \\ h_{22} &= Y_B + \frac{Y_A Y_C}{Y_A + Y_C} \end{aligned} \tag{9.28}$$

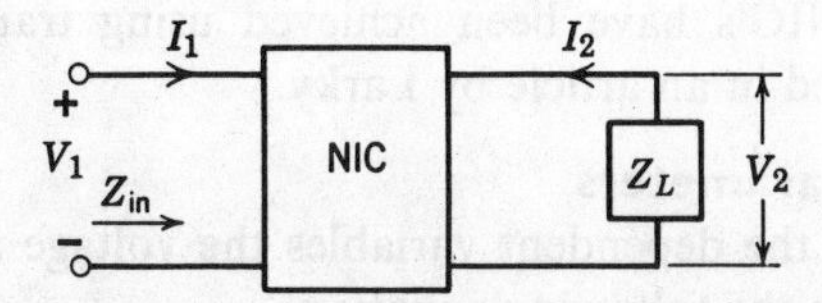

FIG. 9.8. Negative impedance converter with load impedance.

Observe that for the *Pi* circuit, $h_{21} = -h_{12}$. This is the reciprocity condition for the h parameters and can be derived from their relationships to either the z or y parameters.

Next let us consider the h parameters of an ideal device called the *negative impedance converter* (NIC), which converts a positive load impedance into a negative impedance at its input port.[3] Consider the NIC with a load impedance Z_L shown in Fig. 9.8. Its input impedance is

$$Z_{in} = -Z_L \tag{9.29}$$

which can be rewritten as

$$\frac{V_1}{I_1} = \frac{V_2}{I_2} \tag{9.30}$$

The following voltage-current relationships hold for the NIC.

$$\begin{aligned} V_1 &= kV_2 \\ I_1 &= kI_2 \end{aligned} \tag{9.31}$$

If we interpret Eq. 9.31 using h parameters, we arrive at the following conditions.

$$\begin{aligned} h_{11} &= h_{22} = 0 \\ h_{12} &= \frac{1}{h_{21}} = k \end{aligned} \tag{9.32}$$

We see that since $h_{12} \neq -h_{21}$, the NIC is nonreciprocal.

In matrix notation, the h matrix of the NIC is

$$\begin{bmatrix} h_{11} & h_{12} \\ h_{21} & h_{22} \end{bmatrix} = \begin{bmatrix} 0 & k \\ \dfrac{1}{k} & 0 \end{bmatrix} \tag{9.33}$$

The NIC is a convenient device in the modeling of active circuits. It is not, however, a device that exists only in the imagination. Practical

[3] For a lucid discussion of the properties of the NIC, see L. P. Huelsman, *Circuits, Matrices, and Linear Vector Spaces*, McGraw-Hill Company, New York, 1963, Chapter 4.

realizations of NIC's have been achieved using transistors. Some of these are described in an article by Larky.[4]

The *ABCD* parameters

Let us take as the dependent variables the voltage and current at the port 1, and define the following equation.

$$\begin{bmatrix} V_1 \\ I_1 \end{bmatrix} = \begin{bmatrix} A & B \\ C & D \end{bmatrix} \begin{bmatrix} V_2 \\ -I_2 \end{bmatrix} \tag{9.34}$$

This matrix equation defines the A, B, C, D parameters, whose matrix is known as the *transmission matrix* because it relates the voltage and current at the input port to their corresponding quantities at the output. The reason the current I_2 carries a negative sign is that most transmission engineers like to regard their output current as coming *out* of the output port instead of going into the port, as per standard usage.

In explicit form, the *ABCD* parameters can be expressed as

$$\begin{aligned} A &= \left.\frac{V_1}{V_2}\right|_{I_2=0} & B &= -\left.\frac{V_1}{I_2}\right|_{V_2=0} \\ C &= \left.\frac{I_1}{V_2}\right|_{I_2=0} & D &= -\left.\frac{I_1}{I_2}\right|_{V_2=0} \end{aligned} \tag{9.35}$$

From these relations we see that A represents an open-circuit voltage transfer function; B is a short-circuit transfer impedance; C is an open-circuit transfer admittance; and D is a short-circuit current ratio. Note that all four parameters are transfer functions so that the term *transmission matrix* is a very appropriate one. Let us describe the short-circuit transfer functions B and D in terms of y parameters, and the open-circuit transfer functions A and C in terms of z parameters. Using straightforward algebraic operations, we obtain

$$\begin{aligned} A &= \frac{z_{11}}{z_{21}} & B &= -\frac{1}{y_{21}} \\ C &= \frac{1}{z_{21}} & D &= -\frac{y_{11}}{y_{21}} \end{aligned} \tag{9.36}$$

For the *ABCD* parameters, the reciprocity condition is expressed by the equation

$$\det \begin{bmatrix} A & B \\ C & D \end{bmatrix} = AD - BC = 1 \tag{9.37}$$

[4] A. I. Larky, "Negative-Impedance Converters," *Trans. IRE on Circuit Theory*, CT-4, No. 3 (September 1957), 124–131.

Let us find, as an example, the *ABCD* parameter for the *ideal transformer* in Fig. 9.9, whose defining equations are

$$V_1 = nV_2$$

$$I_1 = \frac{1}{n}(-I_2) \tag{9.38}$$

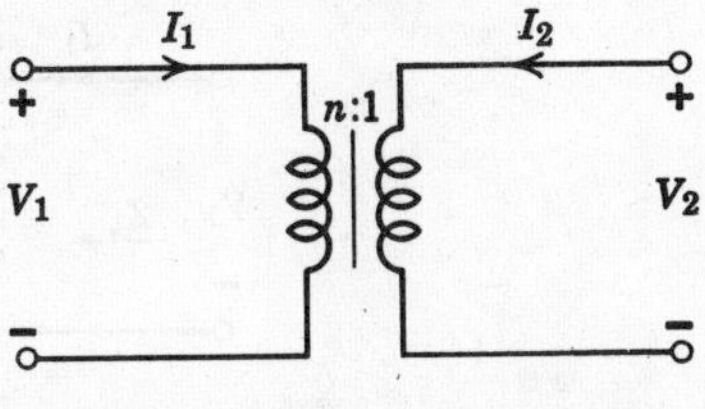

FIG. 9.9. Ideal transformer.

If we express Eq. 9.38 in matrix form, we have

$$\begin{bmatrix} V_1 \\ I_1 \end{bmatrix} = \begin{bmatrix} n & 0 \\ 0 & \frac{1}{n} \end{bmatrix} \begin{bmatrix} V_2 \\ -I_2 \end{bmatrix} \tag{9.39}$$

so that the transmission matrix of the ideal transformer is

$$\begin{bmatrix} A & B \\ C & D \end{bmatrix} = \begin{bmatrix} n & 0 \\ 0 & \frac{1}{n} \end{bmatrix} \tag{9.40}$$

Note, incidentally, that the ideal transformer does *not* possess an impedance or admittance matrix because the self- and mutual inductances are infinite.[5]

For the ideal transformer terminated in a load impedance shown in Fig. 9.10*a*, the following set of equations apply.

$$V_1 = nV_2$$

$$I_1 = \frac{V_2}{nZ_L} \tag{9.41}$$

Taking the ratio of V_1 to I_1, we find the input impedance at port 1 to be

$$Z_1 = \frac{V_1}{I_1} = n^2 Z_L \tag{9.42}$$

Thus we see that an ideal transformer is an *impedance transformer*. If the load element were an inductor L (Fig. 9.10*b*), at port 1 we would see an equivalent inductor of value n^2L. Similarly, a capacitor C at the load would appear as a capacitor of value C/n^2 at port 1 (Fig. 9.10*c*).

As a second example indicating the use of the transmission matrix in

[5] For a detailed discussion concerning ideal transformers, see M. E. Van Valkenburg, *Introduction to Modern Network Synthesis*, John Wiley and Sons, New York, 1960.

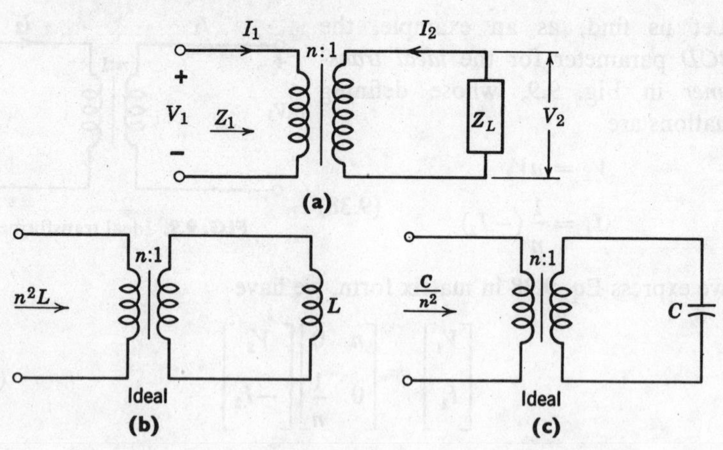

FIG. 9.10. Ideal transformer as an impedance transformer.

network analysis, consider the *ABCD* parameters of the *Pi* circuit in Fig. 9.5.

$$\begin{aligned} A &= \frac{Y_B + Y_C}{Y_C} \\ B &= \frac{1}{Y_C} \\ C &= \frac{Y_A Y_B + Y_B Y_C + Y_A Y_C}{Y_C} \\ D &= \frac{Y_A + Y_C}{Y_C} \end{aligned} \tag{9.43}$$

If we check for reciprocity from Eq. 9.43, we see that

$$\begin{aligned} AD - BC &= \frac{(Y_A + Y_C)(Y_B + Y_C) - (Y_A Y_B + Y_B Y_C + Y_A Y_C)}{Y_C^2} \\ &= \frac{Y_C^2}{Y_C^2} = 1 \end{aligned} \tag{9.44}$$

9.2 RELATIONSHIPS BETWEEN TWO-PORT PARAMETERS

The relationships between two-port parameters are quite easily obtained because of the simple algebraic nature of the two-port equations. For example, we have seen that $h_{11} = 1/y_{11}$ and $h_{22} = 1/z_{22}$. To derive h_{12} in terms of open-circuit parameters, consider the z parameter equations when port 1 is open circuited: $I_1 = 0$.

$$\begin{aligned} V_1 &= z_{12} I_2 \\ V_2 &= z_{22} I_2 \end{aligned} \tag{9.45}$$

Therefore we have
$$h_{12} = \left.\frac{V_1}{V_2}\right|_{I_1=0} = \frac{z_{12}}{z_{22}} \tag{9.46}$$

Similarly, since h_{21} is defined as a short-circuit type parameter, we can derive h_{21} in terms of y parameters as

$$h_{21} = \frac{y_{21}}{y_{11}} \tag{9.47}$$

We can express all the h parameters as functions of the z parameters or y parameters alone. An easy way to accomplish this task is by finding out what the relationships are between the z and y parameters themselves. Certainly, by their very nature, the z and y parameters are not simply reciprocals of each other (as the novice might guess), since one set of parameters is defined for open-circuit conditions and the other for short-circuit.

The z and y relationships can be obtained very easily by using matrix notation. If we define the z matrix as

$$[Z] = \begin{bmatrix} z_{11} & z_{12} \\ z_{21} & z_{22} \end{bmatrix} \tag{9.48}$$

and the y matrix as

$$[Y] = \begin{bmatrix} y_{11} & y_{12} \\ y_{21} & y_{22} \end{bmatrix} \tag{9.49}$$

In simplified notation we can write the two sets of equations as

$$[V] = [Z][I] \tag{9.50}$$

and

$$[I] = [Y][V] \tag{9.51}$$

Replacing $[I]$ in Eq. 9.50 by $[Y][V]$, we obtain

$$[V] = [Z][Y][V] \tag{9.52}$$

so that the product $[Z][Y]$ must yield the unit matrix $[U]$. The matrices $[Y]$ and $[Z]$ must therefore be inverses of each other, that is,

$$[Z]^{-1} = [Y] \quad \text{and} \quad [Y]^{-1} = [Z] \tag{9.53}$$

From the relationship, we can find the relations between the individual z and y parameters.

$$\begin{aligned} y_{11} &= \frac{z_{22}}{\Delta_z} & y_{22} &= \frac{z_{11}}{\Delta_z} \\ y_{12} &= -\frac{z_{12}}{\Delta_z} & y_{21} &= -\frac{z_{21}}{\Delta_z} \end{aligned} \tag{9.54}$$

TABLE 9.1
Matrix Conversion Table

$$\Delta_x = x_{11}x_{22} - x_{12}x_{21}$$

	$[z]$		$[y]$		$[h]$		$[T]$	
$[z]$	z_{11}	z_{12}	$\frac{y_{22}}{\Delta_y}$	$-\frac{y_{12}}{\Delta_y}$	$\frac{\Delta_h}{h_{22}}$	$\frac{h_{12}}{h_{22}}$	$\frac{A}{C}$	$\frac{\Delta_T}{C}$
	z_{21}	z_{22}	$-\frac{y_{21}}{\Delta_y}$	$\frac{y_{11}}{\Delta_y}$	$-\frac{h_{21}}{h_{22}}$	$\frac{1}{h_{22}}$	$\frac{1}{C}$	$\frac{D}{C}$
$[y]$	$\frac{z_{22}}{\Delta_z}$	$-\frac{z_{12}}{\Delta_z}$	y_{11}	y_{12}	$\frac{1}{h_{11}}$	$-\frac{h_{12}}{h_{11}}$	$\frac{D}{B}$	$-\frac{\Delta_T}{B}$
	$-\frac{z_{21}}{\Delta_z}$	$\frac{z_{11}}{\Delta_z}$	y_{21}	y_{22}	$\frac{h_{21}}{h_{11}}$	$\frac{\Delta_h}{h_{11}}$	$-\frac{1}{B}$	$\frac{A}{B}$
$[h]$	$\frac{\Delta_z}{z_{22}}$	$\frac{z_{12}}{z_{22}}$	$\frac{1}{y_{11}}$	$-\frac{y_{12}}{y_{11}}$	h_{11}	h_{12}	$\frac{B}{D}$	$\frac{\Delta_T}{D}$
	$-\frac{z_{21}}{z_{22}}$	$\frac{1}{z_{22}}$	$\frac{y_{21}}{y_{11}}$	$\frac{\Delta_y}{y_{11}}$	h_{21}	h_{22}	$-\frac{1}{D}$	$\frac{C}{D}$
$[T]$	$\frac{z_{11}}{z_{21}}$	$\frac{\Delta_z}{z_{21}}$	$-\frac{y_{22}}{y_{21}}$	$-\frac{1}{y_{21}}$	$-\frac{\Delta_h}{h_{21}}$	$-\frac{h_{11}}{h_{21}}$	A	B
	$\frac{1}{z_{21}}$	$\frac{z_{22}}{z_{21}}$	$-\frac{\Delta_y}{y_{21}}$	$-\frac{y_{11}}{y_{21}}$	$-\frac{h_{22}}{h_{21}}$	$-\frac{1}{h_{21}}$	C	D

where $\Delta_z = z_{11}z_{22} - z_{12}z_{21}$; and

$$z_{11} = \frac{y_{22}}{\Delta_y} \qquad y_{22} = \frac{y_{11}}{\Delta_y}$$
$$z_{12} = -\frac{y_{12}}{\Delta_y} \qquad z_{21} = -\frac{y_{21}}{\Delta_y} \tag{9.55}$$

where $\Delta_y = y_{11}y_{22} - y_{12}y_{21}$. Using these identities we can derive the h or $ABCD$ parameters in terms of either the z or y parameters. Table 9.1 provides a conversion table to facilitate the process. Note that in the table

$$\Delta_T = AD - BC \tag{9.56}$$

9.3 TRANSFER FUNCTIONS USING TWO-PORT PARAMETERS

In this section we will examine how to determine driving-point and transfer functions of a two-port by use of two-port parameters. These functions fall into two broad categories. The first applies to two-ports without load and source impedances. These transfer functions can be described by means of z or y parameters alone. For example, let us derive the expressions for the *open-circuit voltage ratio* V_2/V_1 by using z parameters first and y parameters next. Consider the z parameter equations for the two-port when port 2 is open circuited.

$$\begin{aligned} V_2 &= z_{21}I_1 \\ V_1 &= z_{11}I_1 \end{aligned} \tag{9.57}$$

If we take the ratio of V_2 to V_1, we obtain

$$\frac{V_2}{V_1} = \frac{z_{21}}{z_{11}} \tag{9.58}$$

By letting I_2 of the second y parameter equation go to zero, we derive the open-circuit voltage ratio as

$$\frac{V_2}{V_1} = -\frac{y_{21}}{y_{22}} \tag{9.59}$$

In similar manner we can derive the *short-circuit current ratio* of a two-port as

$$\frac{I_2}{I_1} = \frac{y_{21}}{y_{11}} \tag{9.60}$$

and

$$\frac{I_2}{I_1} = -\frac{z_{21}}{z_{22}} \tag{9.61}$$

The open- and short-circuit transfer functions are not those we usually deal with in practice, since there are frequently source and load impedances to account for. The second category of two-port transfer functions are those including source or load impedances. These transfer functions are functions of the two-port parameters z, h, or y and the source and/or load impedance. For example, let us derive the transfer admittance I_2/V_1 of a two-port network that is terminated in a resistor of R ohms, as given in Fig. 9.11. For this two-port network, the following equations apply.

$$\begin{aligned} I_2 &= y_{21}V_1 + y_{22}V_2 \\ V_2 &= -I_2R \end{aligned} \tag{9.62}$$

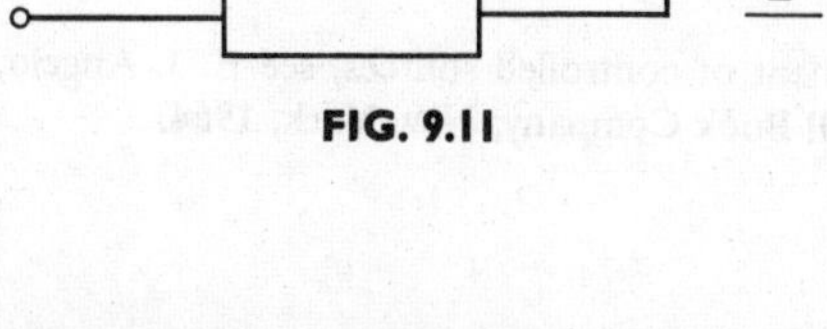

FIG. 9.11

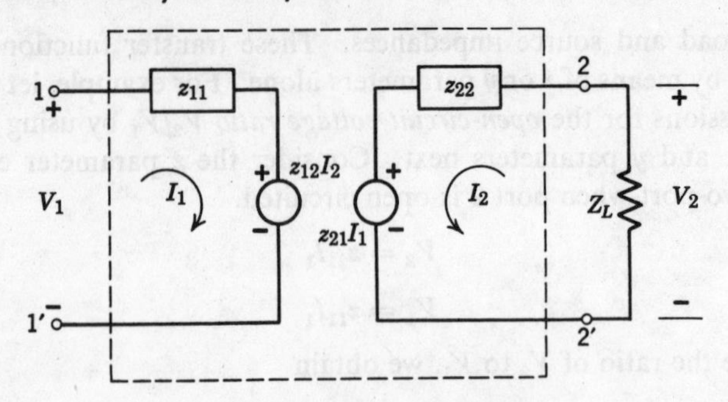

FIG. 9.12. Two-port equivalent.

By eliminating the variable V_2, we obtain

$$Y_{21} = \frac{I_2}{V_1} = \frac{y_{21}/R}{y_{22} + 1/R} \tag{9.63}$$

Note that Y_{21} and y_{21} are not the same. Y_{21} is the transfer admittance of the two-port network terminated in a resistor R, and y_{21} is the transfer admittance when port 2 is short circuited. We must be careful to make this distinction in other cases of a similar nature.

In order to solve for transfer functions of two-ports terminated at either port by an impedance Z_L, it is convenient to use the equivalent circuit of the two-port network given in terms of its z parameters (Fig. 9.12) or y parameters (Fig. 9.13). The equivalent voltage sources $z_{12}I_2$ and $z_{21}I_1$ in Fig. 9.12 are called *controlled sources* because they depend upon a current or voltage somewhere in the network.[6] Similarly, the current sources $y_{12}V_2$ and $y_{21}V_1$ are controlled sources. For the circuit in Fig. 9.12, let us find the transfer impedance $Z_{21} = V_2/I_1$, with port 2

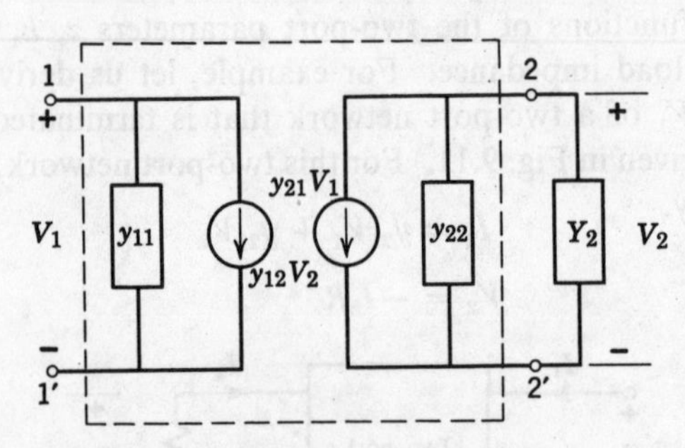

FIG. 9.13. Two-port equivalent.

[6] For a lucid treatment of controlled sources, see E. J. Angelo, *Electronic Circuits, 2nd Ed.*, McGraw-Hill Book Company, New York, 1964.

terminated in a load impedance Z_L. If we write the mesh equation for the I_2 mesh, we have

$$-z_{21}I_1 = (z_{22} + Z_L)\, I_2 \tag{9.64}$$

Since $V_2 = -I_2Z_L$, we readily obtain

$$Z_{21} = \frac{V_2}{I_1} = \frac{z_{21}Z_L}{z_{22} + Z_L} \tag{9.65}$$

It also is clear that the current-ratio transfer function for the terminated two-port network is

$$\frac{I_2}{I_1} = \frac{-z_{21}}{z_{22} + Z_L} \tag{9.66}$$

In similar fashion, we obtain the voltage-ratio transfer function for the circuit represented in Fig. 9.13 as

$$\frac{V_2}{V_1} = -\frac{y_{21}}{Y_2 + y_{22}} \tag{9.67}$$

Next, suppose we are required to find the transfer function V_2/V_g for the two-port network terminated at both ends, as shown in Fig. 9.14. We first write the two mesh equations

$$\begin{aligned} V_g &= (R_1 + z_{11})\, I_1 + z_{12}I_2 \\ 0 &= z_{21}I_1 + (z_{22} + R_2)\, I_2 \end{aligned} \tag{9.68}$$

Next, we solve for I_2 to give

$$I_2 = \frac{V_g z_{21}}{(R_1 + z_{11})(R_2 + z_{22}) - z_{12}z_{21}} \tag{9.69}$$

From the equation $V_2 = -R_2I_2$, we may now arrive at the following solution.

$$\frac{V_2}{V_g} = -\frac{R_2I_2}{V_g} = \frac{z_{21}R_2}{(R_1 + z_{11})(R_2 + z_{22}) - z_{21}z_{12}} \tag{9.70}$$

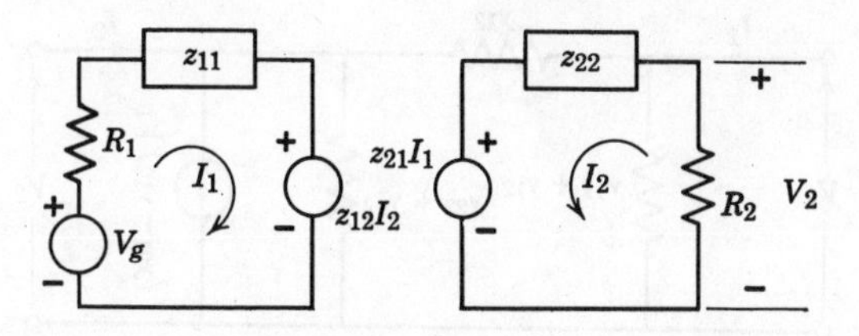

FIG. 9.14

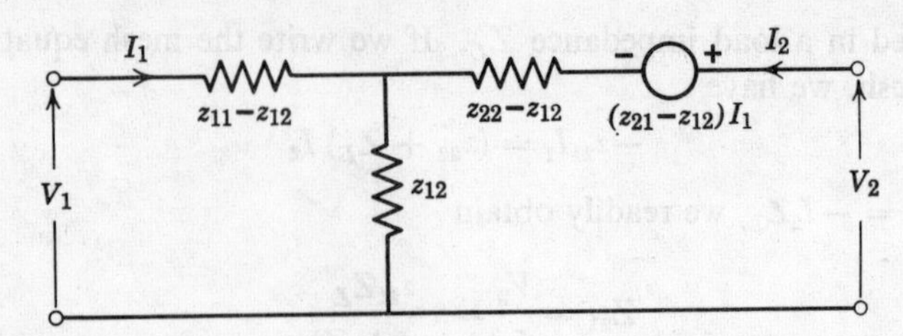

FIG. 9.15. Two-port equivalent circuit with one controlled-voltage source.

Note that the equivalent circuits of the two-ports in Figs. 9.12 and 9.13 are not unique. Two other examples are given in Figs. 9.15 and 9.16. Observe that the controlled sources are nonzero in these equivalent circuits only if the circuit is nonreciprocal.

Finally, let us consider the hybrid equivalent circuit shown in Fig. 9.17. Observe the voltage-controlled source $h_{21}V_2$ at port 1 and the current-controlled source $h_{21}I_1$ at port 2. Let us find the input impedance Z_{in}. The pertinent equations are

$$V_1 = h_{11}I_1 + h_{12}V_2 \tag{9.71}$$

and

$$V_2 = -Z_L I_2 = -(h_{21}I_1 + h_{22}V_2)\,Z_L \tag{9.72}$$

Solving Eq. 9.72 for V_2, we find

$$V_2 = -\frac{h_{21}Z_L I_1}{1 + h_{22}Z_L} \tag{9.73}$$

Substituting V_2 in Eq. 9.73 into Eq. 9.71, we have

$$V_1 = \left(h_{11} - \frac{h_{12}h_{21}Z_L}{1 + h_{22}Z_L}\right) I_1 \tag{9.74}$$

so that

$$Z_{in} = \frac{V_1}{I_1} = h_{11} - \frac{h_{12}h_{21}Z_L}{1 + h_{22}Z_L} \tag{9.75}$$

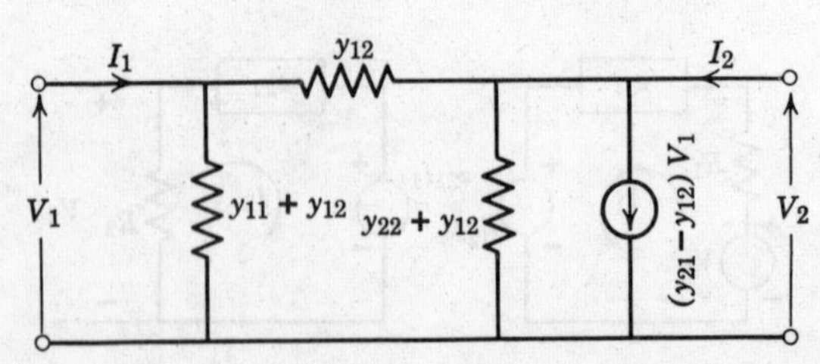

FIG. 9.16. Two-port equivalent circuit with one controlled-current source.

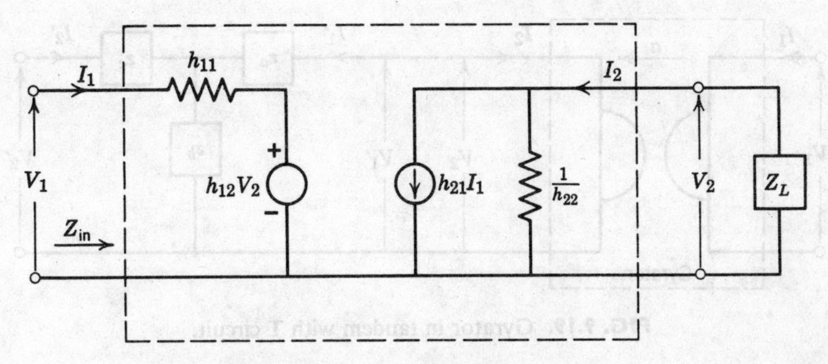

FIG. 9.17. Hybrid equivalent circuit.

We can easily check to see that Eq. 9.75 is dimensionally correct since h_{11} has the dimensions of impedance, h_{22} is an admittance, and h_{12}, h_{21} are dimensionless since they represent voltage and current ratios, respectively.

9.4 INTERCONNECTION OF TWO-PORTS

In this section we will consider various interconnections of two-ports. We will see that when a pair of two-ports are cascaded, the overall transmission matrix is equal to the product of the individual transmission matrices of the two-ports. When two two-ports are connected in series, their z matrices add; when they are connected in parallel, their y matrices add. First let us consider the case in which we connect a pair of two-ports N_a and N_b in cascade or in tandem, as shown in Fig. 9.18. We see that

$$\begin{bmatrix} V_{2a} \\ -I_{2a} \end{bmatrix} = \begin{bmatrix} V_{1b} \\ I_{1b} \end{bmatrix} \tag{9.76}$$

The transmission matrix equation for N_a is

$$\begin{bmatrix} V_1 \\ I_1 \end{bmatrix} = \begin{bmatrix} A_a & B_a \\ C_a & D_a \end{bmatrix} \begin{bmatrix} V_{2a} \\ -I_{2a} \end{bmatrix} \tag{9.77}$$

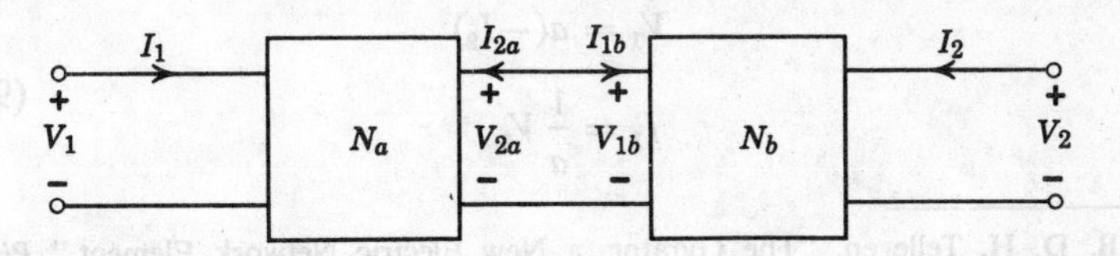

FIG. 9.18. Cascade connection of two ports.

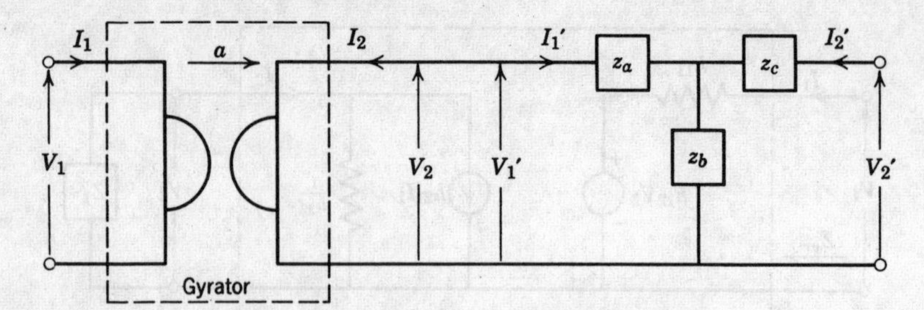

FIG. 9.19. Gyrator in tandem with T circuit.

Correspondingly, for N_b we have

$$\begin{bmatrix} V_{2a} \\ -I_{2a} \end{bmatrix} = \begin{bmatrix} V_{1b} \\ I_{1b} \end{bmatrix} = \begin{bmatrix} A_b & B_b \\ C_b & D_b \end{bmatrix} \begin{bmatrix} V_2 \\ -I_2 \end{bmatrix} \tag{9.78}$$

Substituting the second matrix into the first, we obtain

$$\begin{bmatrix} V_1 \\ I_1 \end{bmatrix} = \begin{bmatrix} A_a & B_a \\ C_a & D_a \end{bmatrix} \cdot \begin{bmatrix} A_b & B_b \\ C_b & D_b \end{bmatrix} \begin{bmatrix} V_2 \\ -I_2 \end{bmatrix} \tag{9.79}$$

We see that the transmission matrix of the overall two-port network is simply the product of the transmission matrices of the individual two-ports.

As an example, let us calculate the overall transmission matrix of a *gyrator*[7] in tandem with a T network shown in Fig. 9.19. The ideal gyrator is an impedance inversion device whose input impedance Z_{in} is related to its load impedance Z_L by

$$\begin{aligned} Z_{\text{in}} &= a^2 Y_L \\ &= \frac{a^2}{Z_L} \end{aligned} \tag{9.80}$$

The constant a in Eq. 9.80 is defined as the *gyration resistance*. If we regard the gyrator as a two-port, its defining equations are

$$\begin{aligned} V_1 &= a(-I_2) \\ I_1 &= \frac{1}{a} V_2 \end{aligned} \tag{9.81}$$

[7] B. D. H. Tellegen, "The Gyrator, a New Electric Network Element," *Phillips Research Repts.*, 3 (April 1948), 81–101, see also Huelsman, *op. cit.*, pp. 140–148.

so that the transmission matrix of the gyrator is

$$\begin{bmatrix} 0 & a \\ \dfrac{1}{a} & 0 \end{bmatrix}$$

We see that for the gyrator

$$AD - BC = -1 \tag{9.82}$$

Therefore the gyrator is a nonreciprocal device, although it is passive.[8]

The overall transmission matrix of the configuration in Fig. 9.19 is obtained by the product of the individual transmission matrices

$$\begin{bmatrix} A & B \\ C & D \end{bmatrix} = \begin{bmatrix} 0 & a \\ \dfrac{1}{a} & 0 \end{bmatrix} \begin{bmatrix} \dfrac{z_a + z_b}{z_b} & \dfrac{z_a z_b + z_b z_c + z_a z_c}{z_b} \\ \dfrac{1}{z_b} & \dfrac{z_b + z_c}{z_b} \end{bmatrix} \tag{9.83}$$

$$= \begin{bmatrix} \dfrac{a}{z_b} & \dfrac{a(z_b + z_c)}{z_b} \\ \dfrac{z_a + z_b}{a z_b} & \dfrac{z_a z_b + z_b z_c + z_a z_c}{a z_b} \end{bmatrix}$$

If we check the configuration in Fig. 9.19 for reciprocity, we see that for transmission matrix in Eq. 9.83

$$AD - BC = -1 \tag{9.84}$$

We thus see that any reciprocal network connected in tandem with a gyrator yields a configuration that is nonreciprocal.

Next consider the situation in which a pair of two-ports N_a and N_b are connected in parallel, as shown in Fig. 9.20. Let us find the

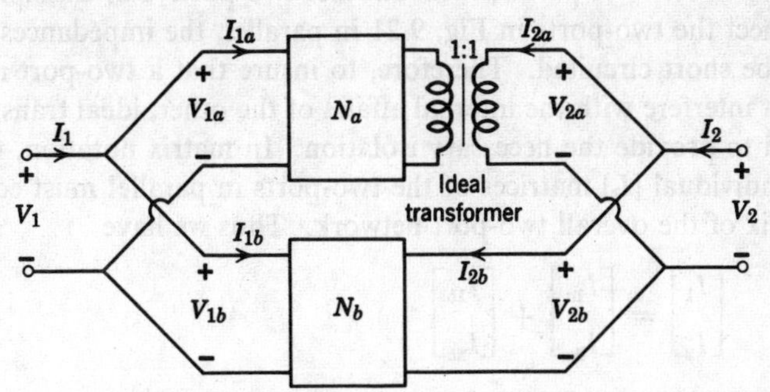

FIG. 9.20. Parallel connection of two ports.

[8] Most gyrators are microwave devices that depend upon the *Hall effect* in ferrites.

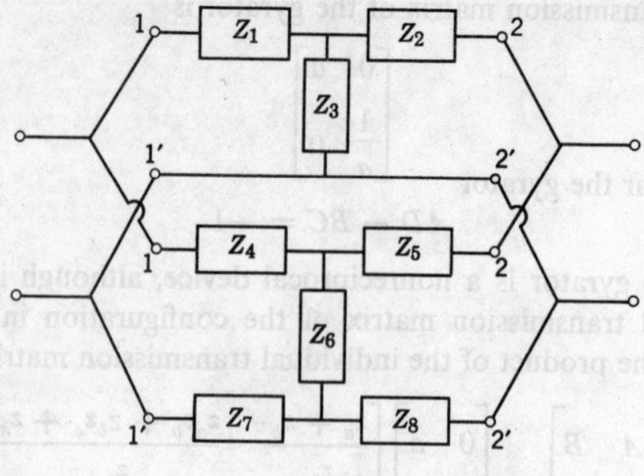

FIG. 9.21

y parameters for the overall two-port network. The matrix equations for the individual two-ports are

$$\begin{bmatrix} I_{1a} \\ I_{2a} \end{bmatrix} = \begin{bmatrix} y_{11a} & y_{12a} \\ y_{21a} & y_{22a} \end{bmatrix} \begin{bmatrix} V_{1a} \\ V_{2a} \end{bmatrix} \tag{9.85}$$

and

$$\begin{bmatrix} I_{1b} \\ I_{2b} \end{bmatrix} = \begin{bmatrix} y_{11b} & y_{12b} \\ y_{21b} & y_{22b} \end{bmatrix} \begin{bmatrix} V_{1b} \\ V_{2b} \end{bmatrix} \tag{9.86}$$

From Fig. 9.20, we see that the following equations must hold.

$$\begin{aligned} V_1 &= V_{1a} = V_{1b} \qquad V_2 = V_{2a} = V_{2b} \\ I_1 &= I_{1a} + I_{1b} \qquad I_2 = I_{2a} + I_{2b} \end{aligned} \tag{9.87}$$

In connecting two-ports in series or in parallel, we must be careful that the individual character of a two-port network is not altered when connected in series or parallel with another two-port. For example, when we connect the two-ports in Fig. 9.21 in parallel, the impedances Z_7 and Z_8 will be short circuited. Therefore, to insure that a two-port network does not interfere with the internal affairs of the other, ideal transformers are used to provide the necessary isolation. In matrix notation, the sum of the individual $[I_i]$ matrices of the two-ports in parallel must equal the $[I]$ matrix of the overall two-port network. Thus we have

$$\begin{aligned} \begin{bmatrix} I_1 \\ I_2 \end{bmatrix} &= \begin{bmatrix} I_{1a} \\ I_{2a} \end{bmatrix} + \begin{bmatrix} I_{1b} \\ I_{2b} \end{bmatrix} \\ &= \left(\begin{bmatrix} y_{11a} & y_{12a} \\ y_{21a} & y_{22a} \end{bmatrix} + \begin{bmatrix} y_{11b} & y_{12b} \\ y_{21b} & y_{22b} \end{bmatrix} \right) \begin{bmatrix} V_1 \\ V_2 \end{bmatrix} \end{aligned} \tag{9.88}$$

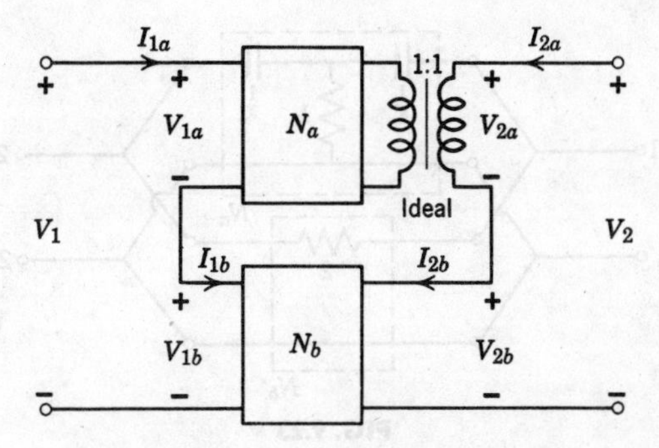

FIG. 9.22. Series connection of two ports.

so that the y parameters of the overall two-port network can be expressed in terms of the y parameters of the individual two-ports as

$$\begin{bmatrix} y_{11} & y_{12} \\ y_{21} & y_{22} \end{bmatrix} = \begin{bmatrix} y_{11a} + y_{11b} & y_{12a} + y_{12b} \\ y_{21a} + y_{21b} & y_{22a} + y_{22b} \end{bmatrix} \tag{9.89}$$

If we connect two-ports in series, as shown in Fig. 9.22, we can express the z parameters of the overall two-port network in terms of the z parameters of the individual two-ports as

$$\begin{bmatrix} z_{11} & z_{12} \\ z_{21} & z_{22} \end{bmatrix} = \begin{bmatrix} z_{11a} + z_{11b} & z_{12a} + z_{12b} \\ z_{21a} + z_{21b} & z_{22a} + z_{22b} \end{bmatrix} \tag{9.90}$$

We may summarize by the following three points.

1. When two-ports are connected in parallel, find the y parameters first, and, from the y parameters, derive the other two-port parameters.
2. When two-ports are connected in series, it is usually easiest to find the z parameters.
3. When two-ports are connected in tandem, the transmission matrix is generally easier to obtain.

As a final example, let us find the y parameters of the bridged-T circuit in Fig. 9.6. We see that the bridged-T circuit would be decomposed into a parallel connection of two-ports, as shown in Fig. 9.23. Our task is to first find the y parameters of the two-ports N_a and N_b. The y parameters of N_b are obtained by inspection and are

$$\begin{aligned} y_{12b} &= y_{21b} = -\tfrac{1}{2} \\ y_{11b} &= y_{22b} = \tfrac{1}{2} \end{aligned} \tag{9.91}$$

N_a is a T circuit so that the z parameters can be obtained by inspection.

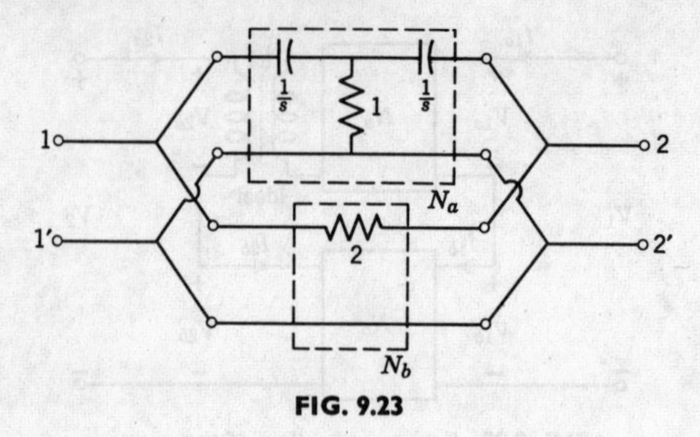

FIG. 9.23

These are

$$z_{11a} = z_{22a} = \frac{s+1}{s}$$
$$z_{12a} = z_{21a} = 1 \tag{9.92}$$

We then find the y parameters from the equations

$$y_{11a} = y_{22a} = \frac{z_{22a}}{\Delta z_a} = \frac{s(s+1)}{2s+1}$$
$$y_{12a} = y_{21a} = \frac{z_{12a}}{\Delta z_a} = -\frac{s^2}{2s+1} \tag{9.93}$$

Since both N_a and N_b are symmetrical two-ports, we know that $y_{11} = y_{22}$ for the overall bridged-T circuit. The y parameters for the bridged-T circuit are now obtained as

$$\begin{aligned} y_{11} &= y_{11a} + y_{11b} \\ &= \frac{s(s+1)}{2s+1} + \frac{1}{2} = \frac{2s^2+4s+1}{2(2s+1)} \\ y_{12} &= y_{12a} + y_{12b} \\ &= -\frac{s^2}{2s+1} - \frac{1}{2} = -\frac{2s^2+2s+1}{2(2s+1)} \end{aligned} \tag{9.94}$$

9.5 INCIDENTAL DISSIPATION

As we have seen, the system function $H(s)$ of an R-L-C network consists of a ratio of polynomials whose coefficients are functions of the resistances, inductances, and capacitances of the network. We have considered, up to this point, that the inductors and capacitors are *dissipationless*; i.e., there are no *parasitic* resistances associated with the L and C. Since, at high frequencies, parasitic losses do play an important role in governing system performance, we must account for this incidental dissipation somehow. An effective way of accounting for parasitic resistances is to "load down" the pure inductances and capacitances with incidental dissipation by associating a resistance r_i in series with every inductor L_i, and, for every capacitor C_i, we associate a resistor whose admittance is g_i, as depicted in Fig. 9.24. Suppose we call the system function of the network without parasitic dissipation (Fig. 9.24*a*) $H(s)$, and the system function of the "loaded" network $H_1(s)$ (Fig. 9.24*b*). Let us consider the the relationship between $H(s)$ and $H_1(s)$ when the dissipation is *uniform*, i.e., in a manner such that

$$\frac{r_i}{L_i} = \frac{g_i}{C_i} = \alpha \tag{9.95}$$

where the constant α is real and positive.

When a network has uniform dissipation or is *uniformly loaded*, the sum of impedances in any mesh of the unloaded network

$$m_{ij} = R_i + sL_i + \frac{1}{sC_i} \tag{9.96}$$

becomes, after loading,

$$\begin{aligned} m'_{ij} &= R_i + sL_i + r_i + \frac{1}{sC_i + g_i} \\ &= R_i + L_i(s + \alpha) + \frac{1}{C_i(s + \alpha)} \end{aligned} \tag{9.97}$$

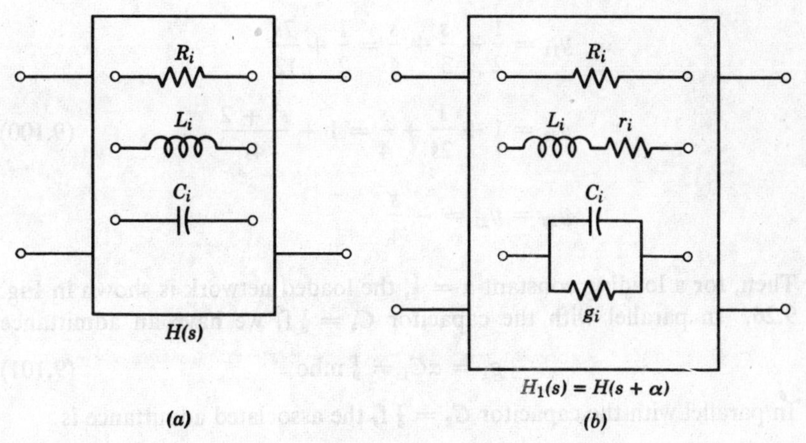

FIG. 9.24. (*a*) Original network. (*b*) Loaded network.

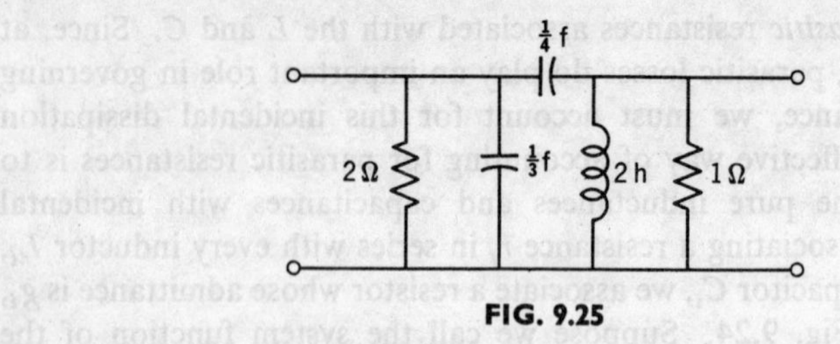

FIG. 9.25

Similarly, on node basis, if the admittance between any two nodes of the original (unloaded) network is

$$n_{ij} = G_i + sC_i + \frac{1}{sL_i} \tag{9.98}$$

then the same node admittance after loading is

$$\begin{aligned} n'_{ij} &= G_i + sC_i + g_i + \frac{1}{sL_i + r_i} \\ &= G_i + C_i(s + \alpha) + \frac{1}{L_i(s + \alpha)} \end{aligned} \tag{9.99}$$

Since any system function can be obtained through mesh or node equations, it is readily seen that the original system function $H(s)$ becomes $H(s + \alpha)$ after the network has been uniformly loaded, that is $H_1(s) = H(s + \alpha)$.

Consider the following example. Let us first find the y parameters for the unloaded network in Fig. 9.25. By inspection we have

$$\begin{aligned} y_{11} &= \frac{1}{2} + \frac{s}{3} + \frac{s}{4} = \frac{1}{2} + \frac{7s}{12} \\ y_{22} &= 1 + \frac{1}{2s} + \frac{s}{4} = 1 + \frac{s^2 + 2}{4s} \\ y_{12} &= y_{21} = -\frac{s}{4} \end{aligned} \tag{9.100}$$

Then, for a loading constant $\alpha = \frac{1}{2}$, the loaded network is shown in Fig. 9.26. In parallel with the capacitor $C_1 = \frac{1}{3}$ f, we have an admittance

$$g_1 = \alpha C_1 = \tfrac{1}{6} \text{ mho} \tag{9.101}$$

In parallel with the capacitor $C_2 = \frac{1}{4}$ f, the associated admittance is

$$g_2 = \alpha C_2 = \tfrac{1}{8} \text{ mho} \tag{9.102}$$

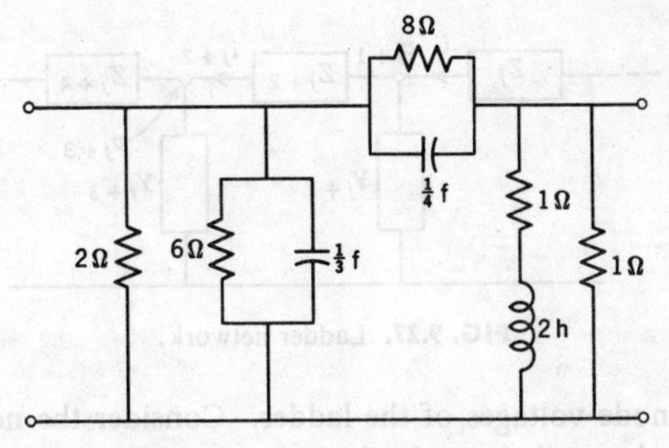

FIG. 9.26. Loaded network $\alpha = \frac{1}{2}$.

In series with the inductor $L = 2$ h, we have a resistor

$$r_3 = \alpha L = 1\ \Omega \tag{9.103}$$

Now we determine the y parameters for the loaded network to be

$$\begin{aligned}
y'_{11} &= \frac{1}{2} + \frac{s}{3} + \frac{1}{6} + \frac{s}{4} + \frac{1}{8} \\
&= \tfrac{1}{2} + \tfrac{1}{3}(s + \tfrac{1}{2}) + \tfrac{1}{4}(s + \tfrac{1}{2}) \\
&= \tfrac{1}{2} + \tfrac{7}{12}(s + \tfrac{1}{2}) \\
y'_{22} &= 1 + \frac{1}{2s + 1} + \frac{s}{4} + \frac{1}{8} \\
&= 1 + \frac{(s + \frac{1}{2})^2 + 2}{4(s + \frac{1}{2})} \\
y'_{12} &= -\left(\frac{s}{4} + \frac{1}{8}\right) = -\frac{1}{4}\left(s + \frac{1}{2}\right)
\end{aligned} \tag{9.104}$$

We see that the y parameters of the loaded network could have been obtained from the y parameters of the unloaded network by the relationship $H_1(s) = H(s + \alpha)$.

We will make use of the uniform loading concept to prove an important theorem concerning network realizability in Chapter 10.

9.6 ANALYSIS OF LADDER NETWORKS

In this section we will consider a simple method of obtaining the network functions of a ladder network in a single operation.[9] This method depends only upon relationships that exist between the branch

[9] F. F. Kuo and G. H. Leichner, "An Iterative Method for Determining Ladder Network Functions," *Proc. IRE*, **47**, No. 10 (Oct. 1959), 1782–1783.

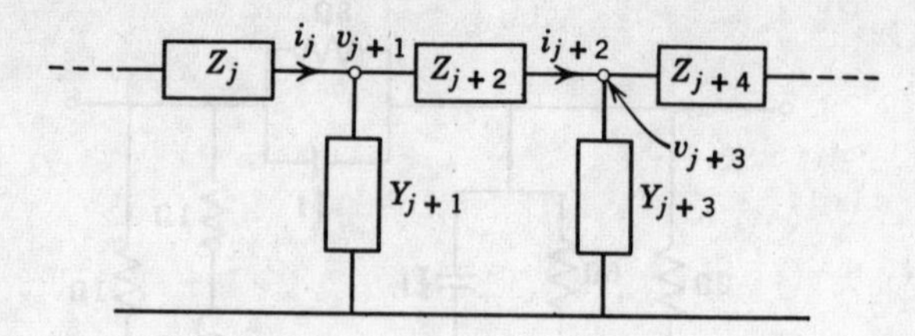

FIG. 9.27. Ladder network.

currents and node voltages of the ladder. Consider the network shown in Fig. 9.27, where all the series branches are given as impedances and all the parallel branches are given as admittances. If the v denote node voltages and i denote branch currents, then the following relationships apply.

$$\begin{aligned} v_{j+1} &= i_{j+2}Z_{j+2} + v_{j+3} \\ i_j &= v_{j+1}Y_{j+1} + i_{j+2} \end{aligned} \tag{9.105}$$

These equations form the basis of the method we discuss here.

To illustrate this method, consider the network in Fig. 9.28, for which the following node voltage and branch current relationships apply.

$$\begin{aligned} V_2(s) &= V_2(s) \\ I_2(s) &= Y_4(s)\,V_2(s) \\ V_a(s) &= I_2(s)\,Z_3(s) + V_2(s) = [1 + Z_3(s)\,Y_4(s)]\,V_2(s) \\ I_1(s) &= Y_2(s)\,V_a(s) + I_2(s) \\ &= \{Y_2(s)[1 + Z_3(s)\,Y_4(s)] + Y_4(s)\}\,V_2(s) \\ V_1(s) &= I_1(s)\,Z_1(s) + V_a(s) \\ &= \{Z_1[Y_2(1 + Z_3Y_4) + Y_4] + (1 + Z_3Y_4)\}\,V_2(s) \end{aligned} \tag{9.106}$$

Upon examining the set of equations given in Eq. 9.106, we see that each equation depends upon the two previous equations only. The first equation, $V_2 = V_2$ is, of course, unnecessary. But, as we shall see later, it is helpful as a starting point. In writing these equations, we begin at

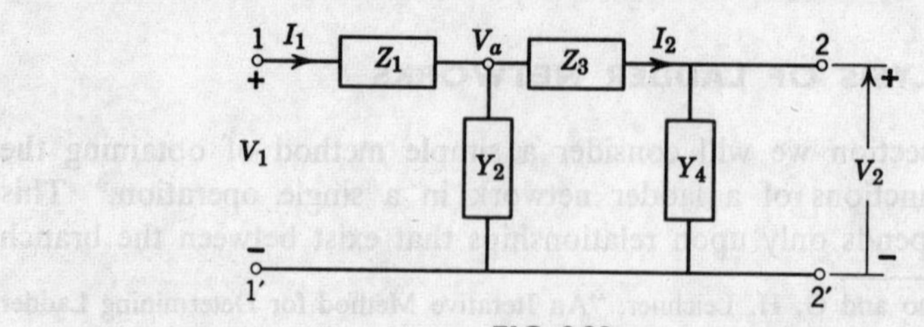

FIG. 9.28

the 2–2′ port of the ladder and work towards the 1–1′ port. Each succeeding equation takes into account one new immittance. We see, further, that with the exception of the first two equations, each subsequent equation is obtained by multiplying the equation just preceding it by the immittance that is next down the line, and then adding to this product the equation twice preceding it. For example, we see that $I_2(s)$ is obtained by multiplying the preceding equation $V_2(s)$ by the admittance $Y_4(s)$. The next immittance is $Z_3(s)$. We obtain $V_a(s)$ by multiplying the previous equation $I_2(s)$ by $Z_3(s)$ to obtain I_2Z_3; then we add to this product the equation twice preceding it, $V_2(s)$, to obtain $V_a = I_2 Z_3 + V_2$. The process is then easily *mechanized* according to the following rules: (1) alternate writing node voltage and branch current equations; (2) the next equation is obtained by multiplying the present equation by the next immittance (as we work from one port to the other), and adding to this product the results of the previous equation.

Using this set of equations, we obtain the input impedance $Z_{in}(s)$ by dividing the equation for $V_1(s)$ by the equation for $I_1(s)$. We obtain the voltage-ratio transfer function $V_2(s)/V_1(s)$ by dividing the first equation $V_2(s)$ by the last equation $V_1(s)$. We obtain other network functions such as transfer immittances and current ratios in similar manner. Note that every equation contains $V_2(s)$ as a factor. In taking ratios of these equations, the $V_2(s)$ term is canceled. Therefore, our analysis can be simplified if we let $V_2(s) = 1$.

If the first equation of the set were a current variable $I_i(s)$ instead of the voltage $V_2(s)$, the subsequent equations would contain the current variable $I_i(s)$ as a factor, which we could also normalize to $I_i(s) = 1$. An example in which the first equation is a current rather than a voltage equation may be seen by determining the y parameters of a two-port network.

Before we embark upon some numerical examples, it is important to note that we must represent the series branches as impedances and the shunt branches as admittances. Suppose the series branch consisted of a resistor $R = 1\ \Omega$ in parallel with a capacitor $C = \frac{1}{2}$ f. Then the impedance of the branch is

$$Z(s) = \frac{R(1/sC)}{1/sC + R} = \frac{2}{s + 2} \tag{9.107}$$

and must be considered as a single entity in writing the equations for the ladder. Similarly, if a shunt branch consists of a resistor $R_1 = 2\ \Omega$ and an inductor $L_1 = 1$ h, then the admittance of the branch is

$$Y(s) = \frac{1}{2 + s} \tag{9.108}$$

The key point in this discussion is that we must use the *total* impedance or admittance of a branch in writing the equations for the ladder.

Example 9.1. Let us find the voltage ratio V_2/V_1, the current ratio I_2/I_1, the input impedance $Z_1 = V_1/I_1$, and the transfer impedance $Z_{21} = V_2/I_1$ for the network in Fig. 9.29. First, we must represent the series branches as impedances and the shunt branches as admittances, as shown in Fig. 9.30. The branch current and node voltage equations for the network are

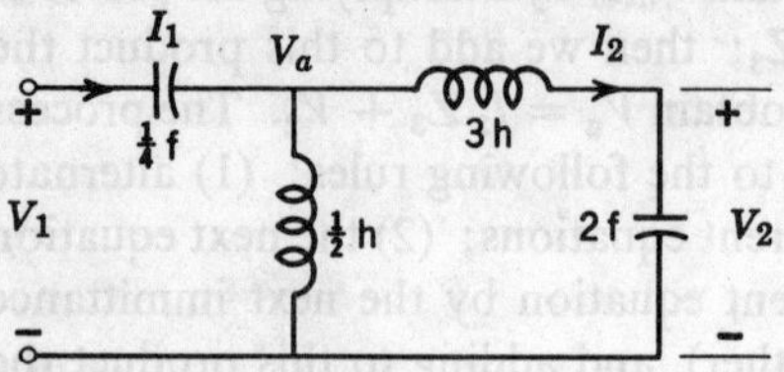

FIG. 9.29

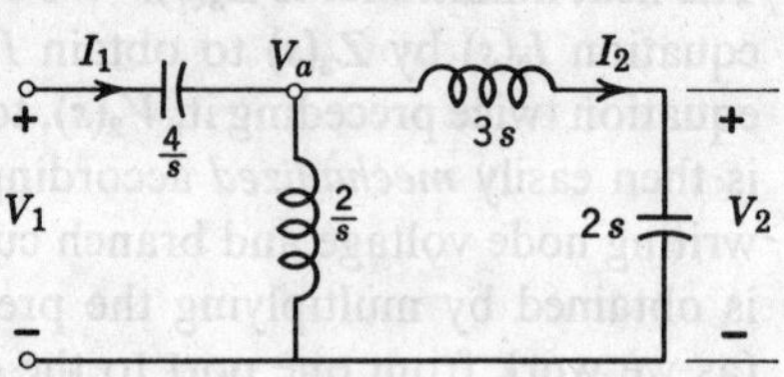

FIG. 9.30

$$
\begin{aligned}
V_2(s) &= 1 \\
I_2(s) &= 2s \\
V_a(s) &= 3s(2s) + 1 = 6s^2 + 1 \\
I_1(s) &= \frac{2}{s}(6s^2 + 1) + 2s = 14s + \frac{2}{s} \\
V_1(s) &= \frac{4}{s}\left(14s + \frac{2}{s}\right) + 6s^2 + 1 = 6s^2 + 57 + \frac{8}{s^2}
\end{aligned}
\tag{9.109}
$$

The various network functions are then obtained.

$$
\begin{aligned}
&(a) \qquad Z_{\text{in}} = \frac{V_1}{I_1} = \frac{6s^4 + 57s^2 + 8}{14s^3 + 2s} \\
&(b) \qquad \frac{V_2}{V_1} = \frac{s^2}{6s^4 + 57s^2 + 8} \\
&(c) \qquad \frac{I_2}{I_1} = \frac{2s^2}{14s^2 + 2} \\
&(d) \qquad Z_{21} = \frac{V_2}{I_1} = \frac{s}{14s^2 + 2}
\end{aligned}
\tag{9.110}
$$

Example 9.2. Find the short-circuit admittance functions y_{11} and y_{21} for the network in Fig. 9.31. To obtain the short-circuit functions, we must short circuit the 2–2′ port. Then we represent the series branches as impedances and the shunt branches as admittances so that the resulting network is given as is

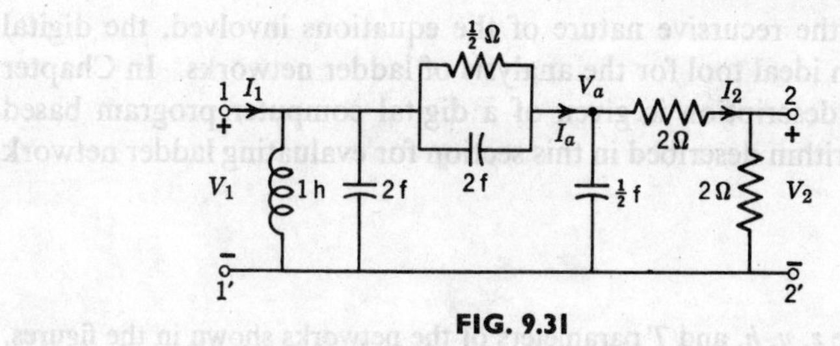

FIG. 9.31

illustrated in Fig. 9.32. The pertinent equations are

$$
\begin{aligned}
I_2 &= 1 \qquad V_a = 2 \\
I_a &= \tfrac{1}{2}s(2) + 1 = s + 1 \\
V_1 &= \frac{1}{2(s+1)}(s+1) + 2 = \frac{5}{2} \\
I_1 &= \frac{2s^2+1}{s}\left(\frac{5}{2}\right) + (s+1) = 6s + 1 + \frac{5/2}{s}
\end{aligned}
\tag{9.111}
$$

We now obtain our short-circuit functions as

$$
y_{11} = \frac{I_1}{V_1} = \frac{12}{5}s + \frac{2}{5} + \frac{1}{s}, \qquad y_{21} = \frac{I_2}{V_1} = -\frac{2}{5} \tag{9.112}
$$

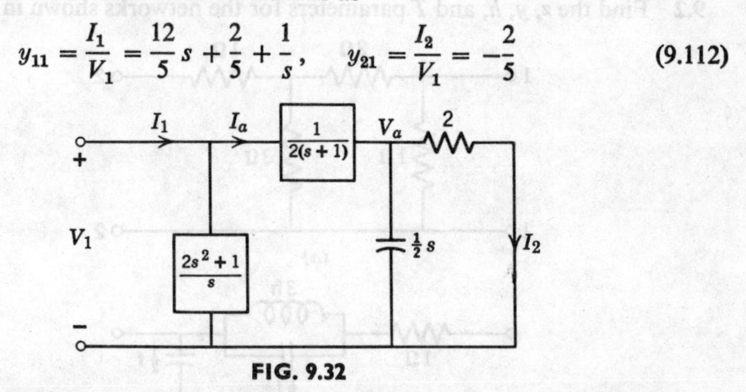

FIG. 9.32

A number of other contributions on ladder networks have appeared in the recent literature. To name but a few, there are the works of Bashkow,[10] O'Meara,[11] Bubnicki,[12] Walker,[13] and Dutta Roy.[14]

[10] T. R. Bashkow, "A Note on Ladder Network Analysis," *IRE Trans. on Circuit Theory*, **CT-8**, (June 1961), 168.

[11] T. R. O'Meara, "Generating Arrays for Ladder Network Transfer Functions," *IEEE Trans. on Circuit Theory*, **CT-10**, (June, 1963), 285.

[12] Z. Bubnicki, "Input Impedance and Transfer Function of a Ladder Network," *IEEE Trans. on Circuit Theory*, **CT-10**, (June, 1963), 286.

[13] F. Walker, "The Topological Analysis of Non-Recurrent Ladder Networks," *Proc. IEEE*, **52**, No. 7, (July, 1964), 860.

[14] S. C. Dutta Roy, "Formulas for the Terminal Impedances and Transfer Functions of General Multimesh Ladder Networks," *Proc. IEEE*, **52**, No. 6, (June, 1964), 738.

Because of the recursive nature of the equations involved, the digital computer is an ideal tool for the analysis of ladder networks. In Chapter 15 a detailed description is given of a digital computer program based upon the algorithm described in this section for evaluating ladder network functions.

Problems

9.1 Find the z, y, h, and T parameters of the networks shown in the figures. Some of the parameters may not be defined for particular circuit configurations.

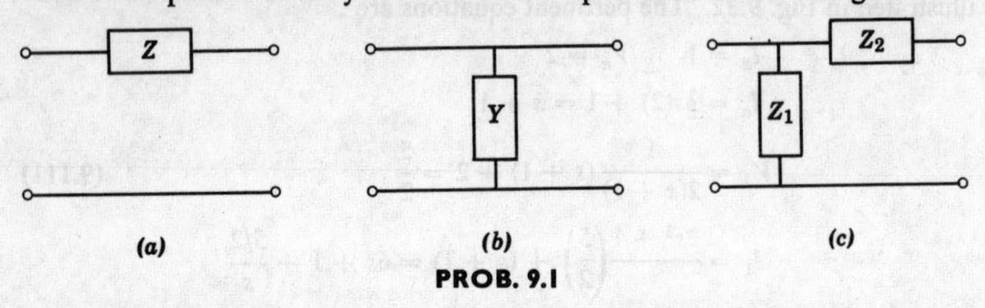

PROB. 9.1

9.2 Find the z, y, h, and T parameters for the networks shown in the figures.

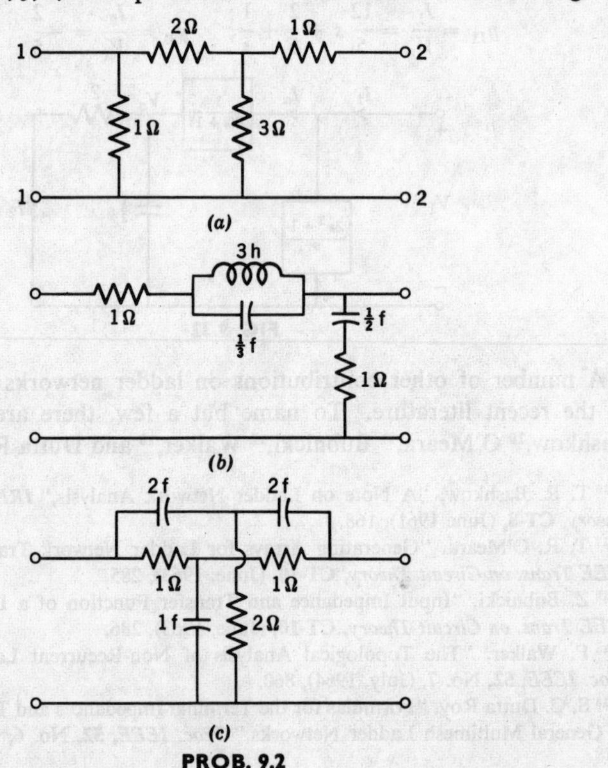

PROB. 9.2

9.3 Find the z parameters for the lattice and bridge circuits in the figures shown. (The results should be identical.)

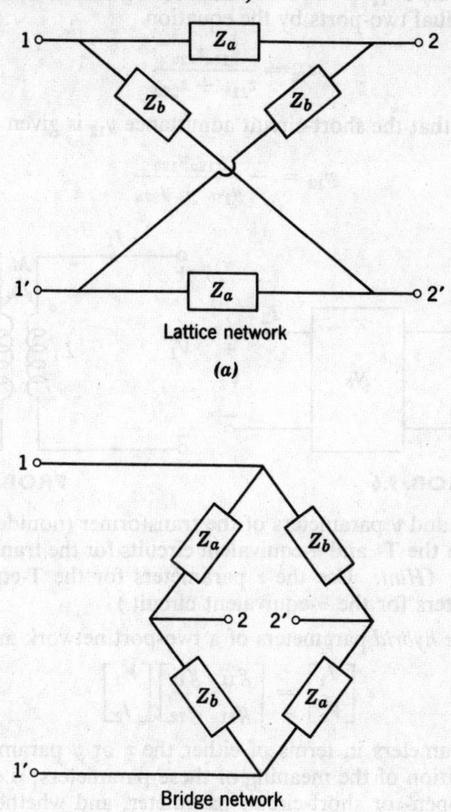

Lattice network

(a)

Bridge network

(b)

PROB. 9.3

9.4 For the lattice and bridge circuits in Prob. 9.3 find the y parameters in terms of the admittances $Y_a = 1/Z_a$ and $Y_b = 1/Z_b$.

9.5 For the circuit shown, find the voltage-ratio transfer function V_2/V_1 and the input impedance $Z_1 = V_1/I_1$ in terms of the z parameters of the two-port network N and the load resistor R_L.

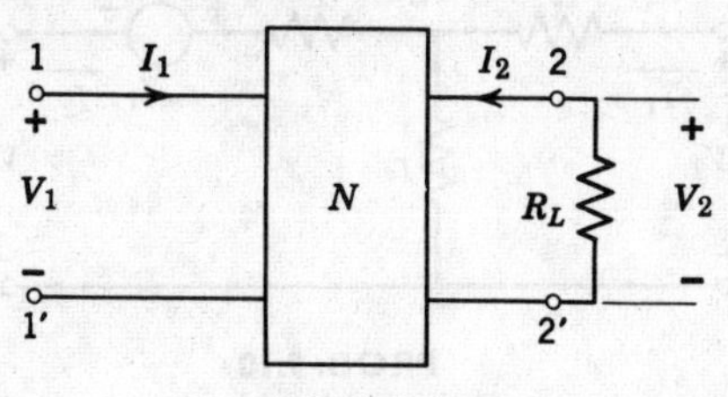

PROB. 9.5

9.6 For the cascade connection of two-ports depicted in the figure, show that the transfer impedance z_{12} of the overall circuit is given in terms of the z parameters of the individual two-ports by the equation

$$z_{12} = \frac{z_{12a}z_{12b}}{z_{11b} + z_{22a}}$$

In addition, show that the short-circuit admittance y_{12} is given by

$$y_{12} = -\frac{y_{12a}y_{12b}}{y_{11b} + y_{22a}}$$

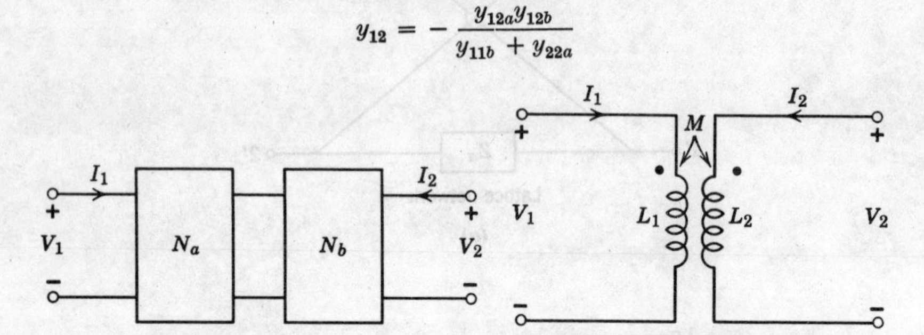

PROB. 9.6　　PROB. 9.7

9.7 Find the z and y parameters of the transformer (nonideal) shown in the figure. Determine the T- and π-equivalent circuits for the transformer in terms of L_1, L_2, and M. (*Hint:* Use the z parameters for the T-equivalent circuit, and the y parameters for the π-equivalent circuit.)

9.8 The *inverse hybrid* parameters of a two-port network are defined by the equation

$$\begin{bmatrix} I_1 \\ V_2 \end{bmatrix} = \begin{bmatrix} g_{11} & g_{12} \\ g_{21} & g_{22} \end{bmatrix} \begin{bmatrix} V_1 \\ I_2 \end{bmatrix}$$

Express the g parameters in terms of either the z or y parameters and give a physical interpretation of the meaning of these parameters; i.e., say whether a parameter is an open- or short-circuit parameter, and whether it is a driving point or transfer function. Finally, derive the conditions of reciprocity for the g parameters.

9.9 Prove that for a passive reciprocal network

$$AD - BC = 1$$

where A, B, C, D are the elements of the transmission matrix.

9.10 Find the z and h parameters of the common emitter transistor represented by its T-circuit model.

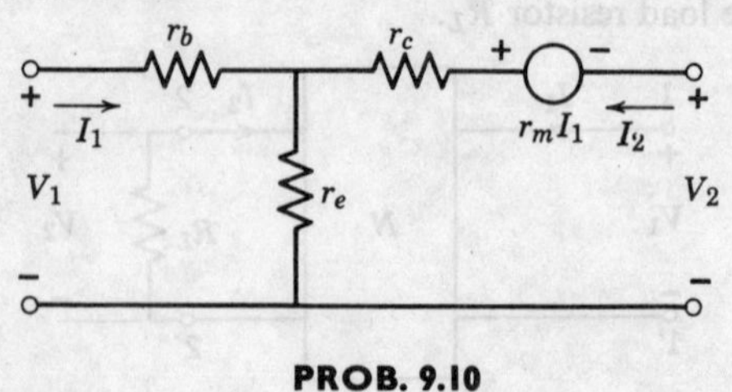

PROB. 9.10

9.11 The circuit in part (a) of the figure is to be described by an *equivalent input circuit* shown in part (b). Determine Z_{eq} in (b) as a function of the elements and voltages in (a).

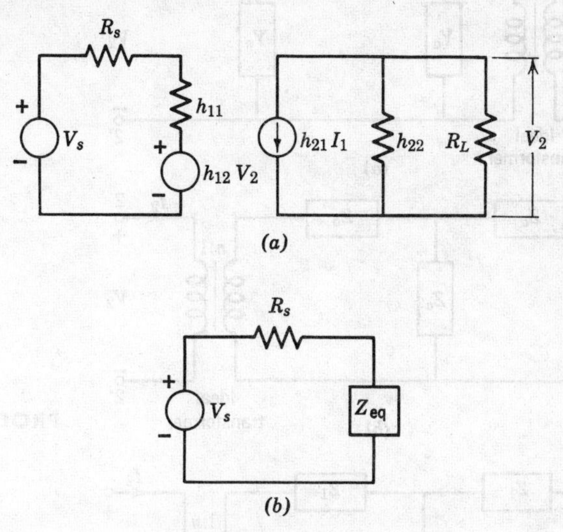

PROB. 9.11

9.12 Find the T parameters of the configurations shown in parts (a) and (b) of the figure.

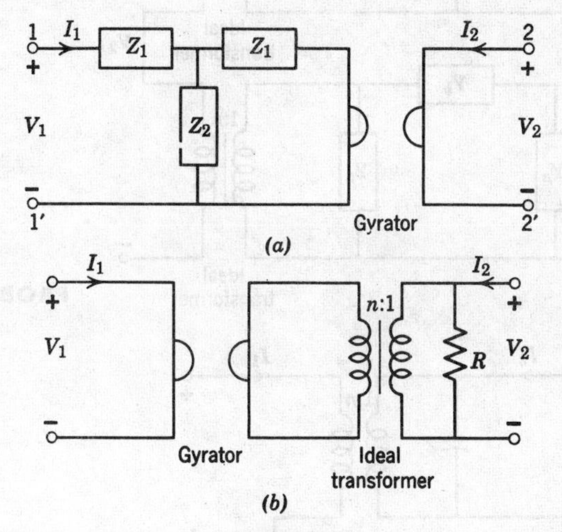

PROB. 9.12

9.13 Find the y parameters of the twin-T circuit in Prob. 9.2c by considering the circuit to be made up of two T circuits in parallel.

9.14 Find the z parameters of the circuits shown.

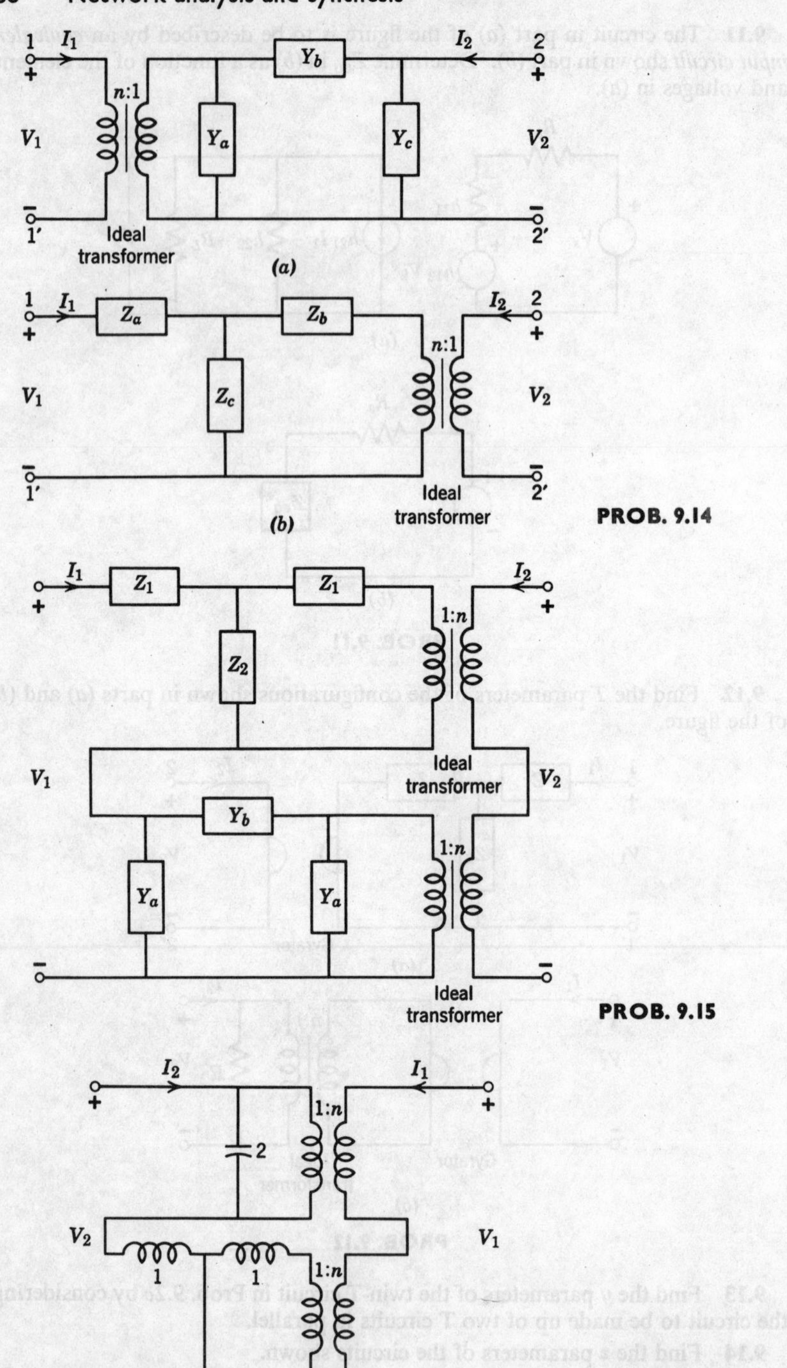

PROB. 9.14

PROB. 9.15

PROB. 9.16

9.15 Find the z parameters of the circuit shown.

9.16 Find the z parameters of the circuit shown.

9.17 For the circuit in Prob. 9.2*b* determine the y parameters of the uniformly loaded circuit derived from the original circuit with the dissipation $\alpha = 0.1$. Plot the poles and zeros of both cases.

9.18 Find the transfer impedance V_2/I_1 for the circuit in part (*a*) and the voltage ratio V_2/V_1 for the circuit in part (*b*). Plot the poles and zeros for the transfer functions obtained.

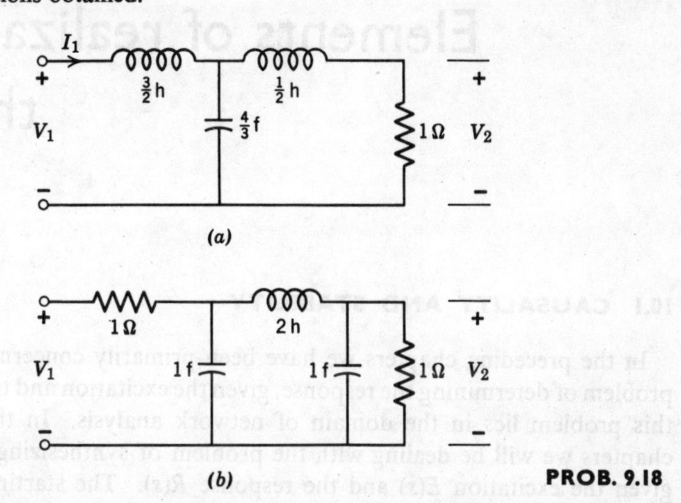

PROB. 9.18

9.19 Find the short-circuit parameters for the ladder network utilizing the method in Section 9.6.

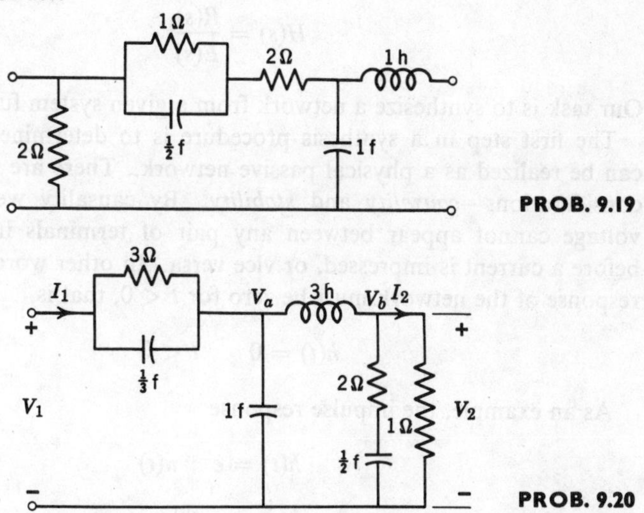

PROB. 9.19

PROB. 9.20

9.20 Determine the voltage ratio V_2/V_1, the current ratio I_2/I_1, the transfer impedance V_2/I_1, and the driving point impedance V_1/I_1 for the network shown.

chapter 10

Elements of realizability theory

10.1 CAUSALITY AND STABILITY

In the preceding chapters we have been primarily concerned with the problem of determining the response, given the excitation and the network; this problem lies in the domain of network analysis. In the next five chapters we will be dealing with the problem of synthesizing a network given the excitation $E(s)$ and the response $R(s)$. The starting point for any synthesis problem is the system function

$$H(s) = \frac{R(s)}{E(s)} \tag{10.1}$$

Our task is to synthesize a network from a given system function.

The first step in a synthesis procedure is to determine whether $H(s)$ can be realized as a physical passive network. There are two important considerations—*causality* and *stability*. By causality we mean that a voltage cannot appear between any pair of terminals in the network before a current is impressed, or vice versa. In other words, the impulse response of the network must be zero for $t < 0$, that is,

$$h(t) = 0 \qquad t < 0 \tag{10.2}$$

As an example, the impulse response

$$h(t) = e^{-at}\,u(t) \tag{10.3a}$$

is causal, whereas

$$h(t) = e^{-a|t|} \tag{10.3b}$$

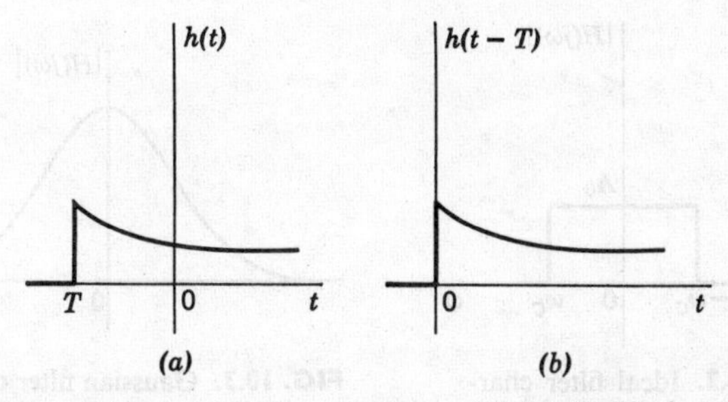

FIG. 10.1. (*a*) Nonrealizable impulse response. (*b*) Realizable impulse response.

is not causal. In certain cases, the impulse response could be made realizable (causal) by delaying it appropriately. For example, the impulse response in Fig. 10.1*a* is not realizable. If we delay the response by T seconds, we find that the delayed response $h(t - T)$ is realizable (Fig. 10.1*b*).

In the frequency domain, causality is implied when the *Paley-Wiener criterion*[1] is satisfied for the amplitude function $|H(j\omega)|$. The Paley-Wiener criterion states that a necessary and sufficient condition for an amplitude function $|H(j\omega)|$ to be realizable (causal) is that

$$\int_{-\infty}^{\infty} \frac{|\log |H(j\omega)||}{1 + \omega^2}\, d\omega < \infty \tag{10.4}$$

The following conditions must be satisfied before the Paley-Wiener criterion is valid: $h(t)$ must possess a Fourier transform $H(j\omega)$; the square magnitude function $|H(j\omega)|^2$ must be integrable, that is,

$$\int_{-\infty}^{\infty} |H(j\omega)|^2\, d\omega < \infty \tag{10.5}$$

The physical implication of the Paley-Wiener criterion is that the amplitude $|H(j\omega)|$ of a realizable network must not be zero over a finite band of frequencies. Another way of looking at the Paley-Wiener criterion is that the amplitude function cannot fall off to zero faster than exponential order. For example, the ideal low-pass filter in Fig. 10.2 is not realizable because beyond ω_C the amplitude is zero. The *Gaussian* shaped curve

$$|H(j\omega)| = e^{-\omega^2} \tag{10.6}$$

shown in Fig. 10.3 is not realizable because

$$|\log |H(j\omega)|| = \omega^2 \tag{10.7}$$

[1] R. E. A. C. Paley and N. Wiener, "Fourier Transforms in the Complex Domain," *Am. Math. Soc. Colloq. Pub.*, **19** (1934), 16–17.

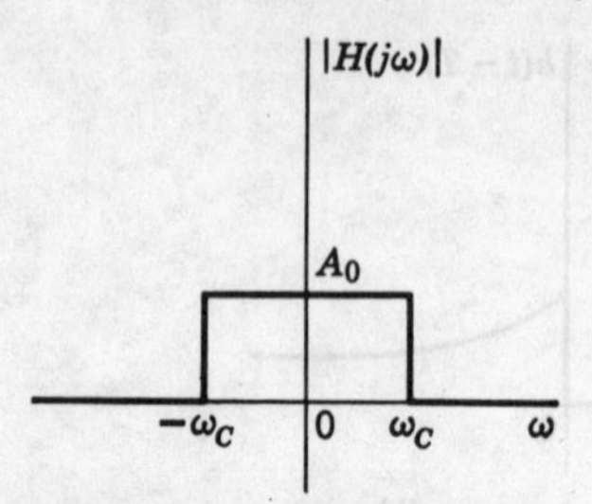

FIG. 10.2. Ideal filter characteristic

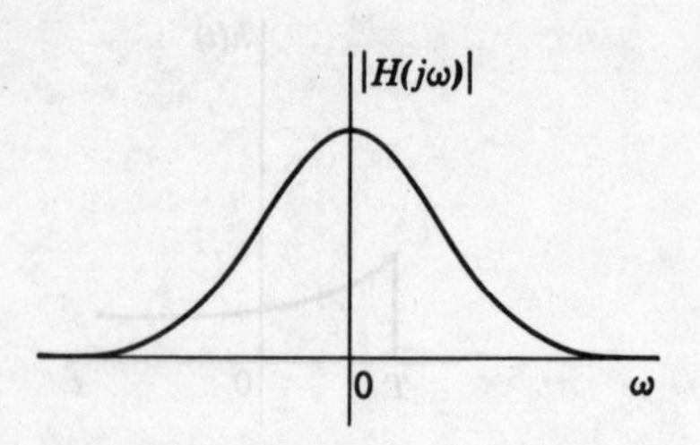

FIG. 10.3. Gaussian filter characteristic.

so that the integral

$$\int_{-\infty}^{\infty} \frac{\omega^2}{1+\omega^2}\, d\omega \tag{10.8}$$

is not finite. On the other hand, the amplitude function

$$|H(j\omega)| = \frac{1}{\sqrt{1+\omega^2}} \tag{10.9}$$

does represent a realizable network. In fact, the voltage-ratio transfer function of the *R-C* network in Fig. 10.4 has an amplitude characteristic given by $|H(j\omega)|$ in Eq. 10.9.

For the ideal filter in Fig. 10.2, the inverse transform $h(t)$ has the form

$$h(t) = \frac{A_0 \sin \omega_C t}{\pi t} \tag{10.10}$$

where A_0 is a constant. From the sin x/x curve in Fig. 3.14, we see that $h(t)$ is nonzero for t less than zero. In fact, in order to make $h(t)$ causal, it must be delayed by an infinite amount. In practice, however, if we delay $h(t)$ by a large but finite amount t_d such that for $t < 0$ the magnitude of $h(t - t_d)$ is less than a very small quantity ϵ, that is,

$$|h(t - t_d)| < \epsilon \qquad t < 0$$

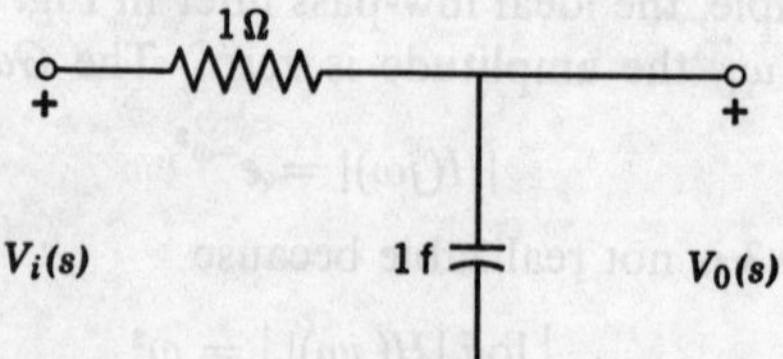

FIG. 10.4

we then can approximate $h(t - t_d)$ by a causal response $h_1(t)$ which is zero for $t < 0$. (For a more detailed discussion of the Paley-Wiener criterion, the reader is referred to an excellent treatment by Wallman.[2])

If a network is stable, then for a bounded excitation $e(t)$ the response (t) is also bounded. In other words, if

$$|e(t)| < C_1 \qquad 0 \leq t < \infty$$

then

$$|r(t)| < C_2 \qquad 0 \leq t < \infty$$

where C_1 and C_2 are real, positive, finite quantities. If a linear system is stable, then from the convolution integral we obtain

$$|r(t)| < C_1 \int_0^\infty |h(\tau)| \, d\tau < C_2 \tag{10.11}$$

Equation 10.11 requires that the impulse response be absolutely integrable, or

$$\int_0^\infty |h(\tau)| \, d\tau < \infty \tag{10.12}$$

One important requirement for $h(t)$ to be absolutely integrable is that the impulse response approach zero as t approaches infinity, that is,

$$\lim_{t \to \infty} h(t) \to 0$$

Generally, it can be said that with the exception of isolated impulses, the impulse response must be bounded for all t, that is,

$$|h(t)| < C \qquad \text{all } t \tag{10.13}$$

where C is a real, positive, finite number.

Observe that our definition of stability precludes such terms as $\sin \omega_0 t$ from the impulse response because $\sin \omega_0 t$ is not absolutely integrable. These undamped sinusoidal terms are associated with simple poles on the $j\omega$ axis. Since pure L-C networks have system functions with simple poles on the $j\omega$ axis, and since we do not wish to call these networks unstable, we say that a system is *marginally* stable if its impulse response is bounded according to Eq. 10.13, but does not approach zero as t approaches infinity.

In the frequency domain, the stability criterion requires that the system

[2] G. E. Valley Jr. and H. Wallman, *Vacuum Tube Amplifiers*, McGraw-Hill Book Company, New York, 1948, Appendix A, pp. 721–727.

function possess poles in the left-half plane or on the $j\omega$ axis only. Moreover, the poles on the $j\omega$ axis must be simple.[3] As a result of the requirement of simple poles on the $j\omega$ axis, if $H(s)$ is given as

$$H(s) = \frac{a_n s^n + a_{n-1} s^{n-1} + \cdots + a_1 s + a_0}{b_m s^m + b_{m-1} s^{m-1} + \cdots + b_1 s + b_0} \tag{10.14}$$

then the order of the numerator n cannot exceed the order of the denominator m by more than unity, that is, $n - m \leq 1$. If n exceeded m by more than unity, this would imply that at $s = j\omega = \infty$, and there would be a multiple pole. To summarize, in order for a network to be stable, the following three conditions on its system function $H(s)$ must be satisfied:

1. $H(s)$ cannot have poles in the right-half plane.
2. $H(s)$ cannot have multiple poles in the $j\omega$ axis.
3. The degree of the numerator of $H(s)$ cannot exceed the degree of the denominator by more than unity.

Finally, it should be pointed out that a rational function $H(s)$ with poles in the left-half plane only has an inverse transform $h(t)$, which is zero for $t < 0$.[4] In this respect, stability implies causality. Since system functions of passive linear networks with lumped elements are rational functions with poles in the left-half plane or $j\omega$ axis only, causality ceases to be a problem when we deal with system functions of this type. We are only concerned with the problem of causality when we have to design a filter for a given amplitude characteristic such as the ideal filter in Fig. 10.2. We know we could never hope to realize exactly a filter of this type because the impulse response would not be causal. To this extent the Paley-Wiener criterion is helpful in defining the limits of our capability.

10.2 HURWITZ POLYNOMIALS

In Section 10.1 we saw that in order for a system function to be stable, its poles must be restricted to the left-half plane or the $j\omega$ axis. Moreover, the poles on the $j\omega$ axis must be simple. The denominator polynomial of the system function belongs to a class of polynomials known as *Hurwitz*

[3] In Chapter 6 it was shown that multiple poles on the $j\omega$ axis gave rise to terms as $A_0 t \sin \omega_0 t$.

[4] G. Raisbeck, "A Definition of Passive Linear Networks in Terms of Time and Energy," *J. Appl. Phys.*, **25** (Dec., 1954), 1510–1514. The proof follows straightforwardly from the properties of the Laplace transform.

polynomials. A polynomial $P(s)$ is said to be Hurwitz if the following conditions are satisfied:

1. $P(s)$ is real when s is real.
2. The roots of $P(s)$ have real parts which are zero or negative.

As a result of these conditions, if $P(s)$ is a Hurwitz polynomial given by

$$P(s) = a_n s^n + a_{n-1}s^{n-1} + \cdots + a_1 s + a_0 \tag{10.15}$$

then all the coefficients a_i must be real; if $s_i = \alpha_i + j\beta$ is a root of $P(s)$, then α_i must be negative. The polynomial

$$P(s) = (s + 1)(s + 1 + j\sqrt{2}\,)(s + 1 - j\sqrt{2}\,) \tag{10.16}$$

is Hurwitz because all of its roots have negative real parts. On the other hand,

$$G(s) = (s - 1)(s + 2)(s + 3) \tag{10.17}$$

is not Hurwitz because of the root $s = 1$, which has a positive real part. Hurwitz polynomials have the following properties:

1. All the coefficients a_i are nonnegative. This is readily seen by examining the three types of roots that a Hurwitz polynomial might have. These are

$$\begin{aligned} s &= -\gamma_i && \gamma_i \text{ real and positive} \\ s &= \pm j\omega_i && \omega_i \text{ real} \\ s &= -\alpha_i \pm j\beta_i && \alpha_i \text{ real and positive} \end{aligned}$$

The polynomial $P(s)$ which contains these roots can be written as

$$P(s) = (s + \gamma_i)(s^2 + \omega_i^2)[(s + \alpha_i)^2 + \beta_i^2] \cdots \tag{10.18}$$

Since $P(s)$ is the product of terms with only positive coefficients, it follows that the coefficients of $P(s)$ must be positive. A corollary is that between the highest order term in s and the lowest order term, none of the coefficients may be zero unless the polynomial is even or odd. In other words, $a_{n-1}, a_{n-2}, \ldots, a_2, a_1$ must not be zero if the polynomial is neither even nor odd. This is readily seen because the absence of a term a_i implies cancellation brought about by a root $s - \gamma_i$ with a positive real part.

2. Both the odd and even parts of a Hurwitz polynomial $P(s)$ have roots on the $j\omega$ axis only. If we denote the odd part of $P(s)$ as $n(s)$ and the even part as $m(s)$, so that

$$P(s) = n(s) + m(s) \tag{10.19}$$

then $m(s)$ and $n(s)$ both have roots on the $j\omega$ axis only. The reader is referred to a proof of this property by Guillemin.[5]

3. As a result of property 2, if $P(s)$ is either even or odd, all its roots are on the $j\omega$ axis.

4. The continued fraction expansion of the ratio of the odd to even parts or the even to odd parts of a Hurwitz polynomial yields all positive quotient terms. Suppose we denote the ratios as $\psi(s) = n(s)/m(s)$ or $\psi(s) = m(s)/n(s)$, then the continued fraction expansion of $\psi(s)$ can be written as

$$\psi(s) = q_1 s + \cfrac{1}{q_2 s + \cfrac{1}{q_3 s + \cfrac{1}{\ddots + \cfrac{1}{q_n s}}}} \tag{10.20}$$

where the quotients $q_1, q_2, \ldots, q_n$ must be positive if the polynomial $P(s) = n(s) + m(s)$ is Hurwitz.[6] To obtain the continued fraction expansion, we must perform a series of long divisions. Suppose $\psi(s)$ is

$$\psi(s) = \frac{m(s)}{n(s)} \tag{10.21}$$

where $m(s)$ is of one higher degree than $n(s)$. Then if we divide $n(s)$ into $m(s)$, we obtain a single quotient and a remainder

$$\psi(s) = q_1 s + \frac{R_1(s)}{n(s)} \tag{10.22}$$

The degree of the term $R_1(s)$ is one lower than the degree of $n(s)$. Therefore if we invert the remainder term and divide, we have

$$\frac{n(s)}{R_1(s)} = q_2 s + \frac{R_2(s)}{R_1(s)} \tag{10.23}$$

Inverting and dividing again, we obtain

$$\frac{R_1(s)}{R_2(s)} = q_3 s + \frac{R_3(s)}{R_2(s)} \tag{10.24}$$

[5] E. Guillemin, *The Mathematics of Circuit Analysis*, John Wiley and Sons, New York, 1949. An excellent treatment of Hurwitz polynomials is given here.

[6] A proof can be undertaken in connection with *L-C* driving-point functions; see M. E. Van Valkenburg, *Introduction to Modern Network Synthesis*, John Wiley and Sons, New York, 1960.

We see that the process of obtaining the continued fraction expansion of $\psi(s)$ simply involves division and inversion. At each step we obtain a quotient term $q_i s$ and a remainder term, $R_{i+1}(s)/R_i(s)$. We then invert the remainder term and divide $R_{i+1}(s)$ into $R_i(s)$ to obtain a new quotient. There is a theorem in the theory of continued fractions which states that the continued fraction expansion of the even to odd or odd to even parts of a polynomial must be finite in length.[7] Another theorem states that, if the continued fraction expansion of the odd to even or even to odd parts of a polynomial yields positive quotient terms, then the polynomial must be Hurwitz to within a multiplicative factor $W(s)$.[8] That is, if we write

$$F(s) = W(s)\, F_1(s) \tag{10.25}$$

then $F(s)$ is Hurwitz, if $W(s)$ and $F_1(s)$ are Hurwitz. For example, let us test whether the polynomial

$$F(s) = s^4 + s^3 + 5s^2 + 3s + 4 \tag{10.26}$$

is Hurwitz. The even and odd parts of $F(s)$ are

$$\begin{aligned} m(s) &= s^4 + 5s^2 + 4 \\ n(s) &= s^3 + 3s \end{aligned} \tag{10.27}$$

We now perform a continued fraction expansion of $\psi(s) = m(s)/n(s)$ by dividing $n(s)$ by $m(s)$, and then inverting and dividing again, as given by the operation

$$\begin{array}{l}
s^3 + 3s\,\overline{)\,s^4 + 5s^2 + 4\,}(\,s \\
\qquad\qquad \underline{s^4 + 3s^2} \\
\qquad\qquad\quad 2s^2 + 4\,)\,s^3 + 3s\,(\,s/2 \\
\qquad\qquad\qquad\qquad\quad \underline{s^3 + 2s} \\
\qquad\qquad\qquad\qquad\qquad\quad s\,)\,2s^2 + 4\,(\,2s \\
\qquad\qquad\qquad\qquad\qquad\qquad\quad \underline{2s^2} \\
\qquad\qquad\qquad\qquad\qquad\qquad\quad 4\,)\,s\,(\,s/4 \\
\qquad\qquad\qquad\qquad\qquad\qquad\qquad\quad \underline{\underline{s}}
\end{array}$$

so that the continued fraction expansion of $\psi(s)$ is

$$\psi(s) = \frac{m(s)}{n(s)} = s + \cfrac{1}{\cfrac{s}{2} + \cfrac{1}{2s + \cfrac{1}{s/4}}} \tag{10.28}$$

---

[7] See Van Valkenburg, *loc. cit.*

[8] $W(s)$ is a common factor in $m(s)$ and $n(s)$.

Since all the quotient terms of the continued fraction expansion are positive, $F(s)$ is Hurwitz.

Example 10.1. Let us test whether the polynomial

$$G(s) = s^3 + 2s^2 + 3s + 6 \tag{10.29}$$

is Hurwitz. The continued fraction expansion of $n(s)/m(s)$ is obtained from the division

$$\begin{array}{r|l} 2s^2 + 6 \overline{)\, s^3 + 3s\,} & s/2 \\ s^3 + 3s & \\ \hline 0 & \end{array}$$

We see that the division has been terminated abruptly by a common factor $s^3 + 3s$. The polynomial can then be written as

$$G(s) = (s^3 + 3s)\left(1 + \frac{2}{s}\right) \tag{10.30}$$

We know that the term $1 + 2/s$ is Hurwitz. Since the multiplicative factor $s^3 + 3s$ is also Hurwitz, then $G(s)$ is Hurwitz. The term $s^3 + 3s$ is the multiplicative factor $W(s)$, which we referred to earlier.

Example 10.2. Next consider a case where $W(s)$ is non-Hurwitz.

$$F(s) = s^7 + 2s^6 + 2s^5 + s^4 + 4s^3 + 8s^2 + 8s + 4 \tag{10.31}$$

The continued fraction expansion of $F(s)$ is now obtained.

$$\frac{n(s)}{m(s)} = \frac{s}{2} + \cfrac{1}{\frac{4}{3}s + \cfrac{1}{\cfrac{\frac{3}{2}s(s^4+4)}{(s^4+4)}}} \tag{10.32}$$

We thus see that $W(s) = s^4 + 4$, which can be factored into

$$W(s) = (s^2 + 2s + 2)(s^2 - 2s + 2) \tag{10.33}$$

It is clear that $F(s)$ is not Hurwitz.

Example 10.3. Let us consider a more obvious non-Hurwitz polynomial

$$F(s) = s^4 + s^3 + 2s^2 + 3s + 2 \tag{10.34}$$

The continued fraction expansion is

$$\begin{array}{rl}
s^3+3s\,\overline{)\,s^4+2s^2+2}\,(s & \\
s^4+3s^2 & \\
\hline
-s^2+2\,\overline{)\,s^3+3s}\,(-s & \\
s^3-2s & \\
\hline
5s\,\overline{)-s^2+2}\,(-s/5 & \\
-s^2 & \\
\hline
2\,\overline{)\,5s}\,(\tfrac{5}{2}s & \\
5s & \\
\hline\hline
\end{array}$$

We see that $F(s)$ is not Hurwitz because of the negative quotients.

Example 10.4. Consider the case where $F(s)$ is an odd or even function. It is impossible to perform a continued fraction expansion on the function as it stands. However, we can test the ratio of $F(s)$ to its derivative, $F'(s)$.[9] If the ratio $F(s)/F'(s)$ gives a continued fraction expansion with all positive coefficients, then $F(s)$ is Hurwitz. For example, if $F(s)$ is given as

$$F(s) = s^7 + 3s^5 + 2s^3 + s \tag{10.35}$$

then $F'(s)$ is

$$F'(s) = 7s^6 + 15s^4 + 6s^2 + 1 \tag{10.36}$$

Without going into the details, it can be shown that the continued fraction expansion of $F(s)/F'(s)$ does not yield all positive quotients. Therefore $F(s)$ is not Hurwitz.

10.3 POSITIVE REAL FUNCTIONS

In this section we will study the properties of a class of functions known as *positive real functions*. These functions are important because they represent physically realizable passive driving-point immittances. A function $F(s)$ is positive real (p.r.) if the following conditions are satisfied:

1. $F(s)$ is real for real s; that is, $F(\sigma)$ is real.
2. The real part of $F(s)$ is greater than or equal to zero when the real part of s is greater than or equal to zero, that is,

$$\text{Re}\,[F(s)] \geq 0 \qquad \text{for} \qquad \text{Re}\, s \geq 0$$

Let us consider a complex plane interpretation of a p.r. function. Consider the s plane and the $F(s)$ plane in Fig. 10.5. If $F(s)$ is p.r., then a point σ_0 on the positive real axis of the s plane would correspond to, or *map onto*, a point $F(\sigma_0)$ which must be on the positive real axis of the $F(s)$ plane. In addition, a point s_i in the right half of the s plane would

[9] See Guillemin, *loc. cit.*

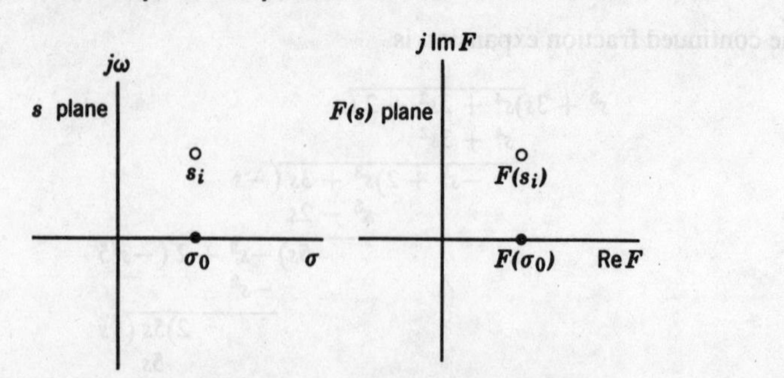

FIG. 10.5. Mapping of s plane onto $F(s)$ plane.

map onto a point $F(s_i)$ in the right half of the $F(s)$ plane. In other words, for a positive real function, the right half of the s plane maps onto the right half of the $F(s)$ plane. The real axis of the s plane maps onto the real axis of the $F(s)$ plane.

A further restriction we will impose is that $F(s)$ be rational. Consider the following examples of p.r. functions:

1. $F(s) = Ls$ (where L is a real, positive number) is p.r. by definition. If $F(s)$ is an impedance function, then L is an inductance.

2. $F(s) = R$ (where R is real and positive) is p.r. by definition. If $F(s)$ is an impedance function, R is a resistance.

3. $F(s) = K/s$ (K real and positive) is p.r. because, when s is real, $F(s)$ is real. In addition, when the real part of s is greater than zero, $\text{Re}\,(s) = \sigma > 0$.

Then
$$\text{Re}\left(\frac{K}{s}\right) = \frac{K\sigma}{\sigma^2 + \omega^2} > 0 \tag{10.37}$$

Therefore, $F(s)$ is p.r. If $F(s)$ is an impedance function, then the corresponding element is a capacitor of $1/K$ farads.

We thus see that the basic passive impedances are p.r. functions. Similarly, it is clear that the admittances

$$\begin{aligned} Y(s) &= K \\ Y(s) &= Ks \\ Y(s) &= \frac{K}{s} \end{aligned} \tag{10.38}$$

are positive real if K is real and positive. We now show that all driving-point immittances of passive networks must be p.r. The proof depends upon the following assertion: for a sinusoidal input, the average power dissipated by a passive network is nonnegative. For the passive network in Fig. 10.6, the average power dissipated by the network is

$$\text{Average power} = \tfrac{1}{2}\,\text{Re}\,[Z_{in}(j\omega)]\,|I|^2 \geq 0 \qquad (10.39)$$

We then conclude that, for any passive network

$$\text{Re}\,[Z_{in}(j\omega)] \geq 0 \qquad (10.40)$$

We can now prove that for $\text{Re}\, s = \sigma \geq 0$, $\text{Re}\, Z_{in}(\sigma + j\omega) \geq 0$. Consider the network in Fig. 10.6, whose driving-point impedance is $Z_{in}(s)$. Let us load the network with incidental dissipation such that if the driving-point impedance of the uniformly loaded network is $Z_1(s)$, then

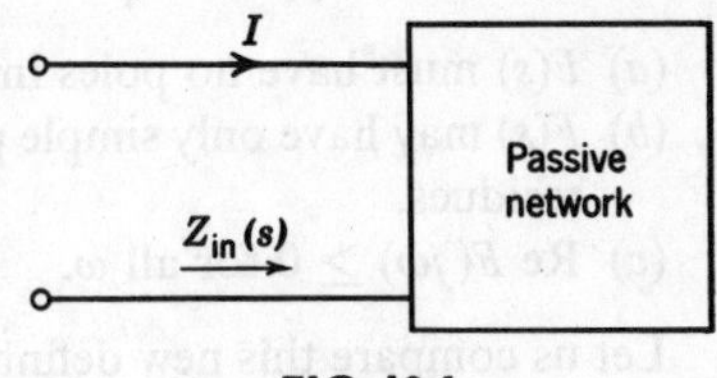

FIG. 10.6

$$Z_1(s) = Z_{in}(s + \alpha) \qquad (10.41)$$

where α, the dissipation constant, is real and positive. Since $Z_1(s)$ is the impedance of a passive network,

$$\text{Re}\, Z_1(j\omega) \geq 0 \qquad (10.42)$$

so that

$$\text{Re}\, Z_{in}(\alpha + j\omega) \geq 0 \qquad (10.43)$$

Since α is an arbitrary real positive quantity, it can he taken to be σ. Thus the theorem is proved.

Next let us consider some useful properties of p.r. functions. The proofs of these properties are given in Appendix D.

1. If $F(s)$ is p.r., then $1/F(s)$ is also p.r. This property implies that if a driving-point impedance is p.r., then its reciprocal, the driving-point admittance, is also p.r.

2. The sum of p.r. functions is p.r. From an impedance standpoint, we see that if two impedances are connected in series, the sum of the impedances is p.r. An analogous situation holds for two admittances in parallel. Note that the difference of two p.r. functions is not necessarily p.r.; for example, $F(s) = s - 1/s$ is not p.r.

3. The poles and zeros of a p.r. function cannot have positive real parts, i.e., they cannot be in the right half of the s plane.

4. Only simple poles with real positive residues can exist on the $j\omega$ axis.

5. The poles and zeros of a p.r. function are real or occur in conjugate pairs. We know that the poles and zeros of a network function are functions of the elements in the network. Since the elements themselves are real, there cannot be complex poles or zeros without conjugates because this would imply imaginary elements.

6. The highest powers of the numerator and denominator polynomials may differ at most by unity. This condition prohibits multiple poles and zeros at $s = \infty$.

7. The lowest powers of the denominator and numerator polynomials may differ by at most unity. This condition prevents the possibility of multiple poles or zeros at $s = 0$.

8. The necessary and sufficient conditions for a rational function with real coefficients $F(s)$ to be p.r. are

(*a*) $F(s)$ must have no poles in the right-half plane.
(*b*) $F(s)$ may have only simple poles on the $j\omega$ axis with real and positive residues.
(*c*) $\text{Re } F(j\omega) \geq 0$ for all ω.

Let us compare this new definition with the original one which requires the two conditions.

1. $F(s)$ is real when s is real.
2. $\text{Re } F(s) \geq 0$, when $\text{Re } s \geq 0$.

In order to test condition 2 of the original definition, we must test every single point in the right-half plane. In the alternate definition, condition (*c*) merely requires that we test the behavior of $F(s)$ along the $j\omega$ axis. It is apparent that testing a function for the three conditions given by the alternate definition represents a considerable saving of effort, except in simple cases as $F(s) = 1/s$.

Let us examine the implications of each criterion of the second definition. Condition (*a*) requires that we test the denominator of $F(s)$ for roots in the right-half plane, i.e., we must determine whether the denominator of $F(s)$ is Hurwitz. This is readily accomplished through a continued fraction expansion of the odd to even or even to odd parts of the denominator. The second requirement—condition (*b*)—is tested by making a partial fraction expansion of $F(s)$ and checking whether the residues of the poles on the $j\omega$ axis are positive and real. Thus, if $F(s)$ has a pair of poles at $s = \pm j\omega_1$, a partial fraction expansion gives terms of the form shown.

$$\frac{K_1}{s - j\omega_1} + \frac{K_1^*}{s + j\omega_1}$$

The residues of complex conjugate poles are themselves conjugates. If the residues are real—as they must be in order for $F(s)$ to be p.r.—then $K_1 = K_1^*$ so that

$$\frac{K_1}{s - j\omega_1} + \frac{K_1^*}{s + j\omega_1} = \frac{2K_1 s}{s^2 + \omega_1^2} \tag{10.44}$$

If K_1 is found to be positive, then $F(s)$ satisfies the second of the three conditions.

In order to test for the third condition for positive realness, we must first find the real part of $F(j\omega)$ from the original function $F(s)$. To do this, let us consider a function $F(s)$ given as a quotient of two polynomials

$$F(s) = \frac{P(s)}{Q(s)} \tag{10.45}$$

We can separate the even parts from the odd parts of $P(s)$ and $Q(s)$ so that $F(s)$ is

$$F(s) = \frac{M_1(s) + N_1(s)}{M_2(s) + N_2(s)} \tag{10.46}$$

where $M_i(s)$ is an even function and $N_i(s)$ is an odd function. $F(s)$ is now decomposed into its even and odd parts by multiplying both $P(s)$ and $Q(s)$ by $M_2 - N_2$ so that

$$\begin{aligned} F(s) &= \frac{M_1 + N_1}{M_2 + N_2}\frac{M_2 - N_2}{M_2 - N_2} \\ &= \frac{M_1 M_2 - N_1 N_2}{M_2^2 - N_2^2} + \frac{M_2 N_1 - M_1 N_2}{M_2^2 - N_2^2} \end{aligned} \tag{10.47}$$

We see that the products $M_1 M_2$ and $N_1 N_2$ are even functions, while $M_1 N_2$ and $M_2 N_1$ are odd functions. Therefore, the even part of $F(s)$ is

$$\text{Ev}\,[F(s)] = \frac{M_1 M_2 - N_1 N_2}{M_2^2 - N_2^2} \tag{10.48}$$

and the odd part of $F(s)$ is

$$\text{Odd}\,[F(s)] = \frac{M_2 N_1 - M_1 N_2}{M_2^2 - N_2^2} \tag{10.49}$$

If we let $s = j\omega$, we see that the even part of any polynomial is real, while the odd part of the polynomial is imaginary, so that if $F(j\omega)$ is written as

$$F(j\omega) = \text{Re}\,[F(j\omega)] + j\,\text{Im}\,[F(j\omega)] \tag{10.50}$$

it is clear that

$$\text{Re}\,[F(j\omega)] = \text{Ev}\,[F(s)]\big|_{s=j\omega} \tag{10.51}$$

and

$$j\,\text{Im}\,[F(j\omega)] = \text{Odd}\,[F(s)]\big|_{s=j\omega} \tag{10.52}$$

Therefore, to test for the third condition for positive realness, we determine the real part of $F(j\omega)$ by finding the even part of $F(s)$ and then letting $s = j\omega$. We then check to see whether $\text{Re}\,F(j\omega) \geq 0$ for all ω.

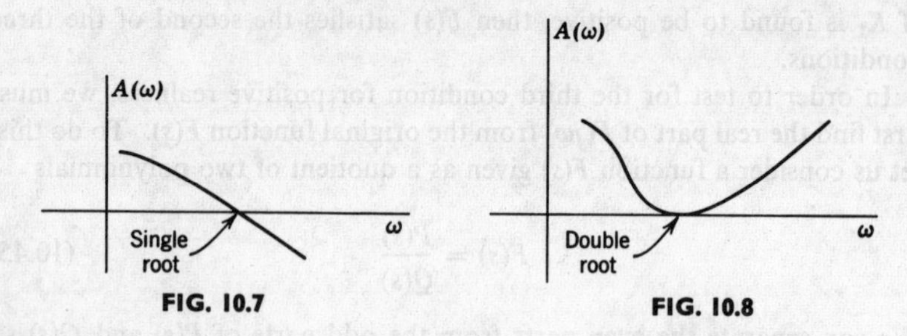

FIG. 10.7

FIG. 10.8

The denominator of Re $F(j\omega)$ is always a positive quantity because

$$M_2(j\omega)^2 - N_2(j\omega)^2 = M_2(\omega)^2 + N_2(\omega)^2 \geq 0 \tag{10.53}$$

That is, there is an extra j or imaginary term in $N_2(j\omega)$, which, when squared, gives -1, so that the denominator of Re $F(j\omega)$ is the sum of two squared numbers and is always positive. Therefore, our task resolves into the problem of determining whether

$$A(\omega) \triangleq M_1(j\omega)\, M_2(j\omega) - N_1(j\omega)\, N_2(j\omega) \geq 0 \tag{10.54}$$

If we call the preceding function $A(\omega)$, we see that $A(\omega)$ must not have positive, real roots of the type shown in Fig. 10.7; i.e., $A(\omega)$ must never have single, real roots of ω. However, $A(\omega)$ may have double roots (Fig. 10.8), because $A(\omega)$ need not become negative in this case.

As an example, consider the requirements for

$$F(s) = \frac{s + a}{s^2 + bs + c} \tag{10.55}$$

to be p.r. First, we know that, in order for the poles and zeros to be in the left-half plane or on the $j\omega$ axis, the coefficients a, b, c must be greater or equal to zero. Second, if $b = 0$, then $F(s)$ will possess poles on the $j\omega$ axis. We can then write $F(s)$ as

$$F(s) = \frac{s}{s^2 + c} + \frac{a}{s^2 + c} \tag{10.56}$$

We will show later that the coefficient a must also be zero when $b = 0$. Let us proceed with the third requirement, namely, Re $F(j\omega) \geq 0$. From the equation

$$M_1(j\omega)\, M_2(j\omega) - N_1(j\omega)\, N_2(j\omega) \geq 0 \tag{10.57}$$

we have

$$a(-\omega^2 + c) + b\omega^2 \geq 0 \tag{10.58a}$$

which simplifies to

$$A(\omega) = (b - a)\omega^2 + ac \geq 0 \tag{10.58b}$$

It is evident that in order to prevent $A(\omega)$ from having positive real roots of ω, b must be greater than or equal to a, that is, $b \geq a$. As a result, when $b = 0$, then $a = 0$. To summarize, the conditions that must be fulfilled in order for $F(s)$ to be positive real are

1. $a, b, c \geq 0$.
2. $b \geq a$.

We see that
$$F_1(s) = \frac{s+2}{s^2+3s+2} \tag{10.59}$$
is p.r., while the functions
$$F_2(s) = \frac{s+1}{s^2+2} \tag{10.60}$$
$$F_3(s) = \frac{s+4}{s^2+2s+1} \tag{10.61}$$
are not p.r. As a second example, let us determine the conditions for the *biquadratic* function
$$F(s) = \frac{s^2+a_1 s+a_0}{s^2+b_1 s+b_0} \tag{10.62}$$
to be p.r. We will assume that the coefficients a_1, a_0, b_1, b_0 are all real, positive constants. Let us test whether $F(s)$ is p.r. by testing each requirement of the second definition.

First, if the coefficients of the denominator b_1 and b_0 are positive, the denominator must be Hurwitz. Second, if b_1 is positive, we have no poles on the $j\omega$ axis. Therefore we can ignore the second condition.

The third condition can be checked by first finding the even part of $F(s)$, which is
$$\begin{aligned}\text{Ev}\,[F(s)] &= \frac{(s^2+a_0)(s^2+b_0)-a_1 b_1 s^2}{(s^2+b_0)^2-b_1{}^2 s^2}\\ &= \frac{s^4+[(a_0+b_0)-a_1 b_1]s^2+a_0 b_0}{(s^2+b_0)^2-b_1{}^2 s^2}\end{aligned} \tag{10.63}$$
The real part of $F(j\omega)$ is then
$$\text{Re}\,[F(j\omega)] = \frac{\omega^4-[(a_0+b_0)-a_1 b_1]\omega^2+a_0 b_0}{(-\omega^2+b_0)^2+b_1{}^2\omega^2} \tag{10.64}$$

We see that the denominator of Re $[F(j\omega)]$ is truly always positive so it remains for us to determine whether the numerator of Re $[F(j\omega)]$ ever goes negative. Factoring the numerator, we obtain
$$\omega_{1,2}^2 = \frac{(a_0+b_0)-a_1 b_1}{2} \pm \frac{1}{2}\sqrt{[(a_0+b_0)-a_1 b_1]^2-4a_0 b_0} \tag{10.65}$$

There are two situations in which Re $[F(j\omega)]$ does not have a simple real root.

1. When the quantity under the radical sign of Eq. 10.65 is zero (double, real root) or negative (complex roots). In other words,

$$[(a_0 + b_0) - a_1b_1]^2 - 4a_0b_0 \leq 0 \tag{10.66}$$

or

$$[(a_0 + b_0) - a_1b_1]^2 \leq 4a_0b_0 \tag{10.67}$$

If

$$(a_0 + b_0) - a_1b_1 \geq 0 \tag{10.68}$$

then

$$(a_0 + b_0) - a_1b_1 \leq 2\sqrt{a_0b_0} \tag{10.69}$$

or

$$a_1b_1 \geq (\sqrt{a_0} - \sqrt{b_0})^2 \tag{10.70}$$

If

$$(a_0 + b_0) - a_1b_1 < 0 \tag{10.71}$$

then

$$a_1b_1 - (a_0 + b_0) \leq 2\sqrt{a_0b_0} \tag{10.72}$$

but

$$(a_0 + b_0) - a_1b_1 < 0 < a_1b_1 - (a_0 + b_0) \tag{10.73}$$

so again

$$a_1b_1 \geq (\sqrt{a_0} - \sqrt{b_0})^2 \tag{10.74}$$

2. The second situation in which Re $[F(j\omega)]$ does not have a simple real root is when $\omega_{1,2}^2$ in Eq. 10.65 is negative so that the roots are imaginary. This situation occurs when

$$[(a_0 + b_0) - a_1b_1]^2 - 4a_0b_0 > 0 \tag{10.75}$$

and

$$(a_0 + b_0) - a_1b_1 < 0 \tag{10.76}$$

From Eq. 10.75 we have

$$a_1b_1 - (a_0 + b_0) > 2\sqrt{a_0b_0} > (a_0 + b_0) - a_1b_1 \tag{10.77}$$

Thus

$$a_1b_1 > (\sqrt{a_0} - \sqrt{b_0})^2 \tag{10.78}$$

We thus see that Eq. 10.70 is a necessary and sufficient condition for a biquadratic function to be positive real. If we have

$$a_1b_1 = (\sqrt{a_0} - \sqrt{b_0})^2 \tag{10.79}$$

then we will have double zeros for Re $[F(j\omega)]$.

Consider the following example:

$$F(s) = \frac{s^2 + 2s + 25}{s^2 + 5s + 16} \tag{10.80}$$

We see that

$$a_1b_1 = 2 \times 5 \geq (\sqrt{a_0} - \sqrt{b_0})^2 = (\sqrt{25} - \sqrt{16})^2 \tag{10.81}$$

so that $F(s)$ is p.r.

The examples just given are, of course, special cases. But they do illustrate the procedure by which functions are tested for the p.r. property. Let us consider a number of other helpful points by which a function might be tested quickly. First, if $F(s)$ has poles on the $j\omega$ axis, a partial fraction expansion will show if the residues of these poles are positive and real. For example,

$$F(s) = \frac{3s^2 + 5}{s(s^2 + 1)} \tag{10.82}$$

has a pair of poles at $s = \pm j1$. The partial fraction expansion of $F(s)$,

$$F(s) = \frac{-2s}{s^2 + 1} + \frac{5}{s} \tag{10.83}$$

shows that the residue of the poles at $s = \pm j$ is negative. Therefore $F(s)$ is not p.r.

Since impedances and admittances of passive time-invariant networks are p.r. functions, we can make use of our knowledge of impedances connected in series or parallel in our testing for the p.r. property. For example, if $Z_1(s)$ and $Z_2(s)$ are passive impedances, then Z_1 connected in parallel with Z_2 gives an overall impedance

$$Z(s) = \frac{Z_1(s)\, Z_2(s)}{Z_1(s) + Z_2(s)} \tag{10.84}$$

Since the connecting of the two impedances in parallel has not affected the passivity of the network, we know that $Z(s)$ must also be p.r. We see that if $F_1(s)$ and $F_2(s)$ are p.r. functions, then

$$F(s) = \frac{F_1(s)\, F_2(s)}{F_1(s) + F_2(s)} \tag{10.85}$$

must also be p.r. Consequently, the functions

$$F(s) = \frac{Ks}{s + \alpha} \tag{10.86}$$

and

$$F(s) = \frac{K}{s + \alpha} \tag{10.87}$$

where α and K are real and positive quantities, must be p.r. We then observe that functions of the type

$$\begin{aligned} F(s) &= \frac{s + \beta}{s + \alpha} \qquad \alpha, \beta \geq 0 \\ &= \frac{s}{s + \alpha} + \frac{\beta}{s + \alpha} \end{aligned} \tag{10.88}$$

must be p.r. also.

Finally, let us determine whether

$$F(s) = \frac{Ks}{s^2 + \alpha} \qquad \alpha, K \geq 0 \tag{10.89}$$

is p.r. If we write $F(s)$ as

$$F(s) = \frac{1}{s/K + \alpha/Ks} \tag{10.90}$$

we see that the terms s/K and α/Ks are p.r. Therefore, the sum of the two terms must be p.r. Since the reciprocal of a p.r. function is also p.r., we conclude that $F(s)$ is p.r.

10.4 ELEMENTARY SYNTHESIS PROCEDURES

The basic philosophy behind the synthesis of driving-point functions is to break up a p.r. function $Z(s)$ into a sum of simpler p.r. functions $Z_1(s), Z_2(s), \ldots, Z_n(s)$, and then to synthesize these individual $Z_i(s)$ as elements of the overall network whose driving-point impedance is $Z(s)$.

$$Z(s) = Z_1(s) + Z_2(s) + \cdots + Z_n(s) \tag{10.91}$$

First, consider the "breaking-up" process of the function $Z(s)$ into the sum of functions $Z_i(s)$. One important restriction is that all $Z_i(s)$ must be p.r. Certainly, if all $Z_i(s)$ were given to us, we could synthesize a network whose driving-point impedance is $Z(s)$ by simply connecting all the $Z_i(s)$ in series. However, if we were to start with $Z(s)$ alone, how would we decompose $Z(s)$ to give us the individual $Z_i(s)$? Suppose $Z(s)$ is given in general as

$$Z(s) = \frac{a_n s^n + a_{n-1}s^{n-1} + \cdots + a_1 s + a_0}{b_m s^m + b_{m-1}s^{m-1} + \cdots + b_1 s + b_0} = \frac{P(s)}{Q(s)} \tag{10.92}$$

Consider the case where $Z(s)$ has a pole at $s = 0$ (that is, $b_0 = 0$). Let us divide $P(s)$ by $Q(s)$ to give a quotient D/s and a remainder $R(s)$, which we can denote as $Z_1(s)$ and $Z_2(s)$.

$$\begin{aligned} Z(s) &= \frac{D}{s} + R(s) \qquad D \geq 0 \\ &= Z_1(s) + Z_2(s) \end{aligned} \tag{10.93}$$

Are Z_1 and Z_2 p.r.? From previous discussions, we know that $Z_1 = D/s$ is p.r. Is $Z_2(s)$ p.r.? Consider the p.r. criteria given previously.

1. $Z_2(s)$ must have no poles in the right-half plane.
2. Poles of $Z_2(s)$ on the imaginary axis must be simple, and their residues must be real and positive.
3. $\text{Re}\,[Z_2(j\omega)] \geq 0$ for all ω.

Let us examine these cases one by one. Criterion 1 is satisfied because the poles of $Z_2(s)$ are also poles of $Z(s)$. Criterion 2 is satisfied by this same argument. A simple partial fraction expansion does not affect the residues of the other poles. When $s = j\omega$, $\text{Re}\,[Z(j\omega) = D/j\omega] = 0$. Therefore we have

$$\text{Re}\, Z_2(j\omega) = \text{Re}\, Z(j\omega) \geq 0 \tag{10.94}$$

From the foregoing discussion, it is seen that if $Z(s)$ has a pole at $s = 0$, a partial fraction expansion can be made such that one of the terms is of the form K/s and the other terms combined still remain p.r.

A similar argument shows that if $Z(s)$ has a pole at $s = \infty$ (that is, $n - m = 1$), we can divide the numerator by the denominator to give a quotient Ls and a remainder term $R(s)$, again denoted as $Z_1(s)$ and $Z_2(s)$.

$$Z(s) = Ls + R(s) = Z_1(s) + Z_2(s). \tag{10.95}$$

Here $Z_2(s)$ is also p.r. If $Z(s)$ has a pair of conjugate imaginary poles on the imaginary axis, for example, poles at $s = \pm j\omega_1$, then $Z(s)$ can be expanded into partial fractions so that

$$Z(s) = \frac{2Ks}{s^2 + \omega_1^{\,2}} + Z_2(s) \tag{10.96}$$

Here

$$\text{Re}\left(\frac{2Ks}{s^2 + \omega_1^{\,2}}\right)_{s=j\omega} = \text{Re}\left(\frac{j2\omega K}{-\omega^2 + \omega_1^{\,2}}\right) = 0 \tag{10.97}$$

so that $Z_2(s)$ is p.r.

Finally, if $\text{Re}\,[Z(j\omega)]$ is minimum at some point ω_i, and if the value of $\text{Re}\, Z(j\omega_i) = K_i$ as shown in Fig. 10.9, we can remove a constant $K \leq K_i$ from $\text{Re}\,[Z(j\omega)]$ so that the remainder is still p.r. This is because $\text{Re}\,[Z(j\omega)]$ will still be greater than or equal to zero for all values of ω.

Suppose we have a p.r. function $Z(s)$, which is a driving-point impedance function. Let $Z(s)$ be decomposed, as before, so that

$$Z(s) = Z_1(s) + Z_2(s) \tag{10.98}$$

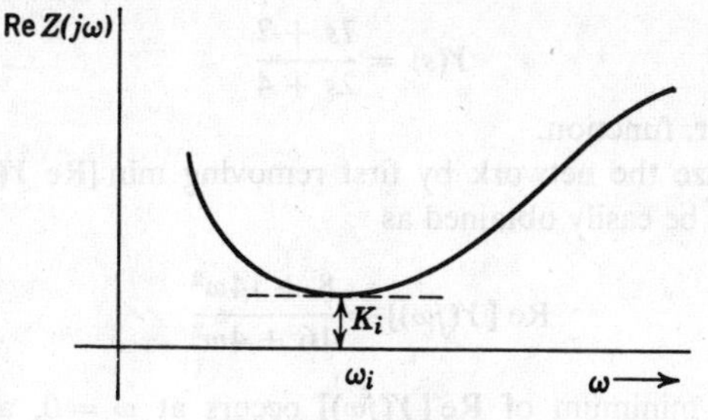

FIG. 10.9

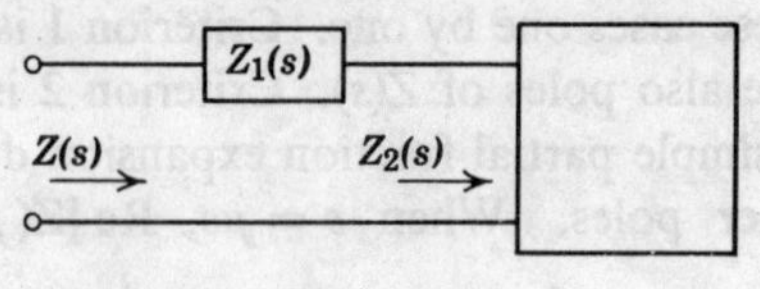

FIG. 10.10

where both Z_1 and Z_2 are p.r. Now let us "remove" $Z_1(s)$ from $Z(s)$ to give us a remainder $Z_2(s)$. This removal process is illustrated in Fig. 10.10 and shows that removal corresponds to synthesis of $Z_1(s)$.

Example 10.5. Consider the following p.r. function

$$Z(s) = \frac{s^2 + 2s + 6}{s(s + 3)} \tag{10.99}$$

We see that $Z(s)$ has a pole at $s = 0$. A partial fraction expansion of $Z(s)$ yields

$$\begin{aligned} Z(s) &= \frac{2}{s} + \frac{s}{s + 3} \\ &= Z_1(s) + Z_2(s) \end{aligned} \tag{10.100}$$

If we remove $Z_1(s)$ from $Z(s)$, we obtain $Z_2(s)$, which can be shown by a resistor in parallel with an inductor, as illustrated in Fig. 10.11.

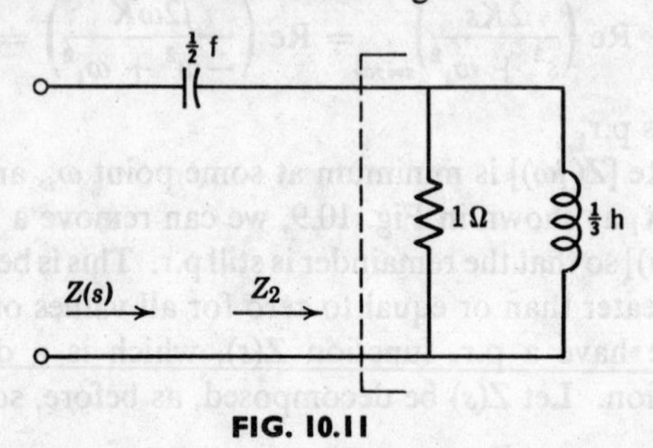

FIG. 10.11

Example 10.6

$$Y(s) = \frac{7s + 2}{2s + 4} \tag{10.101}$$

where $Y(s)$ is a p.r. function.

Let us synthesize the network by first removing min [Re $Y(j\omega)$]. The real part of $Y(j\omega)$ can be easily obtained as

$$\text{Re}\,[Y(j\omega)] = \frac{8 + 14\omega^2}{16 + 4\omega^2} \tag{10.102}$$

We see that the minimum of Re $[Y(j\omega)]$ occurs at $\omega = 0$, and is equal to min [Re $Y(j\omega)$] $= \frac{1}{2}$. Let us then remove $Y_1 = \frac{1}{2}$ mho from $Y(s)$ and denote the remainder as $Y_2(s)$, as shown in Fig. 10.12. The remainder function $Y_2(s)$ is p.r. because we have removed only the minimum real part of $Y_2(j\omega)$. $Y_2(s)$ is obtained as

$$Y_2(s) = Y(s) - \frac{1}{2} = \frac{3s}{s+2} \tag{10.103}$$

It is readily seen that $Y_2(s)$ is made up of a $\frac{1}{3}$-Ω resistor in series with a $\frac{3}{2}$-farad capacitor. Thus the final network is that shown in Fig. 10.13.

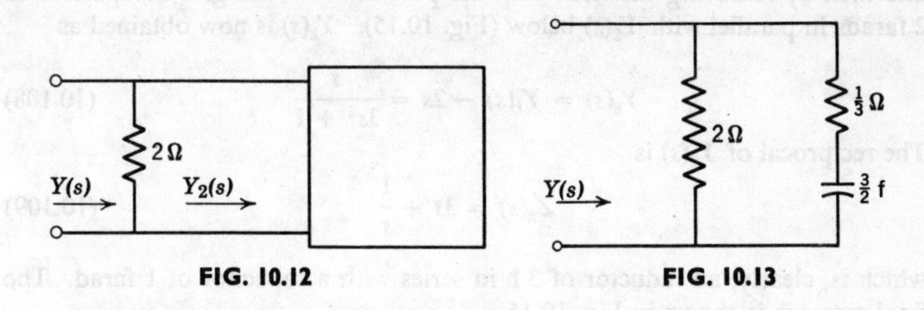

FIG. 10.12 FIG. 10.13

Example 10.7. Consider the p.r. impedance

$$Z(s) = \frac{6s^3 + 3s^2 + 3s + 1}{6s^3 + 3s} \tag{10.104}$$

The real part of the function is a constant, equal to unity. Removing a constant of 1 Ω, we obtain (Fig. 10.14)

$$Z_1(s) = Z(s) - 1 = \frac{3s^2 + 1}{6s^3 + 3s} \tag{10.105}$$

The reciprocal of $Z_1(s)$ is an admittance

$$Y_1(s) = \frac{6s^3 + 3s}{3s^2 + 1} \tag{10.106}$$

which has a pole at $s = \infty$. This pole is removed by finding the partial fraction expansion of $Y_1(s)$;

$$Y_1(s) = 2s + \frac{s}{3s^2 + 1} \tag{10.107}$$

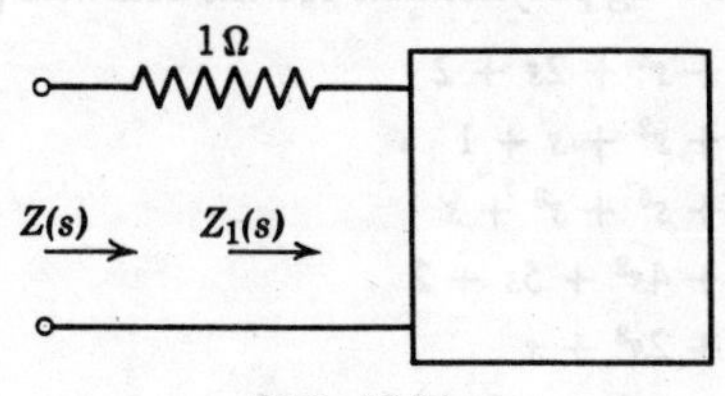

FIG. 10.14

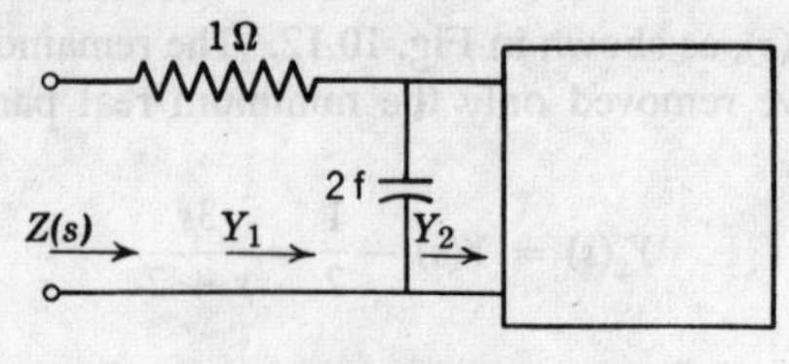

FIG. 10.15

and then by removing the term with the pole at $s = \infty$ to give a capacitor of 2 farads in parallel with $Y_2(s)$ below (Fig. 10.15). $Y_2(s)$ is now obtained as

$$Y_2(s) = Y_1(s) - 2s = \frac{s}{3s^2 + 1} \tag{10.108}$$

The reciprocal of $Y_2(s)$ is

$$Z_2(s) = 3s + \frac{1}{s} \tag{10.109}$$

which is, clearly, an inductor of 3 h in series with a capacitor of 1 farad. The final network is shown in Fig. 10.16.

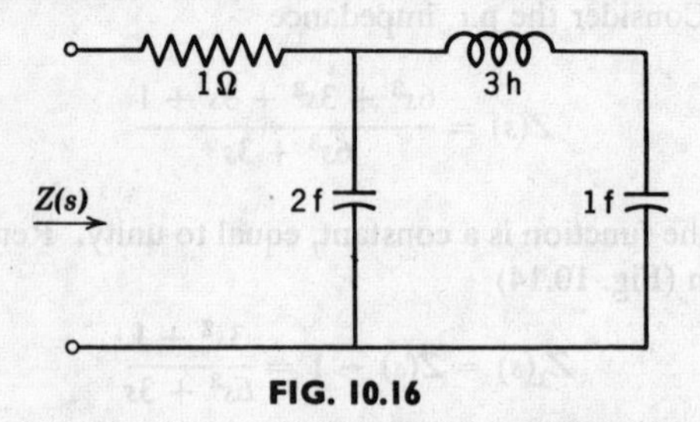

FIG. 10.16

These examples are, of course, special cases of the driving-point synthesis problem. However, they do illustrate the basic techniques involved. In the next chapter, we will discuss the problem of synthesizing a network with two kinds of elements, either *L-C*, *R-C*, or *R-L* networks. The synthesis techniques involved, however, will be the same.

Problems

10.1 Test the following polynomials for the Hurwitz property.

(*a*) $s^3 + s^2 + 2s + 2$

(*b*) $s^4 + s^2 + s + 1$

(*c*) $s^7 + s^5 + s^3 + s$

(*d*) $s^3 + 4s^2 + 5s + 2$

(*e*) $s^5 + 2s^3 + s$

(*f*) $s^7 + 2s^6 + 2s^5 + s^4 + 4s^3 + 8s^2 + 8s + 4$

10.2 Determine whether the following functions are p.r. For the functions with the denominator already factored, perform a partial fraction expansion first.

(*a*)
$$F(s) = \frac{s^2 + 1}{s^3 + 4s}$$

(*b*)
$$F(s) = \frac{2s^2 + 2s + 4}{(s + 1)(s^2 + 2)}$$

(*c*)
$$F(s) = \frac{(s + 2)(s + 4)}{(s + 1)(s + 3)}$$

(*d*)
$$F(s) = \frac{s^2 + 4}{s^3 + 3s^2 + 3s + 1}$$

(*e*)
$$F(s) = \frac{5s^2 + s}{s^2 + 1}$$

10.3 Suppose $F_1(s)$ and $F_2(s)$ are both p.r. Discuss the conditions such that $F(s) = F_1(s) - F_2(s)$ is also p.r.

10.4 Show that the product of two p.r. functions need not be p.r. Also show that the ratio of one p.r. function to another may not be p.r. (Give one example of each.)

10.5 Given $Z(s) = \dfrac{s^2 + Xs}{s^2 + 5s + 4}$:

(*a*) What are the restrictions on X for $Z(s)$ to be a p.r. function?

(*b*) Find X for Re $[Z(j\omega)]$ to have a second-order zero at $\omega = 0$.

(*c*) Choose a numerical value for X and synthesize $Z(s)$.

10.6 Prove that if $Z_1(s)$ and $Z_2(s)$ are both p.r.,

$$Z(s) = \frac{Z_1(s)Z_2(s)}{Z_1(s) + Z_2(s)}$$

must also be positive real.

10.7 $Z(s) = \dfrac{2s^2 + s + 2}{s^2 + s + 1}$ is p.r. Determine min [Re $Z(j\omega)$] and synthesize $Z(s)$ by first removing min [Re $Z(j\omega)$].

10.8 Perform a continued fraction expansion on the ratio

$$Y(s) = \frac{s^3 + 2s^2 + 3s + 1}{s^3 + s^2 + 2s + 1}$$

What does the continued expansion imply if $Y(s)$ is the driving-point admittance of a passive network? Draw the network from the continued fraction.

10.9 The following functions are impedance functions. Synthesize the impedances by successive removals of $j\omega$ axis poles or by removing min [Re $(j\omega)$].

(*a*) $$\frac{s^3 + 4s}{s^2 + 2}$$

(*b*) $$\frac{s + 1}{s(s + 2)}$$

(*c*) $$\frac{2s + 4}{2s + 3}$$

(*d*) $$\frac{s^2 + 3s + 1}{s^2 + 1}$$

chapter 11

Synthesis of one-port networks with two kinds of elements

In this chapter we will study methods for synthesizing one-port networks with two kinds of elements. Since we have three elements to choose from, the networks to be synthesized are either *R-C*, *R-L*, or *L-C* networks. We will proceed according to the following plan. First we will discuss the properties of a particular type of one-port network, and then we will synthesize it. Let us first examine some properties of *L-C* driving-point functions.

11.1 PROPERTIES OF *L-C* IMMITTANCE FUNCTIONS

Consider the impedance $Z(s)$ of a passive one-port network. Let us represent $Z(s)$ as

$$Z(s) = \frac{M_1(s) + N_1(s)}{M_2(s) + N_2(s)} \tag{11.1}$$

where M_1, M_2 are even parts of the numerator and denominator, and N_1, N_2 are odd parts. The average power dissipated by the one-port is

$$\text{Average power} = \tfrac{1}{2} \operatorname{Re} [Z(j\omega)] \, |I|^2 \tag{11.2}$$

where I is the input current. For a pure reactive network, it is known that the power dissipated is zero. We therefore conclude that the real part of $Z(j\omega)$ is zero; that is

$$\operatorname{Re} Z(j\omega) = \operatorname{Ev} Z(j\omega) = 0 \tag{11.3}$$

where
$$\operatorname{Ev} Z(s) = \frac{M_1(s)\, M_2(s) - N_1(s)\, N_2(s)}{M_2^{\,2}(s) - N_2^{\,2}(s)} \tag{11.4}$$

In order for Ev $Z(j\omega) = 0$, that is,

$$M_1(j\omega)\,M_2(j\omega) - N_1(j\omega)\,N_2(j\omega) = 0 \tag{11.5}$$

either of the following cases must hold:

$$\begin{aligned} &(a) \qquad M_1 = 0 = N_2 \\ &(b) \qquad M_2 = 0 = N_1 \end{aligned} \tag{11.6}$$

In case (*a*), $Z(s)$ is

$$Z(s) = \frac{N_1}{M_2} \tag{11.7}$$

and in case (*b*)

$$Z(s) = \frac{M_1}{N_2} \tag{11.8}$$

We see from this development the following two properties of *L-C* functions:

1. $Z_{LC}(s)$ or $Y_{LC}(s)$ is the ratio of even to odd or odd to even polynomials.
2. Since both $M_i(s)$ and $N_j(s)$ are Hurwitz, they have only imaginary roots, and it follows that the poles and zeros of $Z_{LC}(s)$ or $Y_{LC}(s)$ are on the imaginary axis.

Consider the example of an *L-C* immittance function given by

$$Z(s) = \frac{a_4s^4 + a_2s^2 + a_0}{b_5s^5 + b_3s^3 + b_1s} \tag{11.9}$$

Let us examine the constraints on the coefficients a_i and b_j. We know, first of all, that in order for the impedance to be positive real, the coefficients must be real and positive. We also know that an impedance function cannot have multiple poles or zeros on the $j\omega$ axis. Since ∞ is defined to be on the $j\omega$ axis, the highest powers of the numerator and the denominator polynomials can differ by, at most, unity. For example, if the highest order of the numerator is $2n$, then the highest order of the denominator can either be $2n - 1$ so that there is a simple pole at $s = \infty$, or the order can be $2n + 1$ so that there is a simple zero at $s = \infty$. Similarly, the lowest orders of numerator and denominator can differ by at most unity, or else there would be multiple poles or zeros of $Z(s)$ at $s = 0$.

Another property of the numerator and denominator polynomials is that if the highest power of the polynomial is $2n$, for example, the next highest order term must be $2n - 2$, and the succeeding powers must differ by two orders all the way through. There cannot be any missing terms, i.e., no two adjacent terms of either polynomial may differ by more than two powers. For example, for Z(s) given in Eq. 11.9, if $b_3 = 0$, then Z(s) will have poles when

$$b_5 s^5 + b_1 s = 0 \tag{11.10}$$

so that the poles will be at $s = 0$ and at

$$s_k = \left(\frac{b_1}{b_5}\right)^{1/4} e^{j(2k-1)\pi/4} \qquad k = 0, 1, 2, 3 \tag{11.11}$$

It is clearly seen that none of the poles s_k are even on the $j\omega$ axis, thus violating one of the basic properties of an L-C immittance function.

From the properties given in Eq. 11.11, we can write a general L-C impedance or admittance as

$$Z(s) = \frac{K(s^2 + \omega_1^2)(s^2 + \omega_3^2) \cdots (s^2 + \omega_i^2) \cdots}{s(s^2 + \omega_2^2)(s^2 + \omega_4^2) \cdots (s^2 + \omega_j^2) \cdots} \tag{11.12}$$

Expanding Z(s) into partial fractions, we obtain

$$Z(s) = \frac{K_0}{s} + \frac{2K_2 s}{s^2 + \omega_2^2} + \frac{2K_4 s}{s^2 + \omega_4^2} + \cdots + K_\infty s \tag{11.13}$$

where the K_i are the residues of the poles. Since these poles are all on the $j\omega$ axis, the residues must be real and positive in order for Z(s) to be positive real. Letting $s = j\omega$, we see that $Z(j\omega)$ has zero real part, and can thus be written as a pure reactance $jX(\omega)$. Thus we have

$$\begin{aligned} Z(j\omega) &= j\left(-\frac{K_0}{\omega} + \frac{2K_2\omega}{\omega_2^2 - \omega^2} + \cdots + K_\infty\omega\right) \\ &= jX(\omega) \end{aligned} \tag{11.14}$$

Differentiating $X(\omega)$ with respect to ω, we have

$$\frac{dX(\omega)}{d\omega} = \frac{K_0}{\omega^2} + K_\infty + \frac{2K_2(\omega^2 + \omega_2^2)}{(\omega_2^2 - \omega^2)^2} + \cdots \tag{11.15}$$

Since all the residues K_i are positive, it is seen that for an L-C function,

$$\frac{dX(\omega)}{d\omega} \geq 0 \tag{11.16}$$

A similar development shows that the derivative of Im $[Y(j\omega)] = B(\omega)$ is also positive, that is,

$$\frac{dB(\omega)}{d\omega} \geq 0 \tag{11.17}$$

Consider the following example. Z(*s*) is given as

$$Z(s) = \frac{Ks(s^2 + \omega_3^2)}{(s^2 + \omega_2^2)(s^2 + \omega_4^2)} \tag{11.18}$$

Letting $s = j\omega$, we obtain $X(\omega)$ as

$$Z(j\omega) = j\,X(\omega) = +j\,\frac{K\omega(-\omega^2 + \omega_3^2)}{(-\omega^2 + \omega_2^2)(-\omega^2 + \omega_4^2)} \tag{11.19}$$

Let us draw a curve of $X(\omega)$ versus ω. Beginning with the zero at $\omega = 0$, let us examine the sequence of critical frequencies encountered as ω increases. Since the slope of the $X(\omega)$ curve is always positive, the next critical frequency we encounter is when $X(\omega)$ becomes infinitely large or the pole is at ω_2. As we pass ω_2, $X(\omega)$ changes sign and goes from + to −. In general, whenever we pass through any critical frequency, there is always a change of sign, as seen from the way $j\,X(\omega)$ is written in the last equation. After we pass through ω_2, with the slope of $X(\omega)$ always positive, it is easy to see that the next critical frequency is the zero at ω_3. Thus, if an impedance function is an *L-C* immittance, the poles and zeros of the function must *alternate*. The particular $X(\omega)$ under discussion takes the form shown in Fig. 11.1. Since the highest powers of the numerator and the denominator always differ by unity, and the lowest powers also differ by one, we observe that at $s = 0$ and at $s = \infty$, there is always a critical frequency, whether a zero or a pole.

For the example just discussed, there is a zero at $s = 0$ and a zero at $s = \infty$. The critical frequencies at $s = 0$ and $s = \infty$ are called *external* critical frequencies, whereas the remaining finite critical frequencies are referred to as *internal*. Thus, in the previous example, ω_2, ω_3, and ω_4 are internal critical frequencies.

Finally, let us summarize the properties of *L-C* impedance or admittance functions.

1. $Z_{LC}(s)$ or $Y_{LC}(s)$ is the ratio of odd to even or even to odd polynomials.

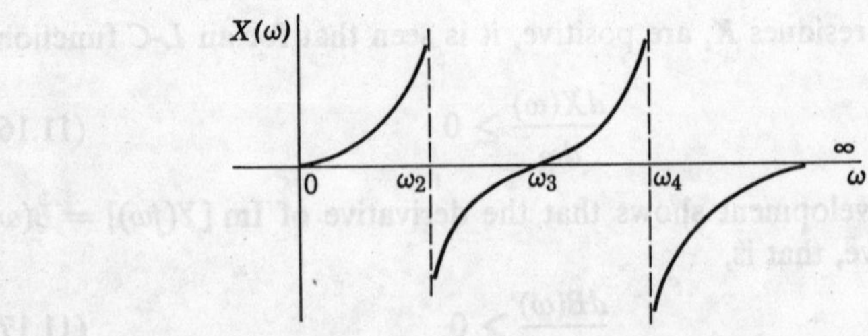

FIG. 11.1

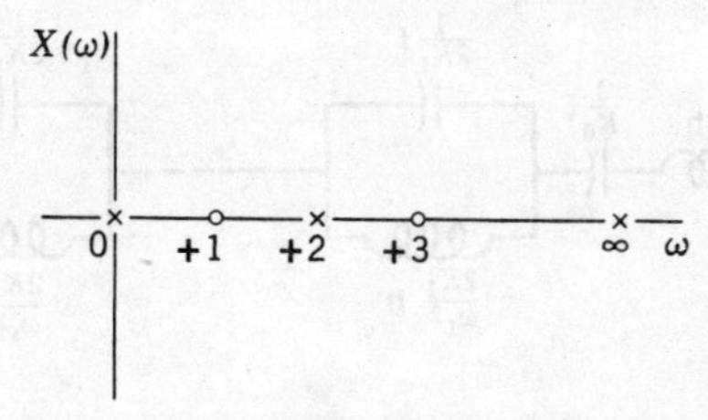

FIG. 11.2

2. The poles and zeros are simple and lie on the $j\omega$ axis.
3. The poles and zeros interlace on the $j\omega$ axis.
4. The highest powers of numerator and denominator must differ by unity; the lowest powers also differ by unity.
5. There must be either a zero or a pole at the origin and infinity.

The following functions are not *L-C* for the reasons listed at the left.

3. $$Z(s) = \frac{Ks(s^2 + 4)}{(s^2 + 1)(s^2 + 3)}$$

2. $$Z(s) = \frac{s^5 + 4s^3 + 5s}{3s^4 + 6s^2} \tag{11.20}$$

1. $$Z(s) = \frac{K(s^2 + 1)(s^2 + 9)}{(s^2 + 2)(s^2 + 10)}$$

On the other hand, the function $Z(s)$ in Eq. 11.21, whose pole-zero diagram is shown in Fig. 11.2, is an *L-C* immittance.

$$Z(s) = \frac{2(s^2 + 1)(s^2 + 9)}{s(s^2 + 4)} \tag{11.21}$$

11.2 SYNTHESIS OF *L-C* DRIVING-POINT IMMITTANCES

We saw in Section 11.1 that an *L-C* immittance is a positive real function with poles and zeros on the $j\omega$ axis only. The partial fraction expansion of an *L-C* function is expressed in general terms as

$$F(s) = \frac{K_0}{s} + \frac{2K_2 s}{s^2 + \omega_2^{\,2}} + \cdots + K_\infty s \tag{11.22}$$

The synthesis is accomplished directly from the partial fraction expansion by associating the individual terms in the expansion with network elements. If $F(s)$ is an impedance $Z(s)$, then the term K_0/s represents a capacitor of $1/K_0$ farads; the term $K_\infty s$ is an inductance of K_∞ henrys, and the term

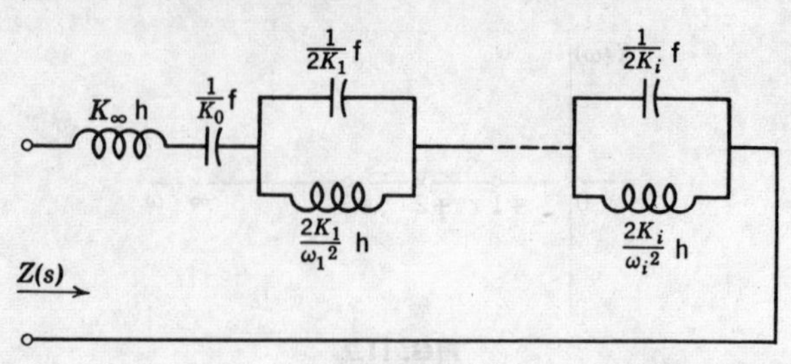

FIG. 11.3

$2K_is/(s^2 + \omega_i^2)$ is a parallel tank circuit that consists of a capacitor of $1/2K_i$ farads in parallel with an inductance of $2K_i/\omega_i^2$. Thus a partial fraction expansion of a general *L-C* impedance would yield the network shown in Fig. 11.3. For example, consider the following *L-C* function.

$$Z(s) = \frac{2(s^2 + 1)(s^2 + 9)}{s(s^2 + 4)} \tag{11.23}$$

A partial fraction expansion of Z(*s*) gives

$$Z(s) = 2s + \frac{\frac{9}{2}}{s} + \frac{\frac{15}{2}s}{s^2 + 4} \tag{11.24}$$

We then obtain the synthesized network in Fig. 11.4.

The partial fraction expansion method is based upon the elementary synthesis procedure of removing poles on the $j\omega$ axis. The advantage with *L-C* functions is that *all* the poles of the function lie on the $j\omega$ axis so that we can remove all the poles simultaneously. Suppose $F(s)$ in Eq. 11.22 is an admittance $Y(s)$. Then the partial fraction expansion of $Y(s)$ gives us a circuit consisting of parallel branches shown in Fig. 11.5. For example,

$$Y(s) = \frac{s(s^2 + 2)(s^2 + 4)}{(s^2 + 1)(s^2 + 3)} \tag{11.25}$$

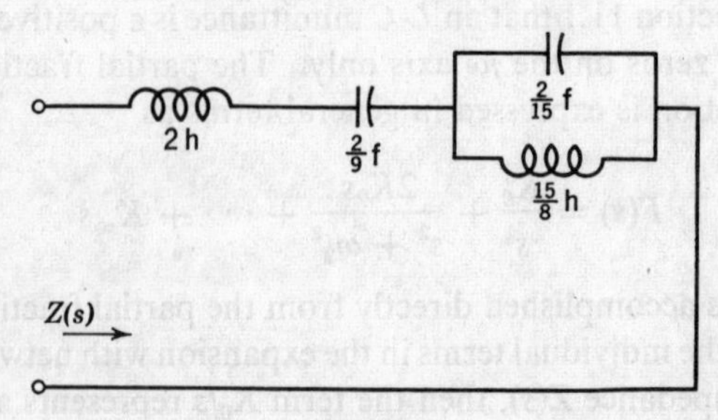

FIG. 11.4

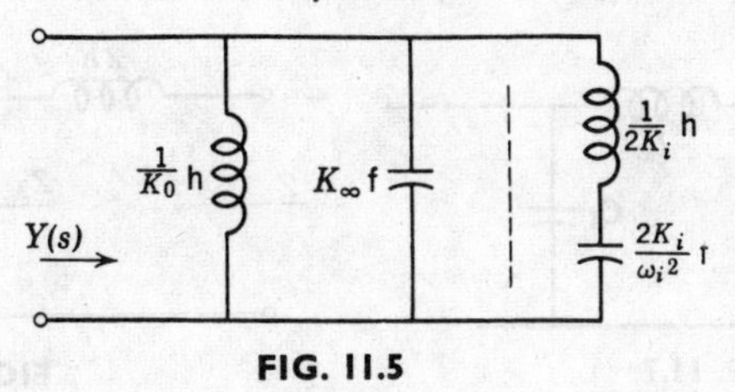

FIG. 11.5

The partial fraction expansion of $Y(s)$ is

$$Y(s) = s + \frac{\frac{1}{2}s}{s^2 + 3} + \frac{\frac{3}{2}s}{s^2 + 1} \tag{11.26}$$

from which we synthesize the network shown in Fig. 11.6. The *L-C* networks synthesized by partial fraction expansions are sometimes called *Foster*-type networks.[1] The impedance form is sometimes called a Foster series network and the admittance form is a Foster parallel network.

A useful property of *L-C* immittances is that the numerator and the denominator always differ in degree by unity. Therefore, there is always a zero or a pole at $s = \infty$. Suppose we consider the case of an *L-C* impedance $Z(s)$, whose numerator is of degree $2n$ and denominator is of degree $2n - 1$, giving $Z(s)$ a pole at $s = \infty$. We can remove this pole by removing an impedance $L_1 s$ so that the remainder function $Z_2(s)$ is still *L-C*:

$$Z_2(s) = Z(s) - L_1 s \tag{11.27}$$

The degree of the denominator of $Z_2(s)$ is $2n - 1$, but the numerator is of degree $2n - 2$, because the numerator and denominator must differ in degree by 1. Therefore, we see that $Z_2(s)$ has a zero at $s = \infty$. If we invert $Z_2(s)$ to give $Y_2(s) = 1/Z_2(s)$, $Y_2(s)$ will have a pole at $s = \infty$, which we can again remove to give a capacitor $C_2 s$ and a remainder $Y_3(s)$, which is

$$Y_3(s) = Y_2(s) - C_2 s. \tag{11.28}$$

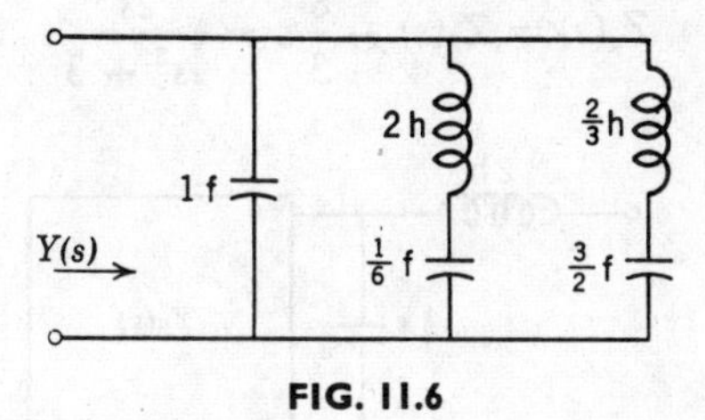

FIG. 11.6

[1] R. M. Foster, "A Reactance Theorem," *Bell System Tech. J.*, No. 3 (1924), 259–267.

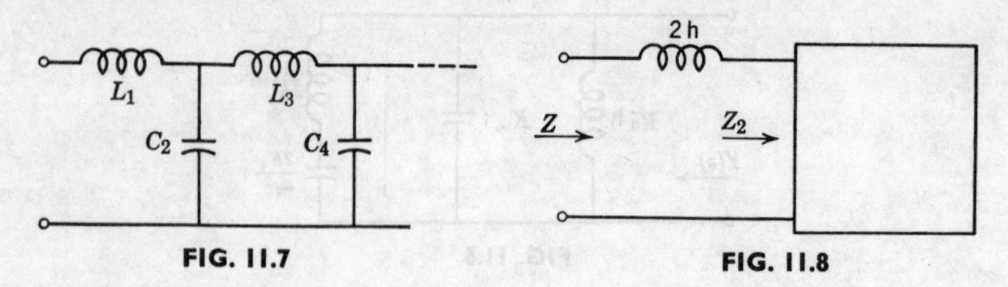

FIG. 11.7 FIG. 11.8

We readily see that $Y_3(s)$ has a zero at $s = \infty$, which we can invert and remove. This process continues until the remainder is zero. Each time we remove a pole, we remove an inductor or a capacitor depending upon whether the function is an impedance or an admittance. Note that the final structure of the network synthesized is a *ladder* whose series arms are inductors and whose shunt arms are capacitors, as shown in Fig. 11.7. Consider the following example.

$$Z(s) = \frac{2s^5 + 12s^3 + 16s}{s^4 + 4s^2 + 3} \tag{11.29}$$

We see that $Z(s)$ has a pole at $s = \infty$, which we can remove by first dividing the denominator into the numerator to give a quotient $2s$ and a remainder $Z_2(s)$, as shown in Fig. 11.8. Then we have

$$Z_2(s) = Z(s) - 2s = \frac{4s^3 + 10s}{s^4 + 4s^2 + 3} \tag{11.30}$$

Observe that $Z_2(s)$ has a zero at $s = \infty$. Inverting $Z_2(s)$, we again remove the pole at infinity. Then we realize a capacitor of $\frac{1}{4}$ farad and a remainder $Y_3(s)$, as may be seen in Fig. 11.9.

$$Y_3(s) = Y_2(s) - \frac{1}{4}s = \frac{\frac{3}{2}s^2 + 3}{4s^3 + 10s} \tag{11.31}$$

Removing the pole at $s = \infty$ of $Z_3(s) = 1/Y_3(s)$, gives a series inductor of $\frac{8}{3}$ h and

$$Z_4(s) = Z_3(s) - \frac{8}{3}s = \frac{2s}{\frac{3}{2}s^2 + 3} \tag{11.32}$$

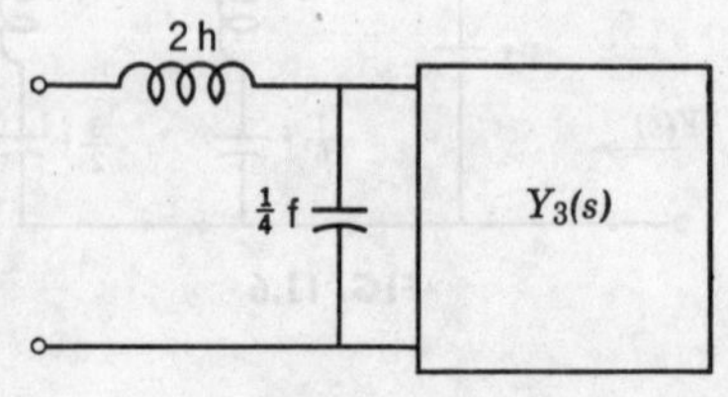

FIG. 11.9

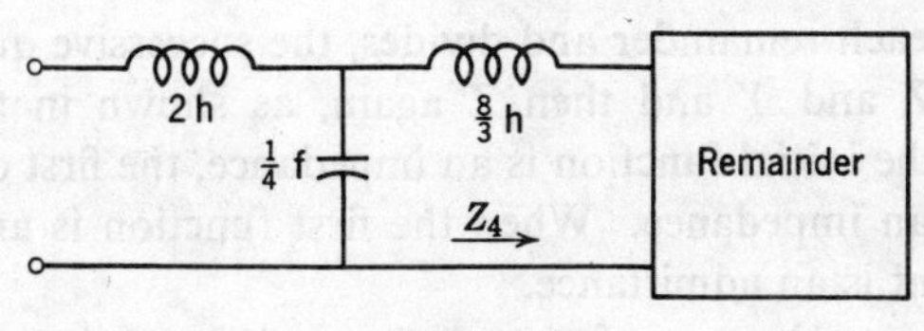

FIG. 11.10

as shown in Fig. 11.10. The admittance $Y_4(s) = 1/Z_4(s)$ has a pole at $s = \infty$, which we remove to give a capacitor of $\frac{3}{4}$ farad and a remainder $Y_5(s) = 3/2s$, which represents an inductor of $\frac{2}{3}$ h. Removing this inductor gives us zero remainder. Our synthesis is therefore complete and the final network is shown in Fig. 11.11.

Since we always remove a pole at $s = \infty$ by inverting the remainder and dividing, we conclude that we can synthesize an L-C ladder network by a continued fraction expansion. The quotients represent the poles at $s = \infty$, which we remove, and we invert the remainder successively until the remainder is zero. For the previous example, the continued fraction expansion is

$$
\begin{array}{l}
s^4 + 4s^2 + 3\overline{)2s^5 + 12s^3 + 16s}(2s \leftrightarrow Z \\
\qquad\qquad\quad \underline{2s^5 + \ \ 8s^3 + \ \ 6s} \\
\qquad\qquad\qquad 4s^3 + 10s)s^4 + 4s^2 + 3(\tfrac{1}{4}s \leftrightarrow Y \\
\qquad\qquad\qquad\qquad\qquad \underline{s^4 + \tfrac{5}{2}s^2} \\
\qquad\qquad\qquad\qquad\qquad\quad \tfrac{3}{2}s^2 + 3)4s^3 + 10s(\tfrac{8}{3}s \leftrightarrow Z \\
\qquad\qquad\qquad\qquad\qquad\qquad\qquad \underline{4s^3 + \ \ 8s} \\
\qquad\qquad\qquad\qquad\qquad\qquad\qquad 2s)\tfrac{3}{2}s^2 + 3(\tfrac{3}{4}s \leftrightarrow Y \\
\qquad\qquad\qquad\qquad\qquad\qquad\qquad\qquad \underline{\tfrac{3}{2}s^2} \\
\qquad\qquad\qquad\qquad\qquad\qquad\qquad\qquad 3)2s(\tfrac{2}{3}s \leftrightarrow Z \\
\qquad\qquad\qquad\qquad\qquad\qquad\qquad\qquad\quad \underline{\underline{2s}}
\end{array}
$$

We see that the quotients of the continued fraction expansion give the elements of the ladder network. Because the continued fraction expansion

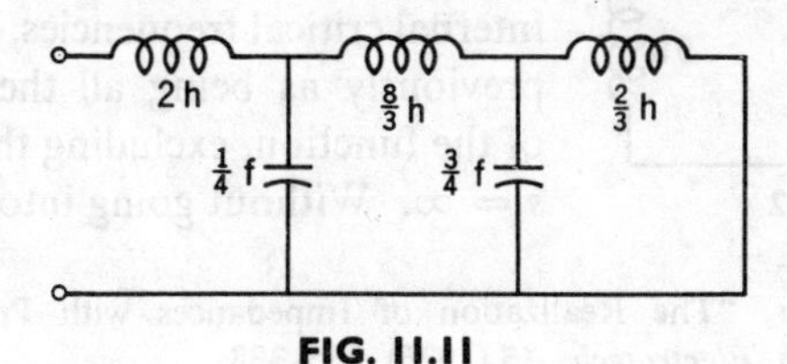

FIG. 11.11

always inverts each remainder and divides, the successive quotients alternate between Z and Y and then Z again, as shown in the preceding expansion. If the initial function is an impedance, the first quotient must necessarily be an impedance. When the first function is an admittance, the first quotient is an admittance.

Since the lowest degrees of numerator and denominator of an L-C admittance must differ by unity, it follows that there must be a zero or a pole at $s = 0$. If we follow the same procedure we have just outlined, and remove successively poles at $s = 0$, we will have an alternate realization in a ladder structure. To do this by continued fractions, we arrange both numerator and denominator in *ascending* order and divide the lowest power of the denominator into the lowest power of the numerator; then we invert the remainder and divide again. For example, in the case of the impedance we have

$$Z(s) = \frac{(s^2 + 1)(s^2 + 3)}{s(s^2 + 2)} \tag{11.33}$$

The continued fraction expansion to give the alternate realization is

$$\begin{array}{l}
2s + s^3\overline{)3 + 4s^2 + s^4}(3/2s \leftrightarrow Z \\
\qquad\quad \underline{3 + \tfrac{3}{2}s^2} \\
\qquad\quad \tfrac{5}{2}s^2 + s^4\overline{)2s + s^3}(4/5s \leftrightarrow Y \\
\qquad\qquad\qquad\quad \underline{2s + \tfrac{4}{5}s^3} \\
\qquad\qquad\qquad\quad \tfrac{1}{5}s^3\overline{)\tfrac{5}{2}s^2 + s^4}(25/2s \leftrightarrow Z \\
\qquad\qquad\qquad\qquad\quad \underline{\tfrac{5}{2}s^2} \\
\qquad\qquad\qquad\qquad\quad s^4\overline{)\tfrac{1}{5}s^3}(1/5s \leftrightarrow Y \\
\qquad\qquad\qquad\qquad\qquad \underline{\underline{\tfrac{1}{5}s^3}}
\end{array}$$

The final synthesized network is shown in Fig. 11.12. The ladder networks realized are called *Cauer* ladder networks because W. Cauer[2] discovered the continued fraction method for synthesis of a passive network.

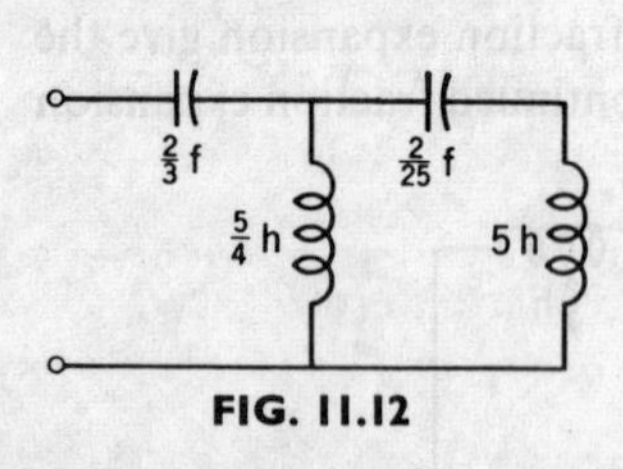

FIG. 11.12

Note that for both the Foster and the Cauer-form realizations, the number of elements is one greater than the number of internal critical frequencies, which we defined previously as being all the poles and zeros of the function, excluding those at $s = 0$ and $s = \infty$. Without going into the proof of the

[2] Wilhelm Cauer, "The Realization of Impedances with Prescribed Frequency Dependence," *Arch. Electrotech.*, **15** (1926), 355–388.

statement, it can be said that both the Foster and the Cauer forms give the minimum number of elements for a specified *L-C* driving-point function. These realizations are sometimes known as *canonical* forms.

11.3 PROPERTIES OF *R-C* DRIVING-POINT IMPEDANCES

The properties of *R-C* driving-point impedances can be derived from known properties of *L-C* functions by a process of mapping the $j\omega$ axis onto the $-\sigma$ axis.[3] We will not resort to this formalism here. Instead, we will assume that all driving-point functions that can be realized with two kinds of elements can be realized in a Foster form. Based upon this assumption we can derive all the pertinent properties of *R-C* or *R-L* driving-point functions. Let us consider first the properties of *R-C* driving-point impedance functions.

Referring to the series Foster form for an *L-C* impedance given in Fig. 11.3, we can obtain a Foster realization of an *R-C* impedance by simply replacing all the inductances by resistances so that a general *R-C* impedance could be represented as in Fig. 11.13. The *R-C* impedance, as seen from Fig. 11.13, is

$$Z(s) = \frac{K_0}{s} + K_\infty + \frac{K_1}{s + \sigma_1} + \frac{K_2}{s + \sigma_2} + \cdots \tag{11.34}$$

where $C_0 = 1/K_0$, $R_\infty = K_\infty$, $C_1 = 1/K_1$, $R_1 = K_1/\sigma_1$, and so on. In order for Eq. 11.34 to represent an *R-C* driving-point impedance, the constants K_i and σ_i must be positive and real. From this development, two major properties of *R-C* impedances are obtained, and are listed in the following.

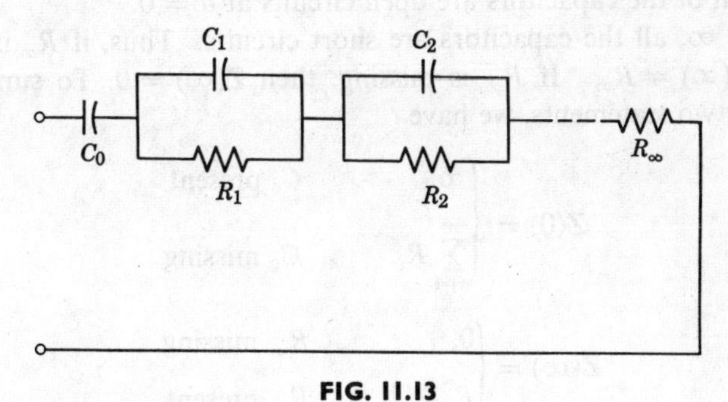

FIG. 11.13

[3] M. E. Van Valkenburg, *Introduction to Modern Network Synthesis*, John Wiley and Sons, New York, 1960, pp. 140–145.

1. The poles of an R-C driving-point impedance are on the negative real $(-\sigma)$ axis. It can be shown from a parallel Foster form that the poles of an R-C admittance function are also on the axis. We can thus conclude that the zeros of an R-C impedance are also on the $-\sigma$ axis.

2. The residues of the poles, K_i, are real and positive. We shall see later that this property does not apply to R-C admittances.

Since the poles and zeros of R-C impedances are on the $-\sigma$ axis, let us examine the slope of $Z(\sigma)$ along the $-\sigma$ axis. To find the slope, $dZ(\sigma)/d\sigma$, we first let $s = \sigma$ in $Z(s)$, and then we take the derivative of $Z(\sigma)$ with respect to σ. Thus we have

$$Z(\sigma) = \frac{K_0}{\sigma} + K_\infty + \frac{K_1}{\sigma + \sigma_1} + \frac{K_2}{\sigma + \sigma_2} + \cdots \tag{11.35}$$

and

$$\frac{dZ(\sigma)}{d\sigma} = -\frac{K_0}{\sigma^2} + \frac{-K_1}{(\sigma + \sigma_1)^2} + \frac{-K_2}{(\sigma + \sigma_2)^2} + \cdots \tag{11.36}$$

It is clear that

$$\frac{dZ(\sigma)}{d\sigma} \leq 0 \tag{11.37}$$

Let us now look at the behavior of $Z(s)$ at the two points where the real axis and the imaginary axis intersect, namely, at $\sigma = \omega = 0$ and at $\sigma = \omega = \infty$. This is readily done by examining the general R-C network in Fig. 11.13 at these two frequencies. At $\sigma = 0$, (d-c), if the capacitor C_0 is in the circuit, it is an open circuit and there is a *pole* of $Z(s)$ at $\sigma = 0$. If C_0 is not in the circuit, then $Z(0)$ is simply the sum of all the resistances in the circuit.

$$Z(0) = R_1 + R_2 + \cdots + R_\infty \tag{11.38}$$

because all of the capacitors are open circuits at $\sigma = 0$.

At $\sigma = \infty$, all the capacitors are short circuits. Thus, if R_∞ is in the circuit, $Z(\infty) = R_\infty$. If R_∞ is missing, then $Z(\infty) = 0$. To summarize these last two statements, we have

$$Z(0) = \begin{cases} \infty, & C_0 \text{ present} \\ \sum_{i=1}^{m} R_i, & C_0 \text{ missing} \end{cases}$$

$$Z(\infty) = \begin{cases} 0, & R_\infty \text{ missing} \\ R_\infty, & R_\infty \text{ present} \end{cases}$$

If we examine the two cases for $Z(0)$ and $Z(\infty)$, we see that

$$Z(0) \geq Z(\infty) \tag{11.39}$$

Next, let us see whether the poles and zeros of an *R-C* impedance function alternate. We have already established that the critical frequency nearest the origin must be a pole and the critical frequency nearest $\sigma = \infty$ must be a zero. Therefore, if $Z(s)$ is given as

$$Z(s) = \frac{(s + \sigma_2)(s + \sigma_4)}{(s + \sigma_1)(s + \sigma_3)} \tag{11.40}$$

Then, if $Z(s)$ is *R-C*, the singularity nearest the origin must be a pole which we will assume to be at $s = -\sigma_1$; the singularity furthest from the origin must be a zero, which we will take to be $s = -\sigma_4$. Let us plot

$$Z(\sigma) = \frac{(\sigma + \sigma_2)(\sigma + \sigma_4)}{(\sigma + \sigma_1)(\sigma + \sigma_3)} \tag{11.41}$$

versus $-\sigma$, beginning at $\sigma = 0$ and extending to $\sigma = -\infty$. At $\sigma = 0$, $Z(0)$ is equal to a positive constant

$$Z(0) = \frac{\sigma_2\sigma_4}{\sigma_1\sigma_2} \tag{11.42}$$

Since the slope of $Z(\sigma)$ is always positive as $-\sigma$ increases, $Z(\sigma)$ must increase until the pole $s = -\sigma_1$ is reached (Fig. 11.14). At $\sigma = -\sigma_1$, $Z(\sigma)$ changes sign, and is negative until the next critical frequency is reached. We see that this next critical frequency must be the zero, $s = -\sigma_2$. Since $Z(\sigma)$ increases for increasing $-\sigma$, the third critical frequency must be the pole $s = -\sigma_3$. Because $Z(\sigma)$ changes sign at $-\sigma_3$, the final critical frequency must be the zero, $s = -\sigma_4$. Beyond $\sigma = -\sigma_4$, the curve becomes asymptotic to $Z(\infty) = 1$. From this analysis we see that the poles and zeros of an *R-C* impedance must alternate so that for

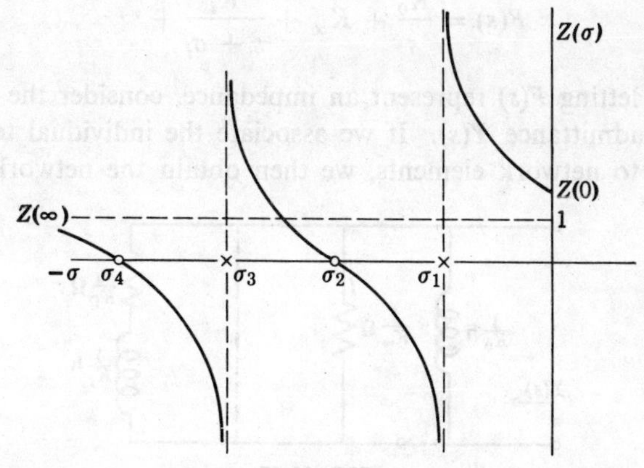

FIG. 11.14

the case being considered

$$\infty > \sigma_4 > \sigma_3 > \sigma_2 > \sigma_1 \geq 0. \tag{11.43}$$

In addition, we see that

$$\frac{\sigma_2\sigma_4}{\sigma_1\sigma_3} > 1 \tag{11.44}$$

which shows that $Z(0) > Z(\infty)$.

To summarize, the three properties we need to recognize an *R-C* impedance are:

1. Poles and zeros lie on the negative real axis, and they alternate.
2. The singularity nearest to (or at) the origin must be a pole whereas the singularity nearest to (or at) $\sigma = -\infty$ must be a zero.
3. The residues of the poles must be real and positive.

An example of an *R-C* impedance is:

$$Z(s) = \frac{(s+1)(s+4)(s+8)}{s(s+2)(s+6)} \tag{11.45}$$

The following impedances are not *R-C*.

$$Z(s) = \frac{(s+1)(s+8)}{(s+2)(s+4)}$$
$$Z(s) = \frac{(s+2)(s+4)}{(s+1)} \tag{11.46}$$
$$Z(s) = \frac{(s+1)(s+2)}{s(s+3)}$$

Let us reexamine the partial fraction expansion of a general *R-C* impedance.

$$F(s) = \frac{K_0}{s} + K_\infty + \frac{K_i}{s+\sigma_i} + \cdots \tag{11.47}$$

Instead of letting $F(s)$ represent an impedance, consider the case where $F(s)$ is an admittance $Y(s)$. If we associate the individual terms in the expansion to network elements, we then obtain the network shown in

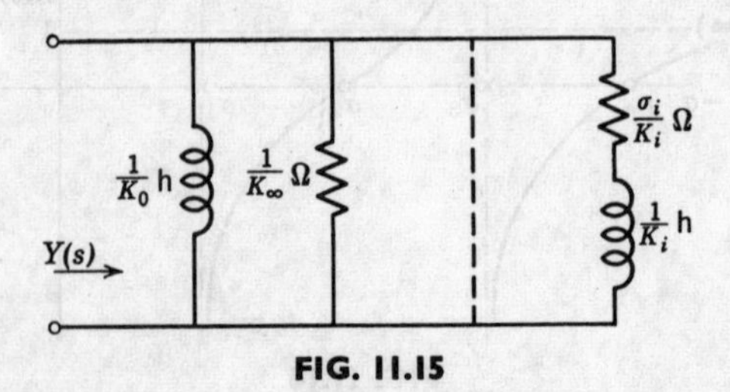

FIG. 11.15

Fig. 11.15. We see that an *R-C* impedance, $Z_{RC}(s)$, also can be realized as an *R-L* admittance $Y_{RL}(s)$. All the properties of *R-L* admittances are the same as the properties of *R-C* impedances. It is therefore important to specify whether a function is to be realized as an *R-C* impedance or an *R-L* admittance.

11.4 SYNTHESIS OF *R-C* IMPEDANCES OR *R-L* ADMITTANCES

We postulated in Section 11.3 that the Foster form realization exists for an *R-C* impedance or an *R-L* admittance. Since Foster networks are synthesized by partial fraction expansions, the synthesis is accomplished with ease. An important point to remember is that we must remove the minimum real part of $Z(j\omega)$ in the partial fraction expansion. It can be shown[4] that min $[\text{Re}\, Z(j\omega)] = Z(\infty)$, so that we have to remove $Z(\infty)$ as a resistor in the partial fraction expansion. In cases where the numerator is of lower degree than the denominator, $Z(\infty) = 0$. When the numerator and the denominator are of the same degree, then $Z(\infty)$ can be obtained by dividing the denominator into the numerator. The quotient is then $Z(\infty)$. Consider the following example.

$$F(s) = \frac{3(s+2)(s+4)}{s(s+3)} \tag{11.48}$$

The partial fraction expansion of the remainder function is obtained as

$$F(s) = \frac{8}{s} + \frac{1}{s+3} + 3 \tag{11.49}$$

where $F(\infty) = 3$. If $F(s)$ is an impedance $Z(s)$, it must be an *R-C* impedance and it is realized in the series Foster form in Fig. 11.16. On the other hand, if $F(s)$ represents an admittance, we realize $Y(s)$ as an *R-L* network in the parallel Foster form (Fig. 11.17).

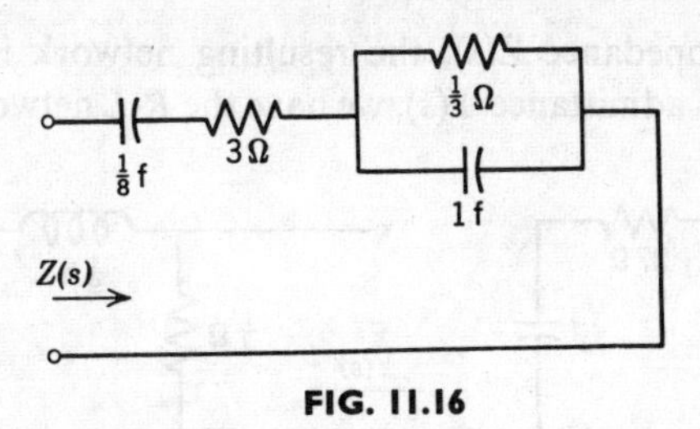

FIG. 11.16

[4] Van Valkenburg, *loc cit.*

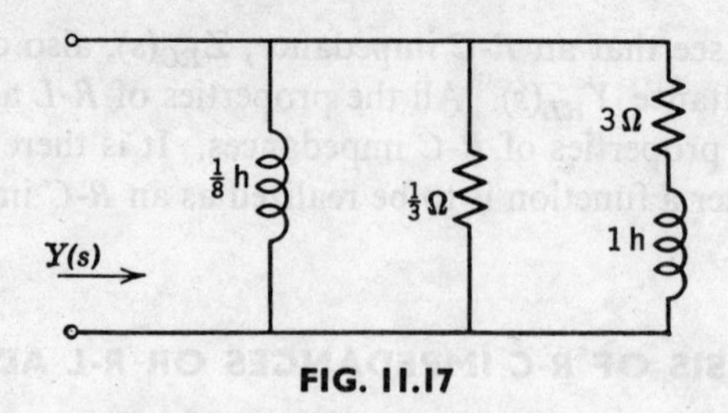

FIG. 11.17

An alternate method of synthesis is based on the following fact. If we remove min Re $[Z(j\omega)] = Z(\infty)$ from $Z(s)$, we create a zero at $s = \infty$ for the remainder $Z_1(s)$. If we invert $Z_1(s)$, we then have a pole at $s = \infty$, which we can remove to give $Z_2(s)$. Since min Re $[Y_2(j\omega)] = Y_2(\infty)$, if we remove $Y_2(\infty)$, we would have a zero at $s = \infty$ again, which we again invert and remove. The process of extracting $Z(\infty)$ or $Y(\infty)$ and the removal of a pole of the reciprocal of the remainder involve dividing the numerator by the denominator. Consequently, we see that the whole synthesis process can be resolved by a continued fraction expansion. The quotients represent the elements of a ladder network.

For example, the continued fraction expansion of $F(s)$ in Eq. 11.48 is

$$s^2 + 3s\overline{)3s^2 + 18s + 24}(3$$
$$\underline{3s^2 + 9s}$$
$$9s + 24\overline{)s^2 + 3s}(\tfrac{1}{9}s$$
$$\underline{s^2 + \tfrac{8}{3}s}$$
$$\tfrac{1}{3}s\overline{)9s + 24}(27$$
$$\underline{9s}$$
$$24\overline{)\tfrac{1}{3}s}(s/72$$
$$\underline{\underline{\tfrac{1}{3}s}}$$

If $F(s)$ is an impedance $Z(s)$, the resulting network is shown in Fig. 11.18. If $F(s)$ is an admittance $Y(s)$, we have the *R-L* network of Fig. 11.19.

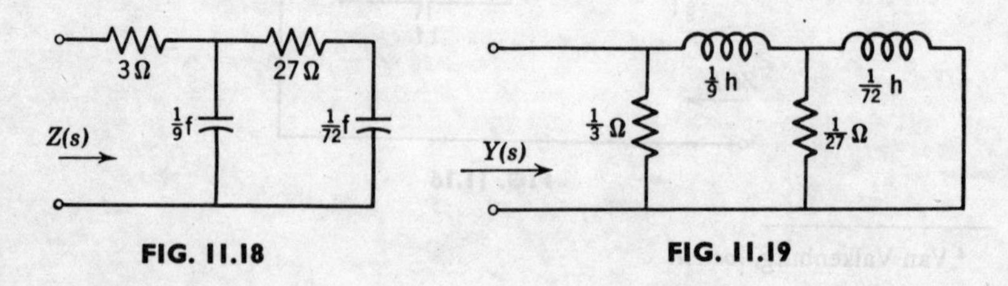

FIG. 11.18

FIG. 11.19

11.5 PROPERTIES OF *R-L* IMPEDANCES AND *R-C* ADMITTANCES

The immittance that represents a series Foster *R-L* impedance or a parallel Foster *R-C* admittance is given as

$$F(s) = K_\infty s + K_0 + \frac{K_i s}{s + \sigma_i} + \cdots \tag{11.50}$$

The significant difference between an *R-C* impedance and an *R-L* impedance is that the partial fraction expansion term for the *R-C* "tank" circuit is $K_i/(s + \sigma_i)$; whereas, for the *R-L* impedance, the corresponding term must be multiplied by an s in order to give an *R-L* tank circuit consisting of a resistor in parallel with an inductor.

The properties of *R-L* impedance or *R-C* admittance functions can be derived in much the same manner as the properties of *R-C* impedance functions. Without going into the derivation of the properties, the more significant ones are given in the following:

1. Poles and zeros of an *R-L* impedance or *R-C* admittance are located on the negative real axis, and they alternate.
2. The singularity nearest to (or at) the origin is a zero. The singularity nearest to (or at) $s = \infty$ must be a pole.
3. The residues of the poles must be real and *negative*.

Because of the third property, a partial fraction expansion of an *R-L* impedance function would yield terms as

$$-\frac{K_i}{s + \sigma_i} \tag{11.51}$$

This does not present any trouble, however, because the term above does not represent an *R-L* impedance at all. To obtain the Foster form of an *R-L* impedance, we will resort to the following artifice. Let us first expand $Z(s)/s$ into partial fractions. If $Z(s)$ is an *R-L* impedance, we will state without proof here that the partial fraction expansion of $Z(s)/s$ yields positive residues.[5] Thus, we have

$$\frac{Z(s)}{s} = \frac{K_0}{s} + K_\infty + \frac{K_i}{s + \sigma_i} + \cdots \tag{11.52}$$

[5] Actually, $Z_{RL}(s)/s$ has the properties of an *R-C* impedance; see Van Valkenburg *loc. cit.*

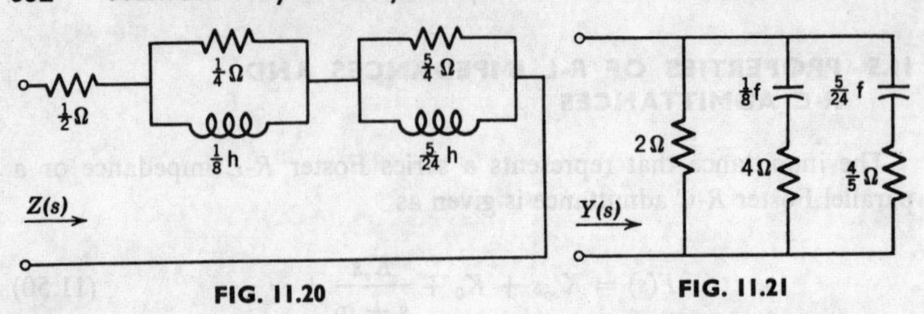

FIG. 11.20

FIG. 11.21

where $K_0, K_1, \ldots, K_\infty \geq 0$. If we multiply both sides by s, we obtain $Z(s)$ in the desired form for synthesis. Consider the following function:

$$F(s) = \frac{2(s+1)(s+3)}{(s+2)(s+6)} \tag{11.53}$$

$F(s)$ represents an R-L impedance or an R-C admittance because it satisfies the first two criteria cited. The partial fraction expansion of $F(s)$ is

$$F(s) = 2 - \frac{\frac{1}{2}}{s+2} - \frac{\frac{15}{2}}{s+6} \tag{11.54}$$

so we see that the residues are negative. The partial fraction expansion of $F(s)/s$, on the other hand, is

$$\frac{F(s)}{s} = \frac{2(s+1)(s+3)}{s(s+2)(s+6)} = \frac{\frac{1}{2}}{s} + \frac{\frac{1}{4}}{s+2} + \frac{\frac{5}{4}}{s+6} \tag{11.55}$$

If we multiply both sides by s, we obtain

$$F(s) = \frac{1}{2} + \frac{\frac{1}{4}s}{s+2} + \frac{\frac{5}{4}s}{s+6} \tag{11.56}$$

If $F(s)$ represents an impedance $Z(s)$, it is synthesized in series Foster form, giving the R-L network in Fig. 11.20. If $F(s)$ is an admittance $Y(s)$, then the resulting network is the R-C network shown in Fig. 11.21.

To synthesize an R-L impedance in ladder form, we make use of the fact that min Re $[Z(j\omega)] = Z(0)$. If we remove $Z(0)$ from $Z(s)$, the remainder function $Z_1(s)$ will have a zero at $s = 0$. After inverting $Z_1(s)$,

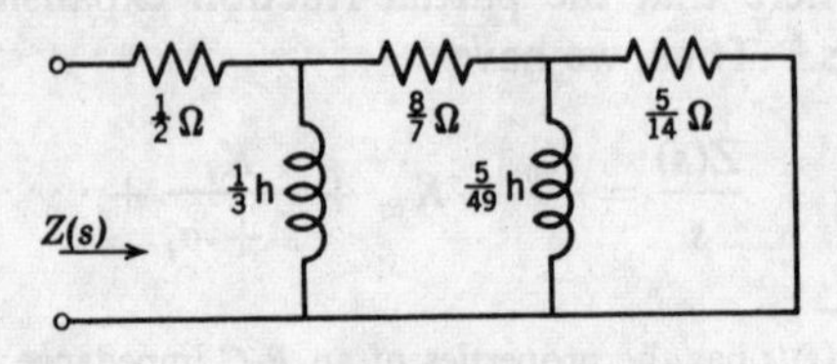

FIG. 11.22

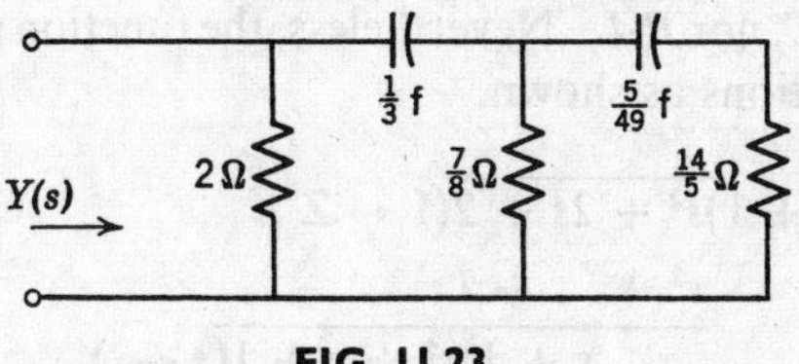

FIG. 11.23

we can then remove the pole at $s = 0$. Since the value $Z(0)$ is obtained by dividing the lowest power of the denominator into the lowest power term of the numerator, the synthesis could be carried out by a continued fraction expansion by arranging the numerator and denominator polynomials in ascending order and then dividing. For example, the following function is either an *R-C* impedance or an *R-C* admittance.

$$F(s) = \frac{2(s+1)(s+3)}{(s+2)(s+6)} = \frac{6 + 8s + 2s^2}{12 + 8s + s^2} \tag{11.57}$$

The continued fraction expansion of $F(s)$ is

$$\begin{array}{l}
12 + 8s + s^2\overline{)6 + 8s + 2s^2}(\tfrac{1}{2} \\
\qquad\qquad\quad 6 + 4s + \tfrac{1}{2}s^2 \\
\qquad\qquad\quad \overline{4s + \tfrac{3}{2}s^2)12 + 8s + s^2}(3/s \\
\qquad\qquad\qquad\qquad\quad 12 + \tfrac{9}{2}s \\
\qquad\qquad\qquad\qquad\quad \overline{\tfrac{7}{2}s + s^2)4s + \tfrac{3}{2}s^2}(\tfrac{8}{7} \\
\qquad\qquad\qquad\qquad\qquad\qquad 4s + \tfrac{8}{7}s^2 \\
\qquad\qquad\qquad\qquad\qquad\qquad \overline{\tfrac{5}{14}s^2)\tfrac{7}{2}s + s^2}(49/5s \\
\qquad\qquad\qquad\qquad\qquad\qquad\qquad\quad \tfrac{7}{2}s \\
\qquad\qquad\qquad\qquad\qquad\qquad\qquad\quad \overline{s^2)\tfrac{5}{14}s^2}(\tfrac{5}{14} \\
\qquad\qquad\qquad\qquad\qquad\qquad\qquad\qquad \tfrac{5}{14}s^2 \\
\qquad\qquad\qquad\qquad\qquad\qquad\qquad\qquad \overline{\overline{}}
\end{array}$$

If $F(s)$ is an impedance function, the resulting network is the *R-L* network shown in Fig. 11.22. If, on the other hand, $F(s)$ is an *R-C* admittance $Y(s)$, the network is synthesized as in Fig. 11.23.

11.6 SYNTHESIS OF CERTAIN *R-L-C* FUNCTIONS

Under certain conditions, *R-L-C* driving-point functions may be synthesized with the use of either partial fractions or continued fractions. For example, the function

$$Z(s) = \frac{s^2 + 2s + 2}{s^2 + s + 1} \tag{11.58}$$

is neither *L-C*, *R-C*, nor *R-L*. Nevertheless, the function can be synthesized by continued fractions as shown.

$$\begin{array}{l} s^2+s+1\overline{)s^2+2s+2(}1 \leftarrow Z \\ \qquad\quad\; \underline{s^2+\;\,s+1} \\ \qquad\qquad\quad s+1\overline{)s^2+s+1(}s \leftarrow Y \\ \qquad\qquad\qquad\quad\;\; \underline{s^2+s} \\ \qquad\qquad\qquad\qquad\qquad 1\overline{)s+1(}s+1 \leftarrow Z \\ \qquad\qquad\qquad\qquad\qquad\qquad \underline{\underline{s+1}} \end{array}$$

The network derived from this expansion is given in Fig. 11.24.

In another case, the poles and zeros of the following admittance are all on the negative real axis, but they do not alternate.

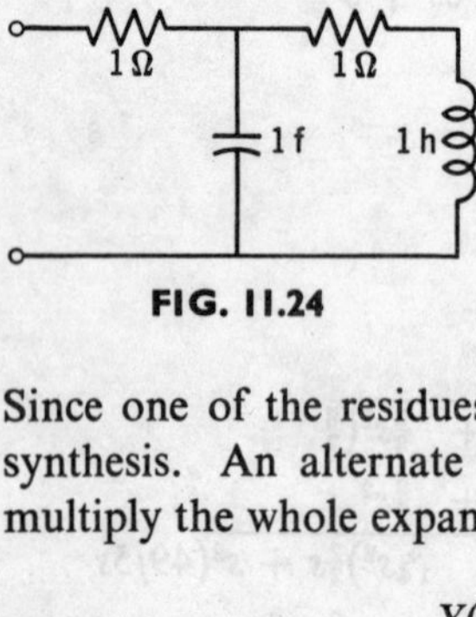

1 Ω 1 Ω 1 f 1 h

FIG. 11.24

$$Y(s)=\frac{(s+2)(s+3)}{(s+1)(s+4)} \tag{11.59}$$

The partial fraction expansion for $Y(s)$ is

$$Y(s)=1+\frac{\frac{2}{3}}{s+1}+\frac{-\frac{2}{3}}{s+4} \tag{11.60}$$

Since one of the residues is negative, we cannot use this expansion for synthesis. An alternate method would be to expand $Y(s)/s$ and then multiply the whole expansion by s.

$$\frac{Y(s)}{s}=\frac{\frac{3}{2}}{s}-\frac{\frac{2}{3}}{s+1}+\frac{\frac{1}{6}}{s+4} \tag{11.61}$$

When we multiply by s, we obtain,

$$Y(s)=\frac{3}{2}-\frac{\frac{2}{3}s}{s+1}+\frac{\frac{1}{6}s}{s+4} \tag{11.62}$$

Note that $Y(s)$ also has a negative term. If we divide the denominator of this negative term into the numerator, we can rid ourselves of any terms with negative signs.

$$\begin{aligned} Y(s)&=\frac{3}{2}-\left(\frac{2}{3}-\frac{\frac{2}{3}}{s+1}\right)+\frac{\frac{1}{6}s}{s+4} \\ &=\frac{5}{6}+\frac{\frac{2}{3}}{s+1}+\frac{\frac{1}{6}s}{s+4} \end{aligned} \tag{11.63}$$

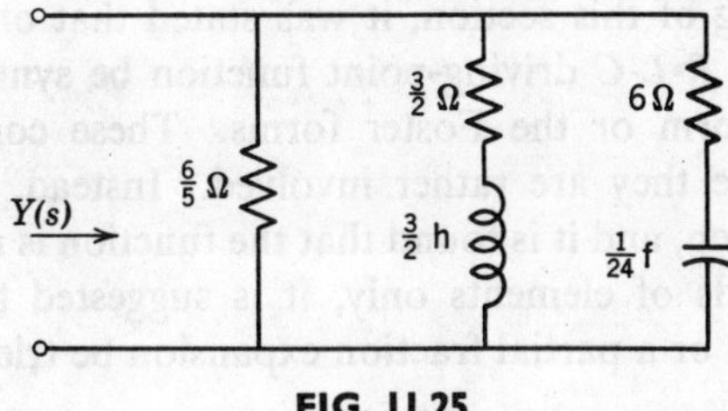

FIG. 11.25

The network that is realized from the expanded function is given in Fig. 11.25.

If we try to expand $Y(s)$ by continued fractions, we see that negative quotients result. However, we can expand $Z(s) = 1/Y(s)$ by continued fractions, although the expansion is not as simple or straightforward as in the case of an *R-C* function, because we sometimes have to reverse the order of division to make the quotients all positive. The continued fraction expansion of $Z(s)$ is

$$
\begin{array}{l}
6 + 5s + s^2 \overline{)\,4 + 5s + s^2}\left(\tfrac{2}{3}\right. \\
\qquad\qquad\quad \underline{4 + \tfrac{10}{3}s + \tfrac{2}{3}s^2} \\
\qquad\qquad\qquad \tfrac{5}{3}s + \tfrac{1}{3}s^2\,)\,6 + 5s + s^2\left(18/5s\right. \\
\qquad\qquad\qquad\qquad\qquad\quad \underline{6 + \tfrac{6}{5}s} \\
\qquad\qquad\qquad\qquad\qquad\qquad \tfrac{19}{5}s + s^2\,)\,\tfrac{1}{3}s^2 + \tfrac{5}{3}s\left(\tfrac{1}{3}\right. \\
\qquad\qquad\qquad\qquad\qquad\qquad\qquad\qquad\quad \underline{\tfrac{1}{3}s^2 + \tfrac{19}{15}s} \\
\qquad\qquad\qquad\qquad\qquad\qquad\qquad\qquad\qquad \tfrac{6}{15}s\,)\,\tfrac{19}{5}s + s^2\left(\tfrac{19}{2}\right. \\
\qquad\qquad\qquad\qquad\qquad\qquad\qquad\qquad\qquad\qquad\quad \underline{\tfrac{19}{5}s} \\
\qquad\qquad\qquad\qquad\qquad\qquad\qquad\qquad\qquad\qquad\quad s^2\,)\,\tfrac{6}{15}s\left(6/15s\right. \\
\qquad\qquad\qquad\qquad\qquad\qquad\qquad\qquad\qquad\qquad\qquad\quad \underline{\underline{\tfrac{6}{15}s}}
\end{array}
$$

As we see, the division process giving the quotient of 1/3 involves a reversal of the order of the polynomials involved. The resulting ladder network is given in Fig. 11.26.

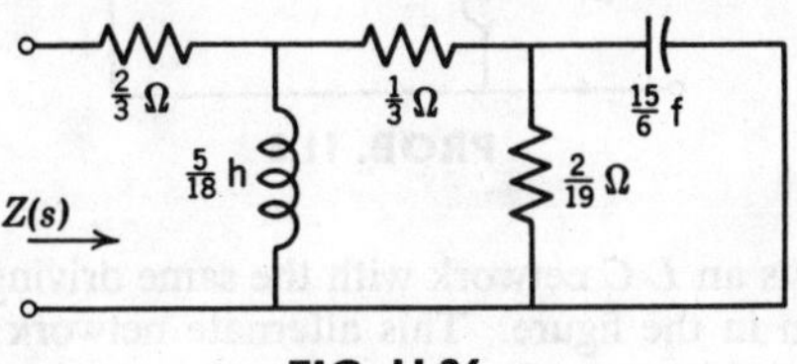

FIG. 11.26

In the beginning of this section, it was stated that only under special conditions can an *R-L-C* driving-point function be synthesized with the use of a ladder form or the Foster forms. These conditions are not given here because they are rather involved. Instead, when a positive real function is given, and it is found that the function is not synthesizable by using two kinds of elements only, it is suggested that a continued fraction expansion or a partial fraction expansion be tried first.

Problems

11.1 (*a*) Which of the following functions are *L-C* driving point impedances? Why?

$$Z_1(s) = \frac{s(s^2+4)(s^2+16)}{(s^2+9)(s^2+25)}, \qquad Z_2(s) = \frac{(s^2+1)(s^2+8)}{s(s^2+4)}$$

(*b*) Synthesize the realizable impedances in a Foster and a Cauer form.

11.2 Indicate the general *form* of the two Foster and the two Cauer networks that could be used to synthesize the following *L-C* impedance.

$$Z(s) = \frac{(s^2+1)(s^2+9)(s^2+25)}{s(s^2+4)(s^2+16)}$$

There is no need to calculate the element values of the four networks.

11.3 Synthesize the *L-C* driving-point impedance

$$Z(s) = \frac{6s^4+42s^2+48}{s^5+18s^3+48s}$$

in the form shown in the figure, i.e., determine the element values of the network in henrys and farads.

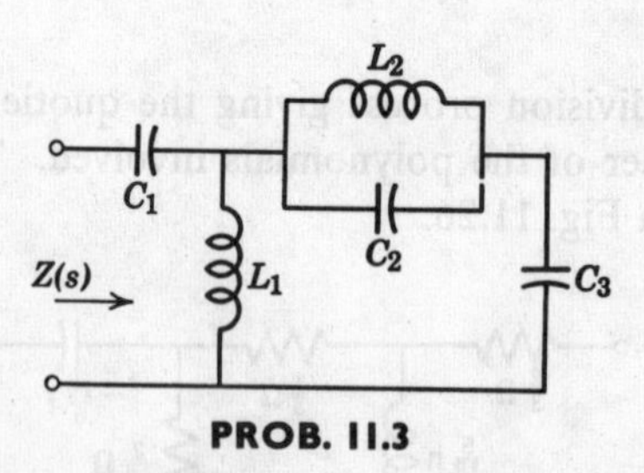

PROB. 11.3

11.4 There exists an *L-C* network with the same driving-point impedance as the network shown in the figure. This alternate network should contain only two elements. Find this network.

PROB. 11.4

11.5 The input impedance for the network shown is

$$Z_{\text{in}} = \frac{2s^2 + 2}{s^3 + 2s^2 + 2s + 2}$$

If Z_0 is an *L-C* network: (*a*) Find the expression for Z_0. (*b*) Synthesize Z_0 in a Foster series form.

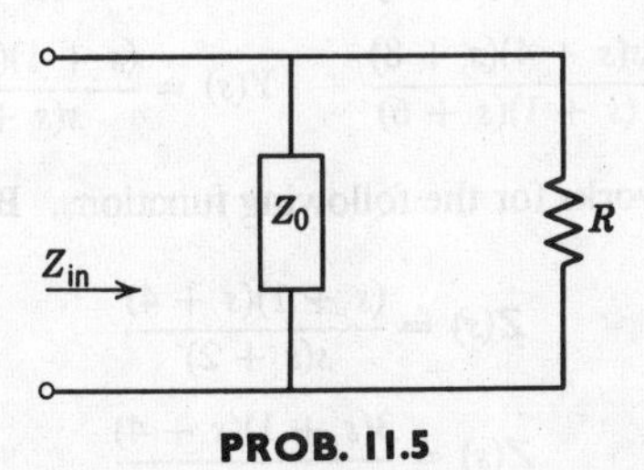

PROB. 11.5

11.6 Indicate which of the following functions are either *R-C*, *R-L*, or *L-C* *impedance* functions.

(*a*) $$Z(s) = \frac{s^3 + 2s}{s^4 + 4s^2 + 3}$$

(*b*) $$Z(s) = \frac{s^2 + 6s + 8}{s^2 + 4s + 3}$$

(*c*) $$Z(s) = \frac{s^2 + 4s + 3}{s^2 + 6s + 8}$$

(*d*) $$Z(s) = \frac{s^2 + 5s + 6}{s^2 + s}$$

(*e*) $$Z(s) = \frac{s^4 + 5s^2 + 6}{s^3 + s}$$

11.7 An impedance function has the pole-zero pattern shown in the figure. If $Z(-2) = 3$, synthesize the impedance in a Foster form and a Cauer form.

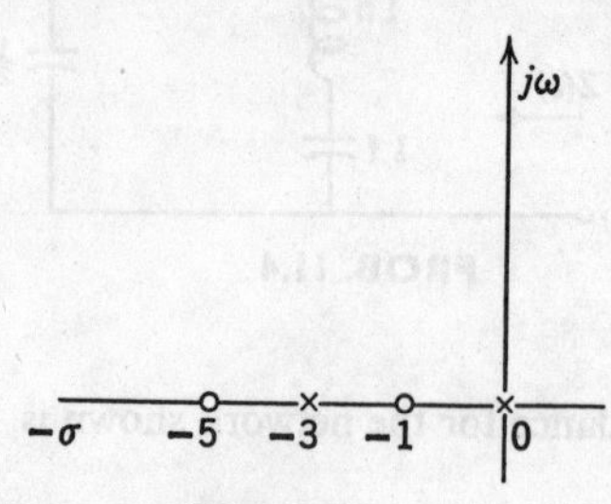

PROB. 11.7

11.8 From the following functions, pick out the ones which are R-C admittances and synthesize in one Foster and one Cauer form.

$$Y(s) = \frac{2(s+1)(s+3)}{(s+2)(s+4)} \qquad Y(s) = \frac{4(s+1)(s+3)}{s(s+2)}$$

$$Y(s) = \frac{s(s+4)(s+8)}{(s+1)(s+6)} \qquad Y(s) = \frac{(s+1)(s+4)}{s(s+2)}$$

11.9 Find the networks for the following functions. Both Foster and ladder forms are required.

(*a*)
$$Z(s) = \frac{(s+1)(s+4)}{s(s+2)}$$

(*b*)
$$Z(s) = \frac{3(s+1)(s+4)}{s+3}$$

11.10 For the network shown, find Y when

$$\frac{V_2}{V_0} = \frac{1}{2+Y}$$

$$= \frac{s(s^2+3)}{2s^3+s^2+6s+1}$$

Synthesize Y as an L-C admittance.

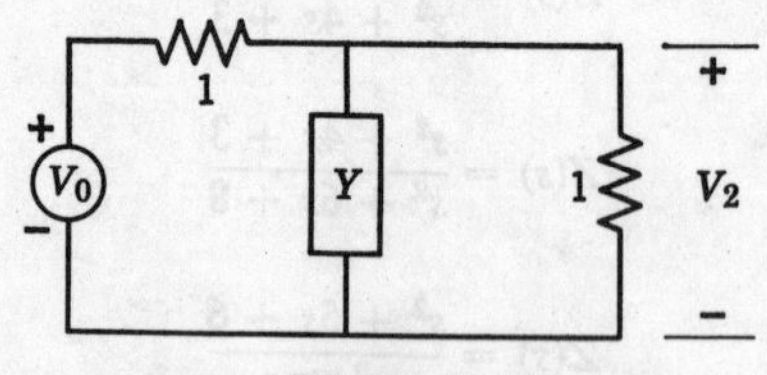

PROB. 11.10

11.11 Synthesize by continued fractions the function

$$Y(s) = \frac{s^3+2s^2+3s+1}{s^3+s^2+2s+1}$$

11.12 Find the networks for the following functions in one Foster and one Cauer form.

$$Y(s) = \frac{(s+1)(s+3)}{(s+2)(s+4)}$$

$$Z(s) = \frac{2(s+0.5)(s+4)}{s(s+2)}$$

11.13 Synthesize the following functions in Cauer form.

$$Z(s) = \frac{s^3 + 2s^2 + s + 1}{s^3 + s^2 + s}$$

$$Z(s) = \frac{s^3 + s^2 + 2s + 1}{s^4 + s^3 + 3s^2 + s + 1}$$

$$Z(s) = \frac{4s^3 + 3s^2 + 4s + 2}{2s^2 + s}$$

11.14 Synthesize $Z(s) = \dfrac{(s+2)(s+4)}{(s+1)(s+5)}$ into the form shown in the figure.

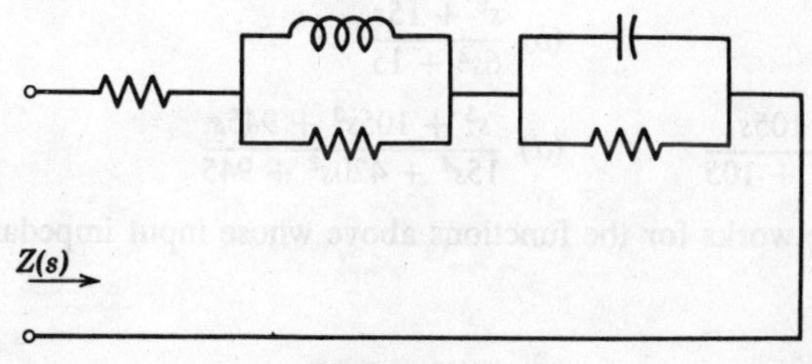

PROB. 11.14

11.15 Of the three pole-zero diagrams shown, pick the diagram that represents an *R-L* impedance function and synthesize in a series Foster form.

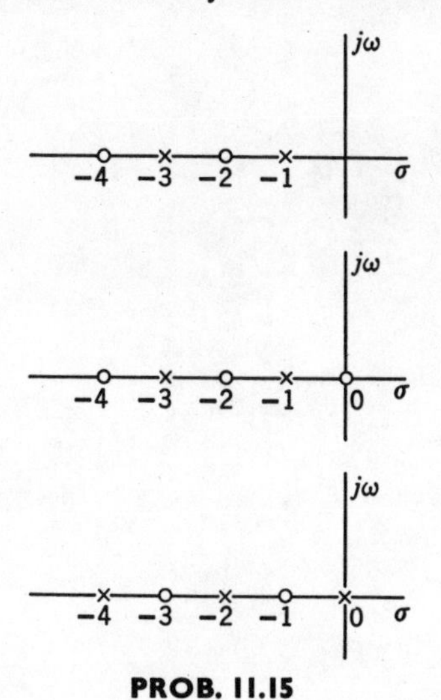

PROB. 11.15

11.16 Synthesize a driving-point impedance with the pole-zero pattern shown in the figure in any form you choose. (*Hint:* Use uniform loading concepts.)

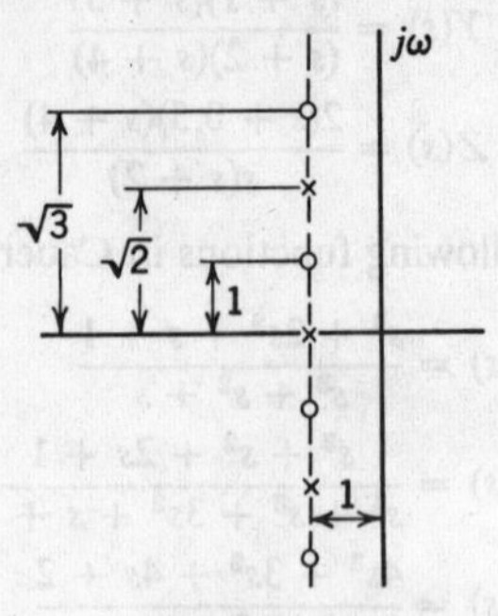

PROB. 11.16

11.17 Following are four successive approximations of tanh s.

(*a*) $\dfrac{3s}{s^2 + 3}$ (*b*) $\dfrac{s^2 + 15s}{6s^2 + 15}$

(*c*) $\dfrac{10s^3 + 105s}{s^4 + 45s^2 + 105}$ (*d*) $\dfrac{s^5 + 105s^3 + 945s}{15s^4 + 420s^2 + 945}$

Synthesize networks for the functions above whose input impedances approximate tanh s.

chapter 12

Elements of transfer function synthesis

12.1 PROPERTIES OF TRANSFER FUNCTIONS

A transfer function is a function which relates the current or voltage at one port to the current or voltage at another port. In Chapter 9 we discussed various descriptions of two-port networks in terms of the open-circuit parameters z_{ij} and the short-circuit parameters y_{ij}. Recall that for the two-port network given in Fig. 12.1, the open-circuit transfer impedances z_{12} and z_{21} were defined as

$$z_{12} = \left.\frac{V_1}{I_2}\right|_{I_1=0} \qquad (12.1)$$

$$z_{21} = \left.\frac{V_2}{I_1}\right|_{I_2=0}$$

In terms of the open-circuit transfer impedances, the voltage-ratio transfer function is given as

$$\frac{V_2}{V_1} = \frac{z_{21}}{z_{11}} \qquad (12.2)$$

In terms of the short-circuit parameters, the voltage ratio is shown to be

$$\frac{V_2}{V_1} = -\frac{y_{21}}{y_{22}} \qquad (12.3)$$

When the network is terminated at port two by a resistor R, as shown in Fig. 12.2, the transfer impedance of the overall network is

$$Z_{21} = \frac{V_2}{I_1} = \frac{z_{21}R}{z_{22} + R} \qquad (12.4)$$

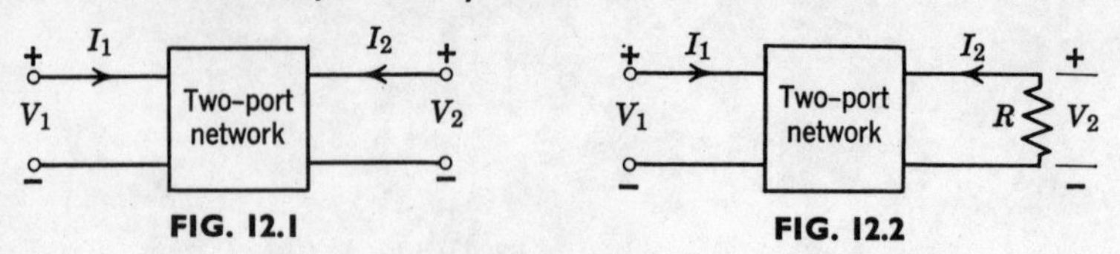

FIG. 12.1 FIG. 12.2

The transfer admittance of the overall structure in Fig. 12.2 is

$$Y_{21} = \frac{I_2}{V_1} = \frac{y_{21}G}{y_{22} + G} \tag{12.5}$$

where $G = 1/R$. When both ports are terminated in resistors, as shown in Fig. 12.3, the voltage-ratio transfer function V_2/V_g is

$$\frac{V_2}{V_g} = \frac{z_{21}R_2}{(z_{11} + R_1)(z_{22} + R_2) - z_{21}z_{12}} \tag{12.6}$$

Other transfer functions such as current-ratio transfer functions can also be described in terms of the open- and short-circuit parameters. In Chapter 10 we discussed the various properties of driving-point impedances such as z_{11} and z_{22}. This chapter deals with the properties of the transfer immittances z_{21} and y_{21} for a passive reciprocal network. First, let us discuss certain properties which apply to all transfer functions of passive linear networks with lumped elements. We denote a transfer function as $T(s)$.

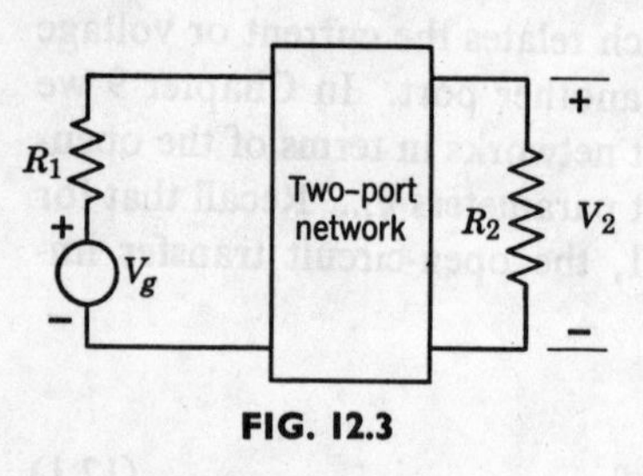

FIG. 12.3

1. $T(s)$ is real for real s. This property is satisfied when $T(s)$ is a rational function with real coefficients.

2. $T(s)$ has no poles in the right-half plane and no multiple poles on the $j\omega$ axis. If $T(s)$ is given as $T(s) = P(s)/Q(s)$, the degree of $P(s)$ cannot exceed the degree of $Q(s)$ by more than unity. In addition, $Q(s)$ must be a Hurwitz polynomial.

3. Suppose $P(s)$ and $Q(s)$ are given in terms of even and odd parts, that is,

$$T(s) = \frac{P(s)}{Q(s)} = \frac{M_1(s) + N_1(s)}{M_2(s) + N_2(s)} \tag{12.7}$$

where $M_i(s)$ is even and $N_i(s)$ is odd. Then $T(j\omega)$ is

$$T(j\omega) = \frac{M_1(j\omega) + N_1(j\omega)}{M_2(j\omega) + N_2(j\omega)} \tag{12.8}$$

The amplitude response of $T(j\omega)$ is

$$|T(j\omega)| = \left[\frac{M_1^2(j\omega) + N_1^2(j\omega)}{M_2^2(j\omega) + N_2^2(j\omega)}\right]^{1/2} \tag{12.9}$$

and is an even function in ω. The phase response is

$$\operatorname{Arg} T(j\omega) = \arctan \left[\frac{N_1(\omega)}{M_1(\omega)}\right] - \arctan \left[\frac{N_2(\omega)}{M_2(\omega)}\right] \tag{12.10}$$

If arg $T(j0) = 0$, we see that the phase response is an odd function in ω.

Now let us discuss some specific properties of the open-circuit and short-circuit parameters.

1. The poles of $z_{21}(s)$ are also the poles of $z_{11}(s)$ and $z_{22}(s)$. However, not *all* the poles of $z_{11}(s)$ and $z_{22}(s)$ are the poles of $z_{21}(s)$. Recall that in Chapter 9 we defined the z parameters in terms of a set of node equations as

$$z_{11}(s) = \frac{\Delta_{11}}{\Delta} \qquad z_{22}(s) = \frac{\Delta_{22}}{\Delta}$$

$$z_{12} = z_{21} = \frac{\Delta_{12}}{\Delta}$$

If there is no cancellation between each numerator and denominator of z_{11}, z_{22}, and z_{12}, then the poles are the roots of the determinant Δ, and all three functions have the same poles. Consider the two-port network described by the black box in Fig. 12.4*a*. Let z'_{11}, z'_{22}, and z'_{12} be the z parameters of the network. Let us examine the case when we attach the impedances Z_1 and Z_2 to ports one and two, as shown in Fig. 12.4*b*. The z parameters for the two-port network in Fig. 12.4*b* are

$$z_{11} = z'_{11} + Z_1$$

$$z_{22} = z'_{22} + Z_2$$

$$z_{12} = z'_{12}$$

It is clear that the poles of z_{11} include the poles of Z_1; the poles of z_{22} include the poles of Z_2. However, the poles of z_{12} include neither the poles of Z_1 nor Z_2. Consequently, we see that all the poles of z_{12} are also poles of z_{11} and z_{22}. The reverse is not necessarily true.

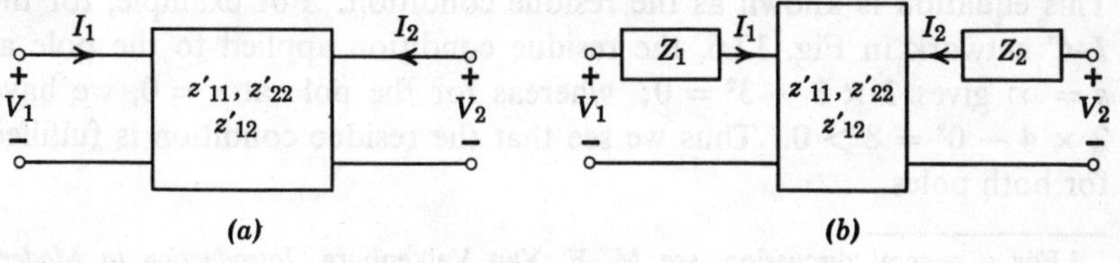

FIG. 12.4

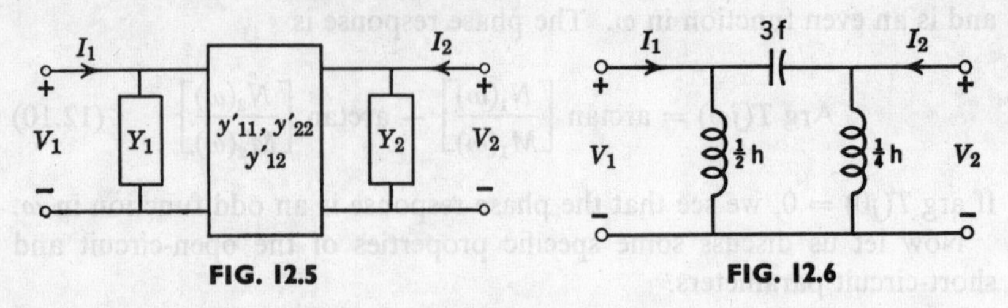

FIG. 12.5

FIG. 12.6

2. The poles of $y_{12}(s)$ are also the poles of $y_{11}(s)$ and $y_{22}(s)$. However, not all of the poles of $y_{11}(s)$ and $y_{22}(s)$ are the poles of $y_{12}(s)$. This property is readily seen when we examine the two-port network in Fig. 12.5. The y parameters are

$$y_{11} = y'_{11} + Y_1$$

$$y_{22} = y'_{22} + Y_2$$

$$y_{12} = y'_{12}$$

Clearly, the poles of $y_{12}(s)$ do not include the poles of either Y_1 and Y_2. Consider the network in Fig. 12.6. The y parameters are

$$y_{11}(s) = \frac{2}{s} + 3s$$

$$y_{22}(s) = \frac{4}{s} + 3s$$

$$y_{12}(s) = -3s$$

Observe that $y_{11}(s)$ and $y_{22}(s)$ have poles at $s = 0$ and $s = \infty$, whereas $y_{12}(s)$ only has a pole at $s = \infty$.

3. Suppose $y_{11}(s)$, $y_{22}(s)$, and $y_{12}(s)$ all have poles at $s = s_1$. Let us denote by k_{11} the residue of the pole at s_1 of the function $y_{11}(s)$. The residue of the pole $s = s_1$ of $y_{22}(s)$ will be denoted as k_{22}, and the residue of the same pole of $y_{12}(s)$ will be denoted as k_{12}. Without going into the proof,[1] a general property of *L-C*, *R-C*, or *R-L* two-port networks is that

$$k_{11}k_{22} - k_{12}^2 \geq 0 \tag{12.11}$$

This equation is known as the residue condition. For example, for the *L-C* network in Fig. 12.6, the residue condition applied to the pole at $s = \infty$ gives $3 \times 3 - 3^2 = 0$; whereas for the pole at $s = 0$, we have $2 \times 4 - 0^2 = 8 > 0$. Thus we see that the residue condition is fulfilled for both poles.

---

[1] For a general discussion, see M. E. Van Valkenburg, *Introduction to Modern Network Synthesis*, John Wiley and Sons, New York, 1960, pp. 305–313.

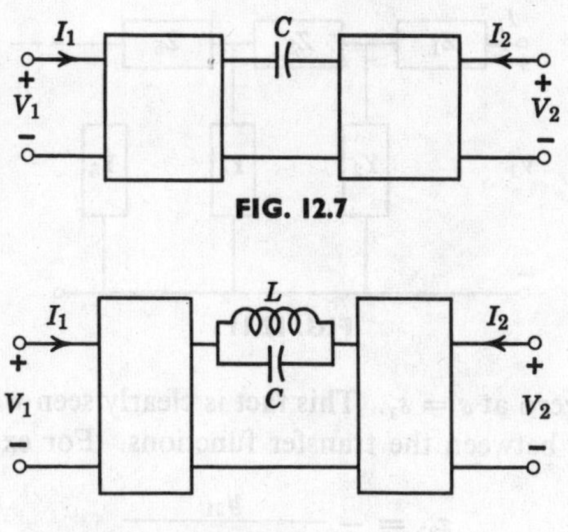

FIG. 12.7

FIG. 12.8

12.2 ZEROS OF TRANSMISSION

A zero of transmission is a zero of a transfer function. At a zero of transmission, there is zero output for an input of the same frequency. For the network in Fig. 12.7, the capacitor is an open circuit at $s = 0$, so there is a zero of transmission at $s = 0$. For the networks in Figs. 12.8 and 12.9, the zero of transmission occurs at $s = \pm j/\sqrt{LC}$. For the network in Fig. 12.10, the zero of transmission occurs at $s = -1/RC$.

In general, all the transfer functions of a given network have the same zeros of transmission, except in certain special cases. For example if $z_{12}(s)$ has a zero of transmission at $s = s_1$, than $y_{12}(s)$, $V_2(s)/V_1(s)$, etc.,

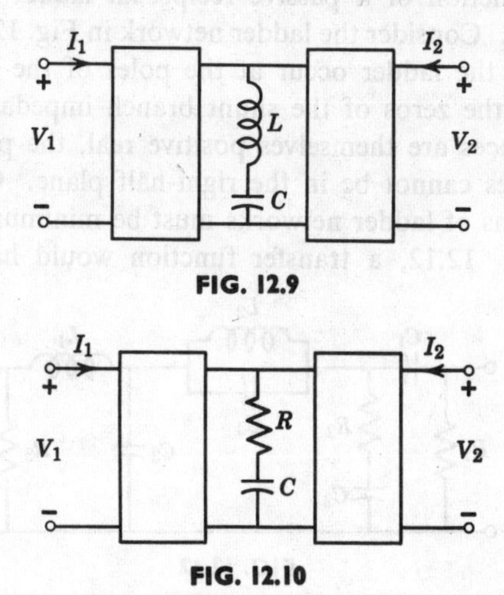

FIG. 12.9

FIG. 12.10

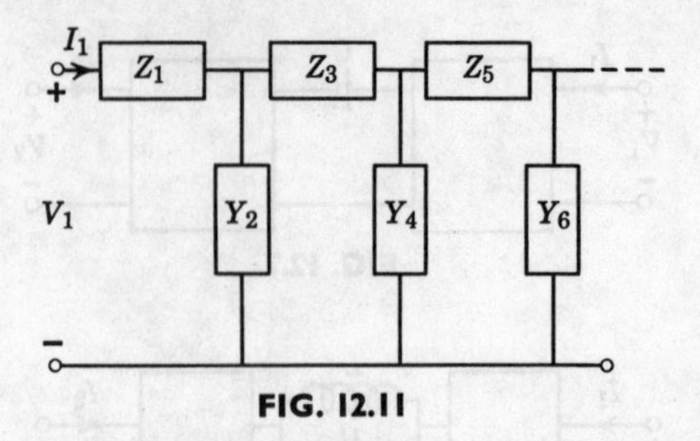

FIG. 12.11

will also have a zero at $s = s_1$. This fact is clearly seen when we examine the relationships between the transfer functions. For example, we have

$$z_{21} = -\frac{y_{21}}{y_{11}y_{22} - y_{12}y_{21}} \tag{12.12}$$

and

$$y_{21} = -\frac{z_{21}}{z_{11}z_{22} - z_{12}z_{21}} \tag{12.13}$$

In addition, the voltage- and current-ratio transfer functions can be expressed in terms of the z and y parameters as

$$\frac{V_2}{V_1} = \frac{z_{21}}{z_{11}}, \qquad \frac{I_2}{I_1} = \frac{y_{21}}{y_{11}} \tag{12.14}$$

In Chapter 8 we saw that transfer functions that have zeros of transmission only on the $j\omega$ axis or in the left-half plane are called *minimum phase* functions. If the function has one or more zeros in the right-half plane, then the function is *nonmimimum phase*. It will be shown now that any transfer function of a passive reciprocal ladder network must be minimum phase. Consider the ladder network in Fig. 12.11. The zeros of transmission of the ladder occur at the poles of the series branch impedances or at the zeros of the shunt branch impedances. Since these branch impedances are themselves positive real, the poles and zeros of these impedances cannot be in the right-half plane. Consequently, the transfer functions of ladder networks must be minimum phase. For the network in Fig. 12.12, a transfer function would have two zeros of

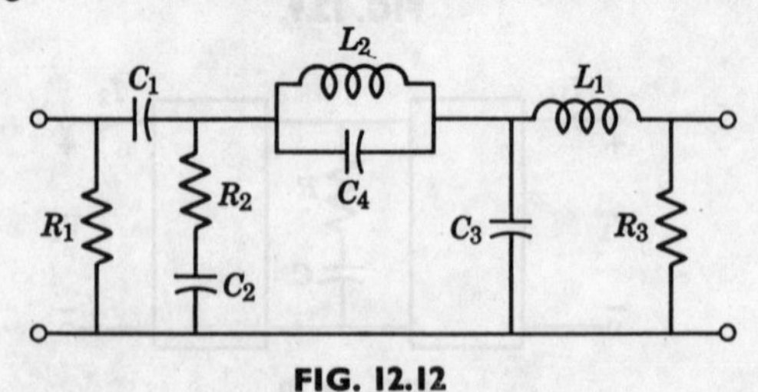

FIG. 12.12

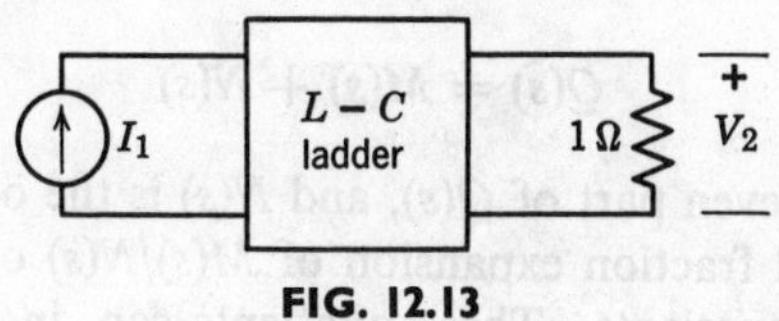

FIG. 12.13

transmission at $s = \infty$ due to the elements L_1 and C_3. It would also have a zero of transmission at $s = 0$ due to C_1, a zero at $s = -1/R_2C_2$ due to the parallel R-C branch, and a zero of transmission at $s = j/(L_2C_4)^{1/2}$ due to the L-C tank circuit. It is seen that none of the transmission zeros are in the right-half plane. We also see that a transfer function may possess multiple zeros on the $j\omega$ axis.

In Section 12.4 it may be seen that lattice and bridge circuits can easily be nonminimum phase. It can also be demonstrated that when two ladder networks are connected in parallel, the resulting structure may have right-half-plane zeros.[2]

12.3 SYNTHESIS OF Y_{21} AND Z_{21} WITH A 1-Ω TERMINATION

In this section we consider the synthesis of an L-C ladder network with a 1-Ω resistive termination to meet a specified transfer impedance Z_{21} or transfer admittance Y_{21}. In terms of the open- and short-circuit parameters of the L-C circuit, $Z_{21}(s)$ can be expressed as

$$Z_{21} = \frac{z_{21}}{z_{22} + 1} \tag{12.15}$$

and $Y_{21}(s)$ is

$$Y_{21} = \frac{y_{21}}{y_{22} + 1} \tag{12.16}$$

as depicted in Figs. 12.13 and 12.14.

Before we proceed with the actual details of the synthesis, it is necessary to discuss two important points. The first deals with the ratio of the odd to even or even to odd parts of a Hurwitz polynomial $Q(s)$. Suppose

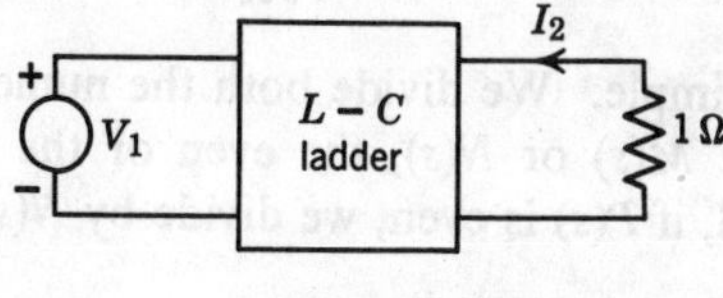

FIG. 12.14

[2] Van Valkenburg, *op. cit.*, Chapter 11.

$Q(s)$ is given as

$$Q(s) = M(s) + N(s) \tag{12.17}$$

where $M(s)$ is the even part of $Q(s)$, and $N(s)$ is the odd part. We know that the continued fraction expansion of $M(s)/N(s)$ or $N(s)/M(s)$ should yield all positive quotients. These quotients can, in turn, be associated with reactances. Therefore it is clear that the ratio of the even to odd or the odd to even parts of a Hurwitz polynomial is an *L-C* driving-point function.

The second point to be discussed is the fact that the open-circuit transfer impedance z_{21} or the short-circuit transfer admittance y_{21} of an *L-C* circuit is an odd function. To show this, we must remember that in an *L-C* circuit with steady-state input, the currents are 90° out of phase with the voltages. Thus the phase shifts between the input currents and output voltages or input voltages and output currents must be 90° out of phase, or

$$\text{Arctan}\,\frac{V_2(j\omega)}{I_1(j\omega)} = \pm\frac{\pi}{2}\text{ rad} \tag{12.18}$$

and

$$\text{Arctan}\,\frac{I_2(j\omega)}{V_1(j\omega)} = \pm\frac{\pi}{2}\text{ rad} \tag{12.19}$$

so that $\text{Re}\, z_{21}(j\omega) = 0 = \text{Re}\, y_{21}(j\omega)$ for an *L-C* network. In order for the real parts to be equal to zero, the functions z_{21} and y_{21} of an *L-C* two-port network must be odd.

Suppose, now, that the transfer admittance Y_{21} is given as the quotient of two polynomials

$$Y_{21} = \frac{P(s)}{Q(s)} = \frac{P(s)}{M(s) + N(s)} \tag{12.20}$$

where $P(s)$ is either even or odd. Now, how do we determine the short-circuit parameters y_{21} and y_{22} from the Eq. 12.20 to get it into the form

$$Y_{21} = \frac{y_{21}}{1 + y_{22}} \tag{12.21}$$

The answer is quite simple. We divide both the numerator $P(s)$ and the denominator $Q(s)$ by $M(s)$ or $N(s)$, the even or the odd part of $Q(s)$. Since y_{21} must be odd, if $P(s)$ is even, we divide by $N(s)$ so that

$$Y_{21} = \frac{P(s)/N(s)}{1 + [M(s)/N(s)]} \tag{12.22}$$

From this we obtain

$$y_{21} = \frac{P(s)}{N(s)}$$
$$y_{22} = \frac{M(s)}{N(s)} \tag{12.23}$$

On the other hand, if $P(s)$ is odd, we divide by $M(s)$ so that

$$Y_{21} = \frac{P(s)/M(s)}{1 + [N(s)/M(s)]} \tag{12.24}$$

and

$$y_{21} = \frac{P(s)}{M(s)}$$
$$y_{22} = \frac{N(s)}{M(s)} \tag{12.25}$$

We assume that $P(s)$, $M(s)$, and $N(s)$ do not possess common roots. For our purposes, we will consider only the synthesis of Y_{21} or Z_{21} with zeros of transmission either at $s = 0$ or $s = \infty$. In a ladder network, a zero of transmission at $s = 0$ corresponds to a single capacitor in a series branch or a single inductor in a shunt branch. On the other hand, a zero of transmission at $s = \infty$ corresponds to an inductor in a series branch or a capacitor in a shunt branch. In terms of the transfer impedance

$$Z_{21}(s) = \frac{P(s)}{Q(s)} = \frac{K(s^n + a_{n-1}s^{n-1} + \cdots + a_1 s + a_0)}{s^m + b_{m-1}s^{m-1} + \cdots + b_1 s + b_0} \tag{12.26}$$

the presence of n zeros of $Z_{21}(s)$ at $s = 0$ implies that the coefficients a_{n-1}, $a_{n-2}, \ldots, a_1, a_0$ are all zero. The number of zeros of $Z_{21}(s)$ at $s = \infty$ is given by the difference between the highest powers of the denominator and the numerator, $m - n$. We know that n can exceed m by at most unity, while m can be greater than n by more than one. For example, if $m - n = 2$, and $n = 3$ with $a_{n-1}, \ldots, a_1, a_0 = 0$, we know that the transfer function has three zeros of transmission at $s = 0$ and two zeros of transmission at $s = \infty$.

We can now proceed with the matter of synthesis. Consider the following example.

$$Z_{21}(s) = \frac{2}{s^3 + 3s^2 + 4s + 2} \tag{12.27}$$

We see that all three zeros of transmission are at $s = \infty$. Since the numerator $P(s)$ is a constant, it must be even, so we divide by the odd part of the denominator $s^3 + 4s$. We then obtain

$$z_{21} = \frac{2}{s^3 + 4s}$$
$$z_{22} = \frac{3s^2 + 2}{s^3 + 4s} \tag{12.28}$$

We see that both z_{21} and z_{22} have the same poles. Our task is thus simplified to the point where we must synthesize z_{22} so that the resulting network has the transmission zeros of z_{21}. This requires that we first examine the possible structures of the networks which have the required zeros of transmission and see if we can synthesize z_{22} in one of those forms. For the example that we are considering, a network which gives us three zeros of transmission at $s = \infty$ is shown in Fig. 12.15. We can synthesize z_{22} to give us this structure by the following continued fraction expansion of $1/z_{22}$.

$$\begin{array}{l} 3s^2 + 2\overline{)s^3 + 4s}(\tfrac{1}{3}s \leftarrow Y \\ \quad\;\; s^3 + \tfrac{2}{3}s \\ \quad\;\; \overline{\tfrac{10}{3}s)3s^2 + 2}(\tfrac{9}{10}s \leftarrow Z \\ \qquad\qquad 3s^2 \\ \qquad\qquad \overline{2)\tfrac{10}{3}s}(\tfrac{5}{3}s \leftarrow Y \\ \qquad\qquad\quad \tfrac{10}{3}s \\ \qquad\qquad\quad \overline{\overline{}} \end{array}$$

Since z_{22} is synthesized from the 1-Ω termination toward the input end, the final network takes the form shown in Fig. 12.16. Examining the network more closely, we see that it takes the form of a *low-pass* filter. Thus the specification of all zeros at $s = \infty$ is equivalent to the specification of a low-pass filter.

As a second example, consider the transfer impedance

$$Z_{21}(s) = \frac{s^3}{s^3 + 3s^2 + 4s + 2} \tag{12.29}$$

Since the numerator of $Z_{21}(s)$ is an odd function, we have to divide both

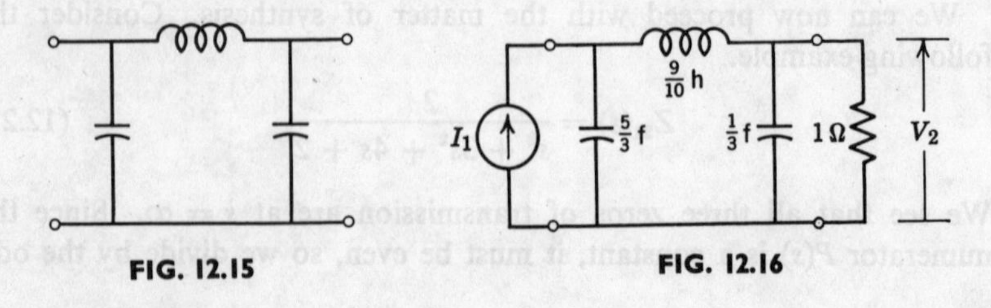

FIG. 12.15

FIG. 12.16

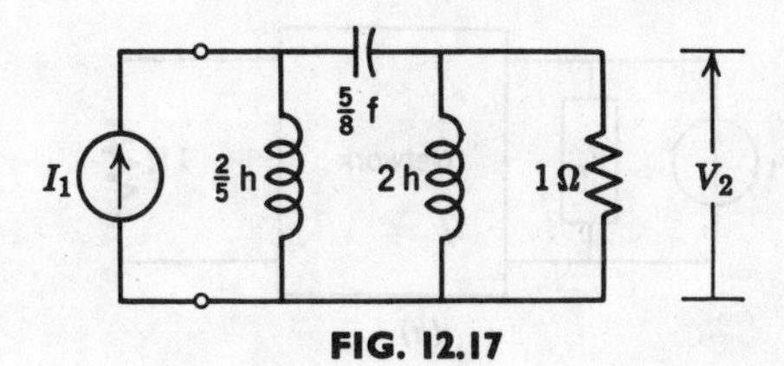

FIG. 12.17

numerator and denominator by the even part of the denominator so that

$$z_{21} = \frac{s^3}{3s^2 + 2} \qquad z_{22} = \frac{s^3 + 4s}{3s^2 + 2} \tag{12.30}$$

The network that gives three zeros of transmission at $s = 0$ is a high-pass structure which is realized by a continued fraction expansion of z_{22}. The final realization is shown in Fig. 12.17.

Finally, consider the transfer admittance

$$Y_{21}(s) = \frac{s^2}{s^3 + 3s^2 + 4s + 2} \tag{12.31}$$

which has two zeros of transmission at $s = 0$, and one zero at $s = \infty$. Since the numerator is even, we divide by $s^3 + 4s$ so that

$$y_{22} = \frac{3s^2 + 2}{s^3 + 4s} \tag{12.32}$$

The question remains as to how we synthesize y_{22} to give a zero of transmission at $s = \infty$ and two zeros at $s = 0$. First, remember that a parallel inductor gives us a zero of transmission at $s = 0$. We can remove this parallel inductor by removing the pole at $s = 0$ of y_{22} to give

$$y_1 = y_{22} - \frac{1}{2s} = \frac{5s/2}{s^2 + 4} \tag{12.33}$$

If we invert y_1, we see that we have a series L-C combination, which gives us another transmission zero at $s = 0$, as represented by the $\frac{5}{8}$-farad capacitor, and we have the zero of transmission at $s = \infty$ also when we remove the inductor of $\frac{2}{5}$ h. The final realization is shown in Fig. 12.18.

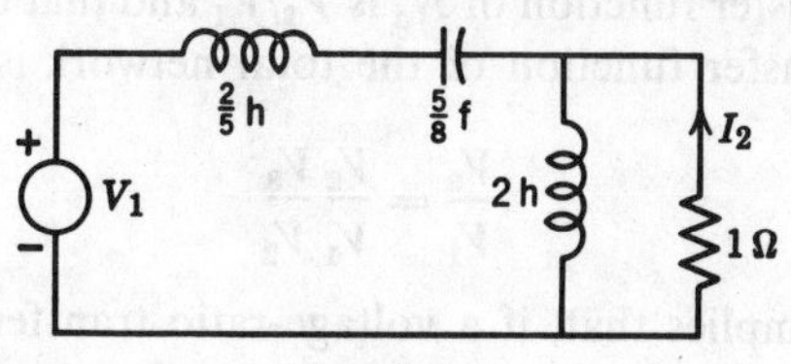

FIG. 12.18

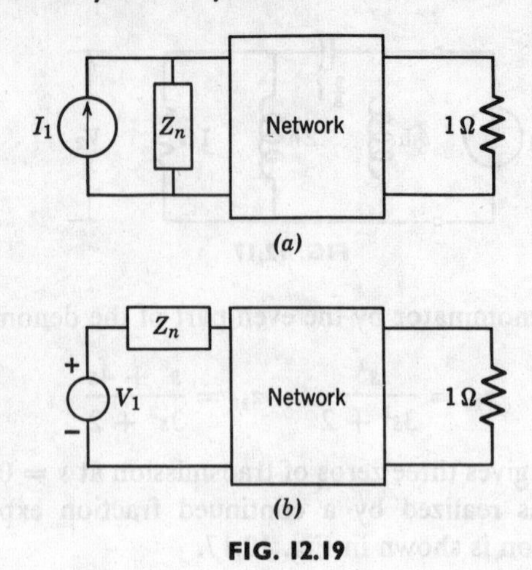

FIG. 12.19

An important point to note in this synthesis procedure is that we must place the last element in series with a voltage source or in parallel with a current source in order for the element to have any effect upon the transfer function. If the last element is denoted as Z_n, then the proper connection of Z_n should be as shown in Fig. 12.19.

12.4 SYNTHESIS OF CONSTANT-RESISTANCE NETWORKS

In this section we will consider the synthesis of *constant-resistance* two-port networks. They derive their name from the fact that the impedance looking in at either port is a constant-resistance R when the other port is terminated in the same resistance R, as depicted in Fig. 12.20. Constant-resistance networks are particularly useful in transfer function synthesis, because when two constant-resistance networks with the same R are connected in tandem, as shown in Fig. 12.21, neither network *loads down* the other. As a result, if the voltage-ratio transfer function of N_a is V_2/V_1 and that of N_b is V_3/V_2, the voltage-ratio transfer function of the total network is

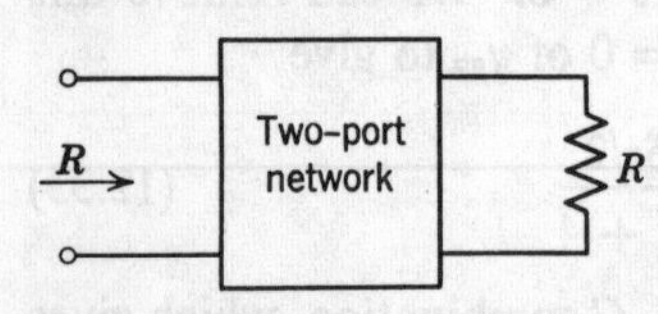

FIG. 12.20. Constant-resistance network.

$$\frac{V_3}{V_1} = \frac{V_2}{V_1}\frac{V_3}{V_2} \tag{12.34}$$

Equation 12.34 implies that, if a voltage-ratio transfer function is to be realized in terms of constant-resistance networks, the voltage ratio could

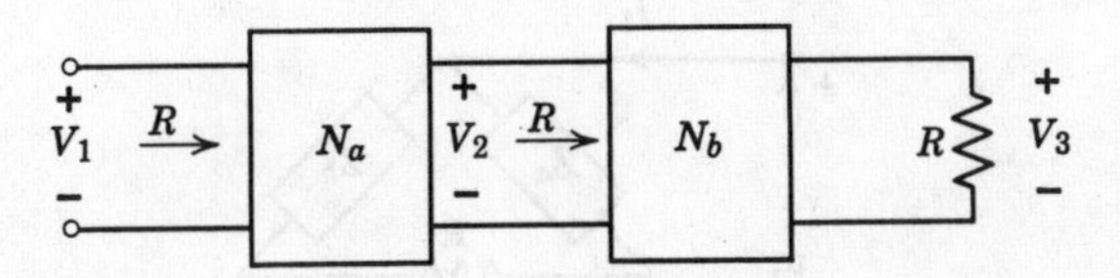

FIG. 12.21. Constant-resistance networks in tandem.

be decomposed into a product of simpler voltage ratios, which could be realized as constant-resistance networks, and then connected in tandem. For example, suppose our objective is to realize

$$\frac{V_b}{V_a} = \frac{K(s - z_0)(s - z_1)(s - z_2)}{(s - p_0)(s - p_1)(s - p_2)} \tag{12.35}$$

in terms of constant-resistance networks. We can first synthesize the individual voltage ratios

$$\begin{aligned} \frac{V_1}{V_a} &= \frac{K_0(s - z_0)}{s - p_0} \\ \frac{V_2}{V_1} &= \frac{K_1(s - z_1)}{s - p_1} \\ \frac{V_b}{V_2} &= \frac{K_2(s - z_2)}{s - p_2} \end{aligned} \tag{12.36}$$

as constant-resistance networks and then connect the three networks in tandem to realize V_b/V_a.

Although there are many different types of constant-resistance networks, we will restrict ourselves to networks of the bridge- and lattice-type structures as shown in Figs. 12.22*a* and 12.22*b*. These networks are *balanced* structures; i.e., the input and output ports do not possess common terminals. Upon a close examination, we see that the bridge and lattice circuits in Figs. 12.22 are identical circuits. The bridge circuit is merely the *unfolded* version of the lattice. Consider the open-circuit parameters of the bridge circuit in Fig. 12.23. First we determine the impedance z_{11} as

$$z_{11} = \frac{Z_a + Z_b}{2} \tag{12.37}$$

Next we determine the transfer impedance z_{21}, which can be expressed as

$$z_{21} = \frac{V_2 - V_2'}{I_1} \tag{12.38}$$

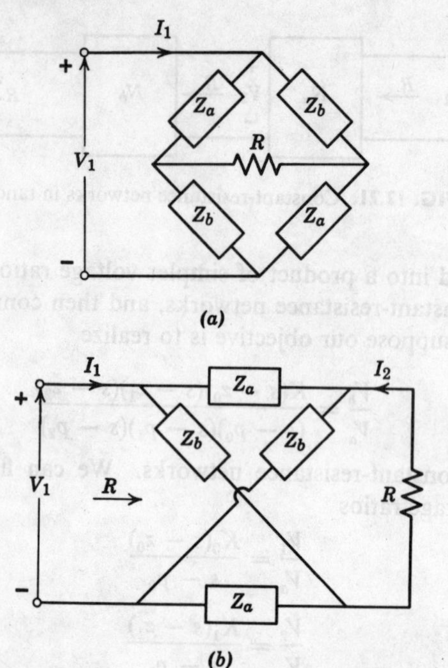

FIG. 12.22. (*a*) Bridge circuit. (*b*) Constant-resistance lattice.

and is obtained as follows. We first obtain the current I' as

$$I' = \frac{V_1}{Z_a + Z_b} = \frac{I_1}{2} \tag{12.39}$$

Next we find that

$$\begin{aligned} V_2 - V_{2'} &= (Z_b - Z_a)I' \\ &= (Z_b - Z_a)\frac{I_1}{2} \end{aligned} \tag{12.40}$$

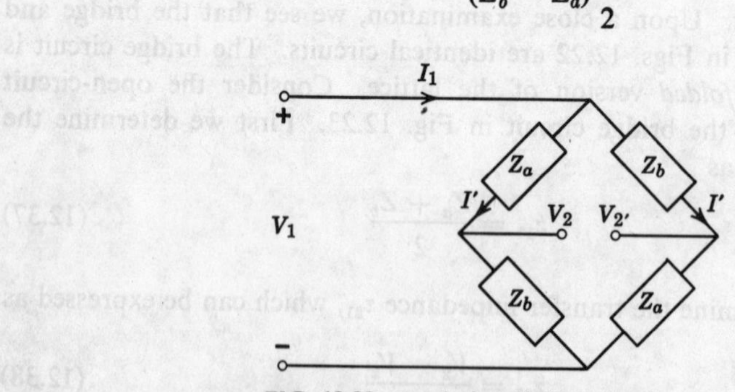

FIG. 12.23. Analysis of bridge circuit.

so that

$$z_{21} = \frac{Z_b - Z_a}{2} \tag{12.41}$$

From the lattice equivalent of the bridge circuit we note that $z_{22} = z_{11}$. Now let us consider the lattice circuit that is terminated in a resistance R, as shown in Fig. 12.22b. What are the conditions on the open-circuit parameters such that the lattice is a constant-resistance network? In other words, what are the conditions upon z_{11} and z_{21} such that the input impedance of the lattice terminated in the resistor R is also equal to R? In Chapter 9 we found that the input impedance could be expressed as

$$Z_{11} = z_{11} - \frac{z_{21}^2}{z_{22} + R} \tag{12.42}$$

Since $z_{22} = z_{11}$ for a symmetrical network, we have

$$Z_{11} = \frac{z_{11}R + z_{11}^2 - z_{21}^2}{z_{11} + R} \tag{12.43}$$

In order for $Z_{11} = R$, the following condition must hold.

$$z_{11}^2 - z_{21}^2 = R^2 \tag{12.44}$$

For the lattice network, we then have

$$\tfrac{1}{4}[(Z_a + Z_b)^2 - (Z_a - Z_b)^2] = R^2 \tag{12.45}$$

which simplifies to give

$$Z_aZ_b = R^2 \tag{12.46}$$

Therefore, in order for a lattice to be a constant-resistance network, Eq. 12.46 must hold.

Next, let us examine the voltage ratio V_2/V_g of a constant-resistance lattice whose source and load impedances are equal to R (Fig. 12.24). From Chapter 9 we can write

$$\frac{V_2}{V_g} = \frac{z_{21}R}{(z_{11} + R)(z_{22} + R) - z_{21}z_{12}} \tag{12.47}$$

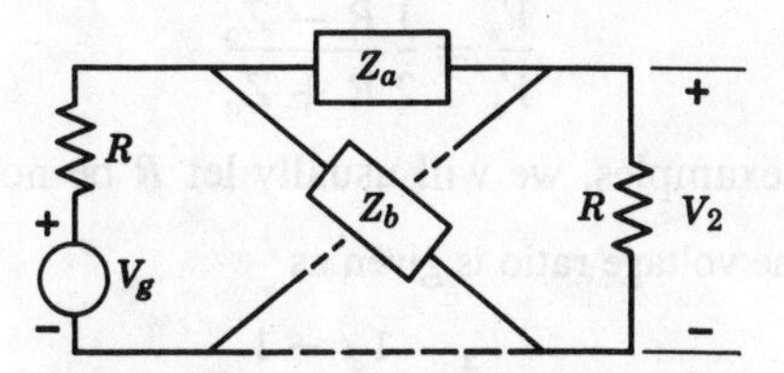

FIG. 12.24. Double-terminated lattice.

which simplifies to
$$\frac{V_2}{V_g} = \frac{z_{21}R}{(z_{11}+R)^2 - z_{21}^{\ 2}}$$
$$= \frac{z_{21}R}{2Rz_{11} + 2R^2} \tag{12.48}$$

In terms of the element values of the lattice, we have

$$\frac{V_2}{V_g} = \frac{\frac{1}{2}(Z_b - Z_a)R}{R(Z_b + Z_a) + 2R^2} \tag{12.49}$$

From the constant-resistance condition in Eq. 12.46, we obtain

$$\frac{V_2}{V_g} = \frac{\frac{1}{2}[Z_b - (R^2/Z_b)]R}{R[Z_b + (R^2/Z_b)] + 2R^2}$$
$$= \frac{\frac{1}{2}(Z_b^{\ 2} - R^2)}{(Z_b^{\ 2} + R^2) + 2RZ_b}$$
$$= \frac{\frac{1}{2}(Z_b^{\ 2} - R^2)}{(Z_b + R)^2}$$
$$= \frac{\frac{1}{2}(Z_b - R)}{Z_b + R} \tag{12.50}$$

In Eq. 12.50, the constant multiplier $\frac{1}{2}$ comes about from the fact that the source resistance R acts as a voltage divider. If we let

$$G(s) \triangleq \frac{2V_2}{V_g} \tag{12.51}$$

we can express Z_b in Eq. 12.50 in terms of G as

$$Z_b = \frac{R[1 + G(s)]}{1 - G(s)} \tag{12.52}$$

In terms of Z_a, the voltage ratio can be given as

$$\frac{V_2}{V_g} = \frac{1}{2}\frac{R - Z_a}{R + Z_a} \tag{12.53}$$

In the following examples, we will usually let R be normalized to unity.

Example 12.1. The voltage ratio is given as

$$\frac{V_2}{V_g} = \frac{1}{2}\frac{s-1}{s+1} \tag{12.54}$$

which, as we recall, is an all-pass transfer function. By associating Eq. 12.54 with Eq. 12.50, we have

$$Z_b = s, \qquad R = 1 \tag{12.55}$$

Since $Z_b Z_a = 1$, we then obtain
$$Z_a = \frac{1}{s} \tag{12.56}$$

We see that Z_b is a 1-h inductor and Z_a is a 1-farad capacitor. The final network is shown in Fig. 12.25.

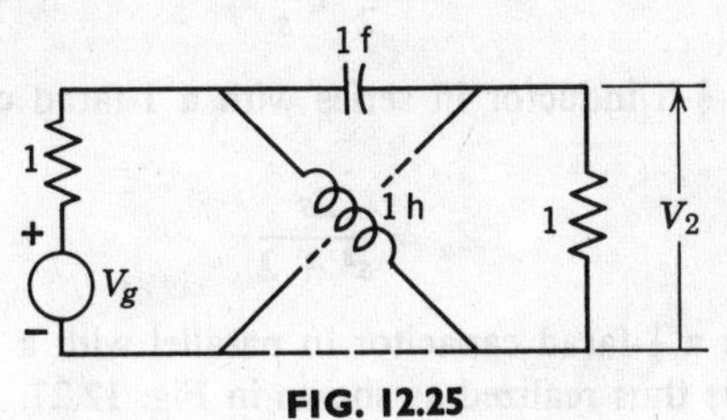

FIG. 12.25

Example 12.2. Let us synthesize the all-pass function

$$\frac{V_2}{V_g} = \frac{1}{2}\frac{s-1}{s+1}\cdot\frac{s^2 - 2s + 2}{s^2 + 2s + 2} \tag{12.57}$$

whose pole-zero diagram is shown in Fig. 12.26. Since the portion

$$\frac{V_2}{V_g} = \frac{1}{2}\frac{s-1}{s+1}$$

has already been synthesized, let us concentrate on synthesizing the function

$$\frac{V_2}{V_a} = \frac{s^2 - 2s + 2}{s^2 + 2s + 2} \tag{12.58}$$

First, we separate the numerator and denominator function into odd and even parts. Thus we have

$$\frac{V_2}{V_a} = \frac{(s^2 + 2) - 2s}{(s^2 + 2) + 2s} \tag{12.59}$$

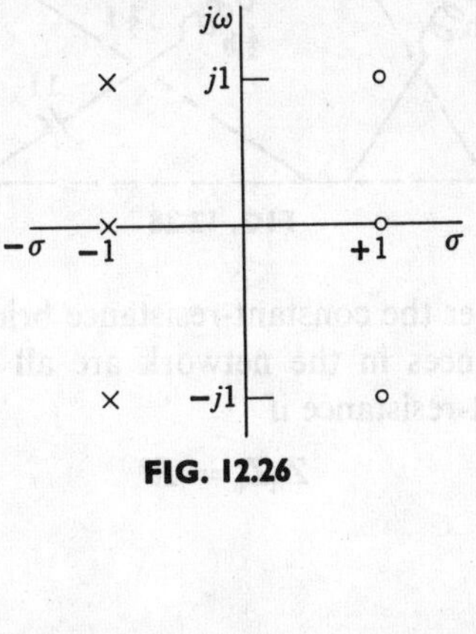

FIG. 12.26

If we divide both numerator and denominator by the odd part $2s$, we obtain

$$\frac{V_2}{V_a} = \frac{[(s^2 + 2)/2s] - 1}{[(s^2 + 2)/2s] + 1} \tag{12.60}$$

We then see that

$$\begin{aligned} Z_b &= \frac{s^2 + 2}{2s} \\ &= \frac{s}{2} + \frac{1}{s} \end{aligned} \tag{12.61}$$

which consists of a $\frac{1}{2}$-h inductor in series with a 1-farad capacitor. The impedance Z_a is then

$$Z_a = \frac{2s}{s^2 + 2} \tag{12.62}$$

and is recognized as a $\frac{1}{2}$-farad capacitor in parallel with a 1-h inductor. The voltage ratio V_2/V_a is thus realized as shown in Fig. 12.27. The structure that

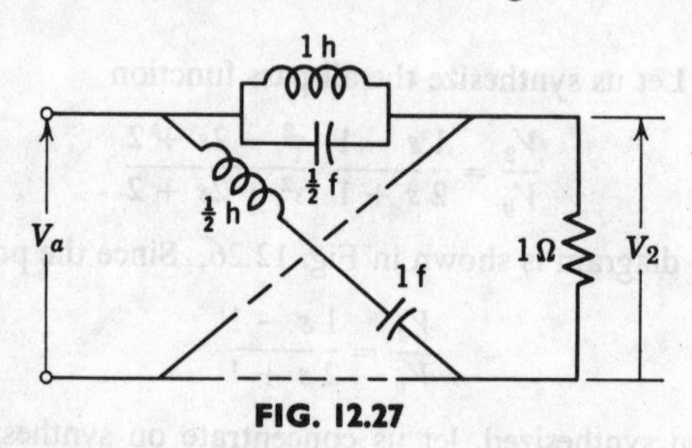

FIG. 12.27

realizes the transfer function V_2/V_g in Eq. 12.57 is formed by connecting the networks in Figs. 12.25 and 12.27 in tandem, as shown in Fig. 12.28. Finally, it should be pointed out that constant-resistance lattices can be used to realize other than all-pass networks.

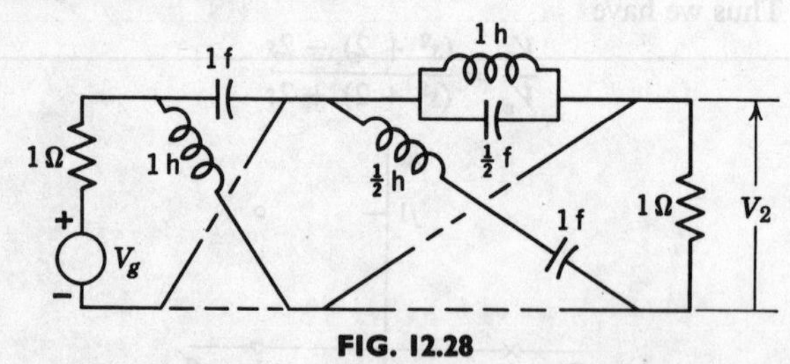

FIG. 12.28

Next, let us consider the constant-resistance bridged-T network in Fig. 12.29. If the resistances in the network are all equal to R ohms, the network has constant-resistance if

$$Z_a Z_b = R^2 \tag{12.63}$$

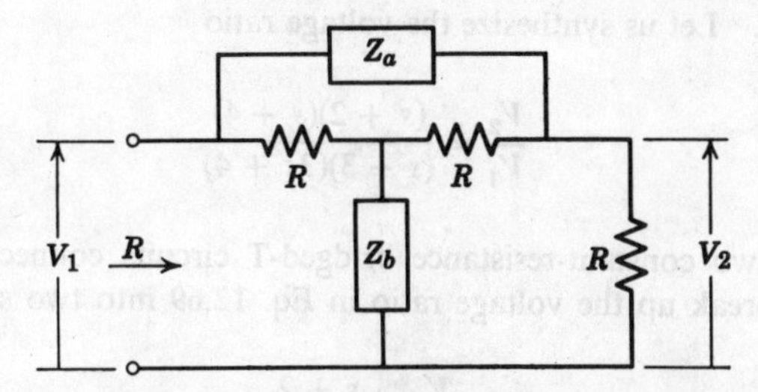

FIG. 12.29. Constant-resistance bridged-T circuit.

Under the constant-resistance assumption, the voltage-ratio transfer function can be given as

$$\frac{V_2}{V_1} = \frac{R}{R + Z_a} = \frac{Z_b}{Z_b + R} \tag{12.64}$$

Example 12.3. Let us synthesize the voltage ratio

$$\frac{V_2}{V_1} = \frac{s^2 + 1}{s^2 + 2s + 1} \tag{12.65}$$

as a constant-resistance bridged-T network terminated in a 1-Ω resistor. First let us write V_2/V_1 as

$$\frac{V_2}{V_1} = \frac{1}{1 + [2s/(s^2 + 1)]} \tag{12.66}$$

so that

$$Z_a = \frac{2s}{s^2 + 1} \tag{12.67}$$

and

$$Z_b = \frac{s^2 + 1}{2s} \tag{12.68}$$

We recognize Z_a as a parallel L-C tank circuit and Z_b as a series L-C tank circuit. The final network is shown in Fig. 12.30.

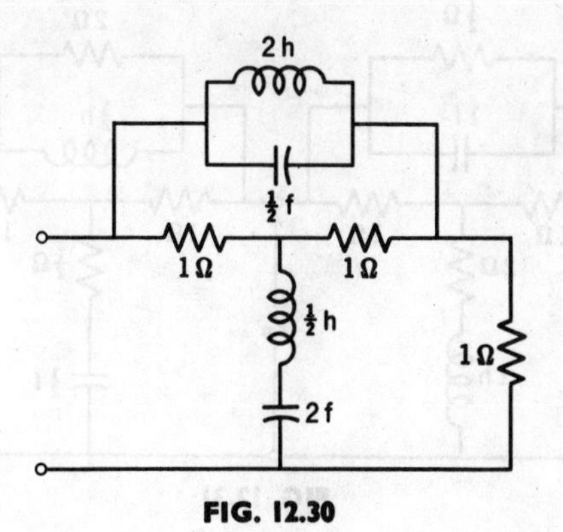

FIG. 12.30

Example 12.4. Let us synthesize the voltage ratio

$$\frac{V_2}{V_1} = \frac{(s+2)(s+4)}{(s+3)(3s+4)} \tag{12.69}$$

in terms of two constant-resistance bridged-T circuits connected in tandem. At first, we break up the voltage ratio in Eq. 12.69 into two separate voltage ratios

$$\frac{V_a}{V_1} = \frac{s+2}{s+3} \tag{12.70}$$

and

$$\frac{V_2}{V_a} = \frac{s+4}{3s+4} \tag{12.71}$$

For the voltage ratio V_a/V_1, we have

$$\frac{s+2}{s+3} = \frac{Z_{b1}}{Z_{b1}+1} \tag{12.72}$$

so that $Z_{b1} = s + 2$ and

$$Z_{a1} = \frac{1}{s+2} \tag{12.73}$$

For the voltage ratio V_2/V_a we have

$$\frac{s+4}{3s+4} = \frac{1}{1+Z_{a2}} \tag{12.74}$$

from which we find

$$Z_{a2} = \frac{2s}{s+4} \tag{12.75}$$

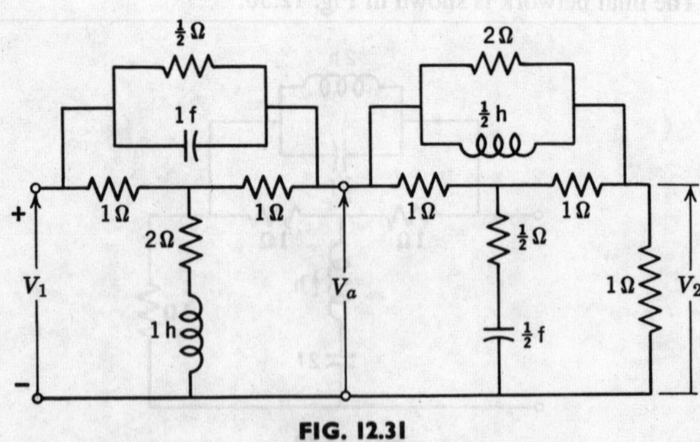

FIG. 12.31

and
$$Z_{b2} = \frac{s+4}{2s} \tag{12.76}$$

The final synthesized network is shown in Fig. 12.31.

Problems

12.1 Give an example of a network where: (*a*) a transfer function has multiple zeros on the $j\omega$ axis; (*b*) the residue of a pole of a transfer function on the $j\omega$ axis is negative.

12.2 Show that the residue condition holds for the networks shown in the figure.

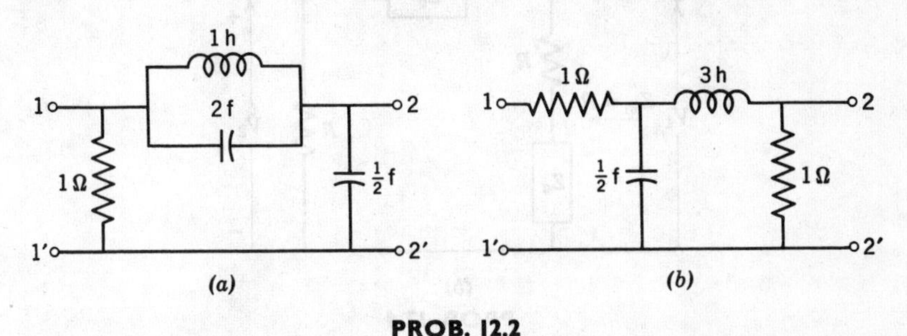

PROB. 12.2

12.3 For the network shown, find by inspection the zeros of transmission and plot on a complex plane.

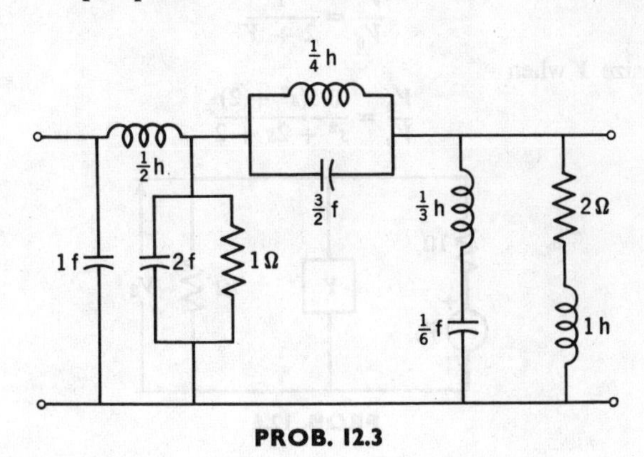

PROB. 12.3

12.4 For the networks in the figure show that the driving point impedances Z_{in} are equal to R when $Z_a Z_b = R^2$.

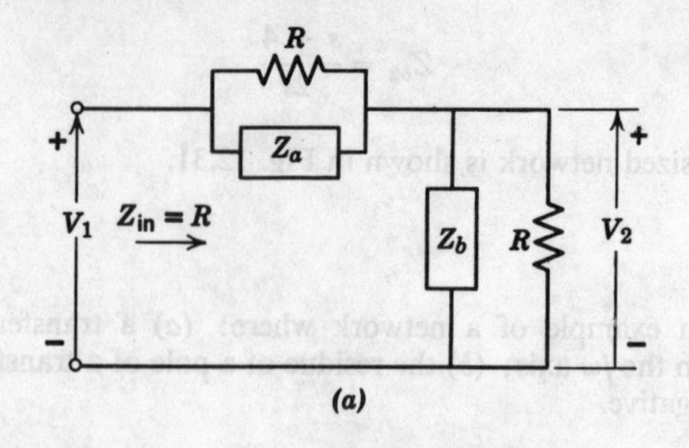

(a)

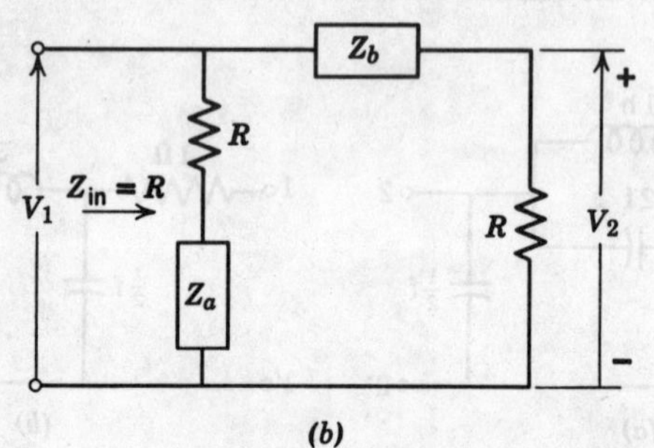

(b)

PROB. 12.4

12.5 For the networks in Prob. 12.4, find the voltage-ratio transfer functions V_2/V_1.

12.6 For the network shown in the figure (*a*) show that

$$\frac{V_2}{V_0} = \frac{1}{2 + Y}$$

(*b*) Synthesize Y when

$$\frac{V_2}{V_0} = \frac{0.5(s^2 + 2)}{s^2 + 2s + 2}$$

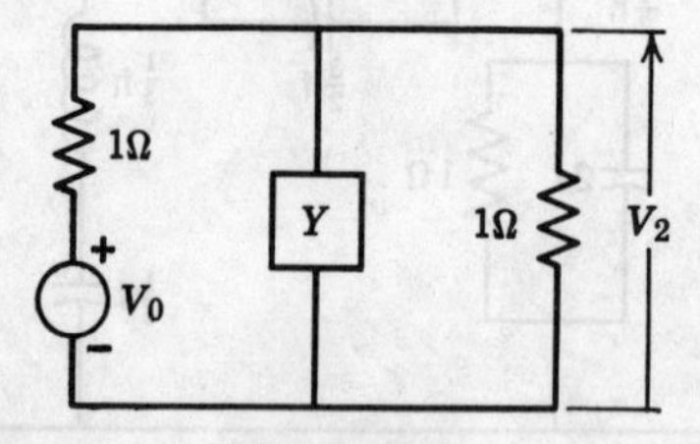

PROB. 12.6

12.7 (*a*) For the constant-resistance bridged-T circuit, show that if $Z_a Z_b = 1$, then

$$\frac{V_2}{V_1} = \frac{1}{1 + Z_a}$$

(*b*) Synthesize Z_a and Z_b if

$$\frac{V_2}{V_1} = \frac{s^2 + 3s + 2}{s^3 + 4s^2 + 5s + 2}$$

12.8 Synthesize the following voltage ratios in one of the forms of the networks in Prob. 12.4

(*a*)
$$\frac{V_2}{V_1} = \frac{s + 2}{s + 3}$$

$$\frac{V_2}{V_1} = \frac{2(s^2 + 3)}{2s^2 + 2s + 6}$$

(*c*)
$$\frac{V_2}{V_1} = \frac{3(s + 0.5)}{4s + 1.5}$$

12.9 Synthesize N_a with termination resistors $R_2 = 4\ \Omega$, $R_1 = 1\ \Omega$ to give

$$\frac{V_2}{V_g} = \frac{12s^2}{15s^2 + 7s + 2}$$

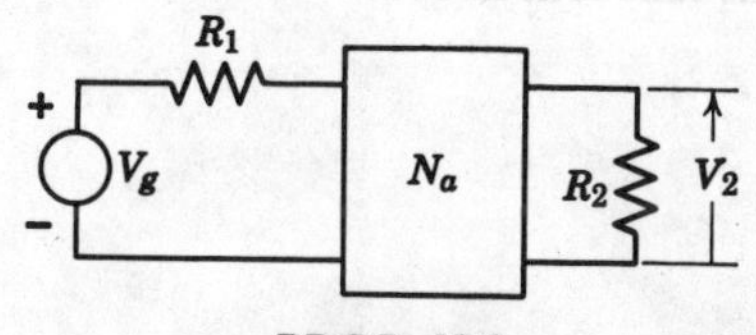

PROB. 12.9

12.10 For the network in Prob. 12.9, realize network N_a to give

$$\frac{V_2}{V_g} = \frac{1}{2}\frac{1}{2s + 3}$$

(*a*) Synthesize N_a as a constant-resistance lattice. ($R = 1\ \Omega$.)

(*b*) Synthesize N_a as a constant-resistance ladder as in Prob. 12.4. ($R = 1\ \Omega$.)

(*c*) Synthesize N_a as a constant-resistance bridged-T circuit. ($R = 1\ \Omega$.)

12.11 Synthesize the following functions into the form shown in the figure

(*a*)
$$Z_{21} = \frac{1}{s^3 + 3s^2 + 3s + 2}$$

(*b*)
$$Z_{21} = \frac{s}{s^3 + 3s^2 + 3s + 2}$$

(*c*)
$$Y_{21} = \frac{s^2}{s^3 + 3s^2 + 3s + 2}$$

(*d*)
$$Y_{21} = \frac{s^3}{s^3 + 3s^2 + 3s + 2}$$

(*e*)
$$Y_{21} = \frac{s^2}{(s + 2)^4}$$

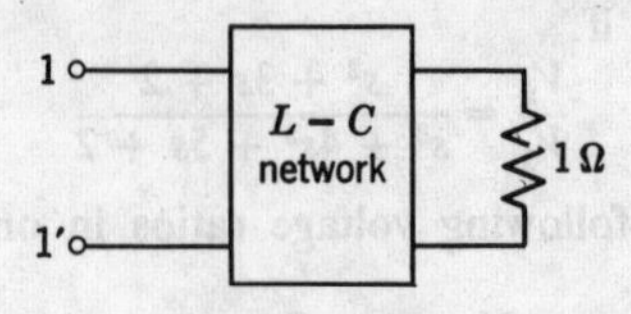

PROB. 12.11

12.12 Synthesize as a constant-resistance lattice terminated in a 1-Ω resistor.

(*a*)
$$\frac{V_2}{V_1} = \frac{s^2 - s + 1}{s^2 + s + 1}$$

(*b*)
$$\frac{I_2}{V_1} = \frac{s^2 - 3s + 2}{s^2 + 3s + 2}$$

(*c*)
$$\frac{V_2(s)}{V_1(s)} = \frac{s^3 - 20s^2 + 5s - 20}{s^3 + 20s^2 + 5s + 20}$$

12.13 Synthesize the functions in Prob. 12.8 as constant-resistance bridged-T circuits.

chapter 13

Topics in filter design

13.1 THE FILTER DESIGN PROBLEM

In the preceding chapters we examined different methods for synthesizing a driving point or transfer function $H(s)$. Most problems have as their initial specification an amplitude or phase characteristic, or an impulse response characteristic instead of the system function $H(s)$. Our problem is to obtain a realizable system function from the given amplitude or phase characteristic. For example, a typical design problem might be to synthesize a network to meet a given low-pass filter characteristic. The specifications might consist of the cutoff frequency ω_C, the maximum allowed deviation from a prescribed amplitude within the pass band, and the rate of *fall off* in the stop band. We must then construct the system function from the amplitude specification. After we obtain $H(s)$, we proceed with the actual synthesis as described in the Chapter 12. Another problem might consist of designing a low-pass filter with a linear phase characteristic within the pass band. Here, both amplitude and phase are specified. We must construct $H(s)$ to meet both specifications. Problems of this nature fall within the domain of *approximation* theory. In this chapter we will consider selected topics in approximation theory and then present examples of filter design where both the approximation and the synthesis problems must be solved.

13.2 THE APPROXIMATION PROBLEM IN NETWORK THEORY

The essence of the problem is the approximation of a given function $f(x)$ by another function $f_a(x;\ \alpha_1, \ldots, \alpha_n)$ in an interval $x_1 \leq x \leq x_2$. The parameters $\alpha_1, \ldots, \alpha_n$ in the approximating function are fixed by the particular error criterion chosen. When we let $\epsilon = f(x) - f_a(x;\ \alpha_1, \ldots, \alpha_n)$, the following error criteria are most common:

1. *Least squares.* The value of $I(\alpha_1, \ldots, \alpha_n)$ is minimized where

$$I(\alpha_1, \ldots, \alpha_n) = \int_{x_1}^{x_2} |\epsilon|^2\, w(x)\, dx$$

and $w(x)$ is a weighting function which stresses the error in certain subintervals.

2. *Maximally flat.* The first $n - 1$ derivatives of ϵ are made to vanish at $x = x_0$.

3. *Chebyshev.* The value of μ is minimized in the interval $x_1 \leq x \leq x_2$ where $\mu = |\epsilon|_{\max}$.

4. *Interpolation.* The value of ϵ is made to vanish at a set of n points in the interval $x_1 \leq x \leq x_2$.

After an error criterion is chosen, we must determine the particular form of the approximating function. This depends upon whether we choose to approximate in the time or frequency domain. Suppose $f(x)$ represents a magnitude function in the frequency domain and the approximating function is to be rational in ω^2; then

$$f_a(x; \alpha_1, \ldots, \alpha_n) = \frac{\alpha_1 + \alpha_3 x + \alpha_5 x^2 + \cdots}{\alpha_2 + \alpha_4 x + \alpha_6 x^2 + \cdots} \tag{13.1}$$

where $x = \omega^2$. In addition, the values of α_k must be restricted to insure that $f_a(x;\ \alpha_1, \ldots, \alpha_n) \geq 0$. In the time domain, $f(x)$ might represent an impulse response of a system to be synthesized. In the case of an *R-C* transfer function, we have

$$f_a(x; \alpha_1, \ldots, \alpha_n) = \alpha_1 \epsilon^{\alpha_2 x} + \alpha_3\, \epsilon^{\alpha_4 x} + \cdots \tag{13.2}$$

where $x = t$. Since an *R-C* transfer function must have its poles on the negative real axis, the values of α_k, k even, are restricted to negative real numbers.

The keystone of any approximation problem lies in the choice of a suitable error criterion subject to realizability restrictions. The problem can be simplified when some of the α's are assigned before applying the error criteria. All the error criteria cited, except the Chebyshev, can then be reduced to a set of linear algebraic equations for the unknowns $\alpha_1, \ldots, \alpha_n$.

Time-domain approximation

The principal problem of time-domain approximation consists of approximating an impulse response $h(t)$ by an approximating function

$h^*(t)$ such that the squared error

$$\epsilon = \int_0^\infty [h(t) - h^*(t)]^2 \, dt$$

is minimum.

A generally effective procedure in time-domain approximation utilizes orthonormal functions $\phi_k(t)$.[1] The approximating function $h^*(t)$ takes the form

$$h^*(t) = \sum_{k=1}^{n} \alpha_k \, \phi_k(t) \tag{13.3}$$

so that the error

$$\epsilon = \int_0^\infty \left[h(t) - \sum_{k=1}^{n} \alpha_k \, \phi_k(t) \right]^2 dt$$

is minimized when

$$\alpha_k = \int_0^\infty h(t) \, \phi_k(t) \, dt \qquad k = 1, 2, \ldots, n \tag{13.4}$$

as we saw in Chapter 3. If the orthonormal set is made up of a sum of exponentials $e^{s_k t}$, then the approximate impulse response

$$h^*(t) = \sum_{k=1}^{n} \alpha_k e^{s_k t} \tag{13.4}$$

has a transform

$$H^*(s) = \sum_{k=1}^{n} \frac{\alpha_k}{s - s_k} \tag{13.5}$$

Realizability is insured if in the orthonormal set $\{\alpha_k e^{s_k t}\}$, Re $s_k \leq 0$; $k = 1, 2, \ldots, n$. Synthesis then proceeds from the system function $H^*(s)$ obtained in Eq. 13.5.

Frequency-domain approximation

In frequency-domain approximation the principal problem is to find a rational function $H(s)$ whose magnitude $|H(j\omega)|$ approximates the ideal low-pass characteristic in Fig. 13.1 according to a predetermined error criterion. In the next few sections we examine several

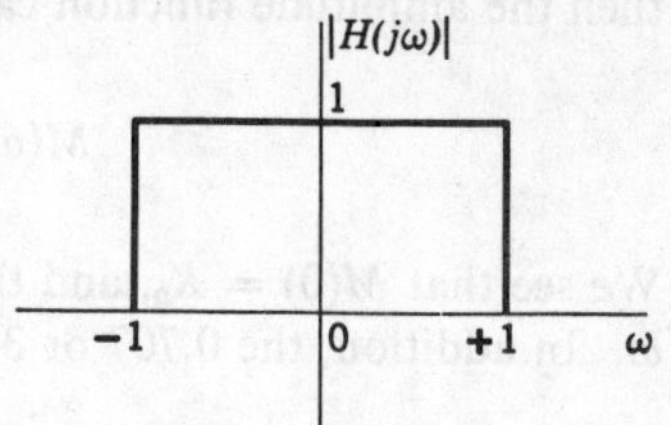

FIG. 13.1. Ideal low-pass filter characteristic.

[1] W. H. Kautz, "Transient Synthesis in the Time Domain," *IRE Trans. on Circuit Theory*, CT-1, No. 3, September 1954, 29–39.

different ways to approximate the ideal low-pass: the *maximally flat* or *Butterworth* approximation, the *equal-ripple* or *Chebyshev* approximation, and the *optimal* or *Legendre* approximation. Another major problem is that of obtaining a transfer function $H_1(s)$, whose phase is approximately linear or whose delay is approximately flat over a given range of frequencies. Here again, there are two different methods: the maximally flat or the equal-ripple methods. Our discussion will center around the maximally flat method. The joint problem of approximating both magnitude and phase over a given frequency range is possible, but will not be discussed here.

13.3 THE MAXIMALLY FLAT LOW-PASS FILTER APPROXIMATION

In Chapter 10 we saw that the ideal low-pass filter in Fig. 13.1 is not realizable because its associated impulse response is not zero for $t < 0$. However, if we use a rational function approximation to this low-pass filter characteristic, the Paley-Wiener criterion will be automatically satisfied. We will therefore restrict ourselves to rational function approximations.

In low-pass filter design, if we assume that all the zeros of the system function are at infinity, the magnitude function takes the general form

$$M(\omega) = \frac{K_0}{[1 + f(\omega^2)]^{1/2}} \tag{13.6}$$

where K_0 is the d-c gain constant and $f(\omega^2)$ is the polynomial to be selected to give the desired amplitude response. For example, if

$$f(\omega^2) = \omega^{2n} \tag{13.7}$$

then the amplitude function can be written as

$$M(\omega) = \frac{K_0}{(1 + \omega^{2n})^{1/2}} \tag{13.8}$$

We see that $M(0) = K_0$, and that $M(\omega)$ is monotonically decreasing with ω. In addition, the 0.707 or 3-decibel point is at $\omega = 1$ for all n, that is,

$$M(1) = \frac{K_0}{\sqrt{2}} \qquad \text{all } n \tag{13.9}$$

The cutoff frequency is thus seen to be $\omega = 1$. The parameter n controls the closeness of approximation in both the pass band and the stop band. Curves of $M(\omega)$ for different n are shown in Fig. 13.2. Observe

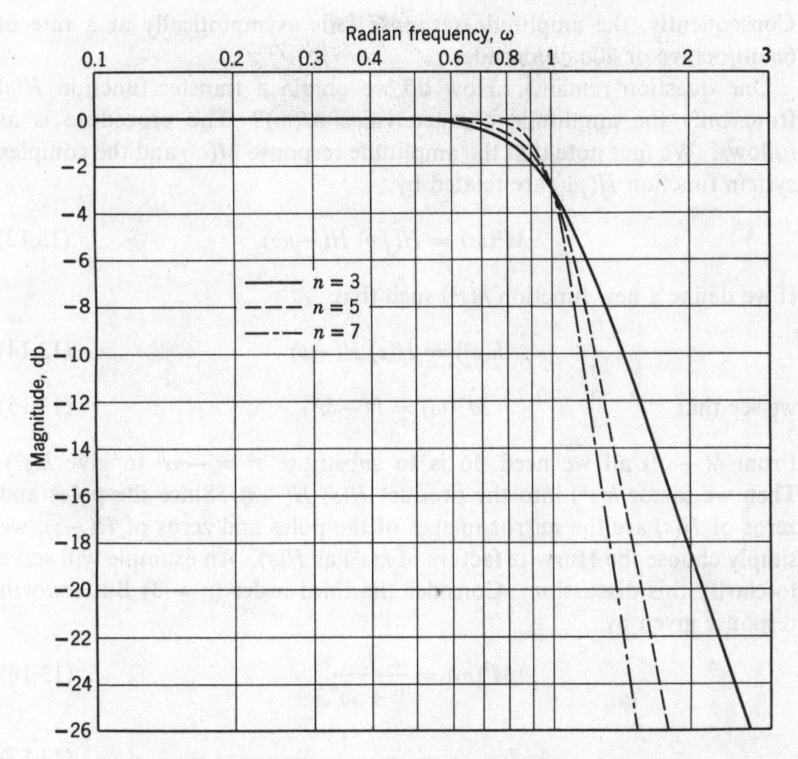

FIG. 13.2. Amplitude response of Butterworth low-pass filters.

that the higher n is, the better the approximation. The amplitude approximation of the type in Eq. 13.8 is called a *Butterworth* or *maximally flat* response. The reason for the term "maximally flat" is that when we expand $M(\omega)$ in a power series about $\omega = 0$, we have

$$M(\omega) = K_0(1 - \tfrac{1}{2}\omega^{2n} + \tfrac{3}{8}\omega^{4n} - \tfrac{5}{16}\omega^{6n} + \tfrac{35}{128}\omega^{8n} + \cdots) \quad (13.10)$$

We see that the first $2n - 1$ derivatives of $M(\omega)$ are equal to zero at $\omega = 0$. For $\omega \gg 1$, the amplitude response of a Butterworth function can be written as (with K_0 normalized to be unity)

$$M(\omega) \simeq \frac{1}{\omega^n} \qquad \omega \gg 1 \quad (13.11)$$

We observe that asymptotically, $M(\omega)$ falls off as ω^{-n} for a Butterworth response. In terms of decibels, the asymptotic slope is obtained as

$$20 \log M(\omega) = -20n \log \omega \quad (13.12)$$

Consequently, the amplitude response falls asymptotically at a rate of $6n$ db/octave or $20n$ db/decade.

One question remains. How do we obtain a transfer function $H(s)$ from only the amplitude characteristics $M(\omega)$? The procedure is as follows. We first note that the amplitude response $M(\omega)$ and the complex system function $H(j\omega)$ are related by

$$M^2(\omega) = H(j\omega)\,H(-j\omega) \tag{13.13}$$

If we define a new function $h(s^2)$ such that

$$h(s^2) = H(s)\,H(-s) \tag{13.14}$$

we see that

$$M^2(\omega) = h(-\omega^2) \tag{13.15}$$

From $h(-\omega^2)$ all we need do is to substitute $s^2 = -\omega^2$ to give $h(s^2)$. Then we factor $h(s^2)$ into the product $H(s)\,H(-s)$. Since the poles and zeros of $H(s)$ are the mirror images of the poles and zeros of $H(-s)$, we simply choose the Hurwitz factors of $h(s^2)$ as $H(s)$. An example will serve to clarify this discussion. Consider the third-order ($n = 3$) Butterworth response given by

$$M^2(\omega) = \frac{1}{1 + \omega^6} \tag{13.16}$$

$$= \frac{1}{1 - (-\omega^2)^3} \tag{13.17}$$

We see that $h(s^2)$ is

$$h(s^2) = \frac{1}{1 - (s^2)^3} \tag{13.18}$$

Factoring $h(s^2)$, we obtain

$$h(s^2) = \frac{1}{1 + 2s + 2s^2 + s^3}\,\frac{1}{1 - 2s + 2s^2 - s^3}$$

$$= H(s)\,H(-s) \tag{13.19}$$

We then have

$$H(s) = \frac{1}{s^3 + 2s^2 + 2s + 1}$$

$$= \frac{1}{(s + 1)(s + \frac{1}{2} + j\sqrt{3}/2)(s + \frac{1}{2} - j\sqrt{3}/2)} \tag{13.20}$$

The poles of $H(s)$ and $H(-s)$ are shown in Fig. 13.3. Observe that the poles of $H(-s)$ are mirror images of the poles of $H(s)$, as given by the theorem on Hurwitz polynomials in the Chapter 2.

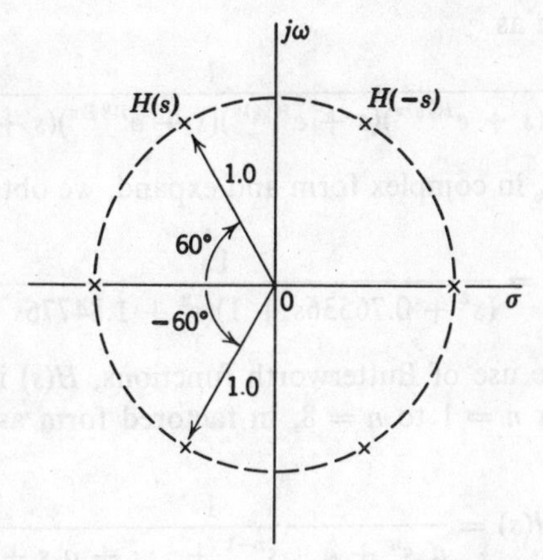

FIG. 13.3. Poles of $H(s)\,H(-s)$ for an $n = 3$ Butterworth filter.

For a Butterworth response, the poles of $H(s)\,H(-s)$ are the roots of

$$\begin{aligned}(-1)^n s^{2n} &= -1 \\ &= e^{j(2k-1)\pi} \qquad k = 0, 1, 2, \ldots, 2n-1\end{aligned} \tag{13.21}$$

The poles s_k are then given by

$$\begin{aligned}s_k &= e^{j[(2k-1)/2n]\pi} \qquad n \text{ even} \\ &= e^{j(k/n)\pi} \qquad n \text{ odd}\end{aligned} \tag{13.22}$$

or simply by $\quad s_k = e^{j[(2k+n-1)/2n]\pi} \qquad k = 0, 1, 2, \ldots, 2n \qquad (13.23)$

Expressing s_k as $s_k = \sigma_k + j\omega_k$, the real and imaginary parts are given by

$$\begin{aligned}\sigma_k &= \cos\frac{2k+n-1}{2n}\pi = \sin\left(\frac{2k-1}{n}\right)\frac{\pi}{2} \\ \omega_k &= \sin\frac{2k+n-1}{2n}\pi = \cos\left(\frac{2k-1}{n}\right)\frac{\pi}{2}\end{aligned} \tag{13.24}$$

It is seen from Eqs. 13.22 and 13.23 that all the poles of $H(s)\,H(-s)$ are located on the unit circle in the s plane, and are symmetrical about both the σ and the $j\omega$ axes. To satisfy realizability conditions, we associate the poles in the right-half plane with $H(-s)$, and the poles in the left-half plane with $H(s)$.

As an example, consider the construction of an $H(s)$ that gives an $n = 4$ Butterworth response. From Eq. 13.23, it is seen that the poles are given by

$$s_k = e^{j[(2k+3)/8]\pi} \tag{13.25}$$

$H(s)$ is then given as

$$H(s) = \frac{1}{(s + e^{j(5/8)\pi})(s + e^{j(7/8)\pi})(s + e^{j(9/8)\pi})(s + e^{j(11/8)\pi})} \quad (13.26)$$

If we express s_k in complex form and expand, we obtain

$$H(s) = \frac{1}{(s^2 + 0.76536s + 1)(s^2 + 1.84776s + 1)} \quad (13.27)$$

To simplify the use of Butterworth functions, $H(s)$ is given in Tables 13.1 and 13.2 for $n = 1$ to $n = 8$, in factored form as in Eq. 13.27, or multiplied out as

$$H(s) = \frac{1}{a_n s^n + a_{n-1}s^{n-1} + \cdots + a_1 s + 1} \quad (13.28)$$

TABLE 13.1
Butterworth Polynomials (Factored Form)

n	
1	$s + 1$
2	$s^2 + \sqrt{2}s + 1$
3	$(s^2 + s + 1)(s + 1)$
4	$(s^2 + 0.76536s + 1)(s^2 + 1.84776s + 1)$
5	$(s + 1)(s^2 + 0.6180s + 1)(s^2 + 1.6180s + 1)$
6	$(s^2 + 0.5176s + 1)(s^2 + \sqrt{2}s + 1)(s^2 + 1.9318s + 1)$
7	$(s + 1)(s^2 + 0.4450s + 1)(s^2 + 1.2456s + 1)(s^2 + 1.8022s + 1)$
8	$(s^2 + 0.3986s + 1)(s^2 + 1.1110s + 1)(s^2 + 1.6630s + 1)(s^2 + 1.9622s + 1)$

TABLE 13.2
Butterworth Polynomials *

n	a_1	a_2	a_3	a_4	a_5	a_6	a_7	a_8
1	1							
2	$\sqrt{2}$	1						
3	2	2	1					
4	2.613	3.414	2.613	1				
5	3.236	5.236	5.236	3.236	1			
6	3.864	7.464	9.141	7.464	3.864	1		
7	4.494	10.103	14.606	14.606	10.103	4.494	1	
8	5.126	13.138	21.848	25.691	21.848	13.138	5.126	1

* $a_0 = 1$.

13.4 OTHER LOW-PASS FILTER APPROXIMATIONS

In Section 13.3, we examined the maximally flat approximation to a low-pass filter characteristic. We will consider other low-pass filter approximants in this section.

The Chebyshev or equal-ripple approximation

We have seen that the maximally flat approximation to the ideal low-pass filter is best at $\omega = 0$, whereas, as we approach the cutoff frequency $\omega = 1$, the approximation becomes progressively poorer. We now consider an approximation which "ripples" about unity in the pass band and falls off rapidly beyond the cutoff $\omega = 1$. The approximation is equally good at $\omega = 0$ and $\omega = 1$, and, as a result is called an *equal-ripple* approximation. The equal-ripple property is brought about by the use of *Chebyshev* cosine polynomials defined as

$$\begin{aligned} C_n(\omega) &= \cos(n \cos^{-1} \omega) \qquad |\omega| \leq 1 \\ &= \cosh(n \cosh^{-1} \omega) \qquad |\omega| > 1 \end{aligned} \tag{13.29}$$

For $n = 0$ we see that
$$C_0(\omega) = 1 \tag{13.30}$$

and for $n = 1$, we have
$$C_1(\omega) = \omega \tag{13.31}$$

Higher order Chebyshev polynomials are obtained through the recursive formula

$$C_n(\omega) = 2\omega\, C_{n-1}(\omega) - C_{n-2}(\omega) \tag{13.32}$$

Thus for $n = 2$, we obtain $C_2(\omega)$ as

$$\begin{aligned} C_2(\omega) &= 2\omega(\omega) - 1 \\ &= 2\omega^2 - 1 \end{aligned} \tag{13.33}$$

In Table 13.3, Chebyshev polynomials of orders up to $n = 10$ are given.

TABLE 13.3

n	Chebyshev polynomials $C_n(\omega) = \cos(n \cos^{-1} \omega)$
0	1
1	ω
2	$2\omega^2 - 1$
3	$4\omega^3 - 3\omega$
4	$8\omega^4 - 8\omega^2 + 1$
5	$16\omega^5 - 20\omega^3 + 5\omega$
6	$32\omega^6 - 48\omega^4 + 18\omega^2 - 1$
7	$64\omega^7 - 112\omega^5 + 56\omega^3 - 7\omega$
8	$128\omega^8 - 256\omega^6 + 160\omega^4 - 32\omega^2 + 1$
9	$256\omega^9 - 576\omega^7 + 432\omega^5 - 120\omega^3 + 9\omega$
10	$512\omega^{10} - 1280\omega^8 + 1120\omega^6 - 400\omega^4 + 50\omega^2 - 1$

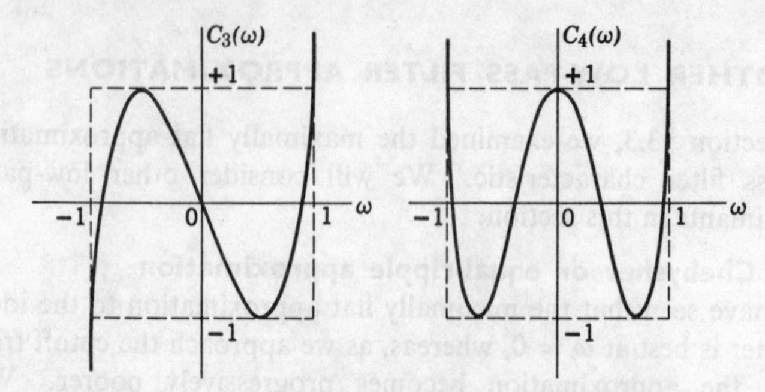

FIG. 13.4. $C_3(\omega)$ and $C_4(\omega)$ Chebyshev polynomials.

The pertinent properties of Chebyshev polynomials used in the low-pass filter approximation are:

1. The zeros of the polynomials are located in the interval $|\omega| \leq 1$, as seen by the plots of $C_3(\omega)$ and $C_4(\omega)$ in Fig. 13.4.
2. Within the interval $|\omega| \leq 1$, the absolute value of $C_n(\omega)$ never exceeds unity; that is, $|C_n(\omega)| \leq 1$ for $|\omega| \leq 1$.
3. Beyond the interval $|\omega| \leq 1$, $|C_n(\omega)|$ increases rapidly for increasing values of $|\omega|$.

Now, how do we apply the Chebyshev polynomials to the low-pass filter approximation? Consider the function $\epsilon^2 C_n^2(\omega)$, where ϵ is real and small compared to 1. It is clear that $\epsilon^2 C_n^2(\omega)$ will vary between 0 and ϵ^2 in the interval $|\omega| \leq 1$. Now we add 1 to this function making it $1 + \epsilon^2 C_n^2(\omega)$. This new function varies between 1 and $1 + \epsilon^2$, a quantity slightly greater than unity, for $|\omega| \leq 1$. Inverting this function, we obtain the function which we will associate with $|H(j\omega)|^2$; thus

$$|H(j\omega)|^2 = \frac{1}{1 + \epsilon^2 C_n^2(\omega)} \tag{13.34}$$

Within the interval $|\omega| \leq 1$, $|H(j\omega)|^2$ oscillates about unity such that the maximum value is 1 and the minimum is $1/(1 + \epsilon^2)$. Outside this interval, $C_n^2(\omega)$ becomes very large so that, as ω increases, a point will be reached where $\epsilon^2 C_n^2(\omega) \gg 1$ and $|H(j\omega)|^2$ approaches zero very rapidly with further increase in ω. Thus, we see that $|H(j\omega)|^2$ in Eq. 13.34 is indeed a suitable approximant for the ideal low-pass filter characteristic.

Figure 13.5 shows a Chebyshev approximation to the ideal low-pass filter. We see that within the pass band $0 \leq \omega \leq 1$, $|H(j\omega)|$ ripples between the value 1 and $(1 + \epsilon^2)^{-1/2}$. The *ripple height* or distance between

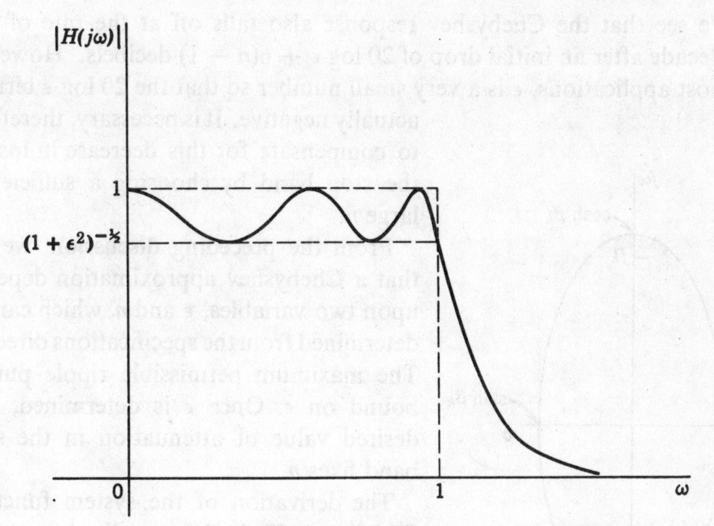

FIG. 13.5. Chebyshev approximation to low-pass filter.

maximum and minimum in the pass band is given as

$$\text{Ripple} = 1 - \frac{1}{(1+\epsilon^2)^{1/2}} \tag{13.35}$$

At $\omega = 1$, $|H(j\omega)|$ is

$$|H(j1)| = \frac{1}{(1+\epsilon^2)^{1/2}} \tag{13.36}$$

because $C_n^2(1) = 1$.

In the stop band, that is, for $|\omega| \geq 1$, as ω increases, we reach a point ω_k, where $\epsilon^2 C_n^2(\omega) \gg 1$ so that

$$|H(j\omega)| \cong \frac{1}{\epsilon C_n(\omega)} \qquad \omega > \omega_k \tag{13.37}$$

The loss in decibels is given as

$$\begin{aligned} \text{Loss} &= -20 \log_{10} |H(j\omega)| \\ &\cong 20 \log \epsilon + 20 \log C_n(\omega) \end{aligned} \tag{13.38}$$

But for large ω, $C_n(\omega)$ can be approximated by its leading term $2^{n-1}\omega^n$, so that

$$\begin{aligned} \text{Loss} &= 20 \log \epsilon + 20 \log 2^{n-1}\omega^n \\ &= 20 \log \epsilon + 6(n-1) + 20n \log \omega \end{aligned} \tag{13.39}$$

We see that the Chebyshev response also falls off at the rate of $20n$ db/decade after an initial drop of $20 \log \epsilon + 6(n - 1)$ decibels. However, in most applications, ϵ is a very small number so that the $20 \log \epsilon$ term is actually negative. It is necessary, therefore, to compensate for this decrease in loss in the stop band by choosing a sufficiently large n.

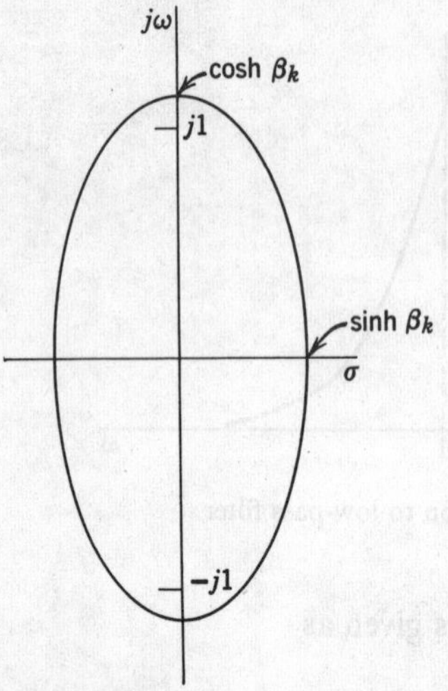

From the preceding discussion, we see that a Chebyshev approximation depends upon two variables, ϵ and n, which can be determined from the specifications directly. The maximum permissible ripple puts a bound on ϵ. Once ϵ is determined, any desired value of attenuation in the stop band fixes n.

The derivation of the system function $H(s)$ from a Chebyshev amplitude approximation $|H(j\omega)|$ is somewhat involved and will not be given here.[2] Instead, we will simply give the results of such a derivation. First we introduce a design parameter.

FIG. 13.6. Locus of poles of Chebshev filter.

$$\beta_k = \frac{1}{n} \sinh^{-1} \frac{1}{\epsilon} \tag{13.40}$$

where n is the degree of the Chebyshev polynomial and ϵ is the factor controlling ripple width. The poles, $s_k = \sigma_k + j\omega_k$, of the equal-ripple approximant $H(s)$ are located on an *ellipse* in the s plane, given by

$$\frac{{\sigma_k}^2}{\sinh^2 \beta_k} + \frac{{\omega_k}^2}{\cosh^2 \beta_k} = 1 \tag{13.41}$$

The major semiaxis of the ellipse is on the $j\omega$ axis and has a value $\omega = \pm\cosh \beta_k$. The minor semiaxis has a value $\sigma = \pm\sinh \beta_k$, and the foci are at $\omega = \pm 1$ (Fig. 13.6). The half-power point of the equal-ripple amplitude response occurs at the point where the ellipse intersects the $j\omega$ axis, i.e., at $\omega = \cosh \beta_k$. Recall that for the Butterworth response, the half-power point occurs at $\omega = 1$. Let us normalize the Chebyshev poles s_k such that the half-power point also falls at $\omega = 1$ instead of at $\omega = \cosh \beta_k$; i.e., let us choose a normalizing factor, $\cosh \beta_k$, such that the

[2] Interested parties are referred to M. E. Van Valkenburg, *Introduction to Modern Network Synthesis*, John Wiley and Sons, New York, 1960, Chapter 13.

normalized pole locations s'_k are given by

$$s'_k = \frac{s_k}{\cosh \beta_k}$$

$$= \frac{\sigma_k}{\cosh \beta_k} + \frac{j\omega_k}{\cosh \beta_k} \tag{13.42}$$

$$\triangleq \sigma'_k + j\omega'_k$$

The normalized pole locations can be derived as

$$\sigma'_k = \tanh \beta_k \sin \left(\frac{2k-1}{n}\right)\frac{\pi}{2}$$

$$\omega'_k = \cos \left(\frac{2k-1}{n}\right)\frac{\pi}{2} \tag{13.43}$$

Comparing the normalized Chebyshev pole locations with the Butterworth pole locations in Eq. 13.24, we see that the imaginary parts are the same, while the real part σ'_k of the Chebyshev pole location is equal to the real part of the Butterworth poles times the factor $\tanh \beta_k$. For example, with $n = 3$ and $\tanh \beta_k = 0.444$, the Butterworth poles are

$$s_1 = -1 + j0$$

$$s_{2,3} = -0.5 \pm j0.866$$

so that the normalized Chebyshev poles are given by

$$s_1 = -1(0.444) + j0$$

$$= -0.444 + j0$$

$$s_{2,3} = -0.5(0.444) \pm j0.866$$

$$= -0.222 \pm j0.866$$

Finally, to obtain the denormalized Chebyshev poles, we simply multiply s'_k by $\cosh \beta_k$, that is,

$$s_k = (\sigma'_k + j\omega'_k) \cosh \beta_k \tag{13.44}$$

There is an easier geometrical method to obtain the Chebyshev poles, given only the semiaxis information and the degree n. First we draw two circles, the smaller of radius $\sinh \beta_k$ and the larger of radius $\cosh \beta_k$, as

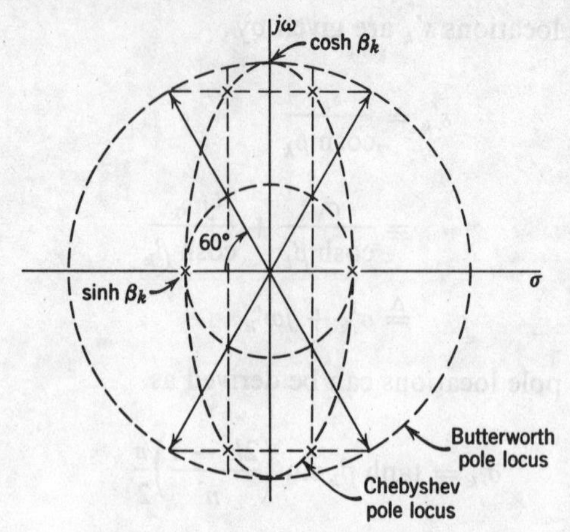

FIG. 13.7. $n = 3$ Chebyshev filter poles.

shown in Fig. 13.7. Next, we draw radial lines according to the angles of the Butterworth poles (Eq. 13.22) as shown. Finally, we draw vertical dashed lines from the intersections of the smaller circle and the radial lines, and horizontal dashed lines from the intersections of the large circle and the radial lines. The Chebyshev poles are located at the intersection of the vertical and horizontal dashed lines, as shown in Fig. 13.7.

Consider the following example. We would like to obtain a system function $H(s)$ that exhibits a Chebyshev characteristic with not more than 1-decibel ripple in the pass band and is down at least 20 decibels at $\omega = 2$.

When we design for 1-decibel ripple, we know that at $\omega = 1$, $|H(j1)|$ is down 1 decibel so that

$$20 \log |H(j1)| = 20 \log \frac{1}{(1 + \epsilon^2)^{1/2}} = -1 \tag{13.45}$$

We then obtain

$$\frac{1}{(1 + \epsilon^2)^{1/2}} = 0.891 \tag{13.46}$$

and

$$\epsilon = 0.509 \tag{13.47}$$

Our next task is to find n from the 20 decibels at $\omega = 2$ specification. From Eq. 13.39 the loss can be given as approximately

$$20 \simeq 20 \log 0.509 + 6(n - 1) + 20n \log 2 \tag{13.48}$$

Solving for n, we obtain $n = 2.65$. Since n must be an integer, we let $n = 3$.

With the specification of n and ϵ, the pole locations are completely specified. Our next task is to determine these pole locations. First we must find β_k. From Eq. 13.40 we have

$$\begin{aligned}\beta_k &= \frac{1}{n}\sinh^{-1}\frac{1}{\epsilon}\\ &= \tfrac{1}{3}\sinh^{-1}1.965 = 0.476\end{aligned} \tag{13.49}$$

In order to find the normalized Chebyshev poles from the Butterworth poles, we must first determine $\tanh \beta_k$. Here we have

$$\tanh \beta_k = \tanh 0.476 = 0.443 \tag{13.50}$$

From Table 13.1, the $n = 3$ Butterworth poles are

$$s_1 = -1.0, \qquad s_{2,3} = -0.5 \pm j0.866 \tag{13.51}$$

Multiplying the real parts of these poles by 0.443, we obtain the normalized Chebyshev poles.

$$s'_1 = -0.443, \qquad s'_{2,3} = -0.222 \pm j0.866$$

Finally, the denormalized Chebyshev poles are obtained by multiplying the normalized ones by $\cosh \beta_k = 1.1155$ so that the denormalized poles are $s_1 = -0.494$ and $s_{2,3} = -0.249 \pm j0.972$.
$H(s)$ is then

$$\begin{aligned}H(s) &= \frac{0.502}{(s + 0.494)(s + 0.249 - j0.972)(s + 0.249 + j0.972)}\\ &= \frac{0.502}{s^3 + 0.992s^2 + 1.255s + 0.502}\end{aligned} \tag{13.52}$$

In Fig. 13.8, the amplitude responses of the Chebyshev and an $n = 3$ Butterworth filter are shown.

Monotonic filters with optimum cutoff

In comparing Butterworth filters with Chebyshev filters, the following can be said. The Butterworth response is a maximally flat, monotonic response, whereas the Chebyshev response is equal ripple in the pass band. In the stop band, the Chebyshev response falls off more rapidly than the Butterworth (except when ϵ is very, very small). In this respect, the Chebyshev filter is a better filter than the Butterworth. However, as we shall see in Section 13.5, the transient response of the Chebyshev filter is very poor. If we require sharp cutoff characteristics for a given degree n, however, the Butterworth filter is quite unsatisfactory. In 1958, Papoulis[3] proposed

[3] A. Papoulis, "Optimum Filters with Monotonic Response," *Proc. IRE*, **46**, No. 3, March 1958, pp. 606–609.

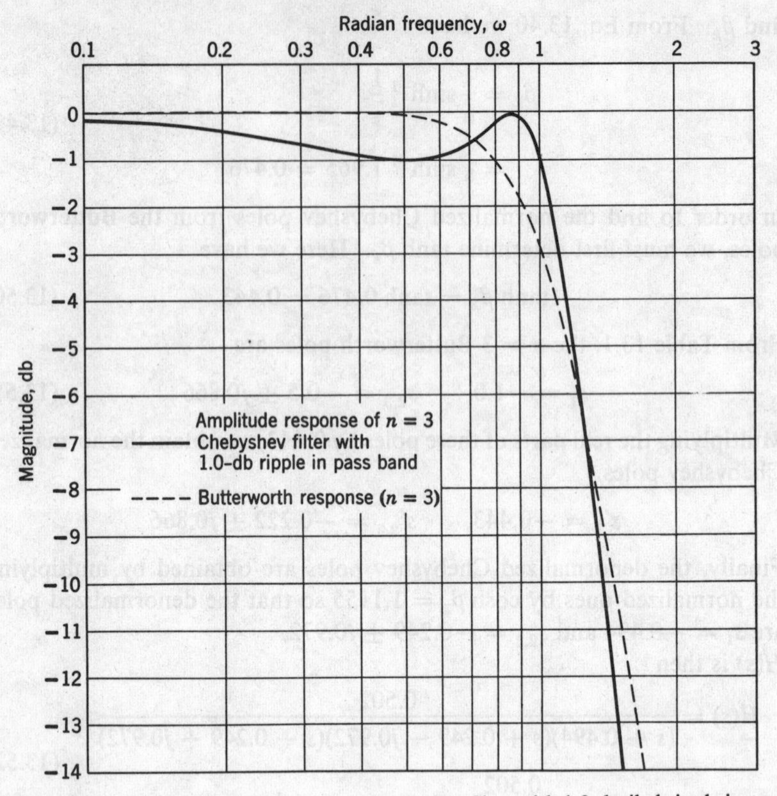

FIG. 13.8. Amplitude response of $n = 3$ Chebyshev filter with 1.0-decibel ripple in pass band and Butterworth response ($n = 3$).

a class of filters called *Optimum* or "L" filters, which have the following properties:

1. The amplitude response is monotonic.
2. The fall-off rate at ω cutoff is the greatest possible, if monotonicity is assumed.
3. The zeros of the system function of the L filter are all at infinity.

Recall that the magnitude response of a low-pass filter with all zeros at infinity can be expressed as

$$M(\omega) = \frac{K_0}{[1 + f(\omega^2)]^{1/2}} \tag{13.53}$$

Let us denote the polynomial generating the L filter by

$$f(\omega^2) = L_n(\omega^2) \tag{13.54}$$

The polynomial $L_n(\omega^2)$ has the following properties:

(*a*) $$L_n(0) = 0$$

(*b*) $$L_n(1) = 1$$

(*c*) $$\frac{dL_n(\omega^2)}{d\omega} \leq 0$$

(*d*) $$\left.\frac{dL_n(\omega^2)}{d\omega}\right|_{\omega=1} = M \qquad (M \text{ maximum})$$

Properties *a*, *b* and *c* are the same as for the Butterworth generating polynomial $f(\omega^2) = \omega^{2n}$. Property *c* insures that the response $M(\omega)$ is monotonic and property *d* requires that the slope of $L_n(\omega^2)$ at $\omega = 1$ be the steepest to insure sharpest cutoff.

Papoulis originally derived the generating equation for the polynomials L_n (for n odd) to be

$$L_n(\omega^2) = \int_{-1}^{2\omega^2-1} \left[\sum_{i=0}^{k} a_i P_i(x)\right]^2 dx \tag{13.55}$$

where $n = 2k + 1$ and the $P_i(x)$ are the Legendre polynomials of the first kind[4]

$$\begin{aligned} P_0(x) &= 1 \\ P_1(x) &= x \\ P_2(x) &= \tfrac{1}{2}(3x^2 - 1) \\ P_3(x) &= \tfrac{1}{2}(5x^3 - 3x) \\ &\cdots \end{aligned} \tag{13.56}$$

and the constants a_i are given by

$$a_0 = \frac{a_1}{3} = \frac{a_2}{5} = \cdots = \frac{a_k}{2k+1} = \frac{1}{\sqrt{2}(k+1)} \tag{13.57}$$

Later Papoulis[5] and, independently, Fukada,[6] showed that the even-ordered L_n polynomials can be given by

$$L_{2k+2}(\omega^2) = \int_{-1}^{2\omega^2-1} (x+1)\left[\sum_{i=0}^{k} a_i P_i(x)\right]^2 dx \tag{13.58}$$

$$n = 2k + 2$$

---

[4] E. Jahnke and F. Emde, *Tables of Functions*, Dover Publications, New York, 1945.

[5] A. Papoulis, "On Monotonic Response Filters," *Proc. IRE*, **47**, February 1959, 332–333.

[6] M. Fukada, "Optimum Filters of Even Orders with Monotonic Response," *Trans. IRE*, **CT-6**, No. 3, September 1959, 277–281.

where the constants a_i are given by:

Case 1 (k even):

$$a_0 = \frac{a_2}{5} = \cdots = \frac{a_k}{2k+1} = \frac{1}{\sqrt{(k+1)(k+2)}} \tag{13.59}$$

$$a_1 = a_3 = \cdots = a_{k-1} = 0$$

Case 2 (k odd):

$$\frac{a_1}{3} = \frac{a_3}{7} = \cdots = \frac{a_k}{2k+1} = \frac{1}{\sqrt{(k+1)(k+2)}} \tag{13.60}$$

$$a_0 = a_2 = \cdots = a_{2k} = 0$$

Fukada tabulated the $L_n(\omega^2)$ polynomials up to $n = 7$ together with $dL_n(\omega^2)/d\omega$ evaluated at $\omega = 1$ to give an indication of the steepness of the cutoff. This is shown in Table 13.4.

To obtain the system function $H(s)$ for the L filter, we must factor the equation for $h(s^2)$ and choose the Hurwitz factors as $H(s)$.

$$h(s^2) = H(s)\,H(-s) = \frac{1}{1 + L_n(-s^2)} \tag{13.61}$$

For example, for $n = 3$, the magnitude response squared is

$$\begin{aligned} M^2(\omega) &= \frac{1}{1 + L_3(\omega^2)} \\ &= \frac{1}{1 + \omega^2 - 3\omega^4 + 3\omega^6} \end{aligned} \tag{13.62}$$

Substituting $-\omega^2 = s^2$, we obtain

$$h(s^2) = H(s)\,H(-s) = \frac{1}{1 - s^2 - 3s^4 - 3s^6} \tag{13.63}$$

TABLE 13.4
$L_n(\omega^2)$ Polynomials

n	$L_n(\omega^2)$	$\dfrac{dL_n(1)}{d\omega}$
2	ω^4	4
3	$3\omega^6 - 3\omega^4 + \omega^2$	8
4	$6\omega^8 - 6\omega^6 + 3\omega^4$	12
5	$20\omega^{10} - 40\omega^8 + 28\omega^6 - 8\omega^4 + \omega^2$	18
6	$50\omega^{12} - 120\omega^{10} + 105\omega^8 - 40\omega^6 + 6\omega^4$	24
7	$175\omega^{14} - 525\omega^{12} + 615\omega^{10} - 355\omega^8 + 105\omega^6 - 15\omega^4 + \omega^2$	32

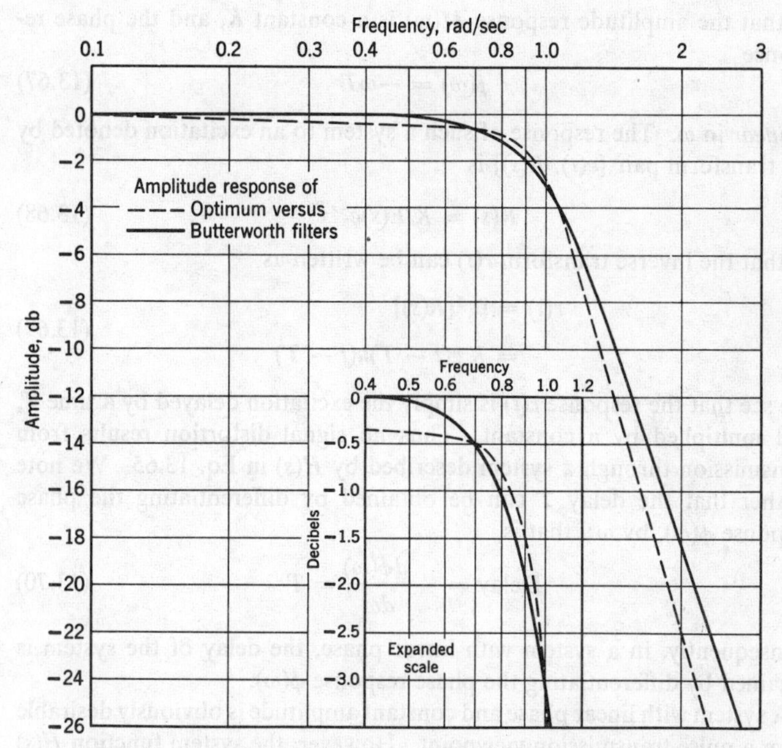

FIG. 13.9. Amplitude response of Optimum versus Butterworth filters.

After we factor $h(s^2)$, we obtain

$$H(s) = \frac{0.577}{s^3 + 1.31s^2 + 1.359s + 0.577} \tag{13.64}$$

where the numerator factor, 0.577, is chosen to let the d-c gain be unity. The poles of $H(s)$ are $s_1 = -0.62$; $s_{2,3} = -0.345 \pm j0.901$. The amplitude response of third-order Optimum (L) and Butterworth filters are compared in Fig. 13.9 Note that the amplitude response of the Optimum filter is not maximally flat, although still monotonic. However, the cutoff characteristic of the Optimum filter is sharper than the cutoff of the Butterworth filter.

Linear phase filters

Suppose a system function is given by

$$H(s) = Ke^{-sT} \tag{13.65}$$

where K is a positive real constant. Then the frequency response of the system can be expressed as

$$H(j\omega) = Ke^{-j\omega T} \tag{13.66}$$

so that the amplitude response $M(\omega)$ is a constant K, and the phase response

$$\phi(\omega) = -\omega T \tag{13.67}$$

is *linear* in ω. The response of such a system to an excitation denoted by the transform pair $\{e(t), E(s)\}$ is

$$R(s) = K\,E(s)e^{-sT} \tag{13.68}$$

so that the inverse transform $r(t)$ can be written as

$$\begin{aligned} r(t) &= \mathcal{L}^{-1}[R(s)] \\ &= K\,e(t-T)u(t-T) \end{aligned} \tag{13.69}$$

We see that the response $r(t)$ is simply the excitation delayed by a time T, and multiplied by a constant. Thus no signal distortion results from transmission through a system described by $H(s)$ in Eq. 13.65. We note further that the delay T can be obtained by differentiating the phase response $\phi(\omega)$, by ω; that is,

$$\text{Delay} = -\frac{d\phi(\omega)}{d\omega} = T \tag{13.70}$$

Consequently, in a system with linear phase, the delay of the system is obtained by differentiating the phase response $\phi(\omega)$.

A system with linear phase and constant amplitude is obviously desirable from a pulse transmission viewpoint. However, the system function $H(s)$ in Eq. 13.65 is only realizable in terms of a lossless transmission line called a *delay line*. If we require that the transmission network be made up of lumped elements, then we must approximate $H(s) = Ke^{-sT}$ by a rational function in s. The approximation method we shall describe here is due to Thomson.[7] We can write $H(s)$ as

$$\begin{aligned} H(s) &= \frac{K_0}{e^{sT}} \\ &= \frac{K_0}{\sinh sT + \cosh sT} \end{aligned} \tag{13.71}$$

where K_0 is chosen such that $H(0) = 1$. Let the delay T be normalized to unity and let us divide both numerator and denominator of $H(s)$ by $\sinh s$ to obtain

$$H(s) = \frac{K_0/\sinh s}{\coth s + 1} \tag{13.72}$$

[7] W. E. Thomson, "Network with Maximally Flat Delay," *Wireless Engrg.*, *29*, Oct. 1952, 256–263.

If sinh s and cosh s are expanded in power series, we have

$$\cosh s = 1 + \frac{s^2}{2!} + \frac{s^4}{4!} + \frac{s^6}{6!} + \cdots$$
$$\sinh s = s + \frac{s^3}{3!} + \frac{s^5}{5!} + \frac{s^7}{7!} + \cdots \tag{13.73}$$

From these series expansions, we then obtain a continued fraction expansion of coth s as

$$\coth s = \frac{1}{s} + \cfrac{1}{\frac{3}{s} + \cfrac{1}{\frac{5}{s} + \cfrac{1}{\frac{7}{s} + \cdots}}} \tag{13.74}$$

If the continued fraction is terminated in n terms, then $H(s)$ can be written as

$$H(s) = \frac{K_0}{B_n(s)} \tag{13.75}$$

where $B_n(s)$ are *Bessel* polynomials defined by the formulas

$$\begin{aligned} B_0 &= 1 \\ B_1 &= s + 1 \\ &\cdots \\ B_n &= (2n-1)B_{n-1} + s^2 B_{n-2} \end{aligned} \tag{13.76}$$

From these formulas, we obtain

$$\begin{aligned} B_2 &= s^2 + 3s + 3 \\ B_3 &= s^3 + 6s^2 + 15s + 15 \end{aligned} \tag{13.77}$$

Higher order Bessel polynomials are given in Table 13.5, and the roots of Bessel polynomials are given in Table 13.6. Note that the roots are all in the left-half plane. A more extensive table of roots of Bessel polynomials is given by Orchard.[8]

The amplitude and phase response of a system function employing an unnormalized third-order Bessel polynomial

$$H(s) = \frac{15}{s^3 + 6s^2 + 15s + 15} \tag{13.78}$$

---

[8] H. J. Orchard, "The Roots of Maximally Flat Delay Polynomials" *IEEE Trans. on Circuit Theory*, **CT-12** No. 3, September 1965, 452–454.

TABLE 13.5

Coefficients of Bessel Polynomials

n	b_0	b_1	b_2	b_3	b_4	b_5	b_6	b_7
0	1							
1	1	1						
2	3	3	1					
3	15	15	6	1				
4	105	105	45	10	1			
5	945	945	420	105	15	1		
6	10,395	10,395	4,725	1,260	210	21	1	
7	135,135	135,135	62,370	17,325	3,150	378	28	1

are given by the solid lines in Figs. 13.10 and 13.11. These are compared with the amplitude and phase of an unnormalized third-order Butterworth function given by the dotted lines. Note that the phase response of the constant-delay function is more linear than the phase of the Butterworth function. Also, the amplitude cutoff of the constant-delay curve is more gradual than that of the Butterworth.

TABLE 13.6

Roots of Bessel Polynomials

n	Roots of Bessel Polynomials
1	$-1.0 + j0$
2	$-1.5 \pm j0.866025$
3	$-2.32219 + j0$ $-1.83891 \pm j1.75438$
4	$-2.89621 \pm j0.86723$ $-2.10379 \pm j2.65742$
5	$-3.64674 + j0$ $-3.35196 \pm j1.74266$ $-2.32467 \pm j3.57102$
6	$-4.24836 \pm j0.86751$ $-3.73571 \pm j2.62627$ $-2.51593 \pm j4.49267$
7	$-4.97179 + j0$ $-4.75829 \pm j1.73929$ $-4.07014 \pm j3.51717$ $-2.68568 \pm j5.42069$

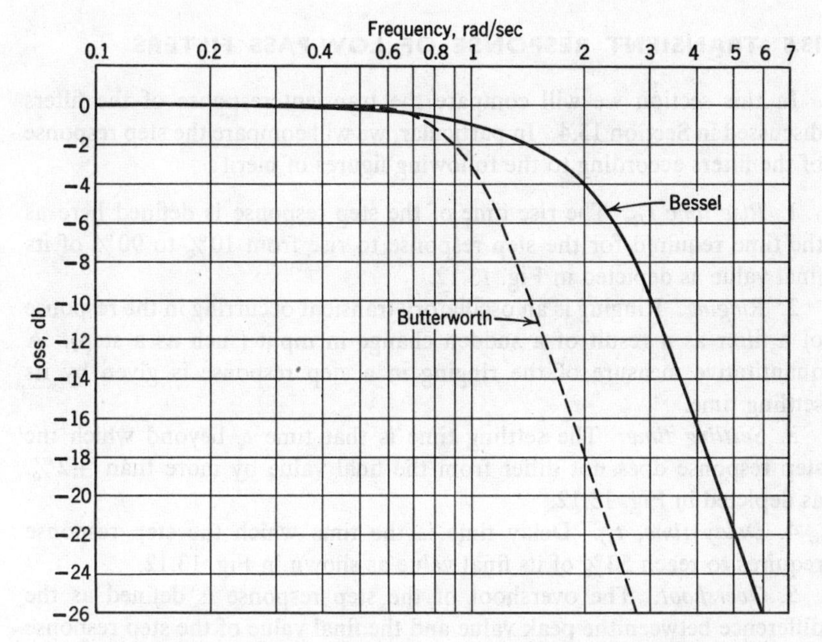

FIG. 13.10. Amplitude response of $n = 3$ Bessel and Butterworth filters.

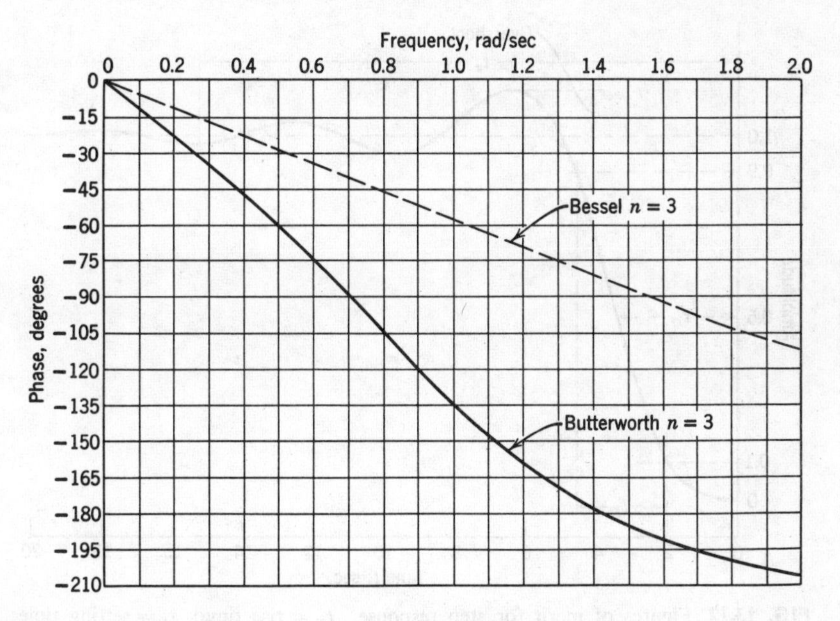

FIG. 13.11. Phase responses of low-pass filters.

13.5 TRANSIENT RESPONSE OF LOW-PASS FILTERS

In this section we will compare the transient response of the filters discussed in Section 13.4. In particular, we will compare the step response of the filters according to the following figures of merit:

1. *Rise time* t_R. The rise time of the step response is defined here as the time required for the step response to rise from 10% to 90% of its final value as depicted in Fig. 13.12.
2. *Ringing*. Ringing is an oscillatory transient occurring in the response of a filter as a result of a sudden change in input (such as a step). A quantitative measure of the ringing in a step response is given by its settling time.
3. *Settling time*. The settling time is that time t_s beyond which the step response does not differ from the final value by more than $\pm 2\%$, as depicted in Fig. 13.12.
4. *Delay time*, t_D. Delay time is the time which the step response requires to reach 50% of its final value as shown in Fig. 13.12.
5. *Overshoot*. The overshoot of the step response is defined as the difference between the peak value and the final value of the step response (see Fig. 13.12) expressed as a percentage of the final value.

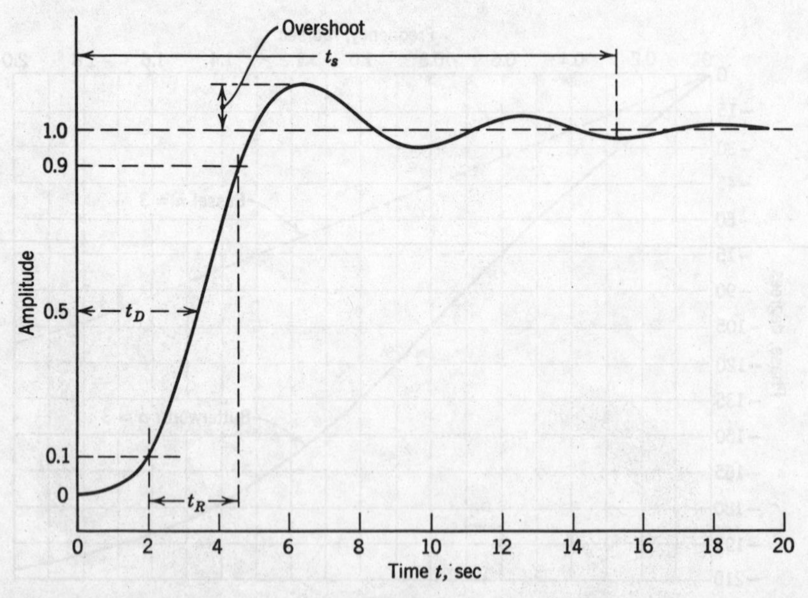

FIG. 13.12. Figures of merit for step response. t_R = rise time; t_s = setting time; t_D = delay time.

Most of the foregoing figures of merit are related to frequency response, particularly bandwidth and phase linearity. Some of the quantities, such as rise time and delay time are intimately related to each other but have rather tenuous ties with overshoot. Let us examine qualitatively the relationships between the transient response criteria just cited and frequency response.

Rise time and bandwidth have an inverse relationship in a filter. The wider the bandwidth, the smaller the rise time; the narrower the bandwidth, the longer the rise time. Physically, the inverse relationship could be explained by noting that the limited performance of the filter at high frequencies slows down the abrupt rise in voltage of the step and prolongs the rise time. Thus we have

$$T_R \times BW = \text{Constant} \tag{13.79}$$

Rise time is a particularly important criterion in pulse transmission. In an article on data transmission,[9] it was shown that in transmitting a pulse of width T_1 through a system with adjustable bandwidths, the following results were obtained:

Bandwidth ($f_c = 1/T_1$)	Rise Time (milliseconds)
f_c	0.5
$2f_c$	0.25
$3f_c$	0.16
$4f_c$	0.12
$5f_c$	0.10

The table shows a definite inverse relationship between rise time and bandwidth.

A definition of time delay is given by Elmore as the first moment or centroid of the impulse response

$$T_D = \int_0^\infty t\, h(t)\, dt \tag{13.80}$$

provided the step response has little or no overshoot. Elmore's definitio of rise time is given as the second moment

$$T_R = \left[2\pi \int_0^\infty (t - T_D)^2\, h(t)\, dt\right]^{1/2} \tag{13.81}$$

[9] R. T. James, "Data Transmission—The Art of Moving Information," *IEEE Spectrum*, January 1965, 65–83.

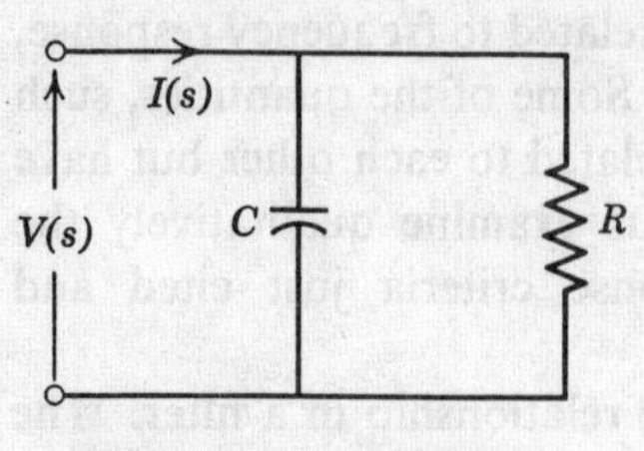

FIG. 13.13. *R-C* network.

These definitions are useful because we can obtain rise time and delay time directly from the coefficients of the system function $H(s)$. Without going into the proof, which is in Elmore and Sands,[10] if $H(s)$ is given as

$$H(s) = \frac{1 + a_1 s + a_2 s^2 + \cdots + a_n s^n}{1 + b_1 s + b_2 s^2 + \cdots + b_m s^m} \tag{13.82}$$

the time delay T_D is

$$T_D = b_1 - a_1 \tag{13.83}$$

and the rise time is

$$T_R = \{2\pi[b_1^2 - a_1^2 + 2(a_2 - b_2)]\}^{1/2} \tag{13.84}$$

For the *R-C* network in Fig. 13.13, $H(s)$ is

$$H(s) = \frac{V(s)}{I(s)} = \frac{R}{1 + sRC} \tag{13.85}$$

so that

$$\begin{aligned} T_D &= RC \\ T_R &= \sqrt{2\pi}RC \end{aligned} \tag{13.86}$$

It should be emphasized that Elmore's definitions are restricted to step responses without overshoot because of the moment definition. The more general definition of rise time is the 10–90% one cited earlier, which has no formal mathematical definition.

Overshoot is generally caused by "excess" gain at high frequencies. By *excess gain* we normally mean a magnitude characteristic with a peak such as the shunt peaked response shown by the dashed curve in Fig. 13.14. A magnitude characteristic with no overshoot is the magnitude characteristic of an *R-C* interstage shown by the solid curve in Fig. 13.14.

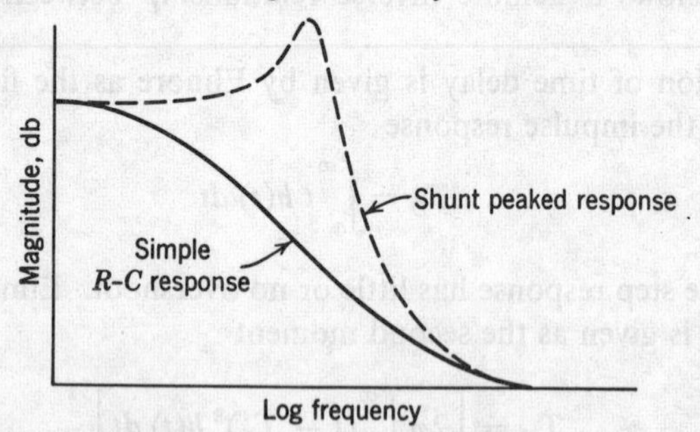

FIG. 13.14. Comparison of shunt-peaked and simple *R-C* magnitudes.

---

[10] W. C. Elmore and M. Sands, *Electronics*, National Nuclear Energy Series, Div. V, 1, McGraw-Hill Book Company, New York, 1949, pp. 137–138.

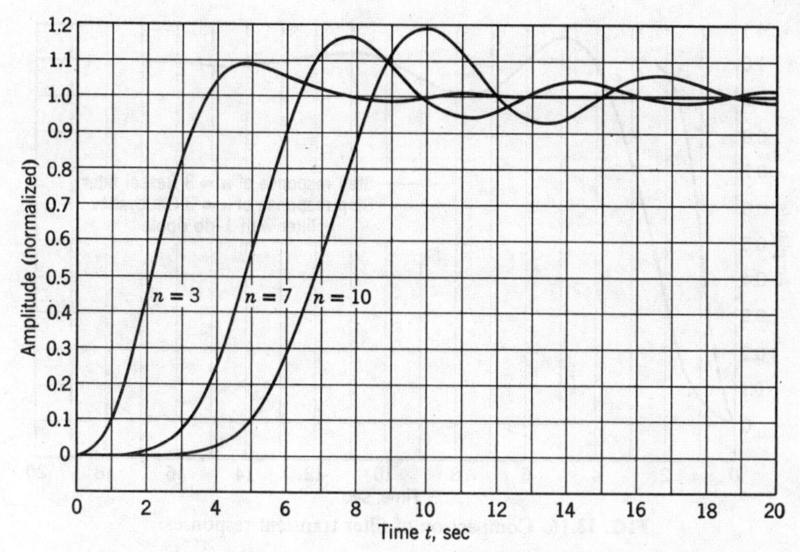

FIG. 13.15. Step response of normalized Butterworth low-pass filters.

The step responses of the $n = 3$, $n = 7$, and $n = 10$ Butterworth filters are shown in Fig. 13.15. Note that as n increases, the overshoot increases. This is because the higher order Butterworth filters have flatter magnitude characteristics (i.e., there is more gain at frequencies just below the cutoff).

Ringing is due to sharp cutoff in the filter magnitude response, and is accentuated by a rising gain characteristic preceding the discontinuity. The step response of an $n = 3$ Bessel (linear phase) filter is compared to the response of an $n = 3$ Chebyshev filter with 1-decibel ripple in Fig. 13.16. We cannot compare their rise times since the bandwidths of the two filters have not been adjusted to be equal. However, we can compare their ringing and settling times. The Chebyshev filter has a sharper cutoff, and therefore has more ringing and longer settling time than the Bessel filter. Note also the negligible overshoot of the Bessel filter that is characteristic of the entire class of Bessel filters.

The decision as to which filter is best depends upon the particular situation. In certain applications, such as for transmission of music, phase is not important. In these cases, the sharpness of cutoff may be the dominant factor so that the Chebyshev or the Optimum filter is better than the others. Suppose we were dealing with a pulse transmission system with the requirement that the output sequence have approximately the same shape as the input sequence, except for a time delay of $T = T_2 - T_1$, as shown in Fig. 13.17*a*. It is clear that a filter with a long rise time is not suitable, because the pulses would "smear" over each other

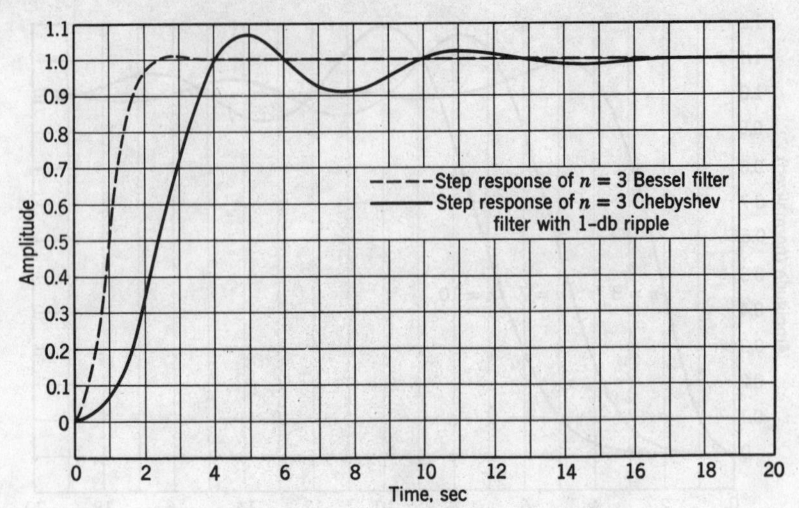

FIG. 13.16. Comparison of filter transient responses.

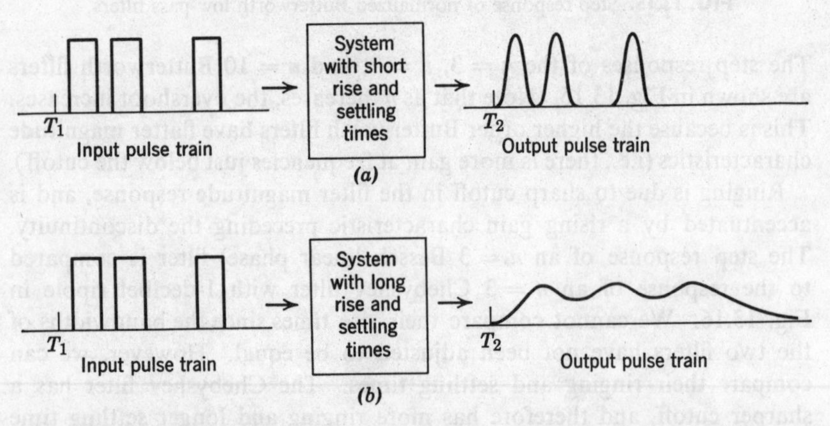

FIG. 13.17. Smearing of pulses in systems with long rise and setting times.

as seen in Fig. 13.17*b*. The same can be said for long settling times. Since a pulse transmission system must have linear phase to insure undistorted harmonic reconstruction at the receiver, the best filter for the system is a linear phase filter with small rise and settling times.

13.6 A METHOD TO REDUCE OVERSHOOT IN FILTERS

We present here a method to reduce the overshoot and ringing of a filter step response. The step response of a tenth-order Butterworth filter is shown in Fig. 13.18. It is seen that the overshoot is about 18%. We

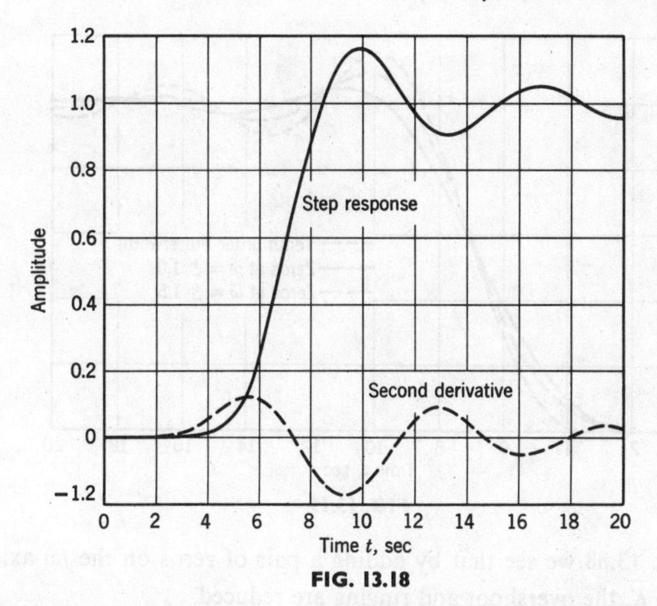

FIG. 13.18

note that after the first peak, the ringing of the step response has an approximate sinusoidal waveshape. Let us now consider the *second derivative* of the step response shown by the dashed curve in Fig. 13.18. Beyond the first peak of the step response, the second derivative is also (approximately) sinusoidal, and is negative when the step response is greater than unity, and positive when the step response is less than unity. If we *add* the second derivative to the step response, we reduce the overshoot and ringing.[11]

Suppose $\alpha(t)$ is the step response and $H(s)$ is the system function of the filter. The corrected step response can be written as

$$\alpha_1(t) = \alpha(t) + K\frac{d^2\alpha(t)}{dt^2} \tag{13.87}$$

where K is a real, positive constant. Taking the Laplace transform of Eq. 13.87, we have

$$\begin{aligned} \mathcal{L}[\alpha_1(t)] &= \frac{H(s)}{s} + Ks^2\left[\frac{H(s)}{s}\right] \\ &= (Ks^2 + 1)\frac{H(s)}{s} \end{aligned} \tag{13.88}$$

[11] F. F. Kuo, "A Method to Reduce Overshoot in Filters," *Trans. IRE on Circuit Theory*, **CT-9,** No. 4, December 1962, 413–414.

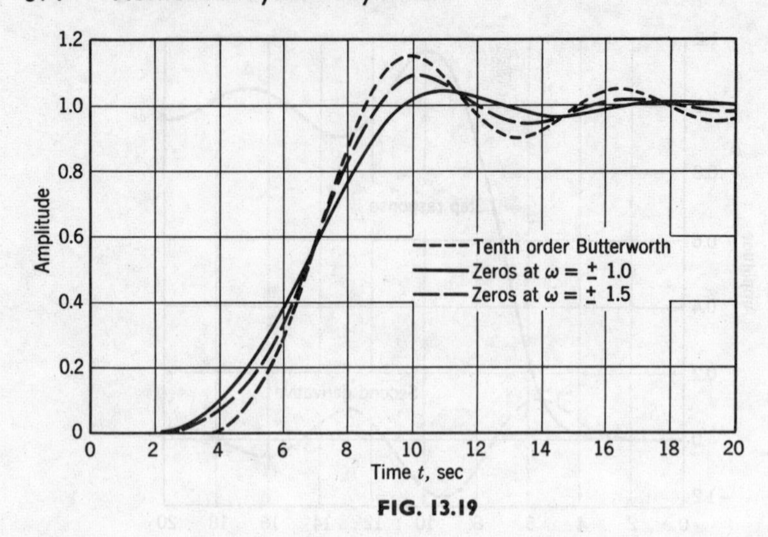

FIG. 13.19

From Eq. 13.88 we see that by adding a pair of zeros on the $j\omega$ axis at $s = \pm j/\sqrt{K}$, the overshoot and ringing are reduced.

For low-pass filters, the factor $1/\sqrt{K}$ must in general be greater than the bandwidth of the system. For normalized Butterworth filters the bandwidth is $\omega = 1$ so that $K \leq 1$. The factor K also controls the amount of overshoot reduction. If K is too small, adding the zeros on the $j\omega$ axis

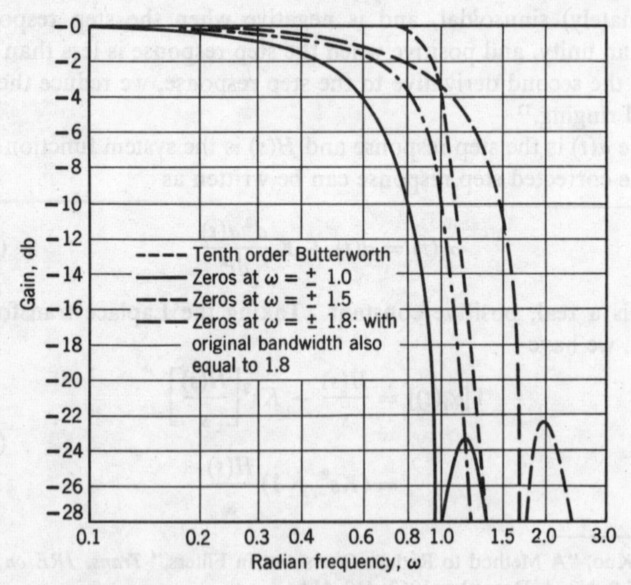

FIG. 13.20

will have negligible effect. Therefore the zeros should be added somewhere near the band edge. Figure 13.19 shows the effects of adding zeros at $\omega = \pm 1$ (i.e., right at the band edge), and at $\omega = 1.5$. We see that the further away the zeros are placed from the band edge, the less effect they will have. The addition of the zeros will decrease the 3-decibel bandwidth of the filter, however, as seen in Fig. 13.20. Therefore, a compromise must be reached between reduction in bandwidth and reduction in overshoot.

An effective way to overcome this difficulty is to scale up the bandwidth by a factor of, for example, 1.8. Then the zeros are placed at $\omega = \pm 1.8$, which will reduce the 3-decibel bandwidth to approximately its original figure $\omega = 1.0$, as shown in Fig. 13.20. The overshoot, however, will be reduced as though the zeros were at the band edge.

13.7 A MAXIMALLY FLAT DELAY AND CONTROLLABLE MAGNITUDE APPROXIMATION

In this section we will examine an interesting result, which is due to Budak.[12] The result deals with linear phase approximation with *controllable* magnitude. In Section 13.4 we discussed approximation of a flat delay using Bessel polynomials. The resulting rational approximant was an all-pole function (all transmission zeros at $s = \infty$), whose denominator was a Bessel polynomial. There was no control of the magnitude using the all-pole approximant. In Budak's method the magnitude *is* controllable, while the *phase is as linear* as the standard Bessel approximation.

Budak's approximation is obtained by introducing the parameter k to split e^{-s} into two parts such that

$$e^{-s} = \frac{e^{-ks}}{e^{-(k-1)s}} \qquad 0 < k \leq 1 \tag{13.89}$$

and then approximate independently e^{-ks} and $e^{-(k-1)s}$ with all-pole Bessel polynomial approximations. Thus the resulting approximation for e^{-s} will have Bessel polynomials for both numerator and denominator. The poles of the $e^{-(k-1)s}$ approximant will be the zeros in the final approximant, while the poles of the e^{-ks} approximant remain as poles in the final approximant. For realizability, the degree of the $e^{-(k-1)s}$ approximant should be less than the degree of the e^{-ks} approximant.

---

[12] A. Budak, "A Maximally Flat Phase and Controllable Magnitude Approximation," *Trans. of IEEE on Circuit Theory*, **CT-12,** No. 2, June 1965, 279.

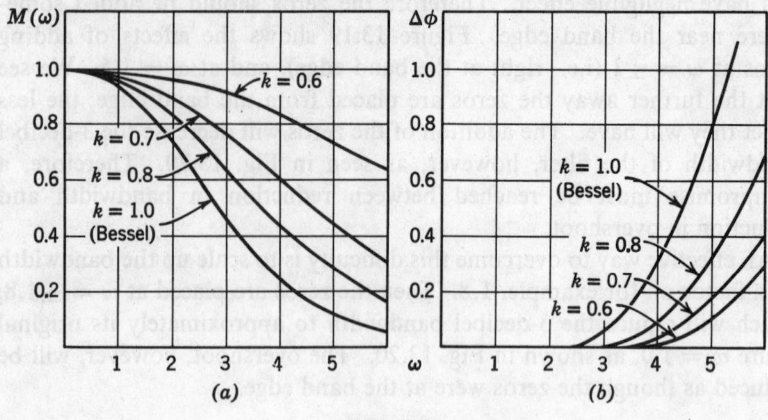

FIG. 13.21

As an example, consider the approximation with three zeros and four poles.

$$e^{-ks} \cong \frac{105}{(ks)^4 + 10(ks)^3 + 45(ks)^2 + 105(ks) + 105} \tag{13.90}$$

$$e^{-(k-1)s} \cong \frac{15}{[(k-1)s]^3 + 6[(k-1)s]^2 + 15[(k-1)s] + 15} \tag{13.91}$$

We then perform the operations as indicated by Eq. 13.89 to obtain

$$e^{-s} \cong \frac{7\{[(k-1)s]^3 + 6[(k-1)s]^2 + 15[(k-1)s] + 15\}}{(ks)^4 + 10(ks)^3 + 45(ks)^2 + 105(ks) + 105} \tag{13.92}$$

In Fig. 13.21*a* the magnitude characteristic of Eq. 13.92 is plotted with k as a parameter. The phase characteristic is given in terms of deviation of phase from linearity $\Delta\phi = \omega - \phi(\omega)$ and is shown in Fig. 13.21*b*. The improvement in phase linearity over the all-pole Bessel approximation $k = 1$ is shown by these curves. Note as the bandwidth is increased (k decreasing), phase linearity is improved. Figure 13.22 shows the step response of Eq. 13.92 also with k as a parameter. Since the effect of decreasing k increases bandwidth, the corresponding effect in the time domain is to decrease rise time.

Budak also observes that as k decreases from unity, the poles and zeros migrate to keep the phase linear. The zeros move inward from infinity along radial lines, while the poles move outward along radial lines.

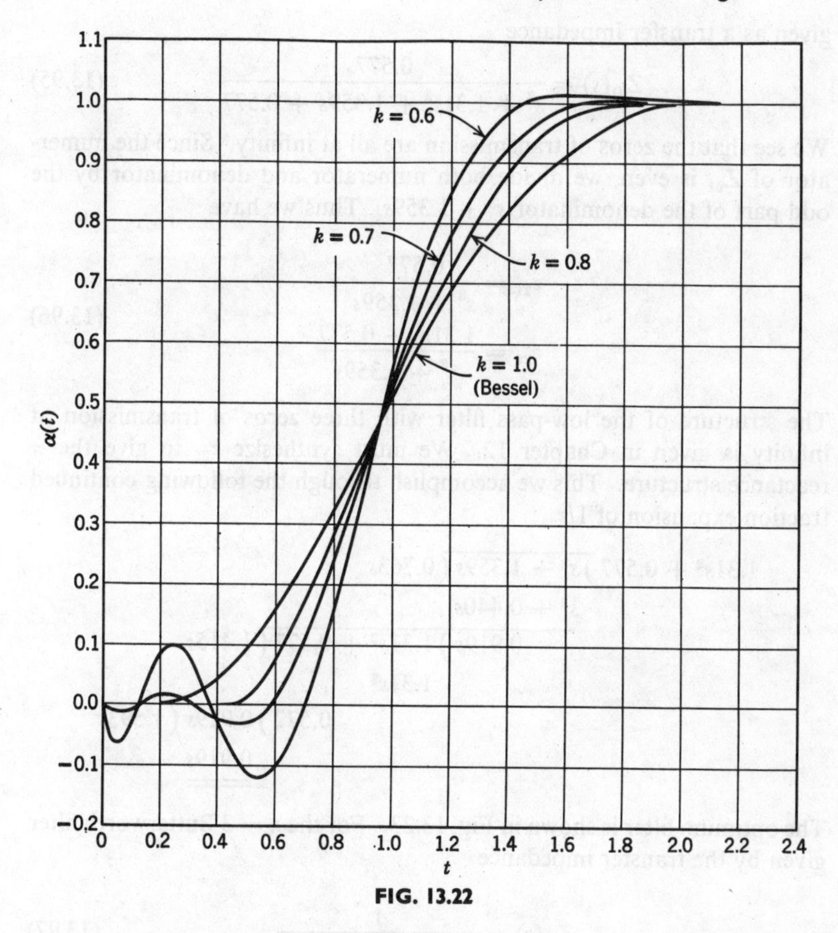

FIG. 13.22

13.8 SYNTHESIS OF LOW-PASS FILTERS

Given the system function of the low-pass filter as derived by the methods described in Section 13.4, we can proceed with the synthesis of the filter network. If we consider the class of filters terminated in a 1-Ω load, and if we let the system function be a transfer impedance,

$$Z_{21}(s) = \frac{z_{21}}{1 + z_{22}} \tag{13.93}$$

or a transfer admittance

$$Y_{21}(s) = \frac{y_{21}}{1 + y_{22}} \tag{13.94}$$

we can synthesize the low-pass filter according to methods given in Chapter 12. For example, consider the $n = 3$ Optimum (L) filter function given as a transfer impedance

$$Z_{21}(s) = \frac{0.577}{s^3 + 1.31s^2 + 1.359s + 0.577} \tag{13.95}$$

We see that the zeros of transmission are all at infinity. Since the numerator of Z_{21} is even, we divide both numerator and denominator by the odd part of the denominator $s^3 + 1.359s$. Thus we have

$$\begin{aligned} z_{21} &= \frac{0.577}{s^3 + 1.359s} \\ z_{22} &= \frac{1.31s^2 + 0.577}{s^3 + 1.359s} \end{aligned} \tag{13.96}$$

The structure of the low-pass filter with three zeros of transmission at infinity is given in Chapter 12. We must synthesize z_{22} to give the π reactance structure. This we accomplish through the following continued fraction expansion of $1/z_{22}$:

$$\begin{array}{l} 1.31s^2 + 0.577\,\overline{)\,s^3 + 1.359s\,(}\,0.763s \\ \qquad\qquad\quad\; \underline{s^3 + 0.440s} \\ \qquad\qquad\qquad 0.919s\,\overline{)\,1.31s^2 + 0.577\,(}\,1.415s \\ \qquad\qquad\qquad\qquad\quad\; \underline{1.31s^2} \\ \qquad\qquad\qquad\qquad\qquad\quad 0.577\,\overline{)\,0.919s\,(}\,1.593s \\ \qquad\qquad\qquad\qquad\qquad\qquad\quad \underline{\underline{0.919s}} \end{array}$$

The optimum filter is shown in Fig. 13.23. For the $n = 3$ Butterworth filter given by the transfer impedance

$$Z_{21}(s) = \frac{1}{s^3 + 2s^2 + 2s + 1} \tag{13.97}$$

we have

$$z_{21}(s) = \frac{1}{s^3 + 2s}\;; \qquad z_{22}(s) = \frac{2s^2 + 1}{s^3 + 2s} \tag{13.98}$$

We then synthesize $z_{22}(s)$ by a continued fraction expansion to give the filter shown in Fig. 13.23.

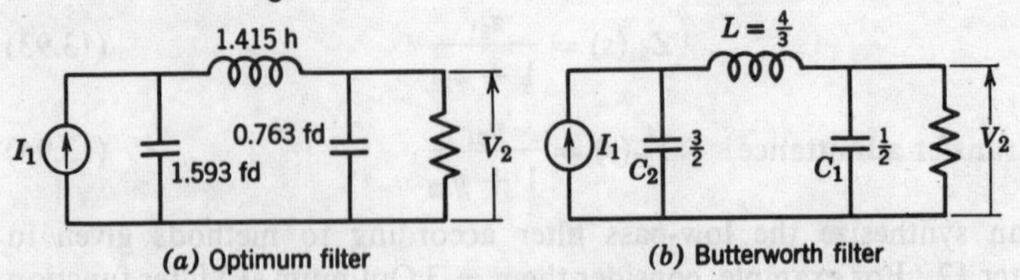

(a) Optimum filter (b) Butterworth filter

FIG. 13.23

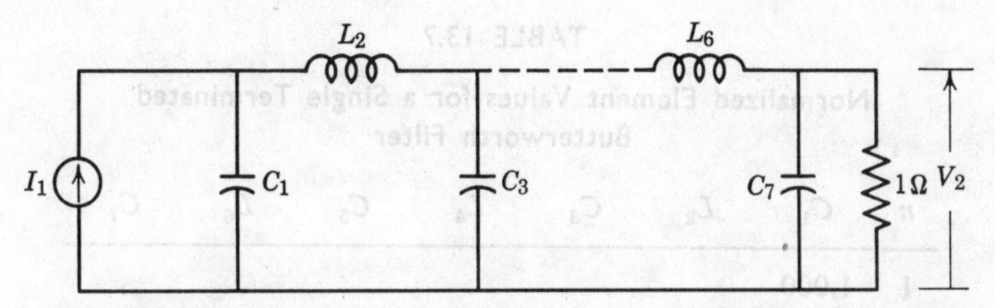

FIG. 13.24. Canonical form for filters described in Tables 13.7, 13.8, and 13.9.

In Tables 13.7, 13.8, 13.9 are listed element values (up to $n = 7$) for single-terminated Butterworth, Chebyshev (1-decibel ripple), and Bessel filters, respectively.[13] These apply to the canonical realization for a transfer impedance $Z_{21}(s)$ shown in Fig. 13.24. If a $Y_{21}(s)$ realization is desired, we simply replace all shunt capacitors by series inductors and vice versa. The element values all carry over.

In Chapter 14, we will consider some examples of synthesis of double terminated filters. To stimulate the curiosity of the reader, note that the voltage-ratio transfer function V_2/V_0 of the network in Fig. 13.25 is precisely the $n = 3$ Butterworth function.

Recall that in Chapter 12, when we cascaded two constant-resistance networks, the overall system function $H_0(s)$ was the product of the individual system functions $H_1(s)\ H_2(s)$. We can apply this property to networks which are not constant-resistance if we place an isolation amplifier between the networks, as shown in Fig. 13.26. Since pentodes provide the necessary isolation, our task is simplified to the design of the individual structures $H_1(s), H_2(s), \ldots, H_n(s)$, which we call *interstage* networks.

Some common interstage structures are shown in Fig. 13.27. In Fig. 13.27*a* a structure known as the *shunt-peaked* network is shown. The transfer impedance of the shunt-peaked network is

$$Z_{21}(s) = \frac{1}{C}\,\frac{s + R/L}{s^2 + sR/L + (1/LC)} \tag{13.99}$$

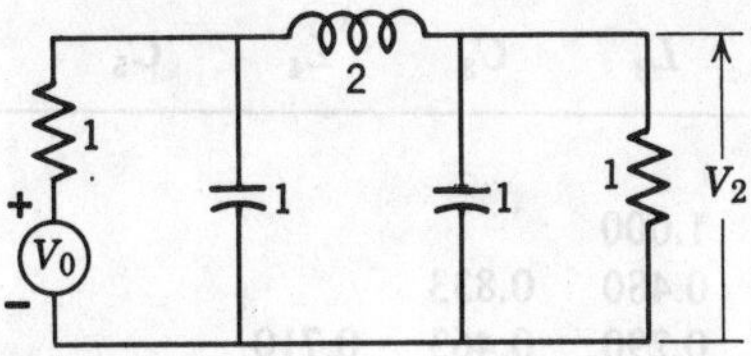

FIG. 13.25. $n = 3$ double-terminated Butterworth filter.

[13] More extensive tables are given in L. Weinberg's excellent book, *Network Analysis and Synthesis*, Chapter 13, McGraw-Hill Book Company, 1962.

TABLE 13.7

Normalized Element Values for a Single Terminated Butterworth Filter

n	C_1	L_2	C_3	L_4	C_5	L_6	C_7
1	1.000						
2	0.707	1.414					
3	0.500	1.333	1.500				
4	0.383	1.082	1.577	1.531			
5	0.309	0.894	1.382	1.694	1.545		
6	0.259	0.758	1.202	1.553	1.759	1.553	
7	0.222	0.656	1.055	1.397	1.659	1.799	1.558

TABLE 13.8

Normalized Element Values for a Single Terminated Chebyshev Filter with 1-db Ripple

n	C_1	L_2	C_3	L_4	C_5	L_6	C_7
1	0.509						
2	0.911	0.996					
3	1.012	1.333	1.509				
4	1.050	1.413	1.909	1.282			
5	1.067	1.444	1.994	1.591	1.665		
6	1.077	1.460	2.027	1.651	2.049	1.346	
7	1.083	1.496	2.044	1.674	2.119	1.649	1.712

TABLE 13.9

Normalized Element Values for a Single Terminated Bessel Filter

n	C_1	L_2	C_3	L_4	C_5	L_6	C_7
1	1.000						
2	0.333	1.000					
3	0.167	0.480	0.833				
4	0.100	0.290	0.463	0.710			
5	0.067	0.195	0.310	0.422	0.623		
6	0.048	0.140	0.225	0.301	0.382	0.560	
7	0.036	0.106	0.170	0.229	0.283	0.349	0.511

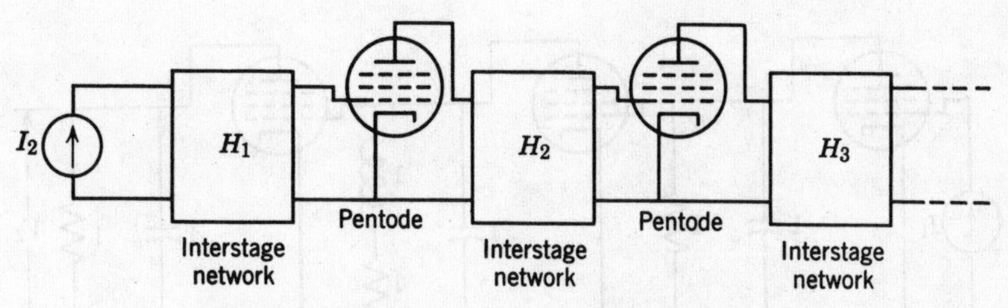

FIG. 13.26. Pentodes used as isolation amplifiers.

We see that $Z_{21}(s)$ has a real zero and a pair of poles which may be complex depending upon the values of R, L, and C. In Fig. 13.27b, a simple R-C interstage is shown, whose transfer impedance is

$$Z_{21}(s) = \frac{1}{C}\frac{1}{s + 1/RC} \tag{13.100}$$

Observe that all the filter transfer functions considered up to this point are made up of pairs of conjugate poles and simple poles on the $-\sigma$ axis. It is clear that if we cascade shunt-peaked stages and R-C stages, we can adjust the R, L, and C elements to give the desired response characteristic. The only problem is to cancel the finite zero of the shunt-peaked stage. For example, if we wish to design an amplifier with an $n = 3$ low-pass Butterworth characteristic, we first break up the system function into complex pole pairs and real pole terms, as given by

$$\begin{aligned} Z_{21}(s) &= \frac{1}{(s^2 + s + 1)(s + 1)} \\ &= \frac{s + 1}{s^2 + s + 1}\frac{1}{s + 1}\cdot\frac{1}{s + 1} \end{aligned} \tag{13.101}$$

We then associate the individual factors with shunt-peaked or simple R-C stages and solve for the element values. The $n = 3$ Butterworth amplifier is given in Fig. 13.28.

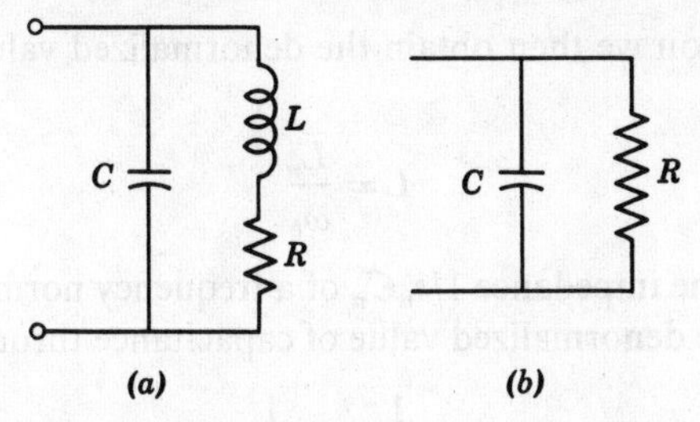

FIG. 13.27. (a) Shunt-peaked interstage. (b) R-C interstage.

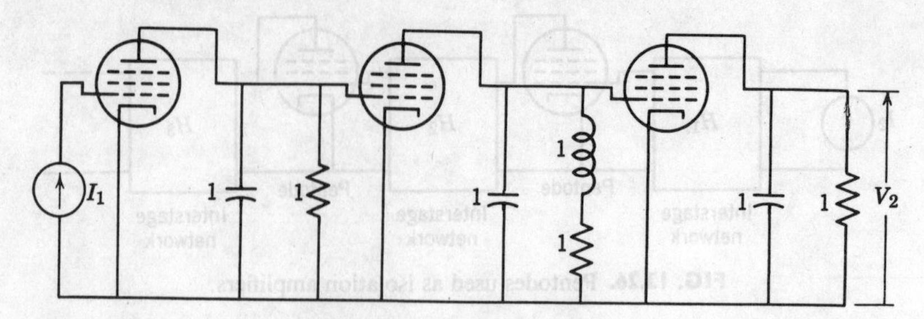

FIG. 13.28. Butterworth amplifier.

13.9 MAGNITUDE AND FREQUENCY NORMALIZATION

In Section 13.8, we discussed the synthesis of low-pass filters with a cutoff frequency of 1 rad/sec and a load impedance of 1 Ω. Filters designed with these restrictions are considered to be *normalized* in both cutoff frequency and impedance level. We will now discuss methods whereby the normalized filters can be converted into filters which meet arbitrary cutoff frequency and impedance level specifications. Let us denote by a subscript n the normalized frequency variable s_n and the normalized element values L_n, R_n, and C_n. The normalized frequency variable s_n is related to the actual frequency s by the relation

$$s_n = \frac{s}{\omega_0} \tag{13.102}$$

where ω_0, the normalizing constant, is dimensionless and is often taken to be the actual cutoff frequency.

Since the impedance of an element remains invariant under frequency normalization, we obtain the actual element values from the normalized values by setting the impedances in the two cases equal to each other. For example, for an inductor, we have

$$s_n L_n = sL = \omega_0 s_n L \tag{13.103}$$

From this equation we then obtain the denormalized value of inductance as

$$L = \frac{L_n}{\omega_0} \tag{13.104}$$

Similarly, from the impedance $1/s_n C_n$ of a frequency normalized capacitor C_n, we obtain the denormalized value of capacitance through the equation

$$\frac{1}{s_n C_n} = \frac{1}{sC} \tag{13.105}$$

so that the actual value of the capacitance is

$$C = \frac{C_n}{\omega_0} \tag{13.106}$$

Since resistances, ideally, are independent of frequency, they are unaffected by frequency normalization.

Consider, next, impedance denormalization. Suppose the actual impedance level should be R_0 ohms instead of 1 Ω. Then a denormalized impedance Z is related to a normalized impedance Z_n by

$$Z = R_0 Z_n \tag{13.107}$$

where R_0 is taken to be dimensionless here. Thus, for a normalized resistor R_n, the denormalized (actual) resistance is

$$R = R_0 R_n \tag{13.108}$$

For an inductance, the corresponding relationship is

$$sL = R_0(sL_n) \tag{13.109}$$

so that the actual inductance value is

$$L = R_0 L_n \tag{13.110}$$

Similarly, for a capacitor we have

$$\frac{1}{sC} = \frac{R_0}{sC_n} \tag{13.111}$$

so that the actual capacitance is

$$C = \frac{C_n}{R_0} \tag{13.112}$$

For combined frequency and magnitude denormalization, we simply combine the two sets of equations to give

$$\begin{aligned} R &= R_0 R_n \\ C &= \frac{C_n}{R_0 \omega_0} \\ L &= \frac{R_0 L_n}{\omega_0} \end{aligned} \tag{13.113}$$

Let us consider an actual example in design. In Section 13.8, we synthesized a transfer impedance Z_{21} with an $n = 3$ Butterworth amplitude characteristic with a cutoff frequency of 1 rad/sec and a load impedance of 1 Ω. Let us redesign this filter for a cutoff frequency of 10^4 rad/sec to work

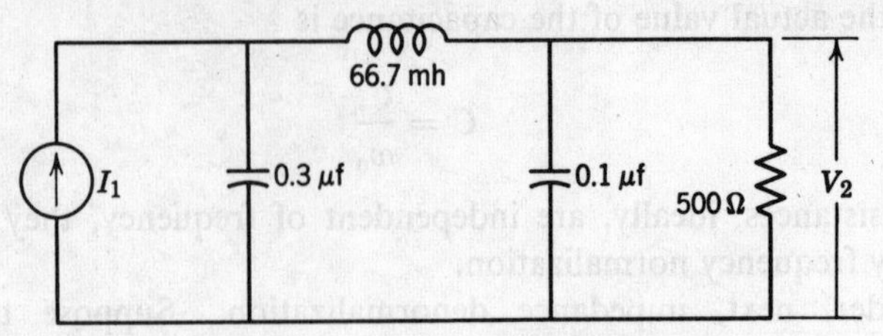

FIG. 13.29. Denormalized low-pass filter.

into a load of 500 Ω. From the original network in Fig. 13.23, we take the element values and denormalize with the normalizing factors, $\omega_0 = 10^4$ and $R_0 = 500$.

Then the denormalized element values are

$$\begin{aligned} R &= 500R_L = 500 \\ C_1 &= \frac{\frac{1}{2}}{500(10^4)} = 0.1\ \mu\text{f} \\ L &= \frac{\frac{4}{3}(500)}{10{,}000} = 0.0067\ \text{h} \\ C_2 &= \frac{\frac{3}{2}}{500(10^4)} = 0.3\ \mu\text{f} \end{aligned} \tag{13.114}$$

The final design is shown in Fig. 13.29.

13.10 FREQUENCY TRANSFORMATIONS

Up to this point, we have discussed only the design of low-pass filters, while neglecting the equally important designs of high-pass, band-pass, and band-elimination filters. We will remedy this situation here, not by introducing new design procedures but through a technique known as a *frequency transformation*, whereby, beginning from a normalized low-pass filter, we can *generate* any other form of filter. Using frequency transformations, the elements of the normalized low-pass filter are changed into elements of a high-pass, band-pass, or band-elimination filter.

Analytically, a frequency transformation simply changes one *L-C* driving-point function into another *L-C* driving-point function. Therefore, the transformation equations must be *L-C* functions themselves. Also, since we proceed from normalized low-pass filters, the transformation equations include built-in frequency denormalization factors so that the resulting networks need only be scaled for impedance level. Consider the simplest transformation equation, that of low-pass to high-pass, which is

$$s = \frac{\omega_0}{s_n} \tag{13.115}$$

where s_n represents the normalized low-pass frequency variable, s is the regular frequency variable, and ω_0 is the cutoff frequency of the high-pass filter. In terms of real and imaginary parts, we have

$$\begin{aligned} \sigma + j\omega &= \frac{\omega_0}{\sigma_n + j\omega_n} \\ &= \frac{\omega_0(\sigma_n - j\omega_n)}{\sigma_n^{\,2} + \omega_n^{\,2}} \end{aligned} \tag{13.116}$$

Since we are interested principally in how the $j\omega_n$ axis maps into the $j\omega$ axis, we let $\sigma_n = 0$ so that

$$\omega = -\frac{\omega_0}{\omega_n} \tag{13.117}$$

which is the equation that transforms normalized low-pass filters to denormalized high-pass filters. From Eq. 13.117 we see that the point $\omega_n = \pm 1$ corresponds to the point $\omega = \pm\omega_0$. It is also clear that the transformation maps the segment $|\omega_n| \leq 1$ on to the segments defined by $\omega_0 \leq |\omega| \leq \infty$, as shown in Fig. 13.30.

Now let us see how the frequency transformations change the network elements. For convenience, let us denote the normalized low-pass network elements with a subscript n, the high-pass elements with a subscript h, the band-pass elements with a subscript b, and the band-elimination elements with a subscript e. For the low-pass to high-pass case, let us first consider the changes for the capacitor C_n. The transformation is given by the

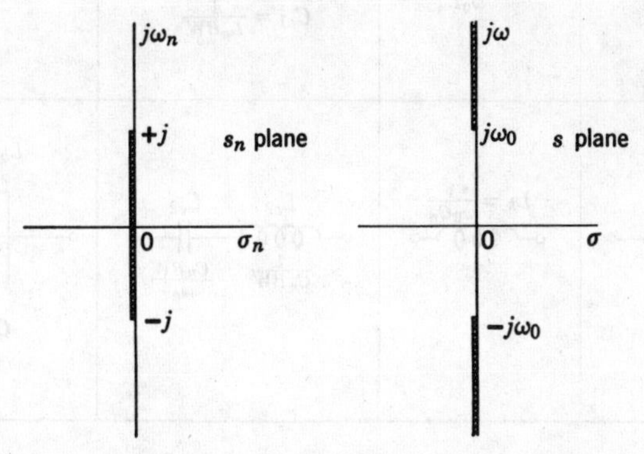

FIG. 13.30. Low-pass to high-pass transformation.

equation

$$\frac{1}{C_n s_n} = \frac{s}{\omega_0 C_n} \overset{\Delta}{=} L_h s \tag{13.118}$$

For the inductor L_n, we have

$$L_n s_n = L_n \frac{\omega_0}{s} \overset{\Delta}{=} \frac{1}{C_h s} \tag{13.119}$$

We observe that a capacitor changes into an inductor and an inductor changes into a capacitor in a low-pass to high-pass transformation (Fig. 13.31). The element values of the high-pass filter are given in terms of the normalized low-pass filter elements as

$$L_h = \frac{1}{\omega_0 C_n} \tag{13.120}$$

and

$$C_h = \frac{1}{\omega_0 L_n} \tag{13.121}$$

Consider the following example. From the normalized third-order Butterworth filter given in Fig. 13.23, let us design a corresponding high-pass filter with its cutoff frequency $\omega_0 = 10^6$ rad/sec and the impedance level of 500 Ω. From the low-pass filter, we can draw by inspection the

Low-pass	High-pass	Band-elimination	Band-pass
L_n	$C_h = \frac{1}{\omega_0 L_n}$	$L_{e1} = \frac{L_n BW}{\omega_0^2}$, $C_{e1} = \frac{1}{L_n BW}$	L_{b1}: $\frac{L_n}{BW}$, C_{b1}: $\frac{BW}{\omega_0^2 L_n}$
C_n	$L_h = \frac{1}{\omega_0 C_n}$	L_{e2}: $\frac{1}{C_n BW}$, C_{e2}: $\frac{C_n BW}{\omega_0^2}$	$L_{b2} = \frac{BW}{\omega_0^2 C_n}$, $C_{b2} = \frac{C_n}{BW}$

FIG. 13.31. Element changes resulting from frequency transformations.

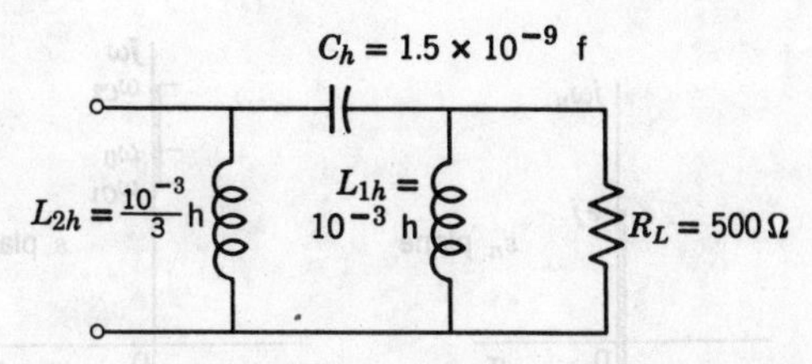

FIG. 13.32. Transformation of low-pass filter in Fig. 13.23 into high-pass filter.

high-pass-filter circuit shown in Fig. 13.32. Its element values are:

$$R_L = 500\ \Omega$$

$$L_{1h} = \frac{500}{10^6(\frac{1}{2})} = 10^{-3}\text{ h}$$

$$C_h = \frac{1}{(500)10^6(\frac{4}{3})} = 1.5 \times 10^{-9}\text{f} \qquad (13.122)$$

$$L_{2h} = \frac{500}{10^6(\frac{3}{2})} = 0.333 \times 10^{-3}\text{ h}$$

Next, let us examine the low-pass to band-pass transformation (also an *L-C* function):

$$s_n = \frac{\omega_0}{BW}\left(\frac{s}{\omega_0} + \frac{\omega_0}{s}\right) \qquad (13.123)$$

where, if ω_{C2} and ω_{C1} denote the upper and lower cutoff frequencies of the band-pass filter, BW is the bandwidth

$$BW = \omega_{C2} - \omega_{C1} \qquad (13.124)$$

and ω_0 is the geometric mean of ω_{C2} and ω_{C1}

$$\omega_0 = \sqrt{\omega_{C2}\omega_{C1}} \qquad (13.125)$$

The low-pass to band-pass transformation maps the segment $|\omega_n| \leq 1$ to the segments $\omega_{C2} \geq |\omega| \geq \omega_{C1}$, shown in Fig. 13.33. The normalized low-pass elements are then modified according to the following equations:

$$L_n s_n = \frac{L_n}{BW}s + \frac{\omega_0{}^2 L_n}{BW s}$$

$$\triangleq L_{b1}s + \frac{1}{C_{b1}s} \qquad (13.126)$$

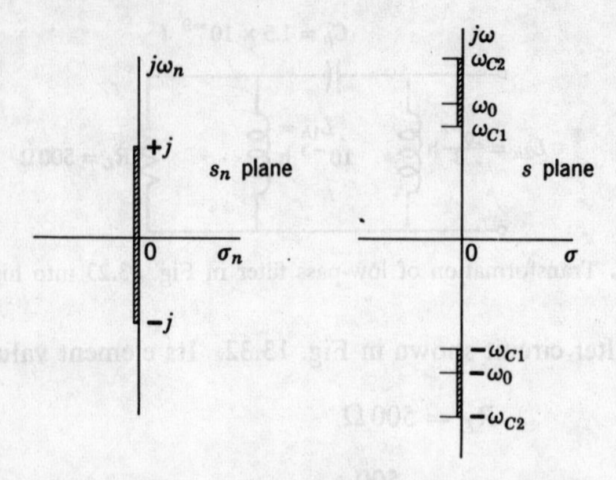

FIG. 13.33. Low-pass to band-pass transformation.

We note that the inductor L_n is transformed into a series-tuned tank, shown in Fig. 13.31, whose elements are given as

$$L_{b1} = \frac{L_n}{BW}$$
$$C_{b1} = \frac{BW}{\omega_0^{\,2} L_n} \tag{13.127}$$

The capacitor C_n is transformed into a parallel-tuned tank (Fig. 13.31), whose elements are

$$L_{b2} = \frac{BW}{\omega_0^{\,2} C_n}$$
$$C_{b2} = \frac{C_n}{BW} \tag{13.128}$$

Let us transform the third-order Butterworth low-pass filter in Fig. 13.23 into a band-pass filter with a 1-Ω impedance level, whose bandwidth is $BW = 6 \times 10^4$ rad/sec, and its band-pass is "centered" at $\omega_0 = 4 \times 10^4$ rad/sec. We draw the band-pass filter shown in Fig. 13.34 by the rules

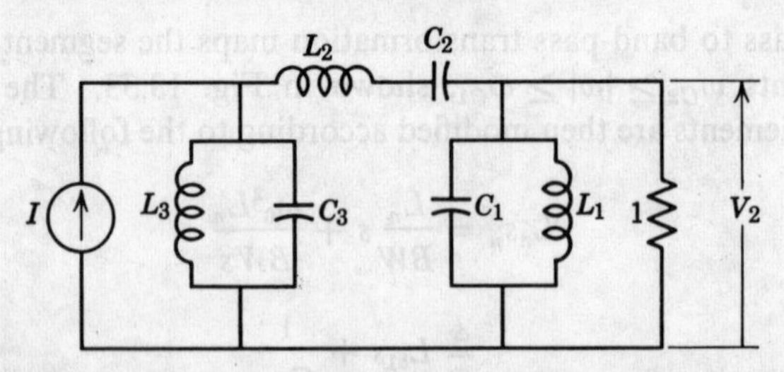

FIG. 13.34. Band-pass filter transformed from low-pass filter in Fig. 13.23.

given above. The element values of the band-pass filter are given in the following equations:

$$
\begin{aligned}
L_1 &= \frac{6 \times 10^4}{(4 \times 10^4)^2(\frac{1}{2})} = 0.75 \times 10^{-4}\text{ h} \\
C_1 &= \frac{\frac{1}{2}}{6 \times 10^4} = \frac{1}{12} \times 10^{-4}\text{ f} \\
L_2 &= \frac{\frac{4}{3}}{6 \times 10^4} = \frac{2}{9} \times 10^{-4}\text{ h} \\
C_2 &= \frac{6 \times 10^4}{(4 \times 10^4)^2(\frac{4}{3})} = \frac{9}{32} \times 10^{-4}\text{ f} \\
L_3 &= \frac{6 \times 10^4}{(4 \times 10^4)^2(\frac{3}{2})} = 0.25 \times 10^{-4}\text{ h} \\
C_3 &= \frac{\frac{3}{2}}{6 \times 10^4} = 0.25 \times 10^{-4}\text{ f}
\end{aligned}
\tag{13.129}
$$

Finally, the band-elimination filter is obtained through the transformation

$$
s_n = \frac{BW}{\omega_0\left(\dfrac{s}{\omega_0} + \dfrac{\omega_0}{s}\right)} \tag{13.130}
$$

where BW and ω_0 are defined in a manner similar to that for the band-pass filter. The transformation maps the segment of the $j\omega_n$ axis in Fig. 13.35*a* onto the segments shown on the $j\omega$ axis in Fig. 13.35*b*. For the low-pass

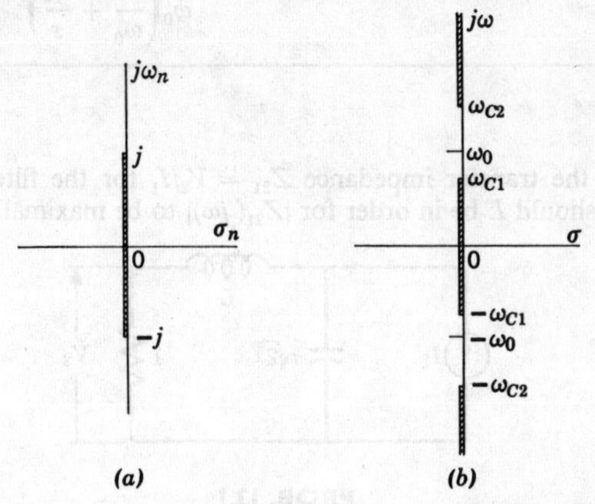

FIG. 13.35. Low-pass to band-elimination transformation.

to band-elimination transformation we, therefore, have the following element changes:

$$
\begin{aligned}
L_n s_n &= \frac{1}{(s/L_n BW) + (\omega_0{}^2/L_n BW s)} \\
&\triangleq \frac{1}{C_{e1}s + (1/L_{e1}s)} \\
\frac{1}{C_n s_n} &= \frac{\omega_0}{C_n BW}\left(\frac{s}{\omega_0} + \frac{\omega_0}{s}\right) \\
&\triangleq L_{e2}s + \frac{1}{C_{e2}s}
\end{aligned}
\tag{13.131}
$$

Observe that the normalized low-pass inductor goes into a parallel-tuned circuit and the capacitor C_n goes into a series-tuned circuit, as shown in Fig. 13.31. In Table 13.10, we have a composite summary of the various transformations.

TABLE 13.10

Table of Various Frequency Transformations

Transformation Low-Pass to	Equation
High-pass	$s_n = \dfrac{\omega_0}{s}$
Band-pass	$s_n = \dfrac{\omega_0}{BW}\left(\dfrac{s}{\omega_0} + \dfrac{\omega_0}{s}\right)$
Band-elimination	$s_n = \dfrac{BW}{\omega_0\left(\dfrac{s}{\omega_0} + \dfrac{\omega_0}{s}\right)}$

Problems

13.1 Find the transfer impedance $Z_{21} = V_2/I_1$ for the filter shown in the figure. What should L be in order for $|Z_{21}(j\omega)|$ to be maximally flat?

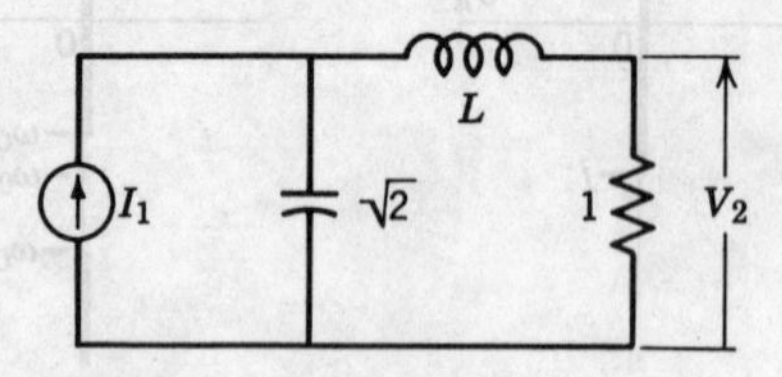

PROB. 13.1

13.2 Find the poles of system functions with $n = 3$, $n = 4$, and $n = 5$ Butterworth characteristics. (Do not use the tables.)

13.3 Show that the half-power point of a Chebyshev low-pass amplitude response is at $\omega = \cosh \beta_k$ for $\epsilon \ll 1$.

13.4 Determine the system function for the following filter specifications:
(*a*) Ripple of $\frac{1}{2}$ db in band $|\omega| \leq 1$;
(*b*) at $\omega = 3$, amplitude is down 30 db.

13.5 Compare the slopes at $\omega = 1$ of the following polynomials (for $n = 3$):

(*a*) $$f(\omega^2) = \omega^{2n}$$

(*b*) $$f(\omega^2) = \tfrac{1}{2}C_n(2\omega^2 - 1) + \tfrac{1}{2} = C_n^2(\omega)$$

(*c*) $$f(\omega^2) = L_n(\omega^2).$$

13.6 Determine the polynomials $L_4(\omega^2)$ and $L_5(\omega^2)$.

13.7 Expand cosh s and sinh s into power series and find the first four terms of the continued fraction expansion of cosh s/sinh s. Truncate the expansion at $n = 4$ and show that $H(s) = K_0/B_4(s)$.

13.8 Synthesize the $n = 3$ linear phase filter as a transfer impedance terminated in a 1-Ω load.

13.9 Synthesize the low-pass filter, which, when terminated in a 1-Ω resistor, will have a transfer admittance whose poles are shown in the figure.

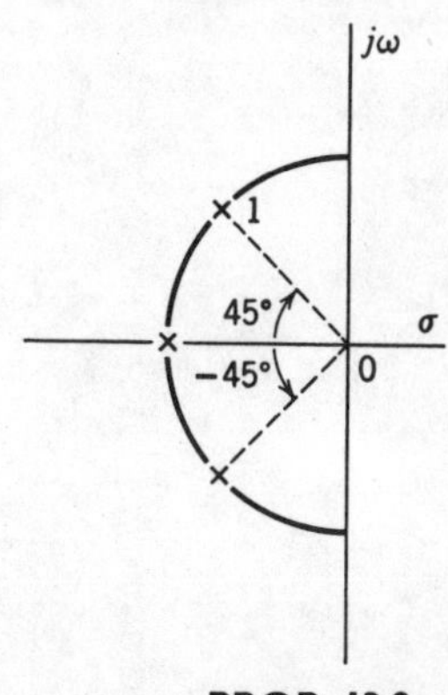

PROB. 13.9.

13.10 Determine the asymptotic rate of falloff in the stop band of: (*a*) optimum filters; (*b*) linear phase filters.

13.11 Synthesize the $n = 3$ and $n = 4$ Butterworth responses as transfer impedances terminated in a load of 600 Ω with a cutoff frequency of 10^6 rad/sec.

13.12 Synthesize a Chebyshev low-pass filter to meet the following specifications:

(*a*) load resistor, $R_L = 600$ Ω
(*b*) $\frac{1}{2}$-db ripple within pass band
(*c*) cutoff frequency $= 5 \times 10^5$ rad/sec
(*d*) at 1.5×10^6 rad/sec, the magnitude must be down 30 db.

13.13 Synthesize $n = 3$ Optimum and linear phase filters to meet the following specifications:

(*a*) load resistor $= 10^3\ \Omega$
(*b*) cutoff frequency $= 10^6$ rad/sec.

13.14 Design an $n = 4$ Butterworth amplifier with the following specifications:

(*a*) impedance level $= 500\ \Omega$
(*b*) cutoff frequency $= 10^3$ rad/sec.

13.15 Synthesize a high-pass filter for a given transfer admittance terminated in a 10^3-Ω load, whose amplitude characteristic is Optimum (L), with a cutoff frequency of $\omega_0 = 10^4$ rad/sec.

13.16 Synthesize: (*a*) a band-pass filter; (*b*) a band-elimination filter, with maximally flat ($n = 4$) amplitude response with $\omega_{c2} = 8 \times 10^4$ and $\omega_{c1} = 2 \times 10^4$.

chapter 14

The scattering matrix

14.1 INCIDENT AND REFLECTED POWER FLOW

In this chapter, we will devote our attention to certain *power* relationships in one- and two-port networks. The characterization of a network in terms of power instead of the conventional voltage-current description is a helpful analytical tool used by transmission engineers. It is especially important in microwave transmission problems where circuits can no longer be given in terms of lumped R, L, and C elements. In the power-flow description, we are concerned with the power into the network, which we call the *incident* power and the power reflected back from the load, which is the *reflected* power. A convenient description of the network in terms of incident and reflected power is given by the scattering matrix, which is the main topic of discussion in this chapter.

It is convenient to think of incident and reflected power when dealing with transmission lines. Therefore, we will briefly review some concepts in transmission line theory. For a more comprehensive treatment of transmission lines, the reader is referred to any standard text on wave propagation.[1] Consider the transmission line shown in Fig. 14.1. The voltage at any point down the line is a function of x, the distance from the source. The parameters which describe the transmission line are given in the following:

R = resistance per unit length
G = conductance per unit length
L = inductance per unit length
C = capacitance per unit length

[1] See, for example, E. C. Jordan, *Electromagnetic Waves and Radiating Systems*, Prentice-Hall, Englewood Cliffs, New Jersey, 1950.

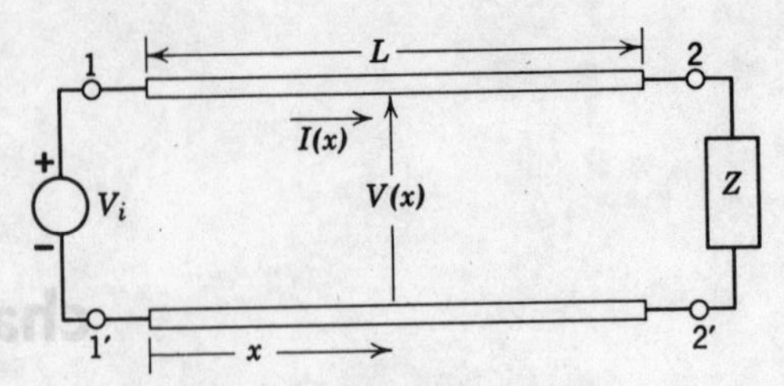

FIG. 14.1. Transmission line.

Given these parameters, we can now define the impedance per unit length as

$$Z = R + j\omega L \tag{14.1}$$

and the admittance per unit length as

$$Y = G + j\omega C \tag{14.2}$$

The *characteristic impedance* Z_0 of the line is given in terms of Z and Y as

$$Z_0 = \sqrt{Z/Y} \tag{14.3}$$

and the *propagation constant* is

$$\gamma = \sqrt{ZY} \tag{14.4}$$

With these definitions in mind, let us turn to the general equations for the current and voltage at any point x down the line

$$\begin{aligned} V(x) &= V_i e^{-\gamma x} + V_r e^{\gamma x} \\ I(x) &= I_i e^{-\gamma x} - I_r e^{\gamma x} \\ &= \frac{V_i}{Z_0} e^{-\gamma x} - \frac{V_r}{Z_0} e^{\gamma x} \end{aligned} \tag{14.5}$$

The terms with the subscript i refer to the incident wave at point x and the terms with subscript r refer to the reflected wave at x. Solving Eqs. 14.5 simultaneously, we obtain explicit expressions for the incident and reflected waves

$$\begin{aligned} V_i e^{-\gamma x} &= \tfrac{1}{2}[V(x) + Z_0 I(x)] \\ V_r e^{\gamma x} &= \tfrac{1}{2}[V(x) - Z_0 I(x)] \end{aligned} \tag{14.6}$$

Consider the case when a transmission line of length L is terminated in its characteristic impedance, that is,

$$\frac{V(L)}{I(L)} = Z_0 \tag{14.7}$$

Then we see that the reflected wave is zero.

$$V_r e^{\gamma L} = 0 \tag{14.8}$$

Since $e^{\gamma L}$ cannot be zero, we see that the coefficient V_r is identically zero for this case. As a result, the reflected wave at any point x is zero. Also, the impedance at any point x down the line is equal to Z_0 as seen from Eq. 14.5 with $V_r = 0$. With these brief thoughts of transmission lines in mind, let us turn our attention to the main topic of this chapter, namely, the *scattering parameters.*

14.2 THE SCATTERING PARAMETERS FOR A ONE-PORT NETWORK

For the one-port network shown in Fig. 14.2*a*, consider the following definitions. The *incident parameter a* is defined as

$$a = \frac{1}{2}\left(\frac{V}{\sqrt{R_0}} + \sqrt{R_0}I\right) \tag{14.9}$$

and the *reflected parameter b*, is defined as

$$b = \frac{1}{2}\left(\frac{V}{\sqrt{R_0}} - \sqrt{R_0}I\right) \tag{14.10}$$

where R_0 is an arbitrary, positive, dimensionless constant called the *reference impedance* factor. For the transmission line described in Section 14.1, if the characteristic impedance $Z_0 = R_0$, then we can describe the incident parameter in terms of the incident voltage as

$$a = \frac{V_i e^{-\gamma x}}{\sqrt{Z_0}} \tag{14.11}$$

Similarly, b can be expressed in terms of the reflected wave as

$$b = \frac{V_r e^{\gamma x}}{\sqrt{Z_0}} \tag{14.12}$$

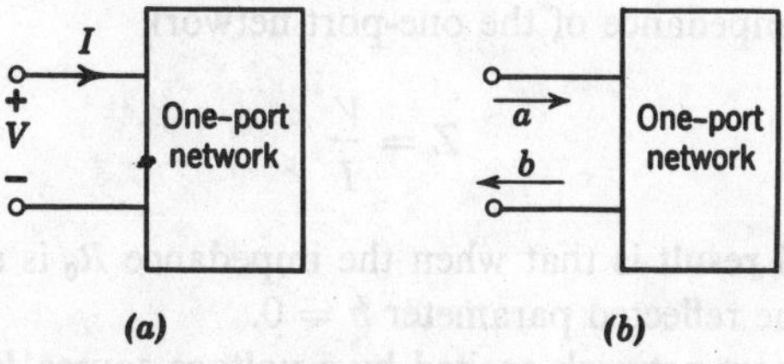

FIG. 14.2. Scattering parameters of a one-port network.

Thus we see that the parameters a and b do indeed describe an incident-reflected wave relationship in a one-port network, as depicted in Fig. 14.2b. To give further meaning to a and b, consider the power dissipated by the one-port network

$$P = \tfrac{1}{2}\text{Re}\, VI^* \tag{14.13}$$

where I^* denotes the complex conjugate of I. From Eqs. 14.9 and 14.10 we solve for V and I in terms of the incident and reflected parameters to give

$$\begin{aligned} V &= (a + b)\sqrt{R_0} \\ I &= \frac{a - b}{\sqrt{R_0}} \end{aligned} \tag{14.14}$$

Then the power dissipated by the one-port network is

$$\begin{aligned} P &= \tfrac{1}{2}(aa^* - bb^*) \\ &= \tfrac{1}{2}(|a|^2 - |b|^2) \end{aligned} \tag{14.15}$$

where, again, the asterisk denotes complex conjugate. The term $\tfrac{1}{2}aa^*$ can be interpreted as the power incident, while $\tfrac{1}{2}bb^*$ can be regarded as the power reflected. The difference yields the power dissipated by the one-port network.

The incident parameter a and the reflected parameter b are related by the equation

$$b = Sa \tag{14.16}$$

where S is called the *scattering element* or, more commonly, the *reflection coefficient*. From the definitions of a and b we can make the following substitution:

$$\left(\frac{V}{\sqrt{R_0}} - \sqrt{R_0}I\right) = S\left(\frac{V}{\sqrt{R_0}} + \sqrt{R_0}I\right) \tag{14.17}$$

Solving for S, we obtain

$$S = \frac{Z - R_0}{Z + R_0} \tag{14.18}$$

where Z is the impedance of the one-port network

$$Z = \frac{V}{I} \tag{14.19}$$

A further useful result is that when the impedance R_0 is set equal to the impedance Z, the reflected parameter $b = 0$.

For the one-port network excited by a voltage source V_g with a source resistor R_g, as depicted in Fig. 14.3, we will show that when we choose the reference impedance R_0 to be equal to R_g

$$a = \frac{V_g}{2\sqrt{R_g}} \qquad (14.20)$$

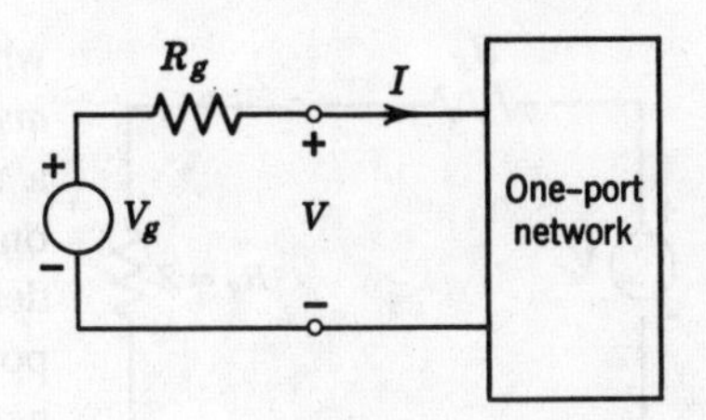

FIG. 14.3

The proof follows. By definition, the incident parameter a is given as

$$a = \frac{1}{2}\left(\frac{V}{\sqrt{R_0}} + \sqrt{R_0}I\right) \qquad (14.21)$$

From Fig. 14.3, we have

$$V_g - IR_g = V \qquad (14.22)$$

and

$$I = \frac{V_g}{R_g + Z} \qquad (14.23)$$

Substituting these equations for V and I into the expression for a in Eq. 14.21, we obtain

$$\begin{aligned} a &= \frac{1}{2\sqrt{R_0}}\left[\left(V_g - \frac{V_gR_g}{R_g + Z}\right) + \frac{R_0V_g}{R_g + Z}\right] \\ &= \frac{V_g}{2\sqrt{R_0}}\left(1 + \frac{R_0 - R_g}{R_g + Z}\right) \\ &= \left.\frac{V_g}{2\sqrt{R_0}}\right|_{R_0=R_g} \end{aligned} \qquad (14.24)$$

Consider once more the expression for the power dissipated in a one-port network:

$$P = \tfrac{1}{2}(aa^* - bb^*) \qquad (14.25)$$

Factoring the term $aa^* = |a|^2$ from within the parentheses, P becomes

$$\begin{aligned} P &= \frac{|a|^2}{2}\left(1 - \frac{|b|^2}{|a|^2}\right) \\ &= \frac{|a|^2}{2}(1 - |S|^2) \end{aligned} \qquad (14.26)$$

When we choose the reference impedance to be equal to the source resistance, i.e., when $R_0 = R_g$, then $S = 0$ and

$$P_A = \frac{|a|^2}{2} = \frac{|V_g|^2}{8R_g} \qquad (14.27)$$

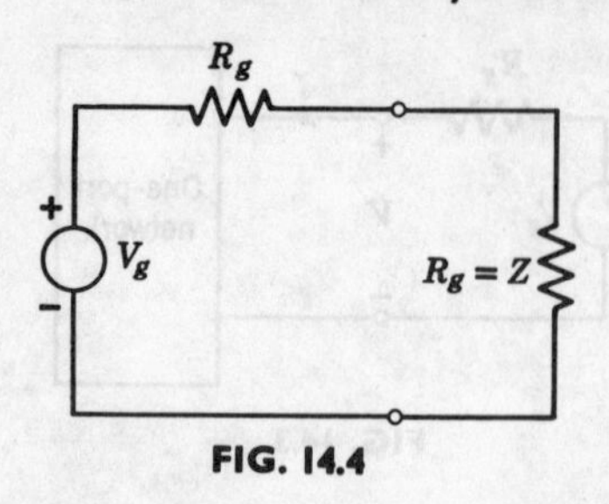

FIG. 14.4

where P_A represents the *available gain* or *available power* of a voltage source V_g with a source resistance R_g. For the case of a one-port network, the available gain is defined as the power dissipated in the one-port network when the impedance of the network Z is equal to the resistance of the source R_g. As a result of this definition, we see that for the one-port network shown in Fig. 14.4, the power dissipated in Z with $Z = R_g$ is

$$P_A = \frac{|V_g|^2}{8R_g} \tag{14.28}$$

The available gain thus represents the *maximum available* power at the terminals of the voltage source.

From this discussion, it is apparent that the value of the reference impedance R_0 should be chosen equal to the source impedance R_g. A standard procedure is to assume a 1-Ω source impedance and denormalize when necessary, that is, let

$$S = \frac{z - 1}{z + 1} \tag{14.29}$$

where

$$z = \frac{Z}{R_0} \tag{14.30}$$

Next, let us briefly consider some of the important properties of the scattering parameter S for a one-port network.

1. The magnitude of S along the $j\omega$ axis is always less or equal to unity for a passive network, that is,

$$|S(j\omega)| \leq 1 \tag{14.31}$$

This property follows from the fact that the power dissipated in a passive network is always greater or equal to zero. Since the power can be expressed as

$$P = \frac{|a|^2}{2}(1 - |S|^2) \geq 0 \tag{14.32}$$

we see that

$$|S(j\omega)|^2 \leq 1 \tag{14.33}$$

2. For a reactive network $|S(j\omega)| = 1$. This property follows from the fact that the power dissipated in a purely reactive network is zero.

3. For an open circuit $S = 1$, and for a short circuit $S = -1$. This is shown to be true from the equation

$$S = \frac{z-1}{z+1} \tag{14.34}$$

For an open circuit $z = \infty$, so that $S = 1$. For a short circuit $z = 0$; therefore $S = -1$.

Before we proceed to the next property, let us consider the following definition.

DEFINITION. A *bounded real function* $F(s)$ is defined by the conditions

(*a*) $|F(s)| \leq K$ for $\text{Re}\, s \geq 0$

(*b*) $F(s)$ is real when s is real.

In (*a*), K denotes any positive real constant.

4. If $z = Z/R_0$ is a positive real function, then S is a bounded real function.

The proof follows from the equation,

$$S = (z-1)/(z+1).$$

From the positive real condition (*a*) $\text{Re}\, z(s) \geq 0$, when $\text{Re}\, s \geq 0$, we see that

$$S = \frac{j\,\text{Im}\, z(s) - [1 - \text{Re}\, z(s)]}{j\,\text{Im}\, z(s) + [1 + \text{Re}\, z(s)]} \tag{14.35}$$

so that

$$|S(s)| = \left\{\frac{\text{Im}^2 z(s) + [1 - \text{Re}\, z(s)]^2}{\text{Im}^2 z(s) + [1 + \text{Re}\, z(s)]^2}\right\}^{1/2} \leq 1 \tag{14.36}$$

when $\text{Re}\, s \geq 0$. (*b*) Where s is real, $z(s)$ is real. Then

$$S = (z-1)/(z+1)$$

must be real. Thus the scattering parameter for a passive network is a bounded real function.

14.3 THE SCATTERING MATRIX FOR A TWO-PORT NETWORK

In this section we will extend the concepts developed for one-port networks discussed in Section 14.2 to two-port networks. In the two-port network shown in Fig. 14.5, we are concerned with two sets of incident and reflected parameters $\{a_1, b_1\}$ at the 1–1′ port, and $\{a_2, b_2\}$ at the 2–2′ port. These parameters are defined in similar manner as for the one-port

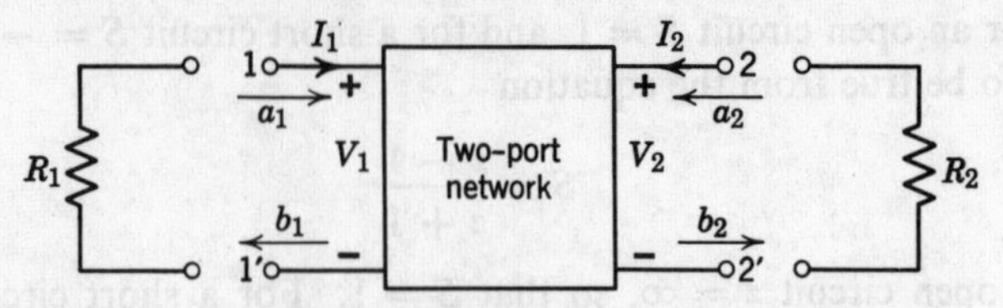

FIG. 14.5. Scattering parameters for a two-port network.

network, that is,

$$
\begin{aligned}
a_1 &= \frac{1}{2}\left(\frac{V_1}{\sqrt{R_{01}}} + \sqrt{R_{01}}I_1\right) \\
b_1 &= \frac{1}{2}\left(\frac{V_1}{\sqrt{R_{01}}} - \sqrt{R_{01}}I_1\right) \\
a_2 &= \frac{1}{2}\left(\frac{V_2}{\sqrt{R_{02}}} + \sqrt{R_{02}}I_2\right) \\
b_2 &= \frac{1}{2}\left(\frac{V_2}{\sqrt{R_{02}}} - \sqrt{R_{02}}I_2\right)
\end{aligned} \tag{14.37}
$$

where R_{01} and R_{02} are the reference impedances at the input and output ports respectively.

The *scattering parameters* S_{ij} for the two-port network are given by the equations

$$
\begin{aligned}
b_1 &= S_{11}a_1 + S_{12}a_2 \\
b_2 &= S_{21}a_1 + S_{22}a_2
\end{aligned} \tag{14.38}
$$

In matrix form the set of equations of Eqs. 14.38 becomes

$$
\begin{bmatrix} b_1 \\ b_2 \end{bmatrix} = \begin{bmatrix} S_{11} & S_{12} \\ S_{21} & S_{22} \end{bmatrix} \begin{bmatrix} a_1 \\ a_2 \end{bmatrix} \tag{14.39}
$$

where the matrix
$$
[S] = \begin{bmatrix} S_{11} & S_{12} \\ S_{21} & S_{22} \end{bmatrix} \tag{14.40}
$$

is called the *scattering matrix* of the two-port network. From Eqs. 14.38, we see that the scattering parameters of the two-port network can be expressed in terms of the incident and reflected parameters as

$$
\begin{aligned}
S_{11} &= \left.\frac{b_1}{a_1}\right|_{a_2=0} \qquad & S_{12} &= \left.\frac{b_1}{a_2}\right|_{a_1=0} \\
S_{21} &= \left.\frac{b_2}{a_1}\right|_{a_2=0} \qquad & S_{22} &= \left.\frac{b_2}{a_2}\right|_{a_1=0}
\end{aligned} \tag{14.41}
$$

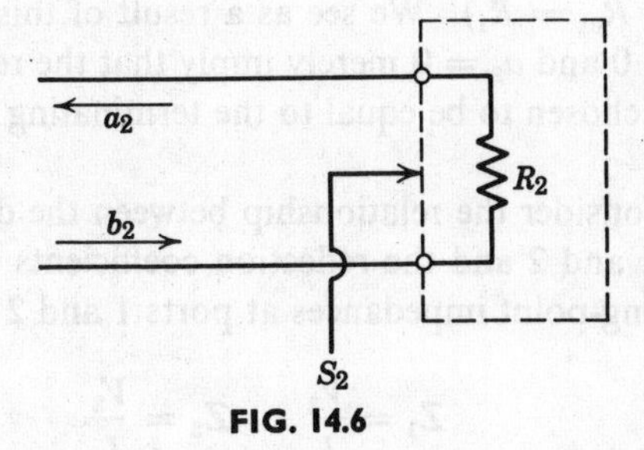

FIG. 14.6

In Eqs. 14.41, the parameter S_{11} is called the *input reflection coefficient*; S_{21} is the *forward transmission coefficient*; S_{12} is the *reverse transmission coefficient*; and S_{22} is the *output reflection coefficient*. Observe that all four scattering parameters are expressed as ratios of reflected to incident parameters.

Now let us examine the physical meaning of these scattering parameters. First, consider the implications of setting the incident parameters a_1 and a_2 to zero in the defining relations in Eqs. 14.41. Let us see what the condition $a_2 = 0$ implies in the definition for the forward reflection coefficient

$$S_{11} = \frac{b_1}{a_1}\bigg|_{a_2=0}$$

Figure 14.6 shows the terminating section of the two-port network of Fig. 14.5 with the parameters a_2 and b_2 of the 2–2′ port shown. If we treat the load resistor R_2 as a one-port network with scattering parameter

$$S_2 = \frac{R_2 - R_{02}}{R_2 + R_{02}} \tag{14.42}$$

where R_{02} is the reference impedance of port 2, then a_2 and b_2 are related by[2]

$$a_2 = S_2 b_2 \tag{14.43}$$

When the reference impedance R_{02} is set equal to the load impedance R_2, then S_2 becomes

$$S_2 = \frac{R_{02} - R_{02}}{R_{02} + R_{02}} = 0 \tag{14.44}$$

so that $a_2 = 0$ under this condition. Similarly, we can show that, when $a_1 = 0$, the reference impedance of port 1 is equal to the terminating

[2] From the viewpoint of the load resistor R_2, the incident parameter is b_2 and the reflected parameter is a_2.

impedance (i.e., $R_{01} = R_1$). We see as a result of this discussion that the conditions $a_1 = 0$ and $a_2 = 0$ merely imply that the reference impedances R_{01} and R_{02} are chosen to be equal to the terminating resistors R_1 and R_2, respectively.

Next, let us consider the relationship between the driving-point impedances at ports 1 and 2 and the reflection coefficients S_{11} and S_{22}. Let us denote the driving-point impedances at ports 1 and 2 as

$$Z_1 = \frac{V_1}{I_1} \qquad Z_2 = \frac{V_2}{I_2} \tag{14.45}$$

From the equation

$$S_{11} = \left.\frac{b_1}{a_1}\right|_{a_2=0} \tag{14.46}$$

We can write

$$S_{11} = \frac{\frac{1}{2}[(V_1/\sqrt{R_{01}}) - \sqrt{R_{01}}I_1]}{\frac{1}{2}[(V_1/\sqrt{R_{01}}) + \sqrt{R_{01}}I_1]} \tag{14.47}$$

which reduces easily to

$$S_{11} = \left.\frac{Z_1 - R_{01}}{Z_1 + R_{01}}\right|_{R_2=R_{02}} \tag{14.48}$$

Similarly, we have

$$S_{22} = \left.\frac{Z_2 - R_{02}}{Z_2 + R_{02}}\right|_{R_1=R_{01}} \tag{14.49}$$

These expressions tell us if we choose the reference impedance at a given port to equal the driving-point impedance at that port, the reflection coefficient of that port will be zero, provided the other port is terminated in its reference impedance.

Next, let us derive some physically meaningful expressions for the forward and reverse transmission coefficients S_{21} and S_{12}. Consider the definition for S_{21}

$$S_{21} = \left.\frac{b_2}{a_1}\right|_{a_2=0} \tag{14.50}$$

As we have just seen, the condition $a_2 = 0$ implies that the reference impedance R_{02} is set equal to the load impedance R_2, as seen in Fig. 14.7. If we connect a voltage source V_{g1} with source impedance $R_{01} = R_1$,

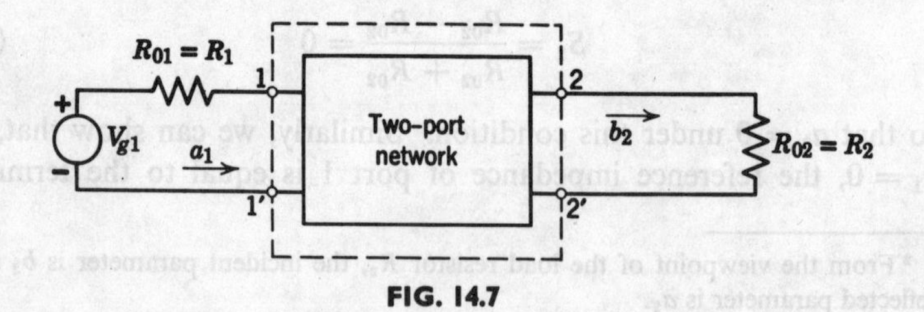

FIG. 14.7

then we can express a_1 as

$$a_1 = \frac{V_{g1}}{2\sqrt{R_1}} \tag{14.51}$$

Since $a_2 = 0$, we have the equation

$$a_2 = 0 = \frac{1}{2}\left(\frac{V_2}{\sqrt{R_2}} + \sqrt{R_2}I_2\right) \tag{14.52}$$

from which we obtain

$$\frac{V_2}{\sqrt{R_2}} = -\sqrt{R_2}I_2 \tag{14.53}$$

Consequently,

$$b_2 = \frac{1}{2}\left(\frac{V_2}{\sqrt{R_2}} - \sqrt{R_2}I_2\right)$$

$$= \frac{V_2}{\sqrt{R_2}} \tag{14.54}$$

Finally, we can express the forward transmission coefficient as

$$S_{21} = \frac{V_2/\sqrt{R_2}}{V_{g1}/2\sqrt{R_1}}$$

$$= \frac{2V_2}{V_{g1}}\sqrt{\frac{R_1}{R_2}}\bigg|_{R_2=R_{02},\,R_1=R_{01}} \tag{14.55}$$

In similar fashion, we find that when port 1 is terminated in $R_{01} = R_1$ and when a voltage source V_{g2} with source impedance R_2 is connected to port 2, then

$$S_{12} = \frac{2V_1}{V_{g2}}\sqrt{\frac{R_2}{R_1}}\bigg|_{R_1=R_{01},\,R_2=R_{02}} \tag{14.56}$$

We see that both S_{12} and S_{21} have the dimensions of a voltage-ratio transfer function. Indeed, if $R_{01} = R_{02}$, then S_{12} and S_{21} are simple voltage ratios. It is seen that for a passive reciprocal network, $S_{21} = S_{12}$.

Now let us consider as an example the scattering matrix of the $n:1$ ratio ideal transformer in Fig. 14.8*a*. Recall that for an ideal transformer

$$V_1 = nV_2, \quad I_1 = -\frac{1}{n}I_2 \tag{14.57}$$

Assuming first that $R_{01} = R_{02} = 1$, let us find S_{11} by terminating the 2–2′ port in a 1-Ω resistor, as shown in Fig. 14.8*b*. Then

$$\frac{V_2}{-I_2} = 1 \tag{14.58}$$

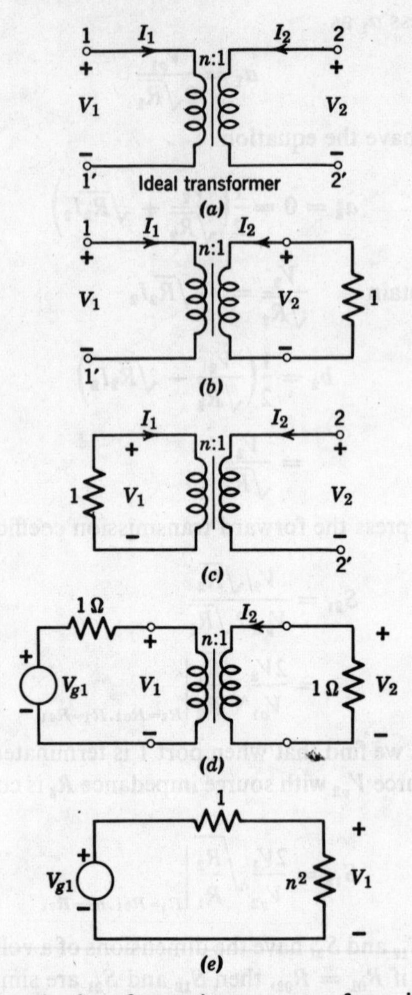

FIG. 14.8. Determination of scattering parameters for an ideal transformer.

so that

$$Z_1 = \frac{V_1}{I_1} = n^2 \tag{14.59}$$

From Eq. 14.48, we have

$$S_{11} = \frac{n^2 - 1}{n^2 + 1} \tag{14.60}$$

Next we terminate the 1–1′ port in a 1-Ω resistor (Fig. 14.8*c*). We obtain, as we did for S_{11},

$$S_{22} = \frac{(1/n^2) - 1}{(1/n^2) + 1} = \frac{1 - n^2}{1 + n^2} \tag{14.61}$$

We obtain S_{21} by connecting a voltage source V_{g1} with a source impedance $R_{01} = 1\ \Omega$ at the 1–1′ port and terminating the 2–2′ port with a resistance $R_{02} = 1\ \Omega$, as seen in Fig. 14.8*d*. Since $V_1/I_1 = n^2$, the equivalent circuit of the ideal transformer as seen from the voltage source as a 1-Ω impedance in series with an n^2-ohm resistance (Fig. 14.8*e*). Then V_1 can be expressed in terms of V_{g1}, as

$$V_1 = \frac{V_{g1}n^2}{n^2+1} \tag{14.62}$$

Since $V_2 = V_1/n$, we have

$$V_2 = \frac{V_{g1}n}{n^2+1} \tag{14.63}$$

Since $R_{01} = R_{02} = 1\ \Omega$, S_{21} is

$$S_{21} = \frac{2V_2}{V_{g1}} = \frac{2n}{n^2+1} \tag{14.64}$$

We can show in similar fashion that

$$S_{12} = \frac{2n}{n^2+1} \tag{14.65}$$

Therefore the scattering matrix for the ideal transformer is given as

$$S = \begin{bmatrix} \dfrac{n^2-1}{n^2+1} & \dfrac{2n}{n^2+1} \\ \dfrac{2n}{n^2+1} & \dfrac{1-n^2}{n^2+1} \end{bmatrix} \tag{14.66}$$

As a second example, let us find the scattering matrix for a lossless transmission line of length L terminated in its characteristic impedance, as shown in Fig. 14.9. If we assume that $R_{01} = R_{02} = Z_0$, then the reflection coefficients are

$$S_{11} = \frac{Z_0 - Z_0}{Z_0 + Z_0} = 0 = S_{22} \tag{14.67}$$

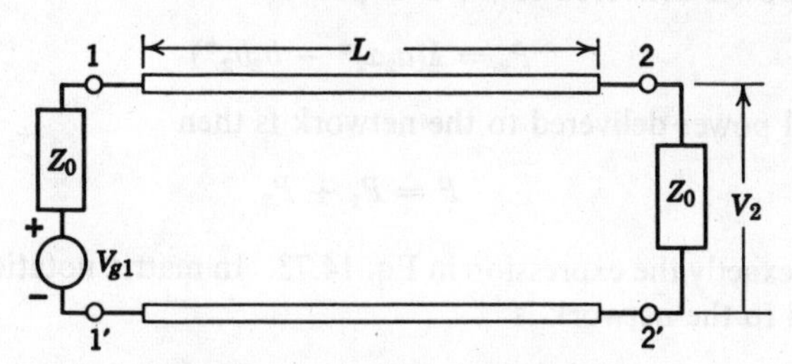

FIG. 14.9. Lossless transmission line.

This result is not implausible, because a transmission line terminated in its characteristic impedance has zero reflected energy. To determine S_{21}, we terminate the line in Z_0 at both ends and connect at the 1–1′ port a voltage source V_{g1}, as depicted in Fig. 14.9. Since the transmission line has zero reflected energy, that is,

$$b_1 = a_2 = 0$$

then

$$V_2 = V_{g1}e^{-\gamma L} \tag{14.68}$$

From the equation

$$S_{21} = 2\left(\frac{R_{01}}{R_{02}}\right)^{1/2}\frac{V_2}{V_{g1}} \tag{14.69}$$

we obtain

$$S_{21} = 2e^{-\gamma L} \tag{14.70}$$

In similar fashion, we find that

$$S_{12} = 2e^{-\gamma L} \tag{14.71}$$

Therefore the scattering matrix for the lossless transmission line is

$$S = \begin{bmatrix} 0 & 2e^{-\gamma L} \\ 2e^{-\gamma L} & 0 \end{bmatrix} \tag{14.72}$$

14.4 PROPERTIES OF THE SCATTERING MATRIX

Having defined the scattering matrix of a two-port network in Section 14.3, let us consider some important properties of the scattering matrix. From the general restriction for a passive network that the net power delivered to all ports must be positive, we obtain the condition

$$P = \tfrac{1}{2}(a_1a_1^* + a_2a_2^* - b_1b_1^* - b_2b_2^*) \geq 0 \tag{14.73}$$

Equation 14.73 follows from the fact that the power delivered to the 1–1′ port is

$$P_1 = \tfrac{1}{2}(a_1a_1^* - b_1b_1^*) \tag{14.74}$$

and the power delivered to the 2–2′ port is

$$P_2 = \tfrac{1}{2}(a_2a_2^* - b_2b_2^*) \tag{14.75}$$

The total power delivered to the network is then

$$P = P_1 + P_2 \tag{14.76}$$

which is exactly the expression in Eq. 14.73. In matrix notation, the power delivered to the network is

$$P = \tfrac{1}{2}\{[a^*]^T[a] - [b^*]^T[b]\} \geq 0 \tag{14.77}$$

where T denotes the transpose operation, and

$$[a] = \begin{bmatrix} a_1 \\ a_2 \end{bmatrix}$$
$$[b] = \begin{bmatrix} b_1 \\ b_2 \end{bmatrix} \tag{14.78}$$

Since $[b] = [S][a]$, then

$$[b^*]^T = [a^*]^T[S^*]^T \tag{14.79}$$

Equation 14.77 can now be rewritten as

$$\begin{aligned} 2P &= \{[a^*]^T[a] - [a^*]^T[S^*]^T[S][a]\} \\ &= [a^*]^T[[u] - [S^*]^T[S]][a] \geq 0 \end{aligned} \tag{14.80}$$

This then implies that the determinant of the matrix $[[u] - [S^*]^T[S]]$ must be greater or equal to zero, that is,

$$\text{Det}\,[[u] - [S^*]^T[S]] \geq 0 \tag{14.81}$$

Consider the special but, nevertheless, important case of a lossless network. In this case $P = 0$, so that

$$[S^*]^T[S] = [u] \tag{14.82}$$

A matrix satisfying the condition in Eq. (14.82) is *unitary*. For a lossless two-port network

$$[S^*]^T[S] = \begin{bmatrix} S_{11}^* & S_{21}^* \\ S_{12}^* & S_{22}^* \end{bmatrix} \begin{bmatrix} S_{11} & S_{12} \\ S_{21} & S_{22} \end{bmatrix} = \begin{bmatrix} 1 & 0 \\ 0 & 1 \end{bmatrix} \tag{14.83}$$

From this equation, we have the following conditions for the scattering matrix

$$S_{11}^*S_{11} + S_{21}^*S_{21} = 1 \tag{14.84}$$

$$S_{12}^*S_{11} + S_{22}^*S_{21} = 0 \tag{14.85}$$

$$S_{11}^*S_{12} + S_{21}^*S_{22} = 0 \tag{14.86}$$

$$S_{12}^*S_{12} + S_{22}^*S_{22} = 1 \tag{14.87}$$

Note that Eqs. 14.85 and 14.86 are conjugates of each other. If the network is reciprocal, then $S_{21} = S_{12}$ and

$$\begin{aligned} |S_{11}(j\omega)|^2 + |S_{21}(j\omega)|^2 = 1 \\ |S_{22}(j\omega)|^2 + |S_{21}(j\omega)|^2 = 1 \end{aligned} \tag{14.88}$$

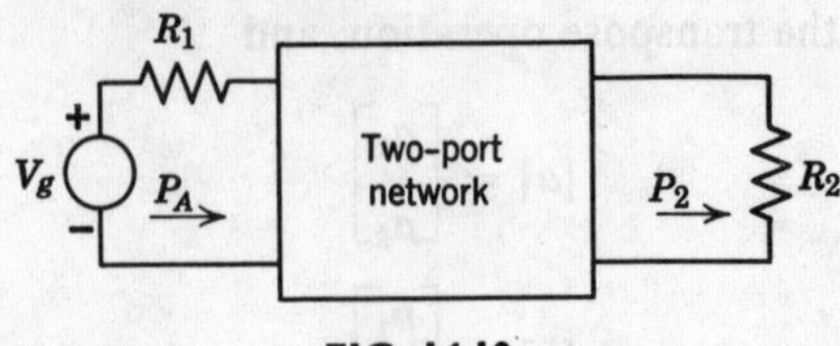

FIG. 14.10

from which it follows that for a lossless reciprocal network $|S_{11}(j\omega)| = |S_{22}(j\omega)| \leq 1$ and $|S_{21}(j\omega)| \leq 1$. Also it is clear that when $|S_{21}(j\omega)| = 0$ (i.e., when there is a zero of transmission), then $|S_{11}(j\omega)| = 1$. This condition states that all the power that has been delivered to the network from port 1–1′ is reflected back to port 1–1′.

At this point, it might be profitable to discuss why we use scattering matrices. What are the advantages of the scattering description over conventional descriptions? Let us discuss three major reasons for the scattering formalism.

1. Many networks do not possess an impedance or admittance matrix. For example, an ideal transformer has no Z or Y matrix because its elements are not finite. However, as we have seen, the ideal transformer can be described by a scattering matrix. Carlin states[3] that all passive networks possess scattering matrices.

2. At high frequencies, incident and reflected parameters play dominant roles in problems of transmission, while voltage-current descriptions are relegated to the background. Then the scattering matrix is necessarily the more powerful description of the system. Note that the voltage standing-wave ratio ($VSWR$) is given in terms of a reflection coefficient S as

$$VSWR = \frac{1 + |S|}{1 - |S|} \tag{14.89}$$

3. In networks where power flow is a prime consideration (e.g., filters), the scattering matrix is very useful. For example, in the network given in Fig. 14.10, if P_A represents the available power from the generator and P_2 is the power dissipated in the load R_2, then we can show that the magnitude-squared forward transmission coefficient is

$$|S_{21}(j\omega)|^2 = \frac{P_2}{P_A} \tag{14.90}$$

We will discuss this point in more detail in Section 14.5.

[3] H. J. Carlin, "The Scattering Matrix in Network Theory," *Trans. IRE*, **CT-3**, No. 2, June 1956, 88–96; see his extensive bibliography.

14.5 INSERTION LOSS

In Section 14.4, we described the forward and reverse transmission coefficients in terms of voltage ratios. Perhaps a more appropriate description of a transmission coefficient is in terms of a power ratio rather than a voltage ratio. In this section we will show that $|S_{21}(j\omega)|^2$ and $|S_{12}(j\omega)|^2$ can be expressed in terms of power ratios. We will then introduce the very important concept of insertion loss and finally relate $|S_{21}(j\omega)|^2$ to the *insertion power ratio*.

Consider the equation for S_{21} in the two-port network shown in Fig. 14.5,

$$S_{21} = \frac{2V_2}{V_{g1}}\left(\frac{R_1}{R_2}\right)^{1/2} \tag{14.91}$$

From the equation $\quad |S_{21}(j\omega)|^2 = S_{21}(j\omega)S_{21}{}^*(j\omega) \quad (14.92)$

we obtain

$$|S_{21}(j\omega)|^2 = \frac{|V_2(j\omega)|^2/2R_2}{|V_{g1}(j\omega)|^2/8R_1}$$

$$= \frac{P_2}{P_{A1}} \tag{14.93}$$

where P_2 is the power dissipated by the load R_2, and P_{A1} is the available gain of the generator V_{g1}. Similarly, we have

$$|S_{12}(j\omega)|^2 = \frac{P_1}{P_{A2}} \tag{14.94}$$

We see that both $|S_{21}(j\omega)|^2$ and $|S_{12}(j\omega)|^2$ are power transfer ratios which relate the power dissipated at a given port to the available power in the other port.

Now let us examine the idea of *insertion loss*. Consider the network shown in Fig. 14.11*a*. Between the terminals 1–1′ and 2–2′ we will insert a two-port network, as shown in Fig. 14.11*b*. Let us denote by V_{20} the voltage across the load resistor R_2 before inserting the two-port network, and by V_2 the voltage across R_2 after inserting the two-port network. A measure of the effect of inserting the two-port network is given by the insertion voltage ratio IVR, which is defined as

$$IVR \triangleq \frac{V_{20}}{V_2} \tag{14.95}$$

Another method of gaging the effect of inserting the two-port network is to measure the power dissipated at the load before and after inserting the

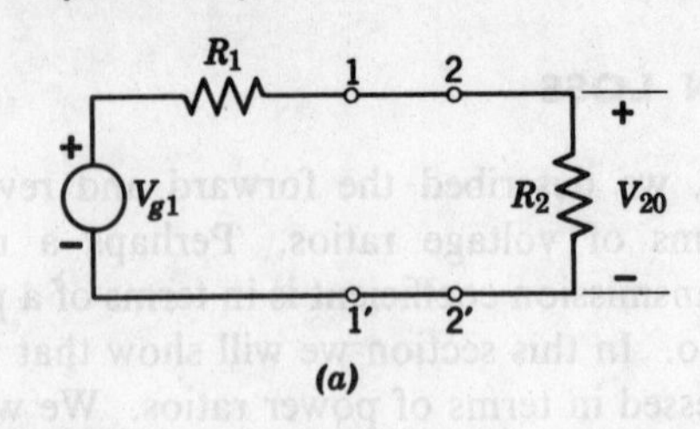

(a)

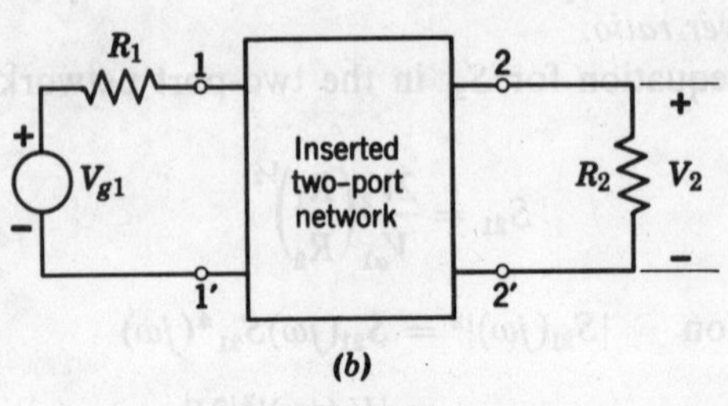

(b)

FIG. 14.11

two-port network. If P_{20} is the power dissipated at the load before the two-port network is inserted, and if P_2 is the power dissipated after insertion, then the *insertion power ratio* of the two-port network is defined as

$$e^{2\alpha} \triangleq \frac{P_{20}}{P_2} \tag{14.96}$$

If we take the logarithm of both sides, we obtain

$$\alpha = 10 \log \frac{P_{20}}{P_2} \tag{14.97}$$

where α is the *insertion loss* of the two-port network. In terms of the circuit given in Fig. 14.11, we can calculate P_{20} from the relation

$$V_{20} = \frac{V_{g1}}{R_1 + R_2} R_2 \tag{14.98}$$

Then P_{20} is

$$P_{20} = \frac{|V_{20}|^2}{2R_2}$$

$$= \frac{R_2 \, |V_{g1}|^2}{2(R_1 + R_2)^2} \tag{14.99}$$

The power dissipated by the load after inserting the two-port network is given by

$$P_2 = \frac{|V_2|^2}{2R_2} \tag{14.100}$$

The insertion power ratio can then be expressed as

$$e^{2\alpha} = \frac{P_{20}}{P_2} = \frac{|V_{g1}|^2}{|V_2|^2} \frac{R_2^{\,2}}{(R_1 + R_2)^2} \tag{14.101}$$

In the special case when the source and load impedances are equal, that is,

$$R_1 = R_2 = R_{01} = R_{02} \tag{14.102}$$

the reciprocal of the squared magnitude of the forward transmission coefficient in Eq. 14.93 is equal to the insertion power ratio

$$\frac{1}{|S_{21}(j\omega)|^2} = \frac{P_{20}}{P_2} \tag{14.103}$$

When $R_1 \neq R_2$, then

$$\frac{P_{20}}{P_2} = \frac{4R_1R_2}{(R_1 + R_2)^2} \frac{1}{|S_{21}(j\omega)|^2} \tag{14.104}$$

In any event, we see that the magnitude-squared transmission coefficients $_1|S_2(j\omega)|_2$ and $|S_{12}(j\omega)|^2$ can be regarded physically as equivalent insertion power ratios. In Section 14.6, we will use this relationship in the synthesis of double-terminated filter networks.

14.6 DARLINGTON'S INSERTION LOSS FILTER SYNTHESIS

In this section, we will consider a filter synthesis procedure first proposed by Darlington in a classic paper in 1939.[4] We will use scattering matrix notation to describe the essence of Darlington's original work. Our coverage will be restricted to the class of low-pass filters which are terminated in equal source and load impedances, $R_{01} = R_{02} = R_0$, as shown in Fig. 14.12. For normalizing purposes we will let R_0 be equal to 1 Ω.

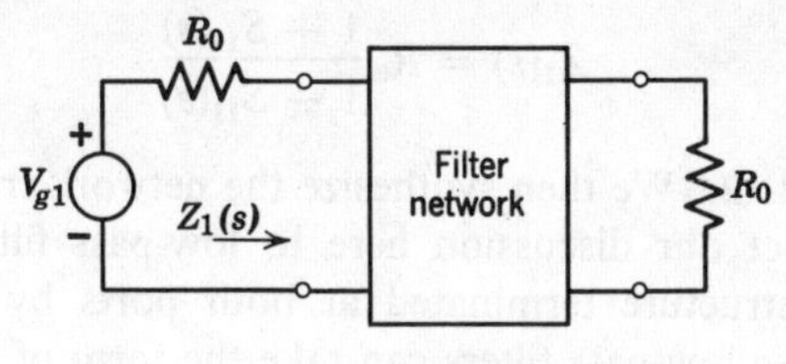

FIG. 14.12

[4] S. Darlington, "Synthesis of Reactance 4-Poles which Produce Prescribed Insertion Loss Characteristics," *J. Math. Phys.*, **18**, 1939, 257–353.

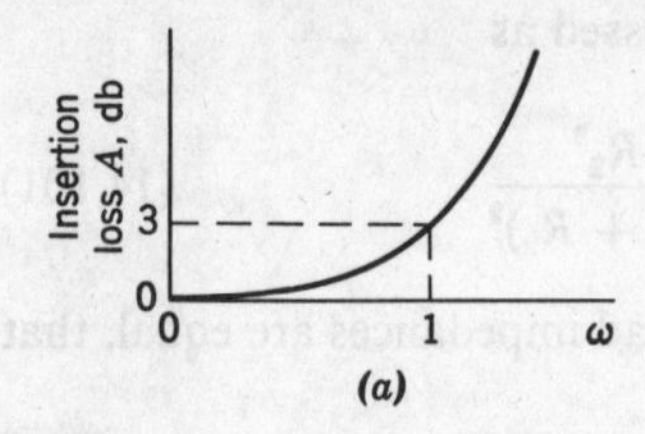

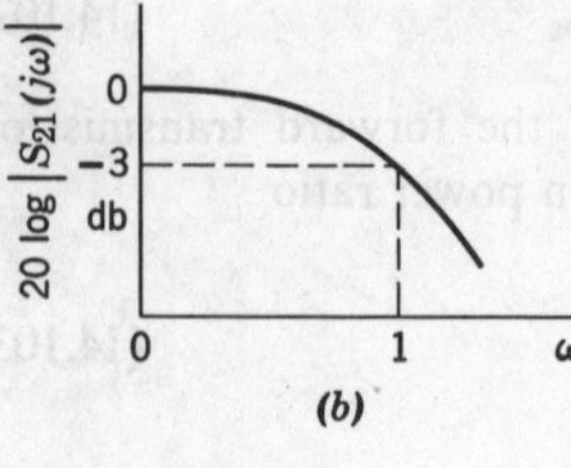

FIG. 14.13

Recall that when the source and load impedances are equal, then the insertion power ratio is equal to the reciprocal of $|S_{21}(j\omega)|^2$, that is,

$$\frac{P_{20}}{P_2} = \frac{1}{|S_{21}(j\omega)|^2} \tag{14.105}$$

Expressed as a loss function, the insertion power ratio is

$$\begin{aligned} A &= 10 \log \frac{P_{20}}{P_2} \\ &= -10 \log |S_{21}(j\omega)|^2 \quad \text{db} \end{aligned} \tag{14.106}$$

In circuit design, the specification of an insertion loss A (Fig. 14.13*a*) is equivalent to the specification of the amplitude-squared transmission coefficient shown in Fig. 14.13*b*. One of the most ingenious techniques given in Darlington's synthesis procedure is the reduction of insertion loss synthesis to an equivalent *L-C* driving-point synthesis problem. This technique can be developed in terms of scattering parameters. Our initial specification is in terms of $|S_{21}(j\omega)|$. For an *L-C* two-port network

$$|S_{11}(j\omega)|^2 = 1 - |S_{21}(j\omega)|^2 \tag{14.107}$$

Next, $S_{11}(s)$ is obtained from the magnitude-squared function

$$S_{11}(s)S_{11}(-s) = 1 - |S_{12}(j\omega)|^2|_{j\omega=s} \tag{14.108}$$

Then from the equation
$$S_{11} = \frac{Z_1 - R_0}{Z_1 + R_0} \tag{14.109}$$

we obtain the driving-point impedance

$$Z_1(s) = R_0 \frac{1 + S_{11}(s)}{1 - S_{11}(s)} \tag{14.110}$$

shown in Fig. 14.12. We then synthesize the network from $Z_1(s)$.

We will restrict our discussion here to low-pass filters given by the lossless ladder structure terminated at both ports by 1-Ω resistors in Fig. 14.14. These low-pass filters can take the form of a Butterworth or Chebyshev specification for $|S_{21}(j\omega)|^2$, that is,

$$|S_{21}(j\omega)|^2 = \frac{1}{1 + \omega^{2n}} \tag{14.111}$$

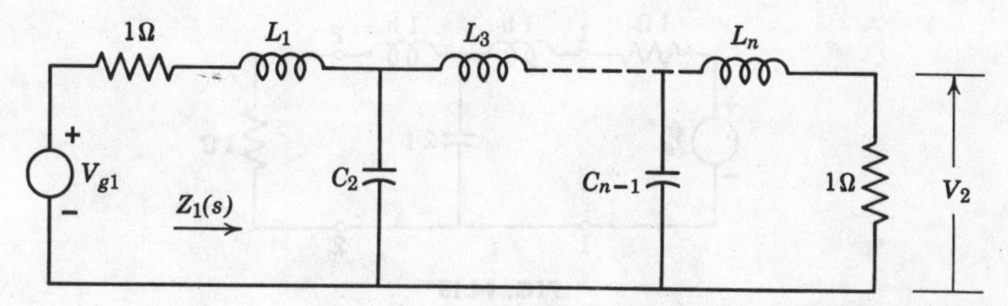

FIG. 14.14. Canonical form for double-terminated low-pass filters.

or

$$|S_{21}(j\omega)|^2 = \frac{1}{1 + \epsilon^2 C_n^{\,2}(\omega)} \tag{14.112}$$

where $C_n(\omega)$ represents an nth-order Chebyshev polynomial.

Example 14.1. Let us synthesize a low-pass filter for the specification

$$|S_{21}(j\omega)|^2 = \frac{1}{1 + \omega^6} \tag{14.113}$$

which represents a third-order Butterworth amplitude characteristic. The load and source impedances are $R_{02} = R_{01} = 1\ \Omega$. First we find $|S_{11}(j\omega)|^2$ as

$$|S_{11}(j\omega)|^2 = 1 - \frac{1}{1 + \omega^6}$$

$$= \frac{\omega^6}{1 + \omega^6} \tag{14.114}$$

Letting $j\omega = s$ in $|S_{11}(j\omega)|^2$, we obtain

$$S_{11}(s)S_{11}(-s) = -\frac{s^6}{1 - s^6} \tag{14.115}$$

which factors into

$$S_{11}(s)\,S_{11}(-s) = \frac{s^3(-s^3)}{(1 + 2s + 2s^2 + s^3)(1 - 2s + 2s^2 - s^3)} \tag{14.116}$$

so that $S_{11}(s)$ is

$$S_{11}(s) = \frac{s^3}{s^3 + 2s^2 + 2s + 1} \tag{14.117}$$

Next, $Z_1(s)$ is obtained from the equation

$$Z_1(s) = \frac{1 + S_{11}(s)}{1 - S_{11}(s)}$$

$$= \frac{2s^3 + 2s^2 + 2s + 1}{2s^2 + 2s + 1} \tag{14.118}$$

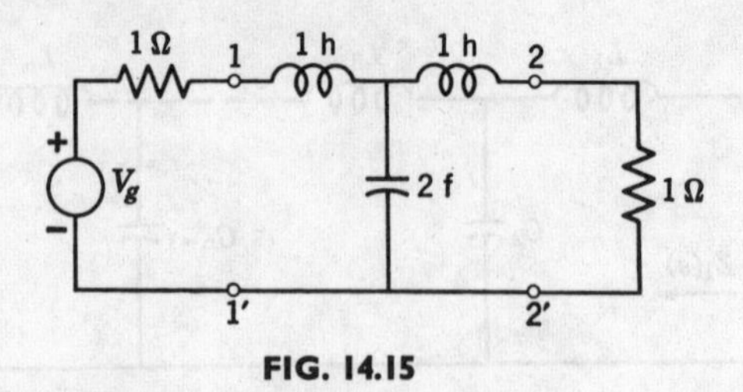

FIG. 14.15

We next perform a Cauer ladder expansion for $Z_1(s)$.

$$
\begin{array}{l}
2s^2 + 2s + 1\,\overline{)\,2s^3 + 2s^2 + 2s + 1\,}(\,s \\
\qquad\qquad\quad\;\; 2s^3 + 2s^2 + s \\
\qquad\qquad\qquad\qquad\quad s + 1\,\overline{)\,2s^2 + 2s + 1\,}(\,2s \\
\qquad\qquad\qquad\qquad\qquad\qquad\; 2s^2 + 2s \\
\qquad\qquad\qquad\qquad\qquad\qquad\qquad\quad 1\,\overline{)\,s + 1\,}(\,s \\
\qquad\qquad\qquad\qquad\qquad\qquad\qquad\qquad\;\; s \\
\qquad\qquad\qquad\qquad\qquad\qquad\qquad\qquad\quad 1\,\overline{)\,1\,}(\,1 \\
\qquad\qquad\qquad\qquad\qquad\qquad\qquad\qquad\qquad\; 1 \\
\qquad\qquad\qquad\qquad\qquad\qquad\qquad\qquad\qquad\; =
\end{array}
$$

The low-pass filter is thus synthesized in the structure shown in Fig. 14.15.

An equivalent realization for the double-terminated filter is obtained if we use the equation

$$S_{11}(s) = \frac{Y_1(s) - G_0}{Y_1(s) + G_0} \tag{14.119}$$

Then, assuming $G_0 = 1$ mho

$$Y_1(s) = \frac{1 + S_{11}(s)}{1 - S_{11}(s)} \tag{14.120}$$

The canonical realization for $Y_1(s)$ is shown in Fig. 14.16. Tables 14.1, 14.2, and 14.3 list element values (up to $n = 7$) for double-terminated Butterworth, Chebyshev (1-db ripple) and Bessel filters, respectively. These apply to the canonical realization for $Y_1(s)$ given in Fig. 14.16. If a $Z_1(s)$ realization in Fig. 14.14 is desired, we simply replace all shunt capacitors by series inductors and vice versa.

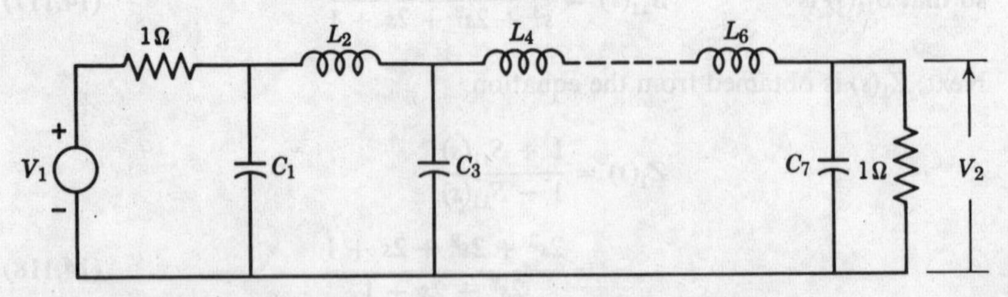

FIG. 14.16. Canonical form for filters in Tables 14.1, 14.2, and 14.3.

TABLE 14.1

Normalized Element Values for a Double-Terminated Butterworth Filter (Equal Terminations)

n	C_1	L_2	C_3	L_4	C_5	L_6	C_7
1	2.000						
2	1.414	1.414					
3	1.000	2.000	1.000				
4	0.765	1.848	1.848	0.765			
5	0.618	1.618	2.000	1.618	0.618		
6	0.518	1.414	1.932	1.932	1.414	0.518	
7	0.445	1.248	1.802	2.000	1.802	1.248	0.445

TABLE 14.2

Normalized Element Values for a Double-Terminated Chebyshev Filter with 1-decibel Ripple (Equal Terminations)

n	C_1	L_2	C_3	L_4	C_5	L_6	C_7
1	1.018						
3	2.024	0.994	2.024				
5	2.135	1.091	3.001	1.091	2.135		
7	2.167	1.112	3.094	1.174	3.094	1.112	2.167

TABLE 14.3

Normalized Element Values for a Double-Terminated Bessel Filter (Equal Terminations)

n	C_1	L_2	C_3	L_4	C_5	L_6	C_7
1	2.000						
2	1.577	0.423					
3	1.255	0.553	0.192				
4	1.060	0.512	0.318	0.110			
5	0.930	0.458	0.331	0.209	0.072		
6	0.838	0.412	0.316	0.236	0.148	0.051	
7	0.768	0.374	0.294	0.238	0.178	0.110	0.038

Note that the even orders for the double-terminated Chebyshev filters are not given. This is because the even-ordered Chebyshev filters do not meet realizability conditions for minimum insertion loss at $s = 0$.[5] We have only given tables for equal source and load terminations. For other possible realizations, the reader should consult L. Weinberg's excellent book.[6]

---

[5] L. Weinberg, *Network Analysis and Synthesis*, McGraw-Hill Book Company, New York, 1962, p. 589.

[6] *Ibid.*, Chapter 13.

Problems

14.1 Determine the reflection coefficient S for the one-port networks shown in the figure.

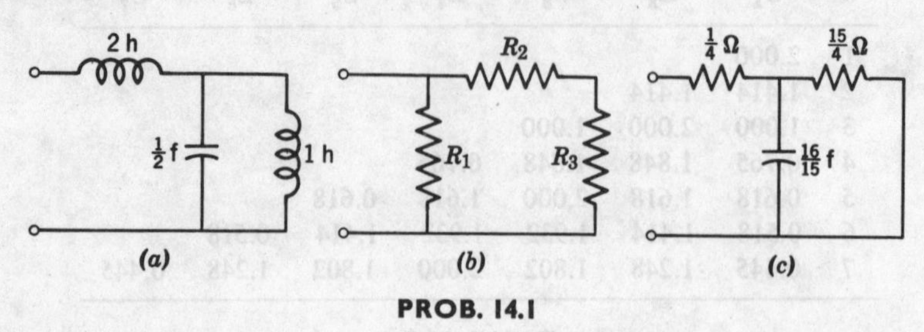

PROB. 14.1

14.2 For the one-port network in Fig. 14.3, let $R_0 = R_g$. If the incident parameter is $a = V_g/2\sqrt{R_0}$, find the reflected parameter b.

14.3 For the network in Prob. 14.1, determine $|S(j\omega)|$. Show that the scattering elements S for the networks in Prob. 14.1 are bounded real functions.

14.4 For each of the networks shown, find the scattering matrix for $R_{01} = R_{02} = 1$.

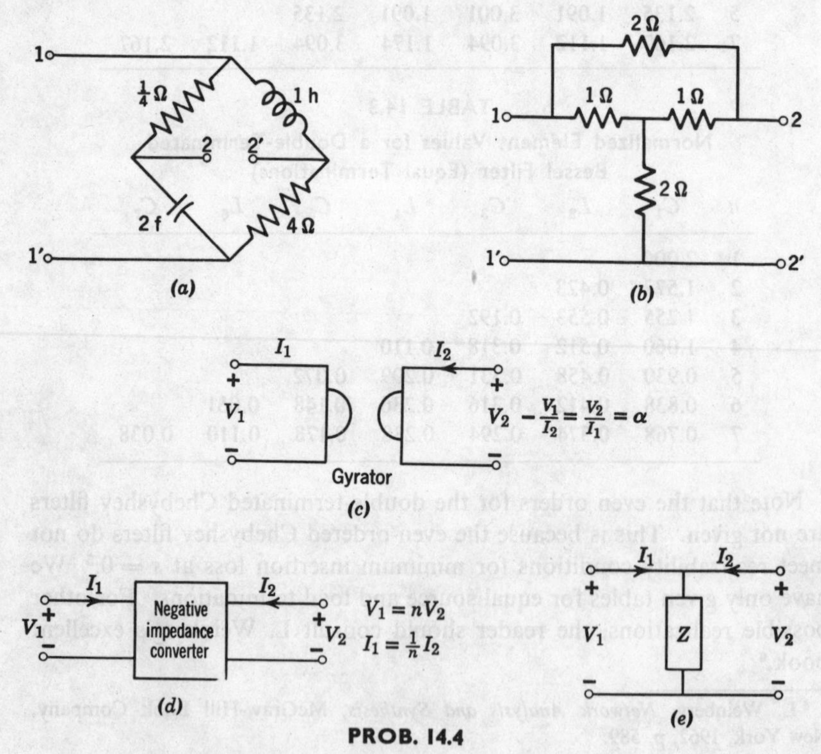

PROB. 14.4

14.5 Find the insertion voltage ratio and insertion power ratios for each of the networks shown. These networks are to be inserted between a source impedance $R_g = 2\ \Omega$ and a load impedance $R_L = 1\ \Omega$. From the insertion power ratio, find $|S_{21}(j\omega)|^2$.

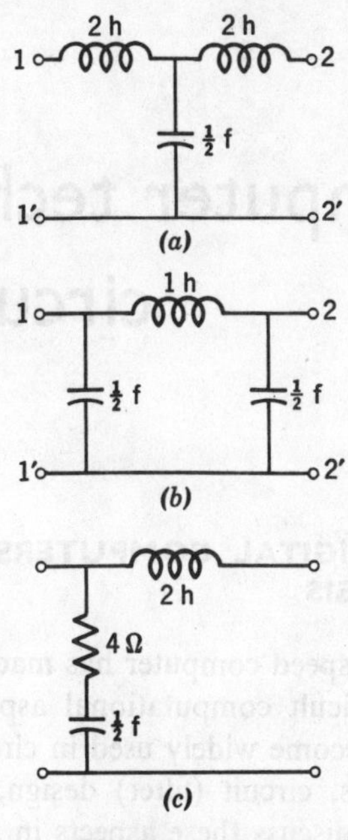

PROB. 14.5

14.6 Synthesize low-pass filters for the specifications

(a) $$|S_{21}(j\omega)|^2 = \frac{1}{1+\omega^4}$$

(b) $$|S_{21}(j\omega)|^2 = \frac{1}{1+\omega^8}$$

14.7 Synthesize an equal-ripple low-pass filter such that $20 \log |S_{21}(j\omega)|$ has at most $\frac{1}{2}$-db ripple in the pass band and an asymptotic falloff of 12 db/octave in the stop band.

chapter 15

Computer techniques in circuit analysis

15.1 THE USES OF DIGITAL COMPUTERS IN CIRCUIT ANALYSIS

The advent of the high-speed computer has made routine many of the formerly tedious and difficult computational aspects of circuit theory. Digital computers have become widely used in circuit analysis, time and frequency-domain analysis, circuit (filter) design, and optimization or iterative design. We will discuss these aspects in general in this section. In succeeding sections, we will discuss some specific circuit-analysis computer programs.

Circuit analysis

The primary objective of a linear circuit-analysis program is to obtain responses to prescribed excitation signals. These programs are based on many different methods: nodal analysis, mesh analysis, topological formulas, and state variables. Most of them can handle active elements such as transistors and diodes by means of equivalent circuit models.

The state-variable programs based upon Bashkow's A matrix formulation[1] perform their calculations directly in the time domain via numerical integration and matrix inversion. The outputs of these programs provide impulse and step response in tabular form. If the excitation signal were given in data form, the state-variable programs would calculate the response directly in the time domain.

---

[1] T. R. Bashkow, "The *A* Matrix—New Network Description," *IRE Trans. on Circuit Theory*, **CT-4**, No. 3, September 1957, 117–119.

The majority of circuit-analysis programs, however, perform their calculations in the frequency domain. The program user is only required to specify the topology of the network, the element values, and what transfer functions he wishes to obtain. The computer does the rest. It calculates the specified transfer functions in polynomial form, calculates the poles and zeros of these functions, and can also provide transient response and steady-state response, if desired. With versatile input-output equipment, the output can also provide a schematic of the original network, as well as plots of time- and frequency-response characteristics.

Time- and frequency-domain analysis

The time- and frequency-domain analysis programs can be used in conjunction with the circuit-analysis programs or independently. The time-domain programs depend upon solving the convolution integral

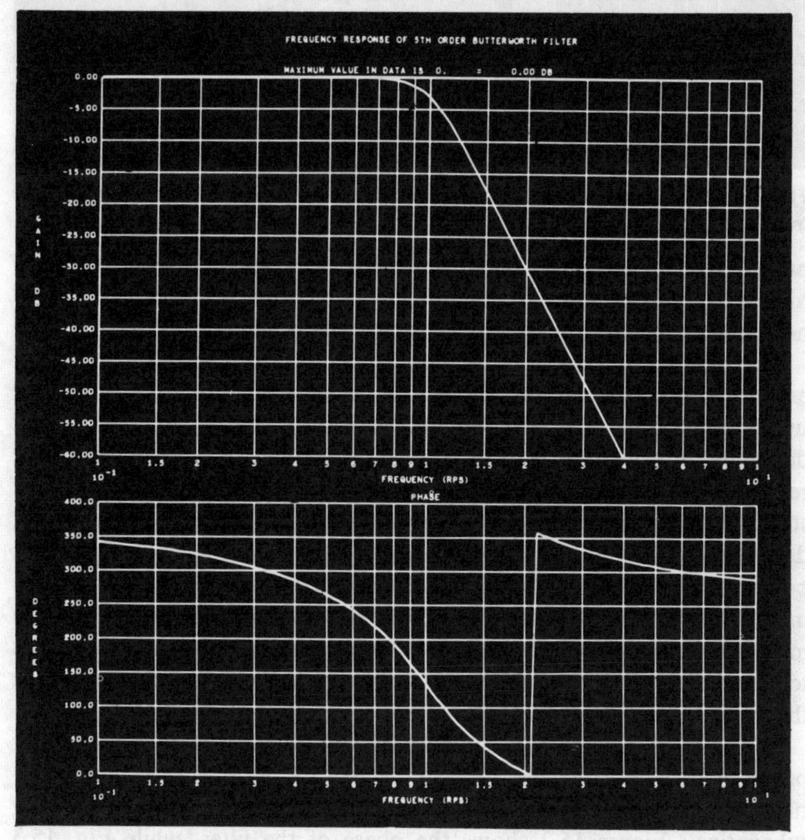

FIG. 15.1. Frequency response of fifth-order Butterworth filter.

FIG. 15.2. Phase response of fifth-order Butterworth filter.

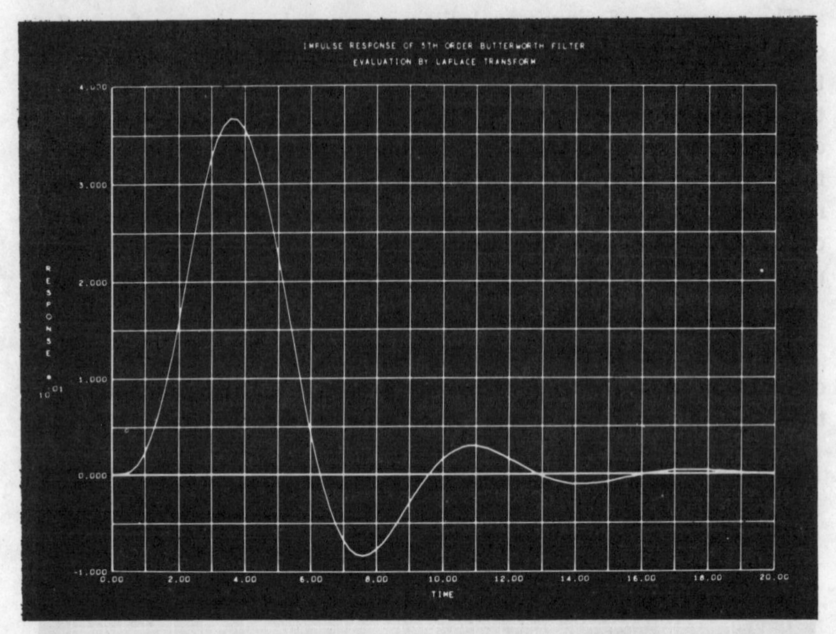

FIG. 15.3. Impulse response of fifth-order Butterworth filter evaluation by Laplace transform.

numerically. This approach obviates the necessity of finding roots of high-order polynomials. It has the advantage that the excitation signal need not be specified analytically, but merely in numerical form.

The frequency-domain programs usually consist of finding transient and steady-state responses, given the transfer function in factored or unfactored form. The program user must specify the numerator and denominator polynomials of the transfer functions, the types of transient response he wishes (i.e., impulse or step response), and the types of steady-state responses he wishes (amplitude, amplitude in decibels, phase, delay, etc.). In addition, he must specify the frequency and time data points at which the calculations are to be performed. This may be done in two ways. If he requires evenly spaced data, he need only specify the minimum point, the increment, and the number of points. If he wishes to obtain data at certain points, he must supply the list of data points at the input.

Examples of outputs of a steady-state and transient analysis computer program are shown in Figs. 15.1, 15.2, and 15.3. In Fig. 15.1, the magnitude of a fifth-order Butterworth filter is plotted via a microfilm plotting subroutine. Figure 15.2 shows the phase of the filter, while Fig. 15.3 shows its impulse response.

In Section 15.2 we will examine further details of a typical steady-state analysis program.

Circuit (filter) design

The filter design programs are probably the most convincing argument for the use of computers in circuit design. Designing insertion loss filters[2] to meet certain amplitude requirements requires considerable numerical calculation even in the simplest cases. The use of digital computers in insertion loss filter design is clearly a logical alternative. The amount of programming time for a general filter synthesis program is considerable. However, the ends certainly justify the means when large numbers of filters must be designed to meet different specifications.

An outstanding example of a digital computer program for filter design is the one written by Dr. George Szentirmai and his associates.[3] The program is complete in that it handles the approximation as well as the synthesis problem. It is capable of dealing with low-, high-, and band-pass filters with prescribed zeros of transmission (also called *attenuation poles* or *loss peaks*). There are provisions for either equal-ripple or maximally flat-type pass-band behavior, for arbitrary ratios of load to source impedances, and for predistortion and incidental dissipation.

In addition, Dr. Szentirmai has a modified program that synthesizes low- and band-pass filters with maximally flat or equal-ripple-type delay in their pass band, and monotonic or equal-ripple-type loss in the stop bands.

In the specifications, the designer could specify both the zeros of transmission (loss peaks) and the network configuration desired.[4] If his specifications include neither, the computer is free to pick both configuration and zeros of transmission. The computer's choice is one in which the inductance values are kept at a minimum. The program was written so that the same network could be synthesized from both ends.

Finally, the computer prints out the network configuration, its dual, the normalized element values, and the denormalized ones. It also provides information such as amplitude and phase response, as well as plots of these responses obtained from a microfilm printer.

Figures 15.4, 15.5, 15.6, 15.7, 15.8, and 15.9 show the results of a band-pass filter synthesis using Dr. Szentirmai's program. Figure 15.4

---

[2] R. Saal and E. Ulbrich, "On the Design of Filters by Synthesis," *Trans. IRE on Circuit Theory*, **CT-5**, December 1958, 287–327.

[3] G. Szentirmai, "Theoretical Basis of a Digital Computer Program Package for Filter Synthesis," *Proceedings of the First Allerton Conference on Circuit and System Theory*, November 1963, University of Illinois.

[4] Saal and Ulbrich, *op. cit.*

```
BAND PASS FILTER SYNTHESIS


CASE NUMBER  3.1

DEGREE ØF FILTER                                = 13
  MULTIPLICITY ØF PEAK AT ZERØ                  =  3
  MULTIPLICITY ØF PEAK AT INFINITY              =  2
  NUMBER ØF FINITE PEAKS BELØW THE BAND         =  1
  NUMBER ØF FINITE PEAKS ABØVE THE BAND         =  3

EQUAL RIPPLE PASS BAND REQUESTED
  PASS BAND RIPPLE MAGNITUDE                    =  0.05000          DB.
  LØWER PASS BAND EDGE FREQUENCY                =  1.0000000E+04 CPS.
  MID-BAND FREQUENCY                            =  1.3416408E+04 CPS.
  UPPER PASS BAND EDGE FREQUENCY                =  1.8000000E+04 CPS.

"ARBITRARY" STØP BAND REQUESTED

NUMBER ØF SPECIAL PØLES = 1      MA =    1.0000000E+00

TERMINATIØNS
  INPUT TERMINATIØN                             =  6.0000000E+02 ØHMS
  ØUTPUT TERMINATIØN                            =  0.            ØHMS

LØWER FINITE STØP BAND PEAKS
  FREQUENCY CPS        NØRMALIZED          VALUE ØF M
   6.2000000E+03       4.6212071E-01        0.6229266         1

UPPER FINITE STØP BAND PEAKS
  FREQUENCY CPS        NØRMALIZED          VALUE ØF M
   2.0700000E+04       1.5428869E+00       2.3788111          1
   2.3100000E+04       1.7217723E+00       1.9296560          2
   3.5200000E+04       2.6236531E+00       1.4968756          3

CØMPUTER WILL SPECIFY CØNFIGURATIØN
  LAST INDUCTØR IS A SERIES BRANCH
```

FIG. 15.4

gives the specification of the problem. The pass-band magnitude is to be equal ripple with ripple magnitude of 0.05 db, and the degree of the filter is to be 13. As we indicated in Chapter 14, odd-degree, equal-ripple filters are nonrealizable. In this example the designer utilized an ingenious device—an extra pole—to accomplish the synthesis. The program logic then provided two extra zeros: one to cancel the extra pole and the other to provide the odd degree. The extra pole is called a *special pole* in Fig. 15.4, and is located at $s = -1.0$.

Further specifications call for the lower band-edge frequency to be 10^4 cycles; the upper, 1.8×10^4 cycles; and the midband frequency to be $\sqrt{1.8 \times 10^4} = 1.3416408 \times 10^4$ cycles. In the stop band, there are to be zeros of transmission at $f = 0$ (three), $f = \infty$ (two), and four finite zeros of transmission: one below the pass band and three above the pass band. The positions of these finite zeros of transmission are chosen by the designer as indicated by the notation "*arbitrary*" *stop band requested.*

```
CASE NUMBER  3.1          FØRWARD REALIZATIØN FRØM A SHØRT CIRCUIT ADMITTANCE

DEGREE ØF FILTER                        = 13
  MULTIPLICITY ØF PEAK AT ZERØ          = 3
  MULTIPLICITY ØF PEAK AT INFINITY =  2

EQUAL RIPPLE PASS BAND                  = 0.0500        DB       "ARBITRARY" STØP BAND
  LØWER PASS-BAND EDGE FREQUENCY        = 1.0000000E+04 CPS
  UPPER PASS-BAND EDGE FREQUENCY        = 1.8000000E+04 CPS
  MID-BAND FREQUENCY                    = 1.3416408E+04 CPS

CØNFIGURATIØN SPECIFIED BY CØMPUTER
   803   501   801   100   802   400   200   100   300

   NØRMALIZED                     UNNØRMALIZED

1.0000000E+00     ....R....      6.0000000E+02        TERMINATIØN
                  .       .
3.0704337E-01     ....C....      6.0706102E-09                        803
                  .       .
                  .     ....
7.8527337E-01     .     L C      5.5892816E-03        PEAK FREQUENCY
1.8499769E-01     .     ....     3.6576228E-09        3.5200000E+04
                  .       .
2.5723015E+00     .       C      5.0857440E-08                        501
                  .       .
1.0085380E+00     ...L.C...      7.1783955E-03        PEAK FREQUENCY
4.6429806E+00     .       .      9.1797212E-08        6.2000000E+03
                  .       .
1.3647491E+00     ....C....      2.6982703E-08                        801
                  .       .
                  .     ....
5.3765272E-01     .     L C      3.8268107E-03        PEAK FREQUENCY
7.8132183E-01     .     ....     1.5447656E-08        2.0700000E+04
                  .       .
7.5623894E-01     .       C      1.4951737E-08                        100
                  .       .
1.4438446E+00     ....C....      2.8546514E-08                        802
                  .       .
                  .     ....
2.7112746E-01     .     L C      1.9297837E-03        PEAK FREQUENCY
1.2441566E+00     .     ....     2.4598445E-08        2.3100000E+04
                  .       .
1.0127197E+01     ....C....      2.0022665E-07                        400
                  .       .
9.5770211E-02     ....L....      6.8165648E-04                        200
                  .       .
1.6378789E+00     .       C      3.2382800E-08                        100
                  .       .
6.2963156E-01     .       L      4.4814816E-03                        300
                  .       .
SHØRT             ....R....      SHØRT                TERMINATIØN
```

FIG. 15.5

The terminations are: input = 60 Ω, output = 0 Ω which means the filter terminates in a short circuit. In this example, the filter configuration is chosen by the computer, and is a *minimum inductance* configuration.[5] In Fig. 15.4 there are, in addition, listings of the finite zeros of transmission (loss peaks), which the designer specified.

Figure 15.5 is a printout of the configuration of the filter as shown by the dotted lines flanked by the associated element values, both normalized (left column) and unnormalized (right column). Since there are four finite zeros of transmission, there must be associated four *L-C* tank circuits.

[5] W. Saraga, "Minimum Inductance or Capacitance Filters," *Wireless Engineer*, **30**, July 1953, 163–175.

FREQUENCY IN CYCLES	VØLTAGE RATIØ IN DB	PHASE IN DEGREES	REAL Z IN ØHMS	IMAG Z IN ØHMS	REAL Y IN MILMHØS	IMAG Y IN MILMHØS
16000	5.718532	145.549717	134.691	56.981	6.2973	-2.6641
16250	5.704379	128.958888	130.664	61.193	6.2766	-2.9395
16500	5.707644	111.534102	125.883	64.762	6.2814	-3.2315
16750	5.729113	92.934808	119.931	67.928	6.3129	-3.5756
17000	5.750516	72.763852	112.687	71.154	6.3445	-4.0061
17250	5.743140	50.659800	104.489	74.696	6.3337	-4.5277
17500	5.708979	26.219204	95.717	77.919	6.2835	-5.1151
17750	5.720983	358.194819	85.165	79.138	6.3011	-5.8552
18000	5.702947	322.320795	67.628	78.767	6.2748	-7.3083
18250	4.131556	272.913485	40.032	87.140	4.3533	-9.4759
18500	-1.119811	225.689589	16.386	111.450	1.2913	-8.7828
18750	-7.753694	195.922988	5.462	139.624	.2797	-7.1512
19000	-14.148566	177.085862	1.725	164.004	.0641	-6.0967
19250	-20.210264	163.748654	.541	184.583	.0158	-5.4176
19500	-26.118735	153.507803	.167	202.605	.0040	-4.9357
19750	-32.100833	145.227029	.049	218.952	.0010	-4.5672
20000	-38.473164	138.296917	.013	234.147	.0002	-4.2708
20250	-45.857895	132.354567	.002	248.501	.0000	-4.0241
20500	-56.325030	127.166241	.000	262.211	.0000	-3.8137
20750	-71.701988	302.572268	.000	275.407	.0000	-3.6310
21000	-59.444189	298.458582	.000	288.178	.0000	-3.4701
21250	-57.483949	294.740796	.000	300.592	.0000	-3.3268
21500	-57.587501	291.354706	.000	312.698	.0000	-3.1980
21750	-58.703992	288.250341	.000	324.536	.0000	-3.0813
22000	-60.546362	285.388013	.000	336.138	.0000	-2.9750
22250	-63.101161	282.735723	.000	347.527	.0000	-2.8775
22500	-66.602572	280.267304	.000	358.726	.0000	-2.7876
22750	-71.867460	277.961066	.000	369.753	.0000	-2.7045
23000	-83.403713	275.798866	.000	380.621	.0000	-2.6273
23250	-80.597817	93.765362	.000	391.345	.0000	-2.5553

FIG. 15.6

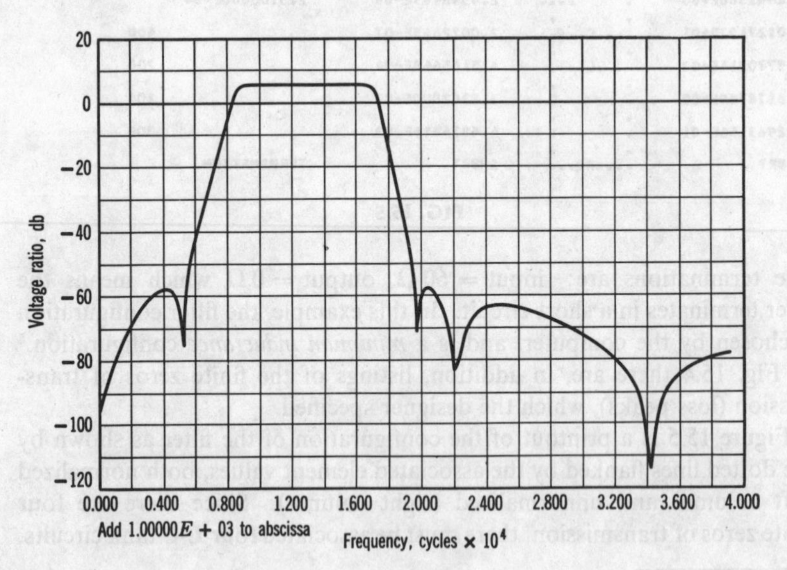

FIG. 15.7

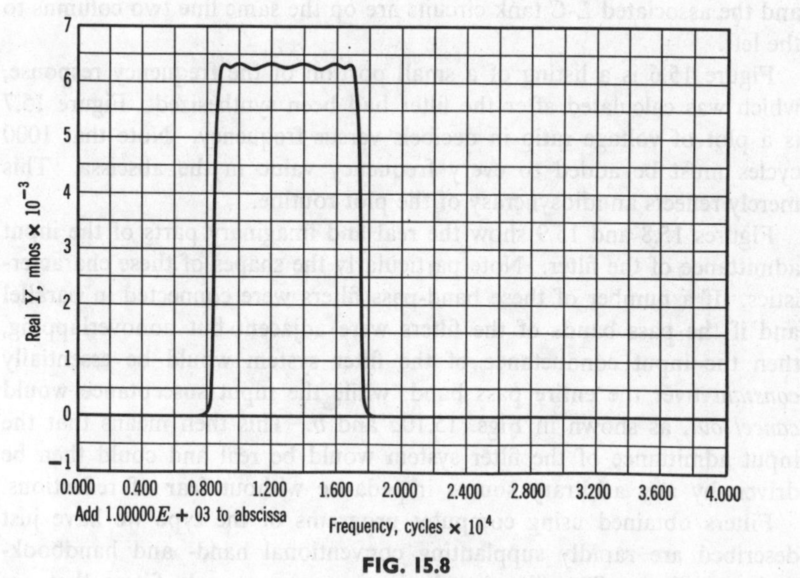

FIG. 15.8

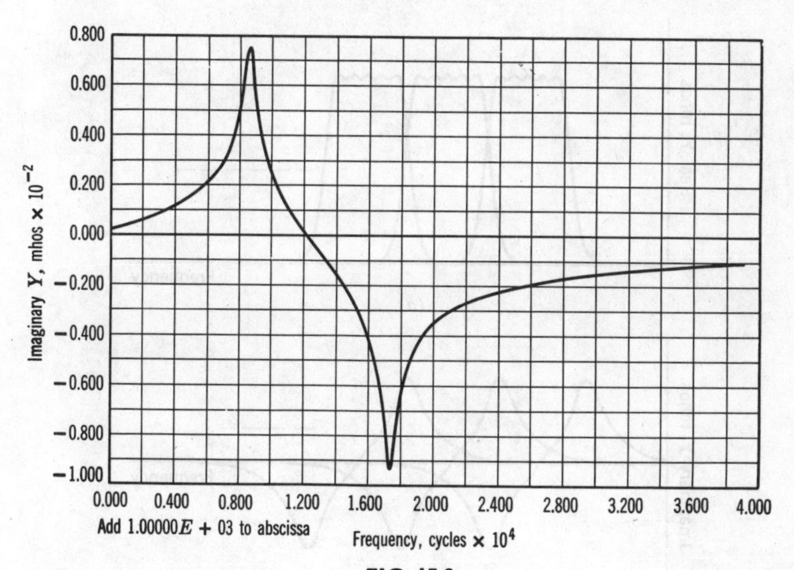

FIG. 15.9

The peak frequencies are listed under the column entitled "termination," and the associated L-C tank circuits are on the same line two columns to the left.

Figure 15.6 is a listing of a small portion of the frequency response, which was calculated after the filter had been synthesized. Figure 15.7 is a plot of voltage ratio in decibels versus frequency. Note that 1000 cycles must be added to every frequency value in the abscissa. This merely reflects an idiosyncrasy of the plot routine.

Figures 15.8 and 15.9 show the real and imaginary parts of the input admittance of the filter. Note particularly the shapes of these characteristics. If a number of these band-pass filters were connected in parallel and if the pass bands of the filters were adjacent but nonoverlapping, then the input conductance of the filter system would be essentially *constant* over the entire pass band, while the input susceptance would *cancel out*, as shown in Figs. 15.10*a* and *b*. This then means that the input admittance of the filter system would be real and could then be driven by any arbitrary source impedance without fear of reflections.

Filters obtained using computer programs of the type we have just described are rapidly supplanting conventional hand- and handbook-designed filters. The filter synthesis programs provide filters that are orders of magnitude more sophisticated than filters designed by the

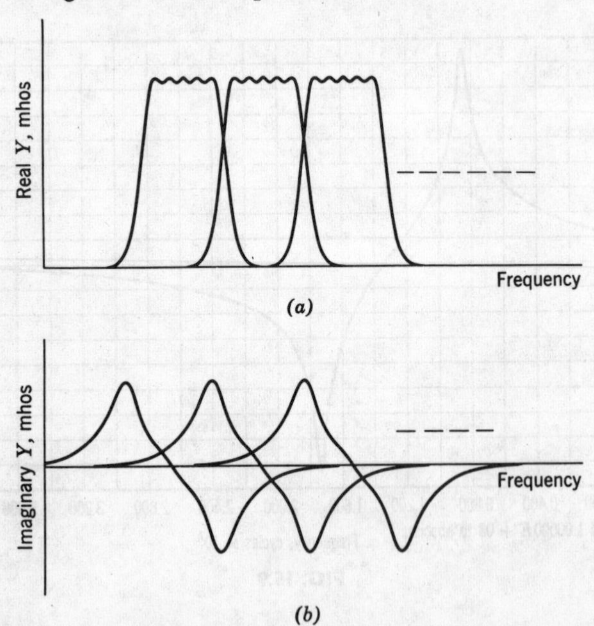

FIG. 15.10. Effects of paralleling several band-pass filters with adjacent pass bands.

conventional manner. Moreover, the designs are completed in minutes rather than days and at a typical cost of twenty dollars rather than two thousand (not including initial programming costs, of course).

Optimization

Network design cannot always be accomplished by way of analytical means. Quite often networks are designed through a trial-and-error process. The network designer begins with a set of specifications. He then selects a network configuration, and makes an initial guess about the element values. Next he calculates or measures the desired responses and compares them with the specifications. If the measured responses differ by a wide margin from the specified responses, the designer changes the values of the elements and compares again. He does this a number of times until (hopefully) the measured responses agree with the specified responses to within a preset tolerance.

This process of cut and try can be made to converge, sometimes quite rapidly, if one uses a method of steepest descent.[6] To get a rough idea of the steepest descent method, suppose there are n parameters in the network. Let us regard each parameter x_i as a dimension in an n-dimensional euclidean space. We define a function $f(x_1, x_2, x_3, \ldots, x_n)$ that assigns a functional value to each point of the euclidean space. We then ask, at point p_i, what direction of motion decreases the value of $f(x_1, x_2, x_3, \ldots, x_n)$ most rapidly? The function $f(x_i)$ may be defined as a least squared error, or as an absolute error between calculated response and specified response. The direction of steepest descent is the direction defined by the gradient

$$\text{grad}\, f = \frac{\partial f}{\partial x_1} x_1 + \frac{\partial f}{\partial x_2} x_2 + \cdots + \frac{\partial f}{\partial x_n} x_n \tag{15.1}$$

Therefore, incremental value of change for each parameter is

$$\delta x_i = -C \frac{\partial f}{\partial x_i} \tag{15.2}$$

where C is a constant. After all parameters have been changed by the incremental value in Eq. 15.2, a new gradient and new incremental values are obtained. The process continues until the optimum is obtained.

An excellent paper by C. L. Semmelman[7] describes a steepest descent

[6] Charles B. Tompkins, "Methods of Steep Descent," in *Modern Mathematics for the Engineer* (Edwin F. Beckenbach, ed.), Chapter 18, McGraw-Hill Book Company, New York, 1956.

[7] C. L. Semmelman, "Experience with a Steepest Descent Computer Program for Designing Delay Networks," *IRE International Convention Rec.*, Part 2, 1962, 206–210.

Fortran program used for designing delay networks. The specifications are the delay values R_j given at the frequency data points f_j. The program successively changes the parameters x_i so that the squared error

$$\epsilon(x_i) = \sum_{j=1}^{n} [R_j - T(x_i, f_j)]^2 \tag{15.3}$$

is minimized. In Eq. 15.3, $T(x_i, f_j)$ represents the delay, at frequency f_j, of the network with parameters x_i.

In order to use the program, the network designer must first select the initial values for the parameters x_i. He must also provide the specified delays R_j and the frequency data points f_j. The program provides for 128 match points and 64 parameter values. It is capable of meeting requirements simultaneously in the time and frequency domains. The designer is not restricted to equal-ripple approximations or infinite Q requirements. He is free to impose requirements such as nonuniform dissipation and range of available element values on the design. For related methods of optimization, the reader should refer to a tutorial paper by M. R. Aaron.[8]

Machine-aided design

The concept of real-time interaction between man and computer holds much promise in the field of network design. Multiple-access computing systems, such as the project MAC of Massachusetts Institute of Technology, make cut and try design procedures practicable. The initials MAC could describe either the term *multiple-access computer* or the term *machine-aided cognition*. In an article describing the MAC computer system,[9] Professor R. M. Fano states "The notion of machine-aided cognition implies an intimate collaboration between a human user and a computer in a real-time dialogue on the solution of a problem, in which the two parties contribute their best capabilities."

A simple multiple-access computer system is shown in Fig. 15.11. There are n data links and n terminals connected to the central processor. Located at each terminal is input-output equipment, such as teletypewriters, teletypes, and oscillographic displays. The sequence in which the central processor accepts programs from the terminals is controlled by a built-in queueing logic.

The main reason a multiple-access computer is effective is that the central processor computes thousands of times faster than the user's reaction time. When a user feeds in a program, it seems only a "moment"

---

[8] M. R. Aaron, "The Use of Least Squares in System Design," *IRE Trans. on Circuit Theory*, CT-3, No. 4, December 1956, 224–231.

[9] R. M. Fano, "The MAC System: The Computer Utility Approach," *IEEE Spectrum*, 2, No. 1, January 1965, 56–64.

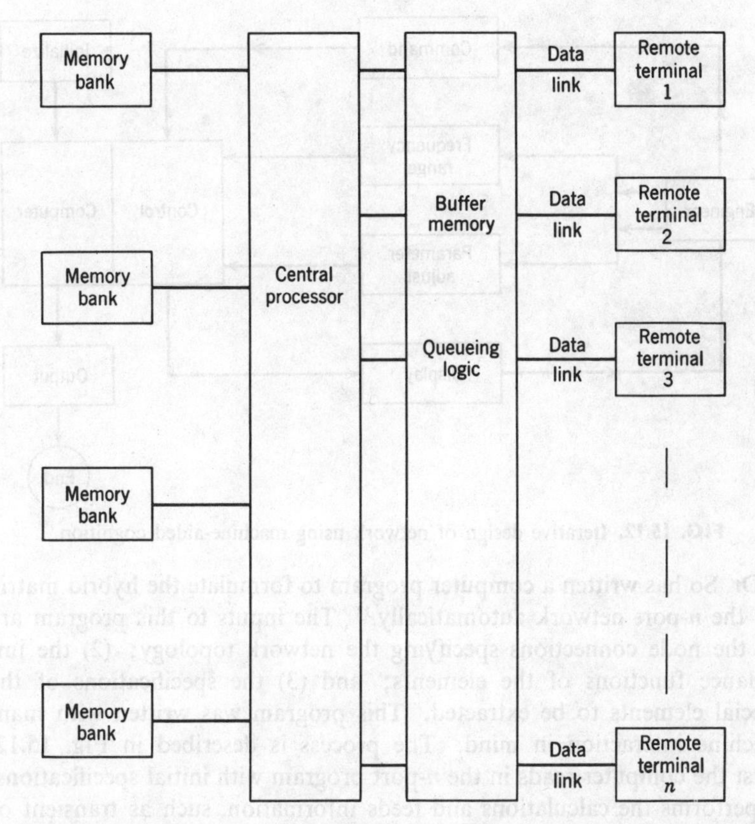

FIG. 15.11. Block diagram showing typical multiple-access computing system.

before he gets the results through the teletypewriter or oscilloscope display. He examines the results, changes some parameters, and feeds the program into the central processor again. The queueing time and processing time on the multiple-access computer may amount to only a minute or two, which, for the user, is probably not significantly long.

Dr. H. C. So has written a paper on a hybrid description of a linear *n*-port network[10] in which he has shown that the hybrid matrix is ideally suited for such problems such as multielement variation studies and iterative design. Suppose there are *n* variable elements in the network. These can be "extracted" and the rest—the bulk of the network—can be described by the *n*-port hybrid matrix.

---

[10] H. C. So, "On the Hybrid Description of a Linear *n*-Port Resulting from the Extraction of Arbitrarily Specified Elements," *Trans. IRE on Circuit Theory*, **CT-12**, No. 3, September 1965, pp. 381–387.

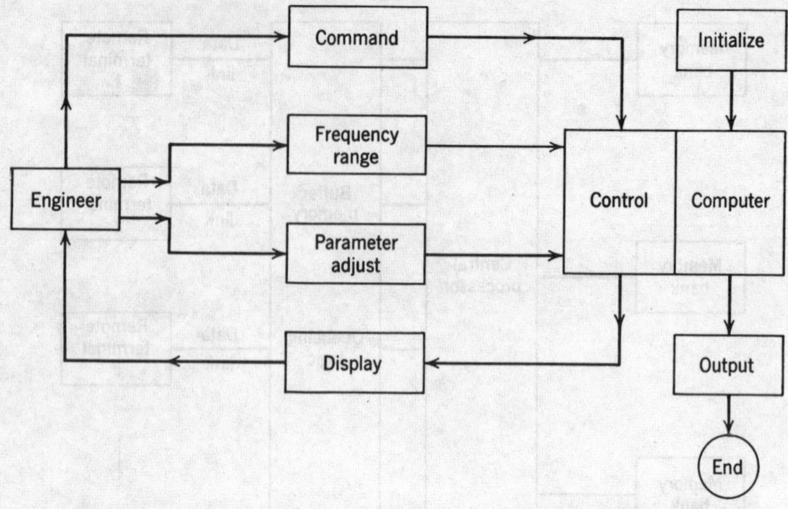

FIG. 15.12. Iterative design of network using machine-aided cognition.

Dr. So has written a computer program to formulate the hybrid matrix for the n-port network automatically.[11] The inputs to this program are (1) the node connections specifying the network topology; (2) the impedance functions of the elements; and (3) the specifications of the special elements to be extracted. This program was written with man-machine interaction in mind. The process is described in Fig. 15.12. First the computer reads in the n-port program with initial specifications. It performs the calculations and feeds information, such as transient or steady-state responses, back to the engineer via a visual display console. The engineer assesses the data and then changes the parameter values of the extracted elements, and, in some instances, the frequency range over which the calculations are made. The process is repeated a number of times until the engineer has obtained the desired results. Such a program could also be controlled by a steepest descent steering program if the engineer wished to obtain his results with less "eyeballing."

Now let us examine some details of specific network analysis programs.

15.2 AMPLITUDE AND PHASE SUBROUTINE

Purpose

Our purpose is to compute the amplitude and phase of a rational function

$$H(s) = \frac{C_n\, E(s)}{C_d\, F(s)} \tag{15.4}$$

[11] H. C. So, unpublished memorandum.

over a set of frequencies ω_i. In Eq. 15.4, C_n and C_d are real constants, and $E(s)$ and $F(s)$ are polynomials in $s = \sigma + j\omega$. The program described here computes the amplitudes $M(\omega_k)$ and phase $\phi(\omega_k)$ over a set of frequencies ω_k, where

$$M(\omega_k) = \frac{|C_n E(j\omega_k)|}{|C_d F(j\omega_k)|} \qquad (15.5)$$

and

$$\phi(\omega_k) = \arctan \frac{E_i(j\omega_k)}{E_r(j\omega_k)} - \arctan \frac{F_i(j\omega_k)}{F_r(j\omega_k)} \qquad (15.6)$$

where in Eq. 15.6 the r and i subscripts indicate real and imaginary parts, respectively. Note that in Eq. 15.6, $\phi(\omega_k)$ is limited to the range $-\pi \leq \phi \leq \pi$ radians. In order to circumvent this restriction on the phase, we use the following method to compute amplitude and phase.

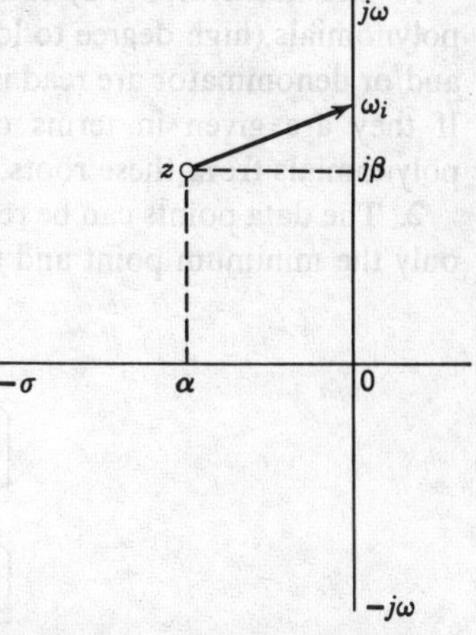

FIG. 15.13. Calculation of magnitude and phase at ω_i due to zero z.

Method used

Let $H(s)$ be factored into poles and zeros such that

$$H(s) = \frac{C_n(s - z_1)(s - z_2)\cdots(s - z_n)}{C_d(s - p_1)(s - p_2)\cdots(s - p_m)} \qquad (15.7)$$

Consider the amplitude and phase due to any pole or zero, for example, $z = -\alpha + j\beta$, shown in Fig. 15.13. The magnitude is

$$M_z(\omega_i) = [\alpha^2 + (\omega_i - \beta)^2]^{1/2} \qquad (15.8)$$

and the phase is

$$\phi_z(\omega_i) = \arctan\left(\frac{\omega_i - \beta}{\alpha}\right) \qquad (15.9)$$

The amplitude and phase of the overall function is

$$M(\omega_i) = \frac{C_n \prod_{k=1}^{n} M_{zk}(\omega_i)}{C_d \prod_{l=1}^{m} M_{pl}(\omega_i)} \qquad (15.10)$$

and

$$\phi(\omega_i) = \sum_{k=1}^{n} \phi_{zk}(\omega_i) - \sum_{l=1}^{m} \phi_{pl}(\omega_i) \qquad (15.11)$$

where the subscripts p and z indicate the contribution due to a pole and zero, respectively.

Input

1. The numerator $E(s)$ and denominator $F(s)$ can be read in either as polynomials (high degree to low) or in terms of their roots. If numerator and/or denominator are read in as polynomials, their roots will be printed. If they are given in terms of their roots, the program could generate polynomials from these roots.

2. The data points can be read in if unequally spaced; if equally spaced, only the minimum point and the spacing need be read in.

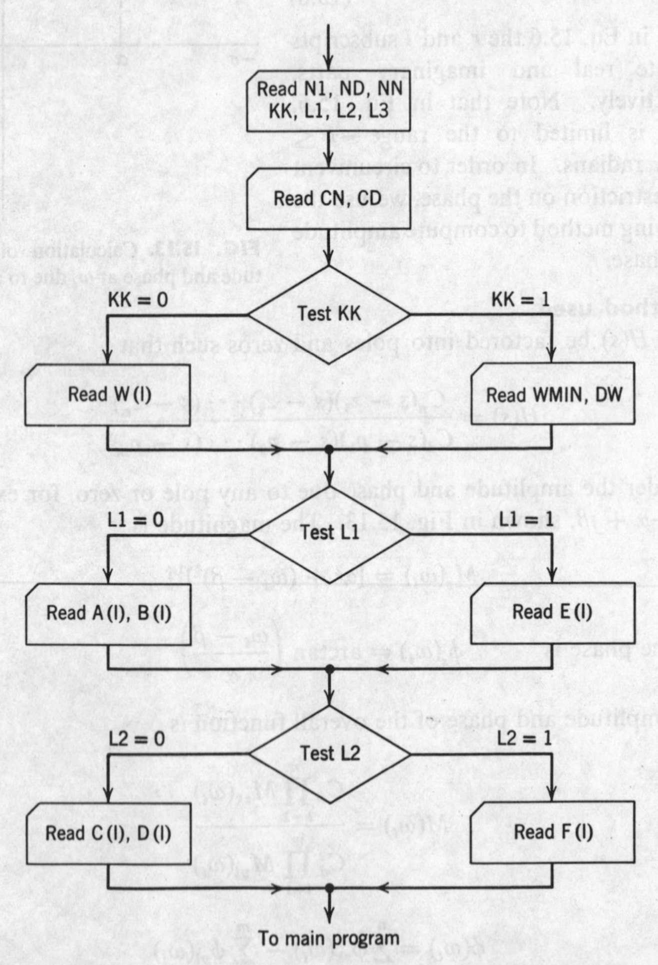

FIG. 15.14. Flow chart of input instructions for magnitude and phase program.

A flow chart listing the input instructions is given in Fig. 15.14. Definitions of the symbols in the flow chart follow.

N1 = number of frequency points
ND = degree of denominator
NN = degree of numerator
KK = 0 indicates equally spaced data points. We then read in only ω_{MIN}, $\Delta\omega$, and number of points.
= 1 indicates unequally spaced data points. We then read in all ω_i.
L1 = 0 Read in zeros (if complex conjugate, both zeros must be read in).
= 1 Read in numerator polynomial, high degree to low.
L2 = 0 Read in poles (if complex conjugate, both poles must be read in).
= 1 Read in denominator polynomial, high degree to low.
CN = numerator multiplier
CD = denominator multiplier

Output

The main output of the program is the frequency response consisting of the following columns: frequency in radians W(I); amplitude, M(W(I)); magnitude in decibels, 20 $\log_{10} M$; and phase in degrees, ϕ(W(I)). A typical printout of an amplitude and phase program is given in Fig. 15.6, which shows the magnitude and phase of the band-pass filter designed using Szentirmai's program.

15.3 A FORTRAN PROGRAM FOR THE ANALYSIS OF LADDER NETWORKS

In this section we will discuss the methods and organization for a Fortran computer program for the analysis of linear ladder networks. The analysis proceeds by computing the voltage and current at the input of successive L sections, beginning with the terminating section.

The program calculates the branch impedances of the ladder by combining *R*, *L*, *C* series impedance arms according to the instructions of the individual programmer.

The poles and zeros and the frequency responses of the input impedance and voltage transfer ratio at the input of each L section are found. In addition, the program provides for the analysis of short- and open-circuited networks as well as for those with normal terminations. Separate problems may be run consecutively if so desired.

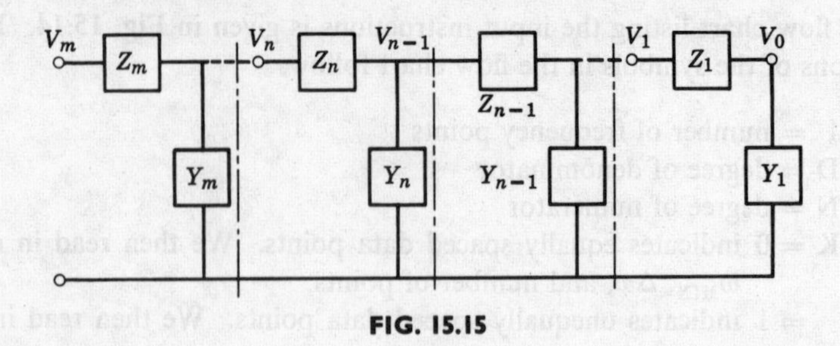

FIG. 15.15

The network is initially decomposed into separate L sections as in Fig. 15.15. These L sections are then added successively to the terminating L section to form the complete network. With the addition of an L section, the voltage and current at the input of the resulting network are computed by the equations

$$\begin{aligned} I_n &= V_{n-1}Y_n + I_{n-1} \\ V_n &= I_nZ_n + V_{n-1} \end{aligned} \tag{15.12}$$

which were originally discussed in Chapter 9 of this book. Thus the program proceeds toward the front of the ladder requiring only the branch immittances of the present L section and the voltage and current of the previous L section to make its calculations.

Suitable assumptions for the initial voltage and current V_0 and I_0 allow for analysis of short- and open-circuited networks as well as for those with normal terminations. These initial values of voltage and current are determined by control instructions.

For the calculation of the voltage and current at an L section, the branch impedances of the section are required. To simplify the preparation of the input data specifying these impedances, the following procedure is used. Each branch impedance is formed by the addition in series or parallel of basic *R*, *L*, *C* impedance arms. A basic *R*, *L*, *C* impedance arm is a series combination of a resistance, an inductance, and a capacitor—any or all of which may be absent. The impedance arms are specified by the element values *R*, *L*, *C* and an instruction indicating how the arm is to be combined.

The elements of the impedance arms are frequency and impedance normalized to aid computation, and then the arms are combined as specified to form the appropriate branch impedance.

The roots of the voltage and current polynomials for each L section are computed by a root-finding routine.

For the normal load and open circuit termination, the frequency responses of the input impedance, and voltage-transfer ratio are computed; for short-circuit termination, the input impedance and current gain are computed. Provision is made for either a *linear* or *logarithmic* increment in frequency. The frequency boundaries and increments are read by the program along with the input data.

According to instructions given by the programmer, the various calculations are printed for each L section or only for the network as a whole.

Program operation

The ladder network-analysis program requires two sets of input data. The first set consists of the control instructions. These tell the computer which of the various options, such as information concerning polynomial roots or frequency responses, are to be exercised.

In addition, the control instructions provide the computer with necessary parameters such as normalizing factors, and the number of L sections in the ladder. The second set of data determines the branch impedances of the network by specifying the R, L, C impedance arms and their combining instructions.

Impedance data

The data specifying the branch impedances consist of an ordered series of instruction cards, each determining an impedance arm and/or a combining instruction. The first arm of any branch impedance requires no combining instruction. The branch impedance is set equal to the impedance of this arm. Subsequent impedance arms are added in parallel or series with the existing branch impedance, according to the combining instructions of the arms. A blank card tells the computer that a complete branch impedance has been formed and that the next arm begins a new branch impedance. If blank cards are not properly inserted, the computer will calculate a single giant branch impedance.

The order of the data determining the branch impedances is Z_1, $1/Y_1$, Z_2, $1/Y_2$, ...

The values of R, L, C and the combining instruction XIN are written on one card in the order XIN, R, L, C.

The values of XIN and the associated operation are shown in the table at the bottom of page 456.

Although the instructions XIN = 1.0, 2.0 will suffice for the construction of many branch impedances, they are inadequate for other impedances.

For example, suppose the series combination of two tank circuits (Fig. 15.16) is desired as a branch impedance, this impedance may only be

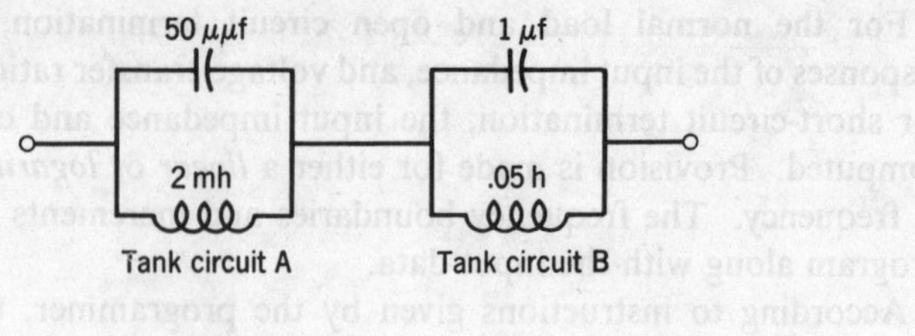

XIN	R	L	C
1.0			$5.00000E-11$
2.0		$2.00000E-03$	
3.0		.05	
2.0			.000001
4.0			
	Blank	Card	

FIG. 15.16. Construction of branch impedances using instructions for XIN.

constructed with use of three more instructions (XIN = 3.0, 4.0, and 5.0) and an additional set of computer storage locations.

Tank circuit A is formed in the usual manner, but must then be temporarily stored during the calculation of tank circuit B. The two tank circuits are then added to form the final branch impedance. Examples of impedances that cannot be formed are shown in Fig. 15.17. Observe that a series branch impedance of zero ohms often simplifies the formation of branch impedances. Two blank cards will indicate a short-circuited impedance.

XIN	Operation
Blank card	A complete branch impedance has been calculated, and the next arm begins a new branch impedance.
1.0	This arm is added in series with the existing branch impedance.
2.0	This arm is paralleled with the existing branch impedance.
3.0	This arm begins a new internal branch impedance. Subsequent XIN = 1.0 and 2.0 refer to this new internal branch.
4.0	Add the two internal branches in series to form the final branch impedance.
5.0	Parallel the internal branches.

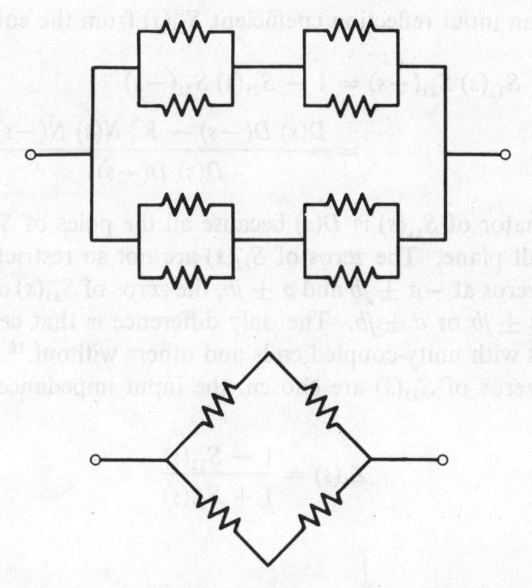

FIG. 15.17. Branch impedances that cannot be formed by program.

Output

The coefficients of the voltage and current polynomials are printed in frequency normalized form. To calculate the denormalized coefficients, divide by the normalizing frequency raised to the power of the corresponding exponent.

The poles and zeros of the voltage and current polynomials are in frequency normalized form to facilitate zero-pole plots.

Frequency responses appear in denormalized form. The frequency variable is in radians per second. The impedance level is in decibels, the phase in degrees. The gain and phase of the transfer ratios are expressed similarly.

15.4 PROGRAMS THAT AID IN DARLINGTON FILTER SYNTHESIS

Two double precision Fortran programs have been written that help in the synthesis of double-terminated filters using the Darlington procedure. The mathematics of the programs is described here.

Given a forward transmission coefficient

$$S_{21}(s) = \frac{KN(s)}{D(s)} \tag{15.13}$$

we generate an input reflection coefficient $S_{11}(s)$ from the equation

$$\begin{aligned} S_{11}(s)\,S_{11}(-s) &= 1 - S_{21}(s)\,S_{21}(-s) \\ &= \frac{D(s)\,D(-s) - K^2\,N(s)\,N(-s)}{D(s)\,D(-s)} \end{aligned} \tag{15.14}$$

The denominator of $S_{11}(s)$ is $D(s)$ because all the poles of $S_{11}(s)$ must be in the left-half plane. The zeros of $S_{11}(s)$ are not so restricted. If $S_{11}(s)$ $S_{11}(-s)$ has zeros at $-a \pm jb$ and $a \pm jb$, the zeros of $S_{11}(s)$ can be chosen as either $-a \pm jb$ or $a \pm jb$. The only difference is that certain choices lead to filters with unity-coupled coils and others without.[12]

Once the zeros of $S_{11}(s)$ are chosen, the input impedance of the filter is then

$$Z_1(s) = \frac{1 - S_{11}(s)}{1 + S_{11}(s)} \tag{15.15}$$

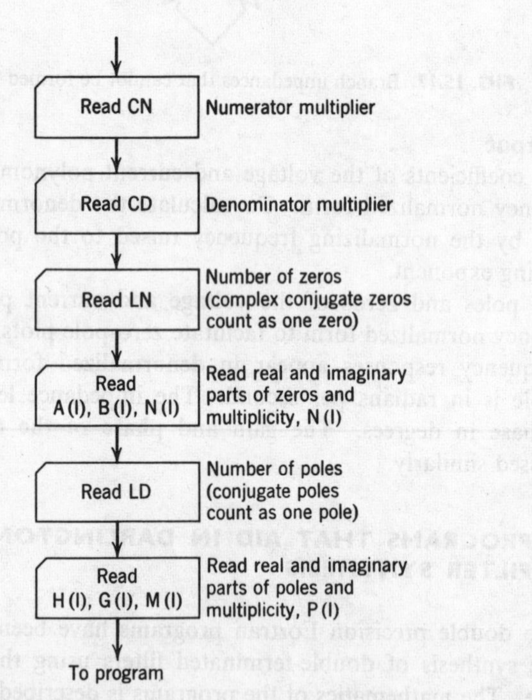

FIG. 15.18. Input to $S_{11}(s)\,S_{11}(-s)$ program.

[12] T. Fujisawa, "Realizability Theorem for Mid-Series or Mid-Shunt Low Pass Ladders without Mutual Induction," *Trans. IRE on Circuit Theory*, **CT-2**, December 1955, 320–325.

Program descriptions

We next proceed with brief descriptions of the two programs. The first program finds the roots of $S_{11}(s)\,S_{11}(-s)$ given the roots of $N(s)$, $N(-s)$, $D(s)$, and $D(-s)$. A flow chart for the input of the program is given in Fig. 15.18. Note that we must read in both the zeros of $N(s)$ and $N(-s)$, although only half of a complex conjugate pair need be read in. The same applies for the roots of $D(s)$ and $D(-s)$. The constant CN is K^2 in Eq. 15.14.

The second program performs the computations in Eq. 15.15 for different combinations of zeros of $S_{11}(s)$, given the fact that the zeros of $S_{11}(s)$ need not be in the left-hand plane. The previous program finds all the zeros of $S_{11}(s)\,S_{11}(-s)$. We must choose only half the number of zero pairs for $S_{11}(s)$. Certain combinations of zeros of $S_{11}(s)\,S_{11}(-s)$ lead to filters with coupled coils; others do not. If one wishes to try a number of combinations of zeros for $S_{11}(s)$, the program has the facility to enable him to do so. He need only supply those zeros of $S_{11}(s)\,S_{11}(-s)$ in the right-hand plane, $\mathrm{C(I)} + j\mathrm{B(I)}$, and a list of constants S(I), which are either 1.0 or -1.0 so that the zeros of $S_{11}(s)$ may be represented as $\mathrm{S(I)} \cdot \mathrm{C(I)} + j\mathrm{B(I)} = \mathrm{A(I)} + j\mathrm{B(I)}$. The routine forms numerator and denominator polynomials of $S_{11}(s)$

$$S_{11}(s) = \frac{E(s)}{F(s)} \tag{15.16}$$

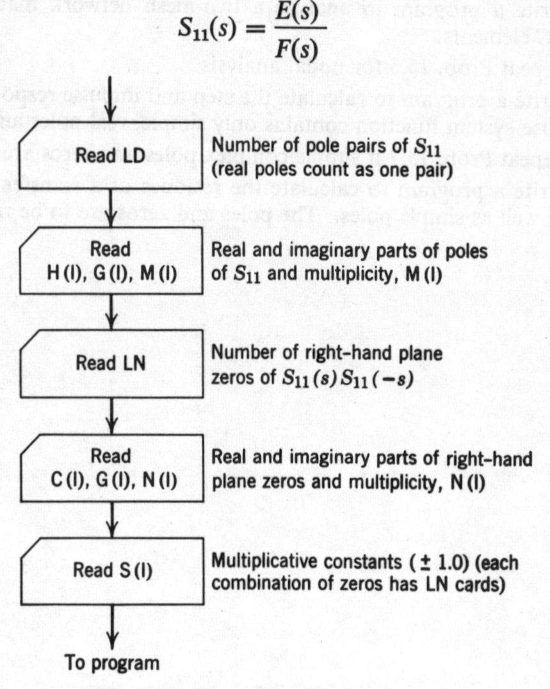

FIG. 15.19. Input to $Z_{\mathrm{in}}(s)$ program.

and then computes the numerator and denominator of $Z_1(s)$ as

$$Z_1(s) = \frac{1 - S_{11}}{1 + S_{11}} = \frac{F(s) - E(s)}{F(s) + E(s)} \tag{15.17}$$

The input instructions are summarized on the flow chart in Fig. 15.19.

Problems

15.1 Organize a flow-chart for computing magnitude and phase given only the unfactored numerator and denominator polynomials. Do *not* use a root-finding subroutine.

15.2 Write a program to calculate the delay of

$$H(s) = \frac{3(s + 1)}{s^2 + 2s + 5}$$

at the points $\omega = 0, 1, 2, \ldots, 10$.

15.3 Repeat Prob. 15.2 for a general transfer function given in terms of its unfactored numerator and denominator polynomials. The frequency points ω_i are to be read in by the computer.

15.4 Suppose you have calculated the phase of a transfer function at points $\omega = 0, 1, 2, \ldots, 50$. Devise an algorithm to test for phase linearity. The deviation from phase linearity is to be called *phase runoff*.

15.5 Write a program to analyze a two-mesh network made up of only *R*, *L*, and *C* elements.

15.6 Repeat Prob. 15.5 for nodal analysis.

15.7 Write a program to calculate the step and impulse response of a linear system whose system function contains only simple, real poles and zeros.

15.8 Repeat Prob. 15.7 if simple complex poles and zeros are allowed.

15.9 Write a program to calculate the residues of a transfer function with multiple as well as simple poles. The poles and zeros are to be real.

appendix A

Introduction to matrix algebra

A.1 FUNDAMENTAL OPERATIONS

Matrix notation is merely a shorthand method of algebraic symbolism that enables one to carry out the algebraic operations more quickly. The theory of matrices originated primarily from the need (1) to solve simultaneous linear equations and (2) to have a compact notation for linear transformations from one set of variables to another.

As an example of (2), consider the set of simultaneous linear equations

$$\begin{aligned} a_{11}x_1 + a_{12}x_2 + \cdots + a_{1n}x_n &= y_1 \\ a_{21}x_1 + a_{22}x_2 + \cdots + a_{2n}x_n &= y_2 \\ a_{m1}x_1 + a_{m2}x_2 + \cdots + a_{mn}x_n &= y_m \end{aligned} \tag{A.1}$$

These may express a general linear transformation from the x_i to the y_i. In general, $m \neq n$. An example where the numbers of variables in the two sets are unequal is that of representing a three-dimensional object in two dimensions (in a perspective drawing). Here, $m = 2$ and $n = 3$.

Definition

A matrix is an ordered rectangular array of numbers, generally the coefficients of a linear transformation. The matrix is defined by giving all its elements, and the location of each.

The matrix of the equations in Eq. A.1 written as

$$\begin{bmatrix} a_{11} & a_{12} & \cdots & a_{1n} \\ a_{21} & a_{22} & & a_{2n} \\ & & \cdots & \\ a_{m1} & a_{m2} & \cdots & a_{mn} \end{bmatrix}$$

is of *order* $m \times n$. (The first number here is the number of rows; the second is the number of columns.) The matrix may be denoted by a single capital letter **A** or by $[a_{ij}]$.

A matrix is a single complete entity, like a position in chess. Two matrices **A** and **B** are equal only if all corresponding elements are the same: $a_{ij} = b_{ij}$ for all i and j. A matrix may consist of a single row or single column. The complete matrix notation applied to Eq. A.1 is

$$\begin{bmatrix} a_{11} & a_{12} & \cdots & a_{1n} \\ a_{21} & a_{22} & \cdots & a_{2n} \\ & & \cdots & \\ a_{m1} & a_{m2} & \cdots & a_{mn} \end{bmatrix} \begin{bmatrix} x_1 \\ x_2 \\ \cdots \\ x_n \end{bmatrix} = \begin{bmatrix} y_1 \\ y_2 \\ \cdots \\ y_m \end{bmatrix} \tag{A.2}$$

which says, if put crudely, that **A** operates on x_i to yield y_i. This emphasizes the similarity between a matrix and a transformation. The **x**- and **y**-matrices are column matrices.

Row and column matrices are called *vectors* (specifically, row vectors and column vectors) and their similarity to the more usual type of vector is discussed later. Here, vectors will be written as small letters, such as **x** or **y**. Note that the elements of vectors need only one subscript, while elements of matrices need two.

A.2 ELEMENTARY CONCEPTS

Square matrix

A square matrix has the same number of rows as columns (i.e., a matrix of order *nn*). The **Y** matrix is an example of a square matrix

$$\mathbf{Y} = \begin{bmatrix} y_{11} & y_{12} \\ y_{21} & y_{22} \end{bmatrix} \tag{A.3}$$

Diagonal matrix

A diagonal matrix is a square matrix whose elements off the main diagonal are zero (i.e., one in which $\alpha_{ij} = 0$ for $i \neq j$). The following matrix is diagonal.

$$\begin{bmatrix} 1 & 0 & 0 \\ 0 & -2 & 0 \\ 0 & 0 & 3 \end{bmatrix} \tag{A.4}$$

Unit matrix

A unit matrix is a diagonal matrix for which $a_{ij} = 1$ for $i = j$, and is denoted as **U**. For example, **U** is

$$\mathbf{U} = \begin{bmatrix} 1 & 0 & 0 \\ 0 & 1 & 0 \\ 0 & 0 & 1 \end{bmatrix} \tag{A.5}$$

Equality

Two matrices are equivalent if they have the same number of rows and columns and if the elements of corresponding orientation are equal Suppose $y_1 = 2$, $y_2 = -3$, and $y_3 = -6$. If we write this set of equations in matrix form, we have

$$\begin{bmatrix} y_1 \\ y_2 \\ y_3 \end{bmatrix} = \begin{bmatrix} 2 \\ -3 \\ -6 \end{bmatrix} \tag{A.6}$$

Transpose

The transpose of a matrix $\mathbf{A}$ denoted as $\mathbf{A}^T$, is the matrix formed by interchanging the rows and columns of $\mathbf{A}$. Thus, if we have

$$\mathbf{A} = \begin{bmatrix} 1 & 0 \\ -6 & 5 \\ 3 & 2 \end{bmatrix} \tag{A.7}$$

then

$$\mathbf{A}^T = \begin{bmatrix} 1 & -6 & 3 \\ 0 & 5 & 2 \end{bmatrix} \tag{A.8}$$

Determinant of a matrix

The determinant of a matrix is defined only for square matrices and is formed by taking the determinant of the elements of the matrix. For example, we have

$$\det \begin{bmatrix} 1 & 2 \\ -5 & 4 \end{bmatrix} = \begin{vmatrix} 1 & 2 \\ -5 & 4 \end{vmatrix} = 14 \tag{A.9}$$

Note that the determinant of a matrix has a particular value, whereas the matrix itself is merely an array of quantities.

Cofactor

The cofactor A_{ij} of a square matrix is the determinant formed by deleting the ith row and jth column, and multiplying by $(-1)^{i+j}$. For example, the cofactor A_{21} of the matrix

$$\mathbf{A} = \begin{bmatrix} 2 & 6 \\ 3 & 4 \end{bmatrix} \tag{A.10}$$

is

$$A_{21} = (-1)^{2+1} \times 6 = -6 \tag{A.11}$$

Adjoint matrix

The adjoint matrix of a square matrix $\mathbf{A}$ is formed by replacing each element of $\mathbf{A}$ by its cofactor and transposing. For example, for the matrix in Eq. A.10, we have

$$\operatorname{adj}\mathbf{A} = \begin{bmatrix} 4 & -3 \\ -6 & 2 \end{bmatrix}^T = \begin{bmatrix} 4 & -6 \\ -3 & 2 \end{bmatrix} \tag{A.12}$$

Singular and nonsingular matrices

A singular matrix is a square matrix $\mathbf{A}$ for which $\det \mathbf{A} = 0$. A nonsingular matrix is one for which $\det \mathbf{A} \neq 0$.

A.3 OPERATIONS ON MATRICES

Addition

Two matrices may be added if both matrices are of the same order. Each element of the first matrix is added to the element of the second matrix, whose row and column orientation is the same. An example of matrix addition is shown in Eq. A.13.

$$\begin{bmatrix} -2 & 4 \\ 3 & 0 \end{bmatrix} + \begin{bmatrix} 3 & 6 \\ -7 & -2 \end{bmatrix} = \begin{bmatrix} 1 & 10 \\ -4 & -2 \end{bmatrix} \tag{A.13}$$

Thus if matrices $\mathbf{A}, \mathbf{B}, \mathbf{C}, \ldots, \mathbf{K}$ are all of the same order, then

$$\mathbf{A} + \mathbf{B} + \cdots + \mathbf{K} = [a_{ij} + b_{ij} + \cdots + k_{ij}] \tag{A.14}$$

The *associative* and *commutative* laws apply.

$$\begin{aligned} &\text{Associative:} \quad \mathbf{A} + (\mathbf{B} + \mathbf{C}) = (\mathbf{A} + \mathbf{B}) + \mathbf{C} \\ &\text{Commutative:} \quad \mathbf{A} + \mathbf{B} = \mathbf{B} + \mathbf{A} \end{aligned} \tag{A.15}$$

Multiplication by a scalar

We define multiplication by a scalar as

$$\lambda\mathbf{A} = \lambda[a_{ij}] = [\lambda a_{ij}] \tag{A.16}$$

Thus to multiply a matrix by a scalar, multiply each of its elements by the scalar.

Example A.1

$$3\begin{bmatrix} 2 & 0 \\ -1 & 1 \\ -3 & 2 \end{bmatrix} = \begin{bmatrix} 6 & 0 \\ -3 & 3 \\ -9 & 6 \end{bmatrix} \tag{A.17}$$

Linear combination of matrices

If the rules of addition and multiplication by a scalar are combined, for two matrices of the same order we have

$$\alpha\mathbf{A} + \beta\mathbf{B} = [\alpha a_{ij} + \beta b_{ij}] \tag{A.18}$$

where α and β are scalars.

Multiplication

In order for matrix multiplication **AB** to be possible, the number of columns of the first matrix **A** must equal the number of rows of the second matrix **B**. The product **C** will have the number of rows of the first and the number of columns of the second matrix. In other words, if **A** has m rows and n columns, and **B** has n rows and p columns, then the product **C** will have m rows and p columns. The individual elements of **C** are given by

$$c_{ij} = \sum_{k=1}^{n} a_{ik} b_{kj} \tag{A.19}$$

Example A.2

$$\begin{bmatrix} 2 & -1 & 0 \\ -1 & 0 & 3 \end{bmatrix} \begin{bmatrix} 4 & 1 \\ 2 & 0 \\ -3 & 2 \end{bmatrix} = \begin{bmatrix} 6 & 2 \\ -13 & 5 \end{bmatrix} \tag{A.20}$$

Example A.3. The system of equations

$$\begin{aligned} z_{11}I_1 + z_{12}I_2 &= V_1 \\ z_{21}I_1 + z_{22}I_2 &= V_2 \end{aligned} \tag{A.21}$$

can be written in matrix notation as

$$\begin{bmatrix} z_{11} & z_{12} \\ z_{21} & z_{22} \end{bmatrix} \begin{bmatrix} I_1 \\ I_2 \end{bmatrix} = \begin{bmatrix} V_1 \\ V_2 \end{bmatrix} \tag{A.22}$$

We see that systems of equations can be very conveniently written in matrix notation.

Matrix multiplication is not generally commutative, that is,

$$\mathbf{A}_{mn}\mathbf{B}_{np} \neq \mathbf{B}_{np}\mathbf{A}_{mn} \tag{A.23}$$

Observe that the product **BA** is not defined unless $p = m$. Even a product of square matrices is generally not commutative, as may be seen in the following example.

$$\begin{bmatrix} 1 & 0 \\ -1 & 2 \end{bmatrix} \begin{bmatrix} -1 & 0 \\ 0 & 2 \end{bmatrix} = \begin{bmatrix} -1 & 0 \\ 1 & 4 \end{bmatrix} \tag{A.24}$$

If we interchange the order of multiplication, we obtain

$$\begin{bmatrix} -1 & 0 \\ 0 & 2 \end{bmatrix} \begin{bmatrix} 1 & 0 \\ -1 & 2 \end{bmatrix} = \begin{bmatrix} -1 & 0 \\ -2 & 4 \end{bmatrix} \tag{A.25}$$

Because of the noncommutative nature of matrix multiplication, we must distinguish between premultiplication and postmultiplication. In **BA**, **A** is premultiplied by **B**; **B** is postmultiplied by **A**. For matrix multiplication, the associative and distributive laws apply.

$$\begin{aligned} \text{Associative:} \quad & \mathbf{A}(\mathbf{BC}) = (\mathbf{AB})\mathbf{C} = \mathbf{ABC} \\ \text{Distributive:} \quad & \mathbf{A}(\mathbf{B} + \mathbf{C}) = \mathbf{AB} + \mathbf{AC} \end{aligned} \tag{A.26}$$

Transpose of a product

The transpose of a product **AB** is equal to the product, in reverse order, of the transpose of the individual matrices **A** and **B**, that is,

$$(\mathbf{AB}) = \mathbf{B}^T\mathbf{A}^T \tag{A.27}$$

and

$$(\mathbf{ABC})^T = \mathbf{C}^T(\mathbf{AB})^T = \mathbf{C}^T\mathbf{B}^T\mathbf{A}^T \tag{A.28}$$

The product $\mathbf{x}^T\mathbf{x}$, if **x** is a column vector, is a scalar number equal to the sum of the squares of the elements of **x**. Thus we have

$$\mathbf{x}^T\mathbf{x} = [x_1 x_2, \ldots, x_n] \begin{bmatrix} x_1 \\ x_2 \\ \cdots \\ x_n \end{bmatrix} \tag{A.29}$$

$$= x_1^2 + x_2^2 + \cdots + x_n^2$$

The product $\mathbf{xx}^T$ is a square matrix **C** such that $\mathbf{C} = \mathbf{C}^T$.

Common expressions in simple notation

(*a*) The sum of products $a_1b_1 + a_2b_2 + \cdots + a_nb_n$, which are the typical element of a product matrix, may be written as $\mathbf{a}^T\mathbf{b}$ or $\mathbf{b}^T\mathbf{a}$ where **a** and **b** are column vectors.

(*b*) The sum of squares $x_1^2 + x_2^2 + \cdots + x_n^2$ is thus $\mathbf{x}^T\mathbf{x}$ (**x**, column vector). It is also $\mathbf{xx}^T$ if **x** is a row vector.

(*c*) The expression $a_{11}x_1^2 + a_{22}x_2^2 + \cdots + a_{nn}x_n^2$ is $\mathbf{x}^T\mathbf{Ax}$ (**x**, column vector), where **A** is a *diagonal* matrix with elements a_{jj}. Similarly, the expression $a_{11}x_1y_1 + a_{22}x_2y_2 + \cdots + a_{nn}x_ny_n$ is $\mathbf{x}^T\mathbf{Ay}$ or $\mathbf{y}^T\mathbf{Ax}$.

(*d*) An expression such as

$$\sum_i \sum_j a_{ij}x_ix_j, \qquad a_{ij} = a_{ji}$$

is a *quadratic form*. This is

$$\begin{aligned} a_{11}x_1^2 + 2a_{12}x_1x_2 + \cdots \qquad \cdots + 2a_{1n}x_1x_n \\ + a_{22}x_2^2 + 2a_{23}x_2x_3 + \cdots + 2a_{2n}x_2x_n \\ + \cdots + a_{nn}x_n^2 = \mathbf{x}^T\mathbf{A}\mathbf{x} \end{aligned} \tag{A.30}$$

Inverse

Division is not defined in matrix algebra. The analogous operation is that of obtaining the *inverse* of a square matrix. The inverse $\mathbf{A}^{-1}$ of a matrix $\mathbf{A}$ is defined by the relation

$$\mathbf{A}^{-1}\mathbf{A} = \mathbf{A}\mathbf{A}^{-1} = \mathbf{U} \tag{A.31}$$

To obtain $\mathbf{A}^{-1}$, we first obtain the adjoint of $\mathbf{A}$, adj $\mathbf{A}$. Then we obtain the determinant of $\mathbf{A}$. The inverse $\mathbf{A}^{-1}$ is equal to adj $\mathbf{A}$ divided by $|\mathbf{A}|$, that is,

$$\mathbf{A}^{-1} = \frac{1}{|\mathbf{A}|} \operatorname{adj} \mathbf{A} \tag{A.32}$$

Example A.4. Let $\mathbf{A}$ be given as

$$\mathbf{A} = \begin{bmatrix} 2 & 1 \\ -1 & 1 \end{bmatrix} \tag{A.33}$$

Its determinant is

$$|\mathbf{A}| = 3 \tag{A.34}$$

and the cofactors are

$$\begin{aligned} A_{11} &= 1 \qquad A_{12} = 1 \\ A_{21} &= -1 \qquad A_{22} = 2 \end{aligned} \tag{A.35}$$

The adjoint matrix is

$$\begin{aligned} \operatorname{adj} \mathbf{A} &= \begin{bmatrix} 1 & 1 \\ -1 & 2 \end{bmatrix}^T \\ &= \begin{bmatrix} 1 & -1 \\ 1 & 2 \end{bmatrix} \end{aligned} \tag{A.36}$$

so that $\mathbf{A}^{-1}$ is

$$\begin{aligned} \mathbf{A}^{-1} &= \tfrac{1}{3} \begin{bmatrix} 1 & -1 \\ 1 & 2 \end{bmatrix} \\ &= \begin{bmatrix} \frac{1}{3} & -\frac{1}{3} \\ \frac{1}{3} & \frac{2}{3} \end{bmatrix} \end{aligned} \tag{A.37}$$

As a check we see that

$$\begin{bmatrix} \frac{1}{3} & -\frac{1}{3} \\ \frac{1}{3} & \frac{2}{3} \end{bmatrix} \begin{bmatrix} 2 & 1 \\ -1 & 1 \end{bmatrix} = \begin{bmatrix} 1 & 0 \\ 0 & 1 \end{bmatrix}$$

$$\begin{bmatrix} 2 & 1 \\ -1 & 1 \end{bmatrix} \begin{bmatrix} \frac{1}{3} & -\frac{1}{3} \\ \frac{1}{3} & \frac{2}{3} \end{bmatrix} = \begin{bmatrix} 1 & 0 \\ 0 & 1 \end{bmatrix} \tag{A.38}$$

If the determinant of the matrix is zero, then the inverse is not defined. In other words, only nonsingular square matrices have inverses.

A.4 SOLUTIONS OF LINEAR EQUATIONS

Consider a set of linear algebraic equations, to be solved simultaneously.

$$\begin{aligned} a_{11}x_1 + a_{12}x_2 + \cdots + a_{1n}x_n &= h_1 \\ a_{21}x_1 + a_{22}x_2 + \cdots + a_{2n}x_n &= h_2 \\ &\cdots \\ a_{n1}x_1 + a_{n2}x_2 + \cdots + a_{nn}x_n &= h_n \end{aligned} \tag{A.39}$$

The h_i are constants. It is desired to solve for the x_j. In matrix notation we have

$$\mathbf{Ax} = \mathbf{h} \tag{A.40}$$

Premultiplying by $\mathbf{A}^{-1}$ gives $\quad \mathbf{x} = \mathbf{A}^{-1}\mathbf{h}$ (A.41)

In expanded form, this is

$$\begin{bmatrix} x_1 \\ x_2 \\ x_3 \end{bmatrix} = \frac{1}{|\mathbf{A}|} \begin{bmatrix} A_{11} & A_{21} & A_{31} \\ A_{12} & A_{22} & A_{32} \\ A_{13} & A_{23} & A_{33} \end{bmatrix} \begin{bmatrix} h_1 \\ h_2 \\ h_3 \end{bmatrix} \tag{A.42}$$

Thus

$$x_1 = \frac{h_1A_{11} + h_2A_{21} + h_3A_{31}}{|\mathbf{A}|} = \frac{\begin{vmatrix} h_1 & a_{12} & a_{13} \\ h_2 & a_{22} & a_{23} \\ h_3 & a_{32} & a_{33} \end{vmatrix}}{|\mathbf{A}|}$$

and similarly for x_2 and x_3. This is the familiar *Cramer's rule* for solving such equations.

Example A.5. Solve for x, y, z.

$$\begin{aligned} x - y + z &= 2 \\ 2x + y &= 1 \\ -x + 3y + z &= -1 \end{aligned} \tag{A.43}$$

In matrix form, these equations are written as $\mathbf{Ax} = \mathbf{h}$ where

$$\mathbf{A} = \begin{bmatrix} 1 & -1 & 1 \\ 2 & 1 & 0 \\ -1 & 3 & 1 \end{bmatrix}, \quad \mathbf{x} = \begin{bmatrix} x \\ y \\ z \end{bmatrix}, \quad \mathbf{h} = \begin{bmatrix} 2 \\ 1 \\ -1 \end{bmatrix} \tag{A.44}$$

Also, $|\mathbf{A}| = 10$. Thus, we have $\mathbf{x} = \mathbf{A}^{-1}\mathbf{h}$

$$\begin{bmatrix} x \\ y \\ z \end{bmatrix} = \tfrac{1}{10} \begin{bmatrix} 1 & 4 & -1 \\ -2 & 2 & 2 \\ 7 & -2 & 3 \end{bmatrix} \begin{bmatrix} 2 \\ 1 \\ -1 \end{bmatrix} = \tfrac{1}{10} \begin{bmatrix} 7 \\ -4 \\ 9 \end{bmatrix} \tag{A.45}$$

Therefore $\quad x = 0.7, y = -0.4, z = 0.9$

A.5 REFERENCES ON MATRIX ALGEBRA

The following is a short list of books on matrices that the reader might wish to examine.

A. C. Aitken, *Determinants and Matrices*, 9th Ed., Interscience Publishers, New York, 1956.

R. Bellman, *Introduction to Matrix Analysis*, McGraw-Hill Book Company, New York, 1960.

R. L. Eisenman, *Matrix Vector Analysis*, McGraw-Hill Book Company, New York, 1963.

D. K. Faddeev and V. N. Faddeeva, *Computational Methods of Linear Algebra*, W. H. Freeman and Company, San Francisco, 1963.

F. R. Gantmacher, *Applications of the Theory of Matrices*, Interscience Publishers, New York, 1959.

F. E. Hohn, *Elementary Matrix Algebra*, The Macmillan Company, New York, 1964.

L. P. Huelsman, *Circuits, Matrices, and Linear Vector Spaces*, McGraw-Hill Book Company, New York, 1963.

P. LeCorbeiller, *Matrix Analysis of Electric Networks*, John Wiley and Sons, New York, 1950.

M. Marcus and H. Minc, *Survey of Matrix Theory and Matrix Inequalities*, Allyn and Bacon, Boston, 1964.

E. D. Nering, *Linear Algebra and Matrix Theory*, John Wiley and Sons, New York, 1964.

S. Perlis, *Theory of Matrices*, Addison-Wesley, Reading, Massachusetts, 1952.

L. A. Pipes, *Matrix Methods for Engineering*, Prentice-Hall, Englewood Cliffs, N.J., 1963.

A. M. Tropper, *Matrix Theory for Electrical Engineering Students*, Harrop, London, 1962.

A. von Weiss, *Matrix Analysis for Electrical Engineers*, D. Van Nostrand, Princeton, N.J., 1964.

appendix B

Generalized functions and the unit impulse

B.1 GENERALIZED FUNCTIONS

The unit impulse, or delta function, is a mathematical anomaly. P. A. M. Dirac, the physicist, first used it in his writings on quantum mechanics.[1] He defined the delta function $\delta(x)$ by the equations

$$\int_{-\infty}^{\infty} \delta(x)\,dx = 1 \qquad \delta(x) = 0 \quad \text{for} \quad x \neq 0 \tag{B.1}$$

Its most important property is

$$\int_{-\infty}^{\infty} f(x)\,\delta(x)\,dx = f(0) \tag{B.2}$$

where $f(x)$ is continuous at $x = 0$. Dirac called the delta function an *improper function*, because there existed no rigorous mathematical justification for it at the time. In 1950 Laurent Schwartz[2] published a treatise entitled *The Theory of Distributions*, which provided, among other things, a fully rigorous and satisfactory basis for the delta function. Distribution theory, however, proved too abstract for applied mathematics and

[1] P. A. M. Dirac, *The Principles of Quantum Mechanics*, Oxford University Press, 1930.

[2] L. Schwartz, *Theorie des Distributions*, Vols. I and II, Hermann et Cie, Paris, 1950 and 1951.

physicists. It was not until 1953, when George Temple produced a more elementary (although no less rigorous) theory through the use of *generalized functions,*[3] that this new branch of analysis received the attention it deserved. Our treatment of generalized functions will be limited to the definition of the generalized step function and its derivative, the unit impulse. The treatment of these functions follows closely the work of Temple[4] and Lighthill.[5]

To get an idea of what a generalized function is, it is convenient to use as an analogy the notion of an irrational number α beng a sequence $\{\alpha_n\}$ of rational numbers α_n such that

$$\alpha = \lim_{n\to\infty} \alpha_n$$

where the limit indicates that the points α_n on the real line converge to the point representing α. All arithmetic operations performed on the irrational number α are actually performed on the sequence $\{\alpha_n\}$ defining α.[6] We can also think of a generalized function as being a sequence of functions, which when multiplied by a test function and integrated over $(-\infty, \infty)$ yields a finite limit. Before we formally define a generalized function, it is important to consider the definition of (1) a testing function and (2) a regular sequence.

DEFINITION B.1 A function $\phi(t)$ of class $C[\phi(t) \in C]$ is one that (1) is differentiable everywhere, any number of times and that (2) when it or any of its derivatives are multiplied by t raised to any power, the limit is

$$\lim_{t\to\pm\infty} [t^m \phi^{(k)}(t)] \to 0 \quad \text{for all} \quad m \,\&\, k \geq 0 \tag{B.3}$$

Any testing function is a function of class C.

Example B.1. The Gaussian function e^{-t^2/n^2} is a function of class C. It is obvious that if a function is of class C then all of its derivatives belong to class C.

---

[3] G. Temple, "Theories and Applications of Generalized Functions," *J. London Math. Soc.*, **28,** 1953, 134–148.

[4] G. Temple, "The Theory of Generalized Functions," Proc. Royal Society, *A*, **228,** 1955, 175–190.

[5] M. J. Lighthill, *Fourier Analysis and Generalized Functions,* Cambridge University Press, 1955. Lighthill dedicated his excellent book to "Paul Dirac, *who saw it must be true,* Laurent Schwartz, *who proved it,* and George Temple, *who showed how simple it could be made.*"

[6] This defines an irrational number according to the Cantor definition. For a more detailed account see any text on real variables such as E. W. Hobson, *The Theory of Functions of a Real Variable,* Vol. I, third edition, Chapter 1, Cambridge University Press, Cambridge, England, 1927.

DEFINITION B.2 A sequence $\{f_n(t)\}$ of functions of class C is said to be *regular* if for any function $\phi(t)$ belonging to C, the limit

$$\lim_{n\to\infty}(f_n, \phi) = \lim_{n\to\infty}\int_{-\infty}^{\infty} f_n(t)\,\phi(t)\,dt \tag{B.4}$$

exists. Note that it is not necessary that the sequence converge pointwise. For example, the sequence $\{e^{-nt^2}(n/\pi)^{1/2}\}$ approaches infinity as $n \to \infty$ at the point $t = 0$. However, the limit $\lim_{n\to\infty}(f_n, \phi)$ exists.

DEFINITION B.3 Two regular sequences $\{f_n\}$ and $\{g_n\}$ are *equivalent* if for all $\phi \in C$

$$\lim_{n\to\infty}(f_n, \phi) = \lim_{n\to\infty}(g_n, \phi) \tag{B.5}$$

Example B.2. The regular sequences $\{e^{-nt^2}(n/\pi)^{1/2}\}$ and $\{e^{-t^2/2n^2}(1/\sqrt{2\pi n})\}$ are equivalent.

DEFINITION B.4 A *generalized function* g is defined as a total, or complete, class of equivalent regular sequences. The term *total* implies here that there exists no other equivalent regular sequence not belonging to this class. Any member of the class, for example, $\{g_n\}$, is sufficient to represent both g and the total class of equivalent regular sequences defining g. We denote this symbolically by the form $g \sim \{g_n\}$.

Example B.3. All of the equivalent, regular sequences

$$[\{e^{-t^2/n^2}\}, \{e^{-t^4/n^4}\}, \{e^{-t^6/n^6}\}, \ldots, \{e^{-t^{2k}/n^{2k}}\}]$$

represent the same generalized function $g \sim \{e^{-t^2/n^2}\}$.

DEFINITION B.5 The inner product (g, ϕ) of a generalized function, g and a function $\phi(t) \in C$ is defined as

$$(g, \phi) = \lim_{n\to\infty}\int_{-\infty}^{\infty} g_n(t)\,\phi(t)\,dt \tag{B.6}$$

The inner product is often given the following symbolic representation.

$$(g, \phi) = \int_{-\infty}^{\infty} g(t)\,\phi(t)\,dt \tag{B.7}$$

Note that the integral here is used symbolically and does not imply actual integration.

DEFINITION B.6 If g and h are two generalized functions represented by the sequences $g \sim \{g_n\}$ and $h \sim \{h_n\}$, the sum $g + h$ is defined by the representation $g + h \sim \{g_n + h_n\}$.

Note that the set of sequences $\{g_n + h_n\}$ represents a total class of equivalent regular sequences made up of the sum of sequences defining g and h; therefore $g + h$ is a properly defined generalized function.

DEFINITION B.7 The product αg of a generalized function $g \sim \{g_n\}$ and a constant α is defined by the representation $\alpha g \sim \{\alpha g_n\}$.

DEFINITION B.8 The derivative g' of a generalized function $g \sim \{g_n\}$ is defined by the representation $g' \sim \{g'_n\}$.

Example B.4. For the generalized function $g_1 \sim \{e^{-t^2/n^2}\}$ the derivative is represented by

$$g'_1 \sim \left\{-\frac{2t}{n^2}e^{-t^2/n^2}\right\}$$

and

$$(g'_1, \phi) = \lim_{n\to\infty}\int_{-\infty}^{\infty}\left(-\frac{2t}{n^2}e^{-t^2/n^2}\right)\phi(t)\,dt \tag{B.8}$$

In Definitions B.6, B.7, and B.8 we have defined the operations of addition, multiplication by a scalar and differentiation. It must be pointed out that the operation of multiplication between two generalized functions is not defined in general.

We next consider an important theorem, whose proof is given in Lighthill,[7] which will enable us to represent any ordinary function, such as a step function by a generalized function equivalent.

Theorem B.1. Given any ordinary function $f(t)$ satisfying the condition

$$\int_{-\infty}^{\infty}\frac{|f(t)|}{(1+t^2)^N}\,dt < \infty \tag{B.9}$$

for some $N \geq 0$, there exists a generalized function[8] $\boldsymbol{f} \sim \{f_n(t)\}$ such that

$$(\boldsymbol{f}, \phi) = \int_{-\infty}^{\infty} f(t)\,\phi(t)\,dt \tag{B.10}$$

for all $\phi \in C$. In other words, an ordinary function satisfying Eq. B.9 is equivalent in terms of inner products to a generalized function. Symbolically, we write $f = \boldsymbol{f}$. If, in addition, f is continuous in an interval, then $\lim_{n\to\infty} f_n = f$ *pointwise* in that interval.

Furthermore, it can be shown that all the operations of addition, scalar multiplication, and differentiation performed on both f and $\boldsymbol{f}$ yield equivalent results, that is,

$$(\alpha f_1 + \beta f_2)' = (\alpha \boldsymbol{f}_1 + \beta \boldsymbol{f}_2)' \tag{B.11}$$

when differentiation is permitted on the ordinary function.

---

[7] Lighthill, *op. cit.*, Section 2.3.

[8] Note that when we represent an ordinary function by generalized function equivalent, we use a bold face italic letter to denote the generalized function.

DEFINITION B.9 The generalized step function $\boldsymbol{u}$ is defined as the total class of equivalent regular sequences $\{u_n(t)\}$ such that

$$\begin{aligned}(\boldsymbol{u}, \phi) &= \lim_{n\to\infty} \int_{-\infty}^{\infty} u_n(t)\,\phi(t)\,dt \\ &= \int_{-\infty}^{\infty} u(t)\,\phi(t)\,dt \\ &= \int_{0}^{\infty} \phi(t)\,dt\end{aligned} \tag{B.12}$$

where $u(t)$ is the unit step defined in Chapter 2. That $\{u_n(t)\}$ exists is guaranteed by the previous theorem allowing representations of ordinary functions by generalized functions. Hence, we write $\boldsymbol{u} = u$.

Example B.5. The sequence

$$\begin{aligned} u_n(t) &= \exp\left[-\frac{1}{n}\left(\frac{k}{t} + t^2\right)\right] \qquad & t > 0 \\ &= 0 & t \le 0 \end{aligned} \tag{B.13}$$

which is plotted in Fig. B.1, is one member of the class of equivalent regular sequences which represents the generalized step function.

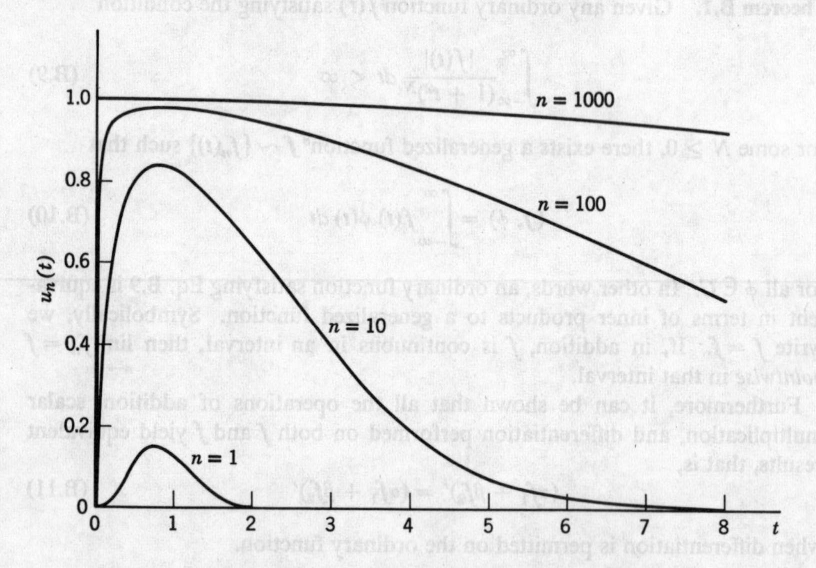

FIG. B.1. The generalized step sequence, $u_n(t)$.

DEFINITION B.10 The unit impulse, or Dirac delta function $\delta(t)$, is defined as the derivative of the generalized step function $\delta(t) \sim \{u'_n(t)\}$.

It should be stressed that $\delta(t)$ is merely the symbolic representation for a total class of equivalent regular sequences represented by $\{u'_n(t)\}$. Thus when we write the integral

$$\int_{-\infty}^{\infty} \delta(t)\, \phi(t)\, dt$$

we actually mean

$$\int_{-\infty}^{\infty} \delta(t)\, \phi(t)\, dt = (\delta, \phi) = \lim_{n\to\infty} \int_{-\infty}^{\infty} u'_n(t)\, \phi(t)\, dt \qquad \text{(B.14)}$$

Example B.6. The sequence

$$u'_n(t) = \left(\frac{k}{t^2} - 2t\right) \exp\left[-\frac{1}{n}\left(\frac{k}{t} + t^2\right)\right] \qquad t > 0$$

$$= 0 \qquad t \leq 0 \qquad \text{(B.15)}$$

in Fig. B.2 is one member of the class of equivalent regular sequences which represents the unit impulse. Other members of the class are the sequences $\{e^{-nt^2}(n/\pi)^{1/2}\}$ and $\{e^{-t^2/2n^2}(1/\sqrt{2\pi n})\}$.

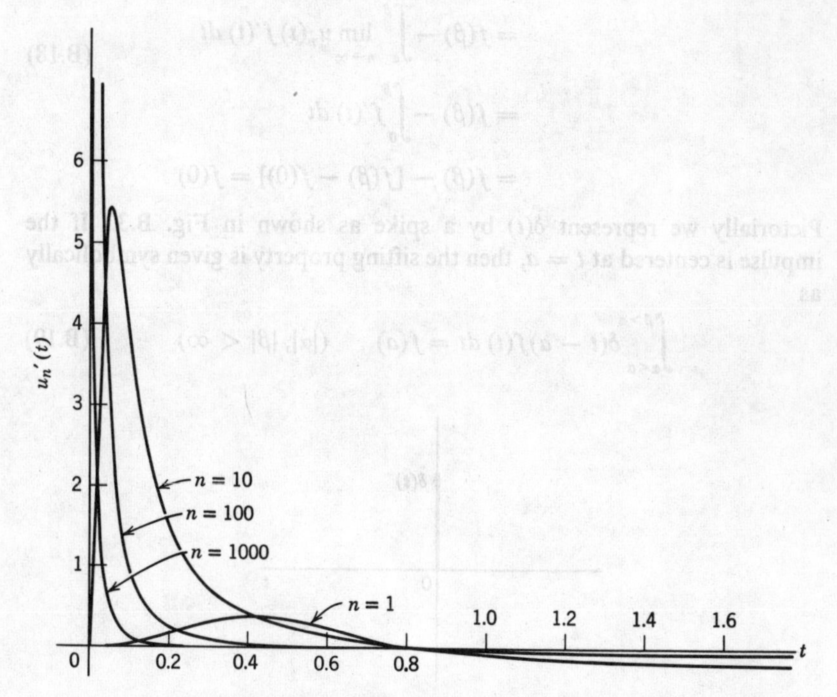

FIG. B.2. The generalized sequence, $u'_n(t)$.

B.2 PROPERTIES OF THE UNIT IMPULSE

Sifting

The most important property of the unit impulse is the sifting property represented symbolically by

$$\int_{\alpha<0}^{\beta>0} \delta(t) f(t)\, dt = f(0); \qquad (|\alpha|, |\beta| < \infty) \tag{B.16}$$

where f is any function differentiable over $[\alpha, \beta]$. The left hand side of Eq. B.16 is defined formally by

$$\int_{\alpha<0}^{\beta>0} \delta(t) f(t)\, dt \equiv \lim_{n\to\infty} \int_{\alpha<0}^{\beta>0} u'_n(t) f(t)\, dt \tag{B.17}$$

The proof of the sifting property is obtained by simply integrating by parts, as follows.

$$\begin{aligned}\lim_{n\to\infty} \int_{\alpha<0}^{\beta>0} u'_n(t) f(t)\, dt &= \lim_{n\to\infty} u_n(t) f(t)\Big|_{\alpha}^{\beta} - \lim_{n\to\infty} \int_{\alpha}^{\beta} u_n(t) f'(t)\, dt \\ &= f(\beta) - \int_{\alpha}^{\beta} \lim_{n\to\infty} u_n(t) f'(t)\, dt \\ &= f(\beta) - \int_{0}^{\beta} f'(t)\, dt \\ &= f(\beta) - [f(\beta) - f(0)] = f(0)\end{aligned} \tag{B.18}$$

Pictorially we represent $\delta(t)$ by a spike as shown in Fig. B.3. If the impulse is centered at $t = a$, then the sifting property is given symbolically as

$$\int_{\alpha<a}^{\beta>a} \delta(t-a) f(t)\, dt = f(a) \qquad (|\alpha|, |\beta| < \infty) \tag{B.19}$$

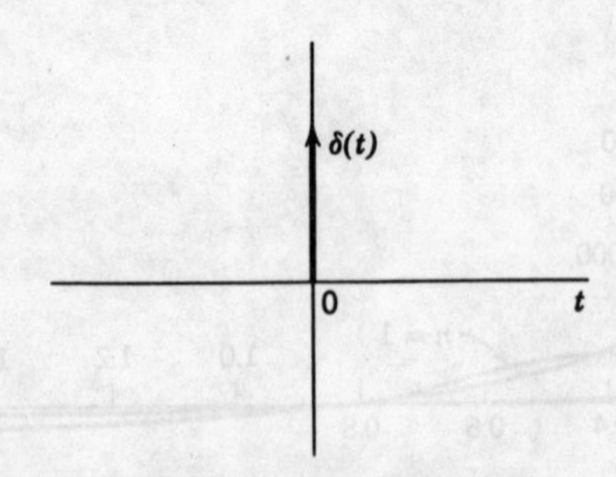

FIG. B.3. The unit impulse.

where $f'(t)$ must exist over $[\alpha, \beta]$. Note that when the limits of integration are infinite, we actually mean

$$\int_{-\infty}^{\infty} \delta(t) f(t)\, dt \equiv \lim_{\substack{\alpha \to -\infty \\ \beta \to \infty}} \int_{\alpha}^{\beta} \delta(t) f(t)\, dt \tag{B.20}$$

In the sifting property, if both $\alpha, \beta > 0$ or $\alpha, \beta < 0$, then

$$\int_{\alpha}^{\beta} \delta(t) f(t)\, dt = 0 \tag{B.21}$$

The proof of this property is similar to the original proof of the sifting property, and will be left as an exercise for the reader.

Integration

The defining equations of the delta function according to Dirac are

$$\int_{\alpha<0}^{\beta>0} \delta(x)\, dx = 1$$
$$\delta(x) = 0; \qquad x \neq 0 \tag{B.22}$$

These are actually *properties* of the delta function as viewed from the generalized function standpoint. The proof can be obtained directly from the sifting property. Suppose we have the integral

$$\int_{\alpha<0}^{\beta>0} \delta(t) f(t)\, dt = f(0) \tag{B.23}$$

and we let $f(t) = 1$. Then we have

$$\int_{\alpha<0}^{\beta>0} \delta(t)\, dt = f(0) = 1 \tag{B.24}$$

If both α, β are greater than zero or both are less than zero, then

$$\int_{\alpha}^{\beta} \delta(t)\, dt = 0 \tag{B.25}$$

This property is stated symbolically by the conditions $\delta(t) = 0$ for $t \neq 0$.

Differentiation across a discontinuity

Consider the function $f(t)$ in Fig. B.4. We see that $f(t)$ has a discontinuity of A at $t = T$. If we let $f_1(t) = f(t)$ for $t < T$, and $f_1(t) = f(t) - A$ for $t \geq T$, then we have

$$f(t) = f_1(t) + A\, u(t - T) \tag{B.26}$$

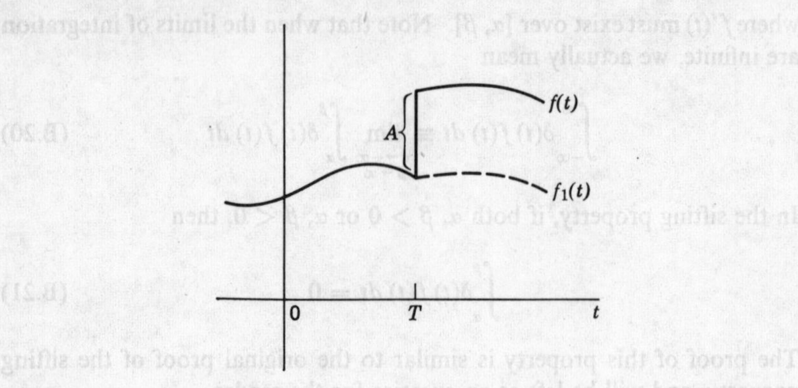

FIG. B.4. Function with discontinuity.

Since $f(t)$, $f_1(t)$, and $u(t)$ satisfy the condition

$$\int_{-\infty}^{\infty} \frac{|g(t)|}{(1 + t^2)^N}\, dt < \infty$$

for some N; we can represent these ordinary functions by generalized functions

$$(ft) = f_1(t) + u(t) \tag{B.27}$$

Taking derivatives on both sides of Eq. B.27 yields

$$f'(t) = f'_1(t) + A\, u'(t) \tag{B.28}$$

which symbolically can be written as

$$f'(t) = f'_1(t) + A\, \delta(t) \tag{B.29}$$

We thus see that whenever we differentiate across a discontinuity, we obtain a delta function times the height of the discontinuity.

Example B.7. The step response of an *R-C* network is given as

$$h(t) = Ae^{-t/T}\, u(t) \tag{B.30}$$

shown in Fig. B.5*a*. The impulse response is

$$h'(t) = A\, \delta(t) - \frac{A}{T} e^{-t/T}\, u(t) \tag{B.31}$$

and is shown in Fig. B.5*b*.

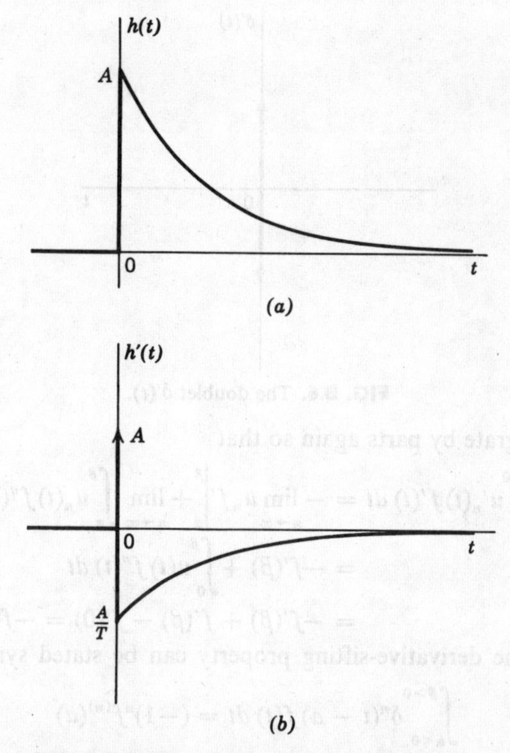

FIG. B.5. Differentiation across a discontinuity.

Differentiation

The derivative of a delta function, which we call a *doublet*, is defined symbolically as $\delta'(t) \sim \{u''_n(t)\}$. It has the following property, where $f''(t)$ exists over $[\alpha, \beta]$.

$$\int_{\alpha<0}^{\beta>0} \delta'(t) f(t)\, dt = -f'(0); \qquad (|\alpha|, |\beta| < \infty) \tag{B.32}$$

The proof is obtained through successive integration by parts.

$$\begin{aligned}\int_{\alpha<0}^{\beta>0} \delta'(t) f(t)\, dt &= \lim_{n\to\infty} \int_{\alpha<0}^{\beta>0} u''_n(t) f(t)\, dt \\ &= \lim_{n\to\infty} u'_n(t) f(t)\Big|_{\alpha<0}^{\beta>0} - \lim_{n\to\infty} \int_{\alpha<0}^{\beta>0} u'_n(t) f'(t)\, dt\end{aligned} \tag{B.33}$$

We see that since $\lim_{n\to\infty} u'_n(\beta) = \lim_{n\to\infty} u'_n(\alpha) = 0$,

$$\lim_{n\to\infty} u'_n(t) f(t)\Big|_{\alpha<0}^{\beta>0} = 0 \tag{B.34}$$

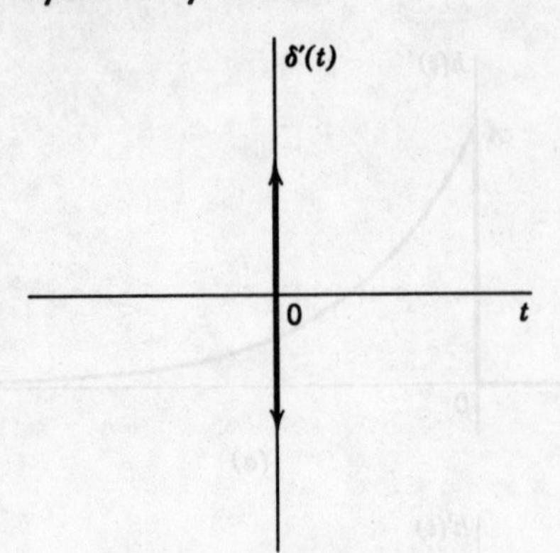

FIG. B.6. The doublet $\delta'(t)$.

We then integrate by parts again so that

$$\begin{aligned} -\lim_{n\to\infty}\int_{\alpha<0}^{\beta>0} u'_n(t)\,f'(t)\,dt &= -\lim_{n\to\infty} u_n f'\Big|_{\alpha}^{\beta} + \lim_{n\to\infty}\int_{\alpha}^{\beta} u_n(t)\,f''(t)\,dt \\ &= -f'(\beta) + \int_0^{\beta} u(t)\,f''(t)\,dt \\ &= -f'(\beta) + f'(\beta) - f'(0) = -f'(0) \end{aligned} \tag{B.35}$$

In general, the derivative-sifting property can be stated symbolically as

$$\int_{\alpha<0}^{\beta>0} \delta^n(t-a)\,f(t)\,dt = (-1)^n f^{(n)}(a) \tag{B.36}$$

where $f^{(n+1)}(t)$ exists over $[\alpha, \beta]$.[9]

The generalized function $\delta'(t)$ is sometimes called a doublet. The pictorial representation of a doublet is given in Fig. B.6.

Other properties of the unit impulse

Dirac and others have obtained a host of identities concerning the unit impulse. We will merely give these here without proof.

$$\delta(-t) = \delta(t) \tag{1}$$
$$\delta'(-t) = -\delta'(t) \tag{2}$$
$$t\,\delta(t) = 0 \tag{3}$$
$$t\,\delta'(t) = -\delta(t) \tag{4}$$
$$\delta(at) = |a|^{-1}\,\delta(t) \tag{5}$$
$$\delta(t^2 - a^2) = \tfrac{1}{2}\,|a|^{-1}\,\{\delta(t-a) + \delta(t+a)\} \tag{6}$$
$$f(t)\,\delta(t-a) = f(a)\,\delta(t-a) \tag{7}$$

The proofs of these properties are obtained through the inner product with a testing function $\phi(t) \in C$.

[9] The condition on $f^{(n+1)}$ is sufficient, but not necessary.

appendix C

Elements of complex variables

C.1 ELEMENTARY DEFINITIONS AND OPERATIONS

A complex variable z is a pair of real variables (x, y) written as

$$z = x + jy \tag{C.1}$$

where j can be thought of as $\sqrt{-1}$.

The variable x is called the real part of z, and y is the imaginary part of z. Written in simpler notation, we have

$$x = \mathrm{Re}\,(z), \qquad y = \mathrm{Im}\,(z) \tag{C.2}$$

The variable z can be plotted on a pair of rectangular coordinates. The abscissa represents the x or real axis, and the ordinate represents the y or imaginary axis. The plane upon which x and y are plotted is called the *complex* plane. Any point on the complex plane, such as $z = 3 + j2$, can be represented in terms of its real and imaginary parts, as shown in Fig. C.1. From the origin of the complex plane, let us draw a vector to any point z. The distance from the origin to z is given by

$$|z| = (x^2 + y^2)^{1/2} \tag{C.3}$$

and is known as the *modulus* of z. The angle which the vector subtends is known as the *argument* of z or

$$\arg z = \tan^{-1} \frac{y}{x} \tag{C.4}$$

Letting $\theta = \arg z$ and $r = |z|$, we can represent z in polar coordinates as

$$z = re^{j\theta} \tag{C.5}$$

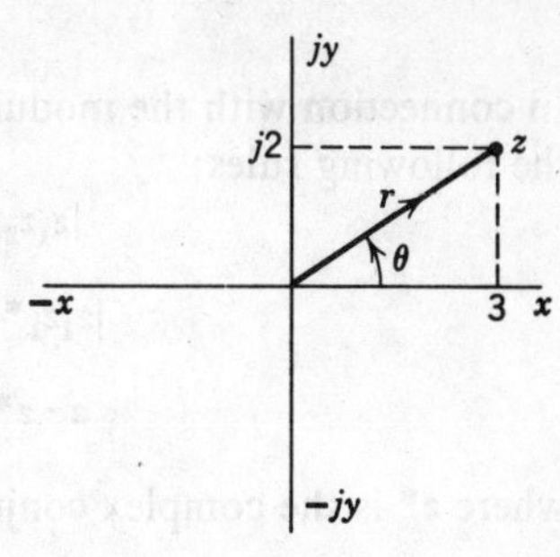

FIG. C.1

Expanding this last equation by Euler's formula, we obtain

$$z = r\cos\theta + jr\sin\theta, \tag{C.6}$$

so that

$$\begin{aligned} x &= r\cos\theta \\ y &= r\sin\theta \end{aligned} \tag{C.7}$$

The rule for addition for two complex numbers is given as

$$(a + jb) + (c + jd) = (a + c) + j(b + d) \tag{C.8}$$

When two complex numbers are multiplied, we have

$$(a + jb)(c + jd) = (ac - bd) + j(ad + bc) \tag{C.9}$$

where $j^2 = -1$. If we express the complex numbers in polar form, we obtain

$$(a + jb) = r_1 e^{j\theta_1} \tag{C.10}$$

and

$$(c + jd) = r_2 e^{j\theta_2} \tag{C.11}$$

When we multiply the two numbers in polar form, then

$$r_1 e^{j\theta_1} r_2 e^{j\theta_2} = r_1 r_2 e^{j(\theta_1 + \theta_2)} \tag{C.12}$$

If we divide these two numbers in polar form, then

$$\frac{r_1 e^{j\theta_1}}{r_2 e^{j\theta_2}} = \frac{r_1}{r_2} e^{j(\theta_1 - \theta_2)} \tag{C.13}$$

In rectangular coordinates, the operation of division can be expressed as

$$\begin{aligned} \frac{a + jb}{c + jd} &= \frac{(a + jb)(c - jd)}{(c + jd)(c - jd)} \\ &= \frac{ac + bd}{c^2 + d^2} + j\frac{bc - ad}{c^2 + d^2} \end{aligned} \tag{C.14}$$

In connection with the modulus of a complex number, it is useful to note the following rules:

$$\begin{aligned} |z_1 z_2| &= |z_1| \cdot |z_2| \\ |z_1 z_1^*| &= |z_1| \cdot |z_1^*| = |z_1|^2 \\ z \cdot z^* &= |z|^2 \end{aligned} \tag{C.15}$$

where z^* is the complex conjugate of z and is defined as

$$z^* = \overline{x + jy} = x - jy \tag{C.16}$$

The following rules deal with operations involving the conjugate definition:

$$\overline{z_1 + z_2} = z_1^* + z_2^*$$

$$\overline{z_1 z_2} = z_1^* \cdot z_2^* \qquad (C.17)$$

$$\overline{z_1/z_2} = z_1^*/z_2^*$$

Finally, if z has a modulus of unity, then

$$z = \frac{1}{z^*} \qquad (C.18)$$

The operations of raising a complex number to the nth power, or taking the nth root of a complex number, can be dealt with most readily by using the polar form of the number. Thus, we have $z^n = (re^{j\theta})^n = r^n e^{jn\theta}$,

and $$z^{1/n} = r^{1/n} e^{j[(\theta+2k\pi)/n]}, \qquad k, 0, 1, \ldots, n-1 \qquad (C.19)$$

C.2 ANALYSIS

If to each $z = x + jy$, we assign a complex number $w = u + jv$, then w is a function of z or

$$w = f(z) \qquad (C.20)$$

The following are examples of complex functions, i.e., functions of a complex variable:

$$w = 2z$$

$$w = \log_e z$$

$$w = 1/z \qquad (C.21)$$

$$w = z^2 + 4$$

$$w = |z|$$

We see that w may be complex, pure real, or pure imaginary, depending upon the particular relationship with z. In general, the real and imaginary parts of w are both functions of x and y. That is, if we let $w = u + jv$, then

$$u = f(x, y) \quad \text{and} \quad v = f(x, y) \qquad (C.22)$$

As an example, let us find u and v for the function $w = z^2 + 4$.

$$w = z^2 + 4 = (x + jy)^2 + 4 \qquad (C.23)$$

Simplifying, we obtain

$$w = (x^2 - y^2 + j2xy) + 4 \qquad (C.24)$$

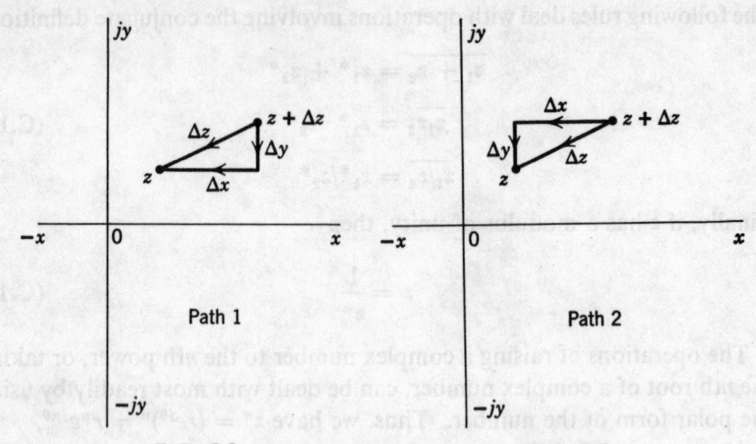

FIG. C.2 FIG. C.3

so that $$u = x^2 - y^2 + 4 \quad \text{and} \quad v = 2xy \tag{C.25}$$

The derivative of a complex function $f(z)$ is defined as

$$f'(z) = \lim_{\Delta z \to 0} \frac{f(z + \Delta z) - f(z)}{\Delta z} \tag{C.26}$$

If one restricts the direction or path along which Δz approaches zero, then we have what is known as a *directional derivative*. However, if a complex function is to possess a derivative at all, the derivative must be the same at any point regardless of the direction in which Δz approaches zero. In other words, in order for $f(z)$ to be differentiable at $z = z_0$, we must have

$$\left.\frac{df(z)}{dz}\right|_{z=z_0} = \text{constant} \tag{C.27}$$

for all directions of approach of Δz.

Consider the two directions in which Δz approaches zero in Figs. C.2 and C.3. For path 1, we have

$$f'(z) = \lim_{\Delta x \to 0} \lim_{\Delta y \to 0} \frac{f(z + \Delta z) - f(z)}{\Delta z} \tag{C.28}$$

If we substitute

$$\Delta z = \Delta x + j\,\Delta y \tag{C.29}$$

into Eq. C.28, we obtain

$$f'(z) = \lim_{\Delta x \to 0} \lim_{\Delta y \to 0} \frac{f[x + \Delta x + j(y + \Delta y)] - f(x + jy)}{\Delta x + j\,\Delta y} \tag{C.30}$$

Since $$f(z) = u + jv \tag{C.31}$$

and $$f(z + \Delta z) = u + \Delta u + j(v + \Delta v) \tag{C.32}$$

we finally arrive at

$$f'(z) = \lim_{\Delta x \to 0} \lim_{\Delta y \to 0} \frac{\Delta u + j\,\Delta v}{\Delta x + j\,\Delta y}$$
$$= \lim_{\Delta x \to 0} \frac{\Delta u + j\,\Delta v}{\Delta x} = \frac{\partial u}{\partial x} + j\,\frac{\partial v}{\partial x} \tag{C.33}$$

For path 2, we have

$$f'(z) = \lim_{\Delta y \to 0} \lim_{\Delta x \to 0} \frac{\Delta u + j\,\Delta v}{\Delta x + j\,\Delta y}$$
$$= \lim_{\Delta y \to o} \frac{\Delta u + j\,\Delta v}{j\,\Delta y} \tag{C.34}$$
$$= \frac{\partial v}{\partial y} - j\,\frac{\partial u}{\partial y}$$

Since we assume that the function $f(z)$ is differentiable, the derivatives must be independent of path. Thus, we have

$$\frac{\partial v}{\partial y} - j\,\frac{\partial u}{\partial y} = \frac{\partial u}{\partial x} + j\,\frac{\partial v}{\partial x} \tag{C.35}$$

From this last equation, we obtain the *Cauchy-Riemann* equations, which are

$$\frac{\partial v}{\partial y} = \frac{\partial u}{\partial x}$$
$$\frac{\partial u}{\partial y} = -\frac{\partial v}{\partial x} \tag{C.36}$$

We have just seen that in order for a function to have a derivative, the *Cauchy-Riemann* equations must hold. A function which is single valued and possesses a unique derivative is called an *analytic* function. A set of sufficient conditions for analyticity is that the Cauchy-Riemann equations are obeyed. For example, consider the function

$$f(z) = z^2 + 4 \tag{C.37}$$

$f(z)$ is analytic because

$$\frac{\partial v}{\partial y} = 2x = \frac{\partial u}{\partial x}$$
$$\frac{\partial u}{\partial y} = -2y = -\frac{\partial v}{\partial x} \tag{C.38}$$

On the other hand, $f(z) = z^*$ is not analytic because

$$u = x \quad \text{and} \quad v = -y$$
$$\frac{\partial u}{\partial x} = +1 \quad \text{and} \quad \frac{\partial v}{\partial y} = -1 \tag{C.39}$$

C.3 SINGULARITIES AND RESIDUES

If $f(z)$ is analytic within a region or *domain* in the complex plane except at a point z_0, then $f(z)$ has an *isolated singularity* at z_0. Suppose $f(z)$ has a singularity at z_0, then we can expand $f(z)$ about z_0 in a *Laurent* series

$$f(z) = \frac{a_{-n}}{(z - z_0)^n} + \cdots + \frac{a_{-1}}{z - z_0} + a_0(z - z_0)^0 + a_1(z - z_0) + \cdots + a_m(z - z_0)^m + \cdots \tag{C.40}$$

In the expansion, if m is finite, then z_0 is called a pole of order m.[1] The term a_{-1} is called the *residue* of the singularity.

Example C.1. Consider the Laurent series for the function $f(z) = e^z/z$ about the pole at the origin. We can expand e^z in a power series to give

$$\begin{aligned} \frac{e^z}{z} &= \frac{1}{z}\left(1 + z + \frac{1}{2!}z^2 + \frac{1}{3!}z^3 + \cdots\right) \\ &= \frac{1}{z} + 1 + \frac{1}{2!}z + \frac{1}{3!}z^2 + \cdots \end{aligned} \tag{C.41}$$

According to the definition, the residue of the pole at $z = 0$ is equal to 1.

Example C.2. Expand the function $f(z) = 1/z(z - 1)^2$ about the pole at $z = 1$, and find the residue of the pole at $z = 1$.

$$\begin{aligned} \frac{1}{z(z-1)^2} &= \frac{1}{(z-1)^2}\frac{1}{1 + (z - 1)} \\ &= \frac{1}{(z-1)^2}[1 - (z - 1) + (z - 1)^2 - (z - 1)^3 + \cdots] \\ &= \frac{1}{(z-1)^2} - \frac{1}{z - 1} + 1 - (z - 1) + (z - 1)^2 + \cdots \end{aligned} \tag{C.42}$$

for $0 < |z - 1| < 1$. Here, the residue of the pole at $z = 1$ is equal to -1.

---

[1] Note that if we have an infinite number of nonzero terms with negative exponents, then z_0 is an *essential singularity*.

Example C.3. Find the residues of the poles at $s = 0$ and $s = -1$ of the function

$$f(s) = \frac{s + 2}{s^2(s + 1)^2} \tag{C.43}$$

To find the residues, we simply perform a partial fraction expansion

$$f(s) = \frac{2}{s^2} - \frac{3}{s} + \frac{1}{(s + 1)^2} + \frac{3}{s + 1} \tag{C.44}$$

Thus the residue of the pole at $s = 0$ is -3, and the residue of the pole at $s = -1$ is $+3$.

C.4 CONTOUR INTEGRATION

In complex integration the integral is taken over a piecewise smooth path C and is defined as the limit of an infinite summation

$$\int_C f(z)\, dz = \lim_{n\to\infty} \sum_{j=1}^{n} f(z_j)\, \Delta z_j \tag{C.45}$$

where z_j lies on C. Unlike the process of differentiation, the path along which we take the integral makes a difference as to the ultimate value of the integral. Thus the integral

$$\int_{z_1}^{z_2} f(z)\, dz \tag{C.46}$$

in general, has different values depending upon whether we choose to integrate along path C_1 or path C_2, as shown in Fig. C.4. If we integrate along a *closed* path, say from a to b and then to a again, we are integrating along a *closed contour*. The path shown in Fig. C.5 is an example of a closed contour. The following theorem, known as *Cauchy's residue theorem* gives a method for rapid evaluation of integrals on closed paths.

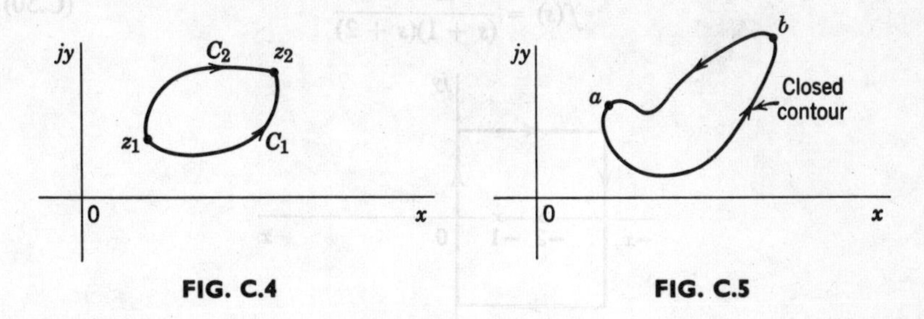

FIG. C.4

FIG. C.5

Theorem C.1. If C is a simple closed curve in a domain D, within which $f(z)$ is analytic except for isolated singularities at $z_1, z_2, \ldots, z_n$, then the integral along the closed path C is

$$\oint_C f(z)\,dz = 2\pi j(K_1 + K_2 + \cdots + K_n) \tag{C.47}$$

where K_i represents the residue of the singularity z_i.

Example C.4. Consider the integral

$$\oint \frac{s+2}{s^2(s+1)^2}\,ds \tag{C.48}$$

along the circle $|s| = 2$, as given in Fig. C.6. Since there are two singularities within the circle, at $s = 0$ and at $s = -1$, whose residues are respectively -3 and $+3$, then the integral along the circle is

$$\oint \frac{s+2}{s^2(s+1)^2}\,ds = 2\pi j(-3+3) = 0 \tag{C.49}$$

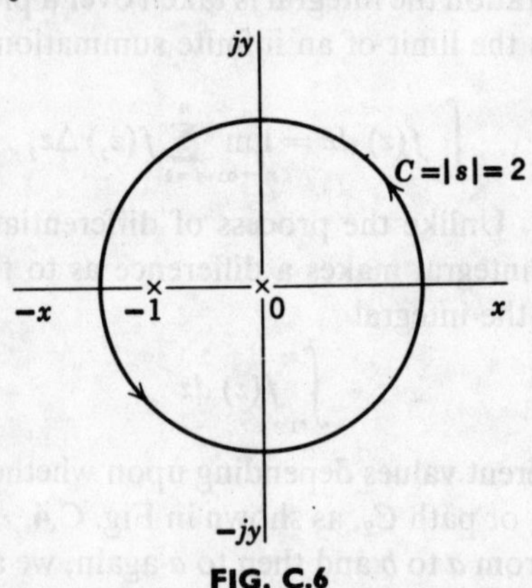

FIG. C.6

Example C.5. Find the integral of $f(s)$ along the closed contour shown in Fig. C.7. The function $f(s)$ is given as

$$f(s) = \frac{3s+5}{(s+1)(s+2)} \tag{C.50}$$

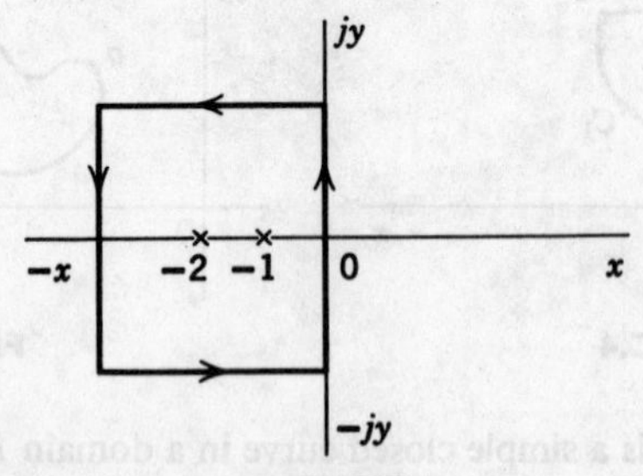

FIG. C.7

A partial fraction expansion of $f(s)$ shows that

$$f(s) = \frac{1}{s+1} + \frac{2}{s+2} \tag{C.51}$$

so that the residues are 1 and 2. The value of the integral along the closed path within which both the singularities lie is then

$$\oint_C f(s)\,ds = 2\pi j(1+2) = 6\pi j \tag{C.52}$$

If $f(z)$ is analytic in a domain with no singularities, then the integral along any closed path is zero, that is,

$$\oint_C f(z)\,dz = 0 \tag{C.53}$$

This result is known as *Cauchy's integral theorem.*

appendix D

Proofs of some theorems on positive real functions

Theorem D.1. If $Z(s)$ and $W(s)$ are both positive real, then $Z(W(s))$ is also positive real.

Proof. When $\operatorname{Re} s \geq 0$, both $\operatorname{Re} Z(s)$ and $\operatorname{Re} W(s) \geq 0$, then $\operatorname{Re} Z(W(s)) \geq 0$, also. When s is real, both $Z(s)$ and $W(s)$ are real, hence $Z(W(s))$ is real. Since $Z(W(s))$ satisfies both conditions of positive realness, it is positive real.

Theorem D.2. If $Z(s)$ is positive real, then $Z(1/s)$ is positive real.

Proof. $W(s) = 1/s$ is positive real, hence $Z(W(s)) = Z(1/s)$ is positive real.

Theorem D.3. If $W(s)$ is positive real, then $1/W(s)$ is also positive real.

Proof. $Z(s) = 1/s$ is positive real, hence $Z(W(s)) = 1/W(s)$ is positive real by Theorem D.1.

Theorem D.4. The sum of positive real functions is positive real.

Proof. Suppose $Z_1(s)$ and $Z_2(s)$ are both positive real. When $\operatorname{Re} s \geq 0$, then

$$\operatorname{Re} Z_1 \geq 0 \quad \text{and} \quad \operatorname{Re} Z_2 \geq 0$$

so that

$$\operatorname{Re} Z_1 + \operatorname{Re} Z_2 = \operatorname{Re} Z \geq 0.$$

Also, when s is real, both Z_1 and Z_2 are real. The sum of two real numbers is a real number. Therefore, $Z_1 + Z_2$ is positive real.

Theorem D.5. The poles and zeros of $Z(s)$ cannot have positive real parts (i.e., lie in the right half of the s plane).

Proof. Suppose there is a pole s_0 in the right-half plane. Let us make a Laurent series expansion about s_0 so that

$$Z(s) = \frac{k_{-n}}{(s - s_0)^n} + \frac{k_{-n+1}}{(s - s_0)^{n-1}} + \cdots + k_1(s - s_0) + \cdots + k_r(s - s_0)^r + \cdots$$

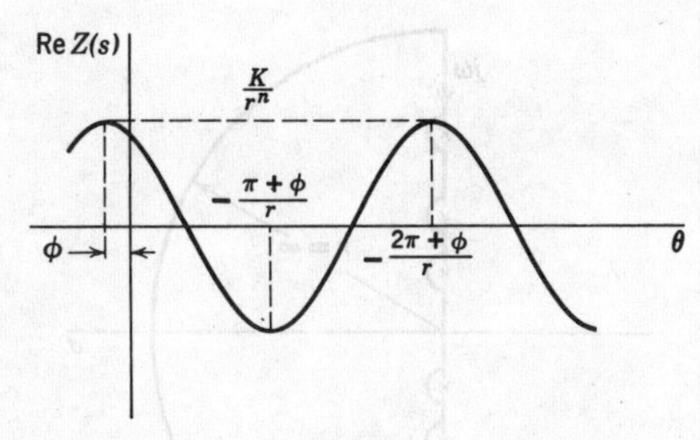

FIG. D.1

where n is real and finite. In the neighborhood of the pole s_0, $Z(s)$ can be approximated by

$$Z(s) \simeq \frac{k_{-n}}{(s - s_0)^n}$$

We can represent $Z(s)$ in polar form by substituting each term by its polar form; i.e., let $(s - s_0)^{+n} = r^n e^{jn\theta}$ and $k_{-n} = Ke^{j\phi}$ so that

$$Z(s) = \frac{K}{r^n} e^{j(\phi - n\theta)}, \qquad \operatorname{Re} Z(s) = \frac{K}{r^n} \cos(\phi - n\theta)$$

which is represented in Fig. D.1. When θ varies from 0 to 2π, the sign of $\operatorname{Re} Z(s)$ will change $2n$ times. Since $\operatorname{Re} Z(s) \geq 0$ when $\operatorname{Re} s \geq 0$, it is seen that any change of sign of $\operatorname{Re} Z(s)$ in the right-half plane will show that the function is not positive real. Therefore, we cannot have a pole in the right-half plane. Since the function $1/Z(s)$ is positive real if $Z(s)$ is positive real, it is obvious that there cannot be any zeros in the right-half plane also.

Theorem D.6. Only simple poles with positive real residues can exist on the $j\omega$ axis.

Proof. As a consequence of the derivation of Theorem D.5, it is seen that poles may exist on the $j\omega$ axis if $n = 1$, and $\phi = 0$. The condition $n = 1$ implies that the pole is simple and the condition $\phi = 0$ implies that the residue is positive and real. It is readily seen that zeros on the $j\omega$ axis must also be simple.

Theorem D.7. The poles and zeros of $Z(s)$ are real or occur in conjugate pairs.

Proof. If a complex pole or zero exists without its conjugate, $Z(s)$ cannot be real when s is real. As a result of this theorem and Theorem D.5, it is seen that both the numerator and denominator polynomials of $Z(s)$ must be Hurwitz.

Theorem D.8. The highest powers of the numerator polynomial and the denominator polynomial of $Z(s)$ may differ by at most unity.

Proof. Let $Z(s)$ be written as

$$Z(s) = \frac{a_n s^n + a_{n-1} s^{n-1} + \cdots + a_1 s + a_0}{b_m s^m + b_{m-1} s^{m-1} + \cdots + b_1 s + b_0} = \frac{P(s)}{Q(s)}$$

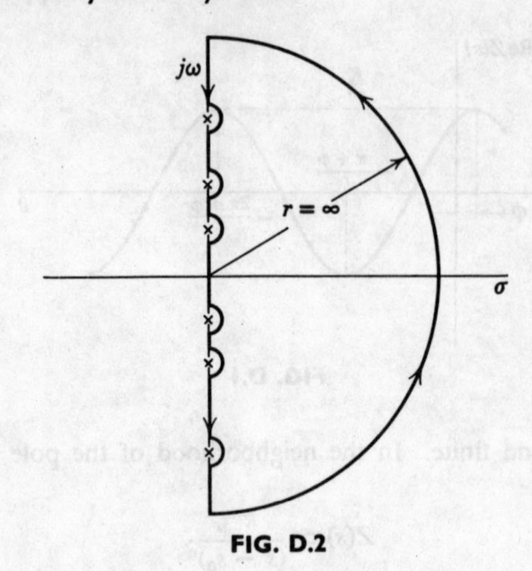

FIG. D.2

If $m - n \geq 2$, when $s = \infty$, $Z(s)$ will have a zero of order 2 or greater at $s = \infty$, which is on the $j\omega$ axis. Similarly, if $n - m \geq 2$, then, at $s = \infty$, $Z(s)$ will have a pole of order 2 or more at $s = \infty$. Since $Z(s)$ cannot have multiple poles or zeros on the $j\omega$ axis, these situations cannot exist; therefore $|n - m| \leq 1$.

Theorem D.9. The lowest powers of $P(s)$ and $Q(s)$ may differ by at most unity.

Proof. The proof is obtained as in Theorem D.8 by simply substituting $1/s$ for s and proceeding as described.

Theorem D.10. A rational function $F(s)$ with real coefficients is positive real if:

(*a*) $F(s)$ is analytic in the right-half plane.

(*b*) If $F(s)$ has poles on the $j\omega$ axis, they must be simple and have real, positive residues.

(*c*) $\text{Re}\, F(j\omega) \geq 0$ for all ω.

Proof. We need only show that these three conditions fulfill the same requirements as $\text{Re}\, Z(s) \geq 0$ for $\text{Re}\, s \geq 0$. We will make use of the *minimum modulus* theorem which states that if a function is analytic within a given region, the minimum value of the real part of the function lies on the boundary of that region. The region with which we are concerned is the right-half plane which is bounded by a semicircle of infinite radius and the imaginary axis with small indentations for the $j\omega$ axis poles. If the minimum value on the $j\omega$ axis is greater than zero, then $\text{Re}\, Z(s)$ must be positive over the entire right-half plane (Fig. D.2).

appendix E

An aid to the improvement of filter approximation

E.1 INTRODUCTION

The introduction of an additional pole and zero in the second quadrant of the complex frequency plane, and at their conjugate locations, can give amplitude or phase corrections to a filter approximant over some desired band of frequencies without significantly changing the approximant at other frequencies. However, a cut-and-try procedure for finding the best positions for such a pole-zero pair can be tedious. A visual aid is presented herein which reduces the amount of labor required to make modest corrections of this type.

Constant phase and constant logarithmic gain contours for the correction by a pole-zero pair[1] are plotted on transparent overlays. One of these may be placed over a suitably scaled sheet of graph paper representing the complex frequency plane. Then the pair-shaped phase and gain corrections along the $j\omega$ axis are indicated by the intersections of the overlay contours with this axis. Corrections which best reduce the errors in the original approximant are then sought by variation of the overlay position and orientation. Either phase or amplitude may be corrected. However, it is not always possible to simultaneously improve both the phase and the amplitude characteristics of an approximant by a single pair-shaped correction.

F. F. Kuo and M. Karnaugh, reprinted from the *IRE Transactions on Circuit Theory*, **CT-9,** No. 4, December 1962, pp. 400–404.

[1] This will be called, hereafter, a pair-shaped correction.

E.2 CONSTANT LOGARITHMIC GAIN CONTOURS

Suppose we begin with a transfer function $G(s)$ with certain deficiencies in its amplitude or phase response. Let us consider the transfer function of a corrective network $G_1(s)$ such that the product

$$G_2(s) = G_1(s)\,G(s) \tag{E.1}$$

will have better gain or phase characteristics. For the purposes of this paper, we will restrict $G_1(s)$ to have the form

$$G_1(s) = C\,\frac{(s-q)(s-\bar{q})}{(s-p)(s-\bar{p})} \tag{E.2}$$

where C is a constant, and q and p are a zero and a pole, respectively, in the second quadrant of the complex frequency plane. If the correction is to be applied at sufficiently high frequencies such that $s - \bar{q} \cong s - \bar{p}$ then

$$G_1(s) \cong \frac{C(s-q)}{s-p} \tag{E.3}$$

Let us consider the effect when the pole-zero pair in Eq. E.3 is used to augment any given rational function in the complex frequency plane.

The added gain, in decibels, due to this pole-zero pair is

$$D = 20 \log_{10} \left| \frac{s-q}{s-p} \right| \tag{E.4}$$

where we have neglected the effect of the constant C in Eq. E.3. If we let

$$k = (10)^{D/20} \tag{E.5}$$

then

$$k = \left| \frac{s-q}{s-p} \right| \tag{E.6}$$

For fixed k, this is the equation of a circle with inverse points[2] at q and p. Its radius is

$$\rho = \frac{k\,|p-q|}{|1-k^2|} \tag{E.7}$$

and its center is at

$$s_c = \frac{q - k^2 p}{1 - k^2} \tag{E.8}$$

---

[2] E. C. Titchmarsh, *The Theory of Functions*, Oxford University Press, Oxford, England, second edition 1939, pp. 191–192.

Let
$$p = s_0 + \frac{\epsilon}{2}$$
$$q = s_0 - \frac{\epsilon}{2} \tag{E.9}$$

Then
$$\rho = \frac{k\,|\epsilon|}{|1 - k^2|} \tag{E.10}$$
$$s_c = s_0 - \frac{1 + k^2}{1 - k^2}\left(\frac{\epsilon}{2}\right) \tag{E.11}$$

Here, s_0 is the midpoint between the pole and zero, and their separation is ϵ. It is easy to see from Eq. E.11 that the center of each circle of constant gain is externally collinear with the pole and zero. For the purpose of drawing the family of constant gain contours, we may let s_0 be the origin of coordinates and the scale factor ϵ may be set equal to unity.

Furthermore, only half the pattern need be drawn because the function D has negative symmetry with respect to a reflection about the perpendicular bisector of the pole-zero pair.

E.3 CONSTANT PHASE CONTOURS

Figure E.1*a* represents a pole at p, a zero at q and an arbitrary point s in the complex frequency plane. When the pole and zero are used to correct a given phase characteristic, the added phase at s is

$$\phi = \theta_q - \theta_p \tag{E.12}$$

where θ_q and θ_p are measured with respect to a single arbitrary reference.

In Fig. E.1*b*, a circle has been drawn through p, q, and s. Angle ϕ is

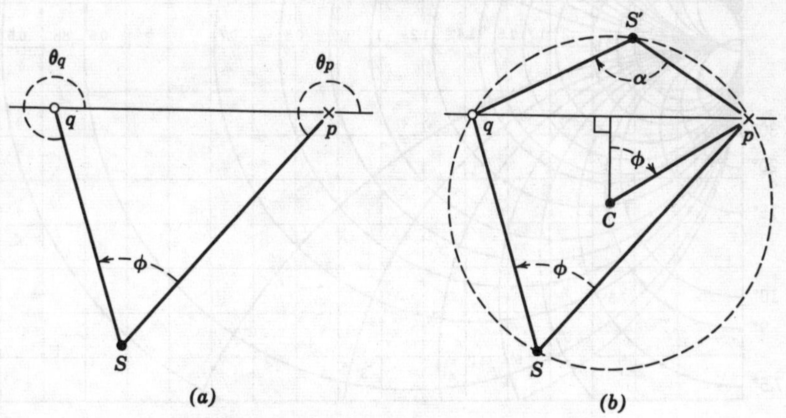

FIG. E.1. Derivation of constant phase contours.

equal to one half the subtended arc $ps'q$. Therefore, the arc qsp is a constant phase contour. The angle at the center of the circle between cp and the perpendicular bisector of chord pq is also equal to ϕ. All of the circular contours of constant phase have their centers on this perpendicular bisector.

Note that minor arc $ps'q$ is also a contour of constant phase, but the phase angle

$$\alpha = \phi - \pi \tag{E.13}$$

is negative, as indicated by the clockwise rotation from $s'p$ to $s'q$. The convention for positive rotation is herein taken to be counterclockwise.

E.4 CONTOUR DRAWINGS

Sets of constant phase and constant logarithmic gain contours are drawn in Figs. E.2 and E.3. The curves are symmetric about the zero-decibel line, except that the gain curves are of opposite sign. Therefore, only one half of each figure has been drawn.

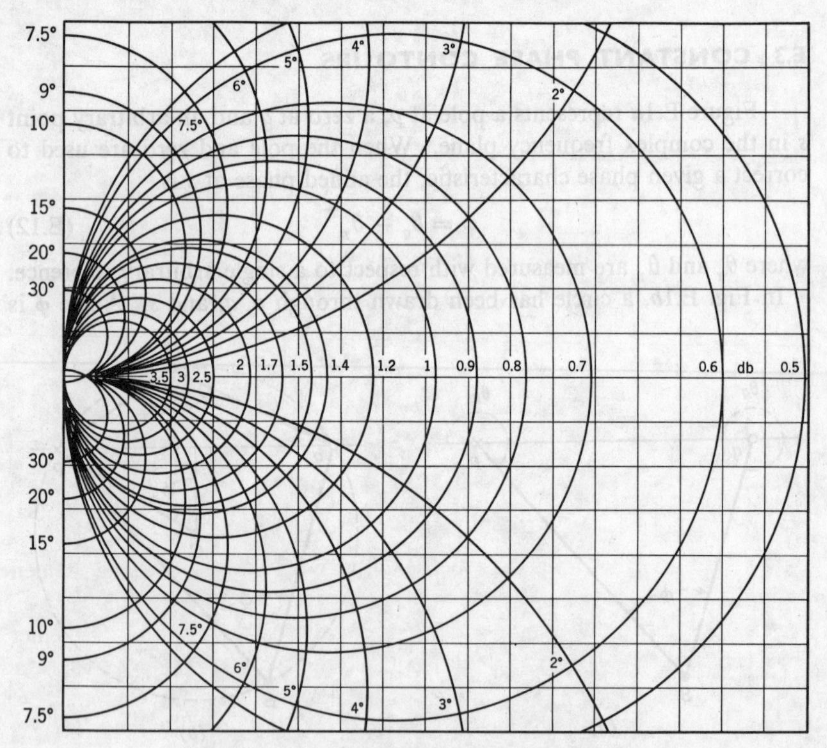

FIG. E.2. Constant amplitude and phase contours.

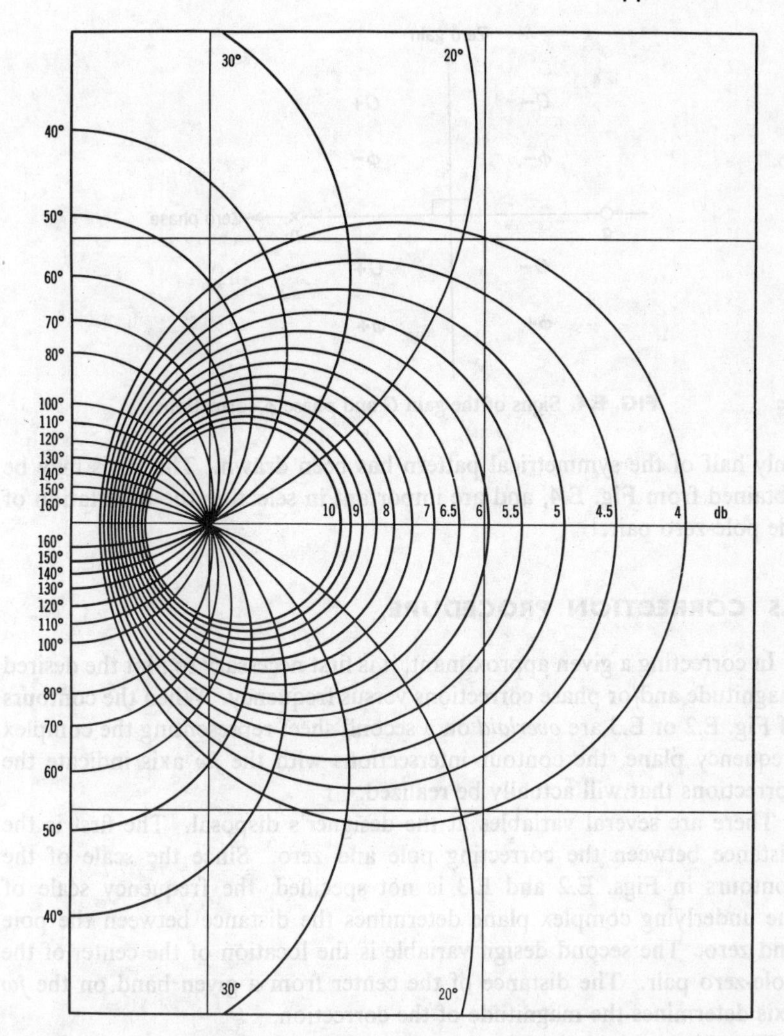

FIG. E.3. Constant amplitude and phase contours.

The zero-decibel line is the perpendicular bisector of the line segment joining the pole and zero. It is a gain contour of infinite radius.

The line through the pole and zero is a zero phase contour. Together, these perpendicular axes form a useful reference system. The signs of the phase and gain in the four quadrants formed by these axes are shown in Fig. E.4.

The corrections in Figs. E.2 and E.3 do not carry algebraic signs because

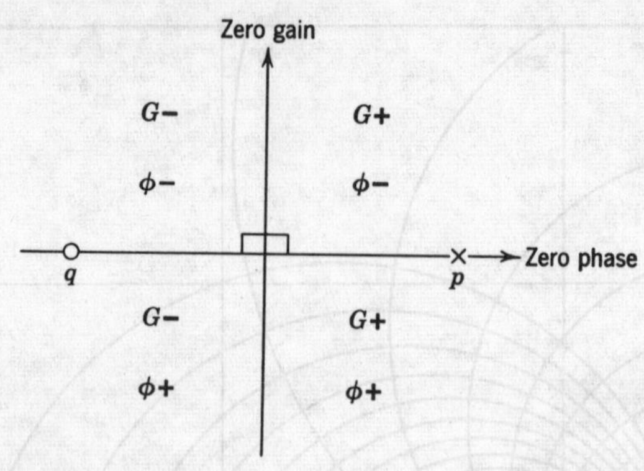

FIG. E.4. Signs of the gain G and phase ϕ corrections.

only half of the symmetrical pattern has been drawn. The signs may be obtained from Fig. E.4, and are important in selecting the orientation of the pole-zero pair.

E.5 CORRECTION PROCEDURE

In correcting a given approximant, it is first necessary to plot the desired magnitude and/or phase corrections versus frequency. When the contours of Fig. E.2 or E.3 are *overlaid* on a second sheet representing the complex frequency plane, the contour intersections with the $j\omega$ axis indicate the corrections that will actually be realized.

There are several variables at the designer's disposal. The first is the distance between the correcting pole and zero. Since the scale of the contours in Figs. E.2 and E.3 is not specified, the frequency scale of the underlying complex plane determines the distance between the pole and zero. The second design variable is the location of the center of the pole-zero pair. The distance of the center from a given band on the $j\omega$ axis determines the magnitude of the correction.

The third variable is the orientation of the pole-zero pair. Figure E.4 shows how the orientation affects the gain and phase corrections. As a simple example, suppose one wishes to have zero phase correction at $\omega = 1.0$, negative phase correction above and positive correction below that frequency. The attack would be to point the zero phase axis at $\omega = 1.0$ with the pole nearest to the $j\omega$ axis. If it is desired to have equal phase correction above and below $\omega = 1.0$, the zero phase axis should be oriented parallel to the real axis of the complex frequency plane. If one wishes to have more phase correction above $\omega = 1.0$ and less below, the zero phase axis should be rotated clockwise with respect to the σ (real) axis.

We thus see that by varying the frequency scale of the complex frequency plane, the position of the center of the pole-zero pair, and the orientation of the pole-zero pair, the pair-shaped correction can be made to approximate the desired correction.

It must be emphasized that the method suggested herein is an aid to cut-and-try correction. As such, it is easier to use the method than to precisely set down rules for applying it. However, a few rather general statements may be helpful.

Unless the pole and zero both lie on the real axis, one must remember that another pole and zero are located at conjugate positions. The contributions from both pole-zero pairs may be added algebraically. In most practical cases, the desired correction will have a band-pass character. Therefore, only one pole-zero pair will normally contribute significantly at any frequency.

The shape and magnitude of the desired correction will dictate the way in which the $j\omega$ axis must intersect the correction contours. The broadness of the desired correction will dictate the proper scaling of the $j\omega$ axis. Usually, only a few trials are needed to fix the pole and zero locations for the best fit.

It will be found that a worthwhile correction can be made in either the phase or the gain characteristic. Only fortuitously can they be improved simultaneously by a single pair-shaped correction.

Example E.1. The amplitude response of a third-order Butterworth filter is given by the solid curve in Fig. E.5. It is desired to steepen the gain roll-off near the cutoff frequency $\omega_c = 1$. This is done by increasing the gain just below $\omega = \omega_c$ and decreasing it above that frequency. Figure E.6 illustrates the type of correction desired. Figure E.7 shows a pole-zero pair that achieves this type of correction. The gain at $\omega = 1$ remains unchanged by this particular choice. Other pole-zero pairs that "aim" the zero-decibel line at $\omega = 1$ but give asymmetrical corrections about that point might also be used. The dashed curve in Fig. E.5 shows the corrected gain.

A different pole-zero pair, also shown in Fig. E.7, has been chosen to minimize the deviation of the slope of the phase response from its slope at $\omega = 0$. This pair-shaped correction decreases the phase for $\omega < 0.7$ and increases it for $\omega > 0.7$.

Figure E.8 shows the deviation of the phase responses from linear phase. It is clear from Figs. E.5 and E.8 that the gain corrected approximant has a poorer phase response than the original, while the phase corrected approximant has a gentler gain roll-off than the original.

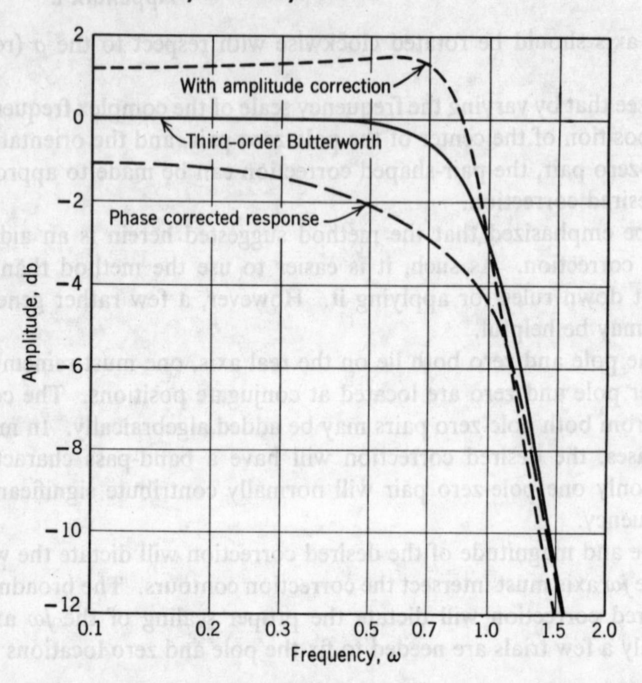

FIG. E.5. Amplitude response of corrected and uncorrected Butterworth filters.

This will not surprise the experienced filter designer. It is possible, however, to achieve moderate corrections in both gain and phase by using two pair-shaped corrections. A useful approach to this objective lies in localizing the gain correction further out of band, and the phase correction further in-band than in the separate corrections just discussed. This can be done by shifting the pole-zero centers to higher or lower frequencies and also by experimenting with nonsymmetrical corrections.

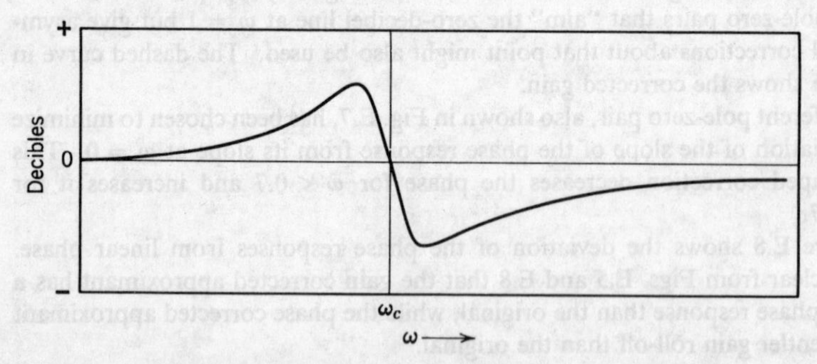

FIG. E.6. Amplitude correction to steepen fall-off of third-degree Butterworth filter.

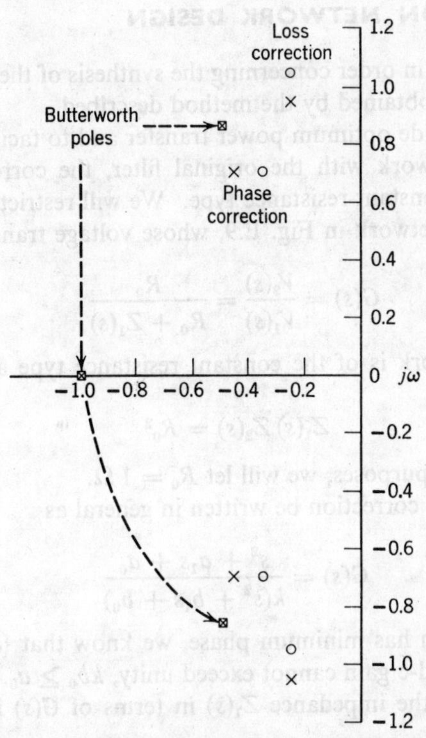

FIG. E.7. Poles and zeros of the original filter and correction equalizers.

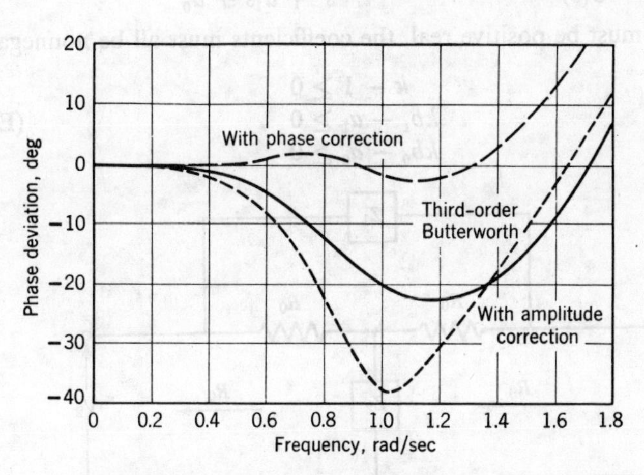

FIG. E.8. Phase deviation from slope at zero frequency.

E.6 CORRECTION NETWORK DESIGN

A few words are in order concerning the synthesis of the equalizer from the pole-zero pair obtained by the method described.

In order to provide optimum power transfer and to facilitate cascading the correction network with the original filter, the correction network should be of the constant-resistance type. We will restrict our discussion to the bridged-T network in Fig. E.9, whose voltage transfer function is given as

$$G(s) = \frac{V_2(s)}{V_1(s)} = \frac{R_0}{R_0 + Z_1(s)} \tag{E.14}$$

provided the network is of the constant resistance type as given by the equation

$$Z_1(s)\,Z_2(s) = R_0^2 \tag{E.15}$$

For normalization purposes, we will let $R_0 = 1\ \Omega$.

Let the pole-zero correction be written in general as

$$G(s) = \frac{s^2 + a_1 s + a_0}{k(s^2 + b_1 s + b_0)} \tag{E.16}$$

Since the correction has minimum phase, we know that $\{a_i, b_i \geq 0\}$. In addition, since the d-c gain cannot exceed unity, $kb_0 \geq a_0$.

We can express the impedance $Z_1(s)$ in terms of $G(s)$ in Eq. E.14 as

$$Z_1(s) = \frac{1}{G(s)} - 1 = \frac{(k-1)s^2 + (kb_1 - a_1)s + (kb_0 - a_0)}{s^2 + a_1 s + a_0} \tag{E.17}$$

Since $Z_1(s)$ must be positive real, the coefficients must all be nonnegative so that

$$\begin{aligned} k - 1 &\geq 0 \\ kb_1 - a_1 &\geq 0 \\ kb_0 - a_0 &\geq 0 \end{aligned} \tag{E.18}$$

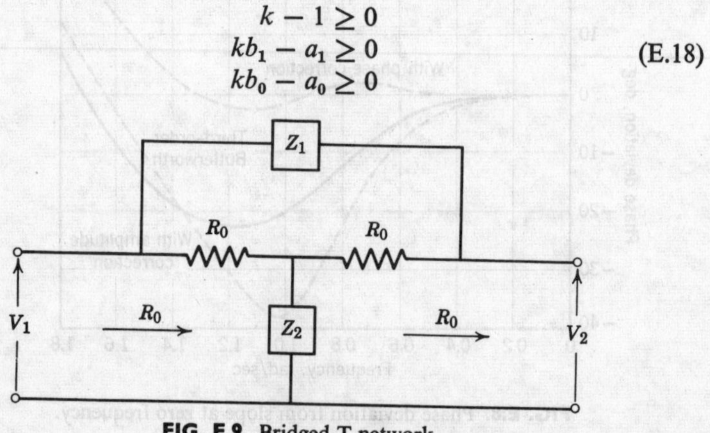

FIG. E.9. Bridged-T network.

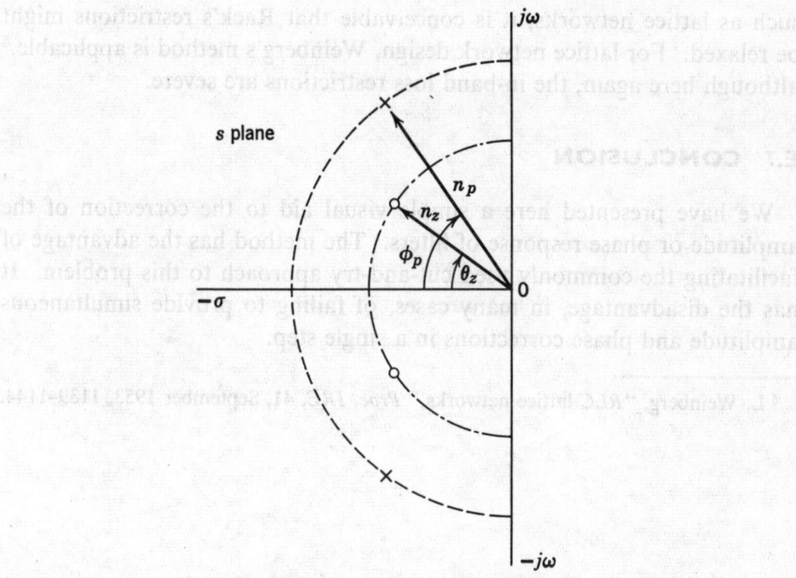

FIG. E.10. Poles and zeros of $Z_1(s)$.

Moreover, in order for a biquadratic driving-point immittance to be positive real, the following condition must apply.[3]

$$a_1 \left[\frac{kb_1 - a_1}{k - 1}\right] \geq \left[\left(\frac{kb_0 - a_0}{k - 1}\right)^{1/2} - a_0^{1/2}\right]^2 \tag{E.19}$$

Details concerning the synthesis of $Z_1(s)$ are also given in Seshu's paper.

Let us plot the pole-zero pair of $Z_1(s)$ as in Fig. E.10. We can represent the locations of the poles and zeros in terms of the polar angles ϕ_p and θ_z and their distances from the origin, n_p and n_z. Rack[4] has shown that in order for in-band loss to be finite

$$\frac{n_p}{n_z} < 2 \tag{E.20}$$

In addition he has shown that for an unbalanced bridged-T circuit, if $\alpha = \max\,[\theta_z, \phi_p]$ then the larger the angle α, the larger the in-band loss. In particular, α should be less than 70° to restrict the in-band loss to reasonable proportions. If one considers other network configurations

[3] S. Seshu, Minimal Realizations of the Biquadratic Minimum Function, *IRE Trans. on Circuit Theory*, **CT-6,** December 1959, 345–350.

[4] A. J. Rack, private communication.

such as lattice networks, it is conceivable that Rack's restrictions might be relaxed. For lattice network design, Weinberg's method is applicable,[5] although here again, the in-band loss restrictions are severe.

E.7 CONCLUSION

We have presented here a simple visual aid to the correction of the amplitude or phase response of filters. The method has the advantage of facilitating the commonly used cut-and-try approach to this problem. It has the disadvantage, in many cases, of failing to provide simultaneous amplitude and phase corrections in a single step.

---

[5] L. Weinberg, "*RLC* lattice networks," *Proc. IRE*, **41**, September 1953, 1139–1144.

Bibliography

SIGNAL ANALYSIS

Goldman, S., *Frequency Analysis, Modulation and Noise*, McGraw-Hill, New York, 1948.

Javid, M. and E. Brenner, *Analysis, Transmission and Filtering of Signals*, McGraw-Hill, New York, 1963.

Lathi, B. P., *Signals, Systems and Communications*, John Wiley and Sons, New York, 1965.

Lighthill, M. J., *Introduction to Fourier Analysis and Generalized Functions*, Cambridge University Press, New York, 1958.

Mason, S. J. and H. J. Zimmerman, *Electronic Circuits, Signals and Systems*, John Wiley and Sons, New York, 1960.

Papoulis, A., *The Fourier Integral and Its Applications*, McGraw-Hill, New York, 1962.

Rowe, H. E., *Signals and Noise in Communications Systems*, D. Van Nostrand, Princeton, N.J. 1965.

Schwartz, M., *Information Transmission, Modulation and Noise*, McGraw-Hill, New York, 1959.

NETWORK ANALYSIS

Balabanian, N., *Fundamentals of Circuit Theory*, Allyn and Bacon, Boston, 1961.

Bohn, E. V., *The Transform Analysis of Linear Systems*, Addison-Wesley, Reading, Massachusetts, 1963.

Brenner, E. and M. Javid, *Analysis of Electric Circuits*, McGraw-Hill, New York, 1959.

Brown, R. G. and J. W. Nilsson, *Introduction to Linear Systems Analysis*, John Wiley and Sons, New York, 1962.

Brown, W. M., *Analysis of Linear Time-Invariant Systems*, McGraw-Hill, New York, 1963.

Carlin, H. J. and A. Giordano, *Network Theory*, Prentice-Hall, Englewood Cliffs, N.J., 1964.

Cassell, W. L., *Linear Electric Circuits*, John Wiley and Sons, New York, 1964.

Chen, W. H., *The Analysis of Linear Systems*, McGraw-Hill, New York, 1963.

dePian, L., *Linear Active Network Theory*, Prentice-Hall, Englewood Cliffs, N.J. 1962.

Friedland, B., O. Wing, and R. B. Ash, *Principles of Linear Networks*, McGraw-Hill, New York, 1961.

Gardner, M. and J. L. Barnes, *Transients in Linear Systems*, Vol. 1, John Wiley and Sons, New York, 1942.

Guillemin, E. A., *Introductory Circuit Theory*, John Wiley and Sons, New York, 1953.

Guillemin, E. A., *Theory of Linear Physical Systems*, John Wiley and Sons, New York, 1963.

Harman, W. W. and D. W. Lytle, *Electrical and Mechanical Networks*, McGraw-Hill, New York, 1962.

Hayt, W. H., Jr. and J. E. Kemmerly, *Engineering Circuit Analysis*, McGraw-Hill, New York, 1962.

Huelsman, L. P., *Circuits, Matrices and Linear Vector Spaces*, McGraw-Hill, New York, 1963.

Kim, W. H. and R. T. Chien, *Topological Analysis and Synthesis of Communication Networks*, Columbia University Press, New York, 1962.

Ku, Y. H., *Transient Circuit Analysis*, D. Van Nostrand, Princeton, N.J., 1961.

Legros, R. and A. V. J. Martin, *Transform Calculus for Electrical Engineers*, Prentice-Hall, Inc., Englewood Cliffs, N.J., 1961.

LePage, W. R. and S. Seely, *General Network Analysis*, McGraw-Hill, New York, 1952.

Ley, B. J., S. G. Lutz, and C. F. Rehberg, *Linear Circuit Analysis*, McGraw-Hill, New York, 1959.

Lynch, W. A. and J. G. Truxal, *Introductory System Analysis*, McGraw-Hill, New York, 1961.

Paskusz, G. F. and B. Bussell, *Linear Circuit Analysis*, Prenctice-Hall, Inc. Englewood Cliffs, N.J., 1964.

Pearson, S. I. and G. J. Maler, *Introductory Circuit Analysis*, John Wiley and Sons, New York, 1965.

Pfeiffer, P. E., *Linear System Analysis*, McGraw-Hill, New York, 1961.

Reza, F. M. and S. Seely, *Modern Network Analysis*, McGraw-Hill, New York, 1959.

Sanford, R. S., *Physical Networks*, Prentice-Hall, Englewood Cliffs, N.J., 1965.

Schwarz, R. J. and B. Friedland, *Linear Systems*, McGraw-Hill, New York, 1965.

Scott, R. E., *Elements of Linear Circuits*, Addison-Wesley, Reading, Massachusetts, 1965.

Seely, S., *Dynamic Systems Analysis*, Reinhold, New York, 1964.

Seshu, S. and N. Balabanian, *Linear Network Analysis*, John Wiley and Sons, New York, 1959.

Skilling, H. H., *Electrical Engineering Circuits*, Second Edition, John Wiley and Sons, New York, 1965.

Van Valkenburg, M. E., *Network Analysis*, Second Edition, Prentice-Hall, Englewood Cliffs, N.J., 1964.

Weber, E., *Linear Transient Analysis*, John Wiley and Sons, New York, 1954.

Zadeh, L. A. and C. A. Desoer, *Linear Systems Theory*, McGraw-Hill, New York, 1963.

NETWORK SYNTHESIS

Balabanian, N., *Network Synthesis*, Prentice-Hall, Englewood Cliffs, N.J., 1958.

Bode, H. W., *Network Analysis and Feedback Amplifier Design*, D. Van Nostrand, Princeton, N.J., 1945.

Calahan, D. A., *Modern Network Synthesis*, Hayden, New York, 1964.

Chen, W. H., *Linear Network Design and Synthesis*, McGraw-Hill, New York, 1964.

Geffe, P. R., *Simplified Modern Filter Design*, John F. Rider, New York, 1963.

Guillemin, E. A., *Synthesis of Passive Networks*, John Wiley and Sons, New York, 1957.

Guillemin, E. A., *The Mathematics of Circuit Analysis*, John Wiley and Sons, New York, 1949.

Hazony, D., *Elements of Network Synthesis*, Reinhold, New York, 1963.

Kuh, E. S. and D. O. Pederson, *Principles of Circuit Synthesis*, McGraw-Hill, New York, 1959.

Matthaei, G. L., L. Young, and E. M. T. Jones, *Microwave Filters, Impedance-Matching Networks and Coupling Structures*, McGraw-Hill, New York, 1964.

Saal, R., *Der Entwurf von Filtern mit Hilfe des Kataloges Normierter Tiefpässe*, Telefunken GMBH, 1961.

Skwirzynski, J. K., *Design Theory and Data for Electrical Filters*, D. Van Nostrand, Princeton, 1965.

Storer, J. E., *Passive Network Synthesis*, McGraw-Hill, New York, 1957.

Truxal, J. G., *Control System Synthesis*, McGraw-Hill, New York, 1955.

Tuttle, D. F., *Network Synthesis*, Vol. 1, John Wiley and Sons, New York, 1958.

Van Valkenburg, M. E., *Introduction to Modern Network Syntheses*, John Wiley and Sons, New York, 1960.

Weinberg, L., *Network Analysis and Synthesis*, McGraw-Hill, New York, 1962.

Yengst, W. C., *Procedures of Modern Network Synthesis*, Macmillan, New York, 1964.

Name Index

Subject Index